Image Analysis and
Mathematical Morphology

IMAGE ANALYSIS AND MATHEMATICAL MORPHOLOGY

by J. SERRA

École Nationale Supérieure des Mines de Paris, Centre de Géostatistique et de Morphologie Mathématique, Fontainebleau, France

1982

ACADEMIC PRESS
A Subsidiary of Harcourt Brace Jovanovich, Publishers
London · New York
Paris · San Diego · San Francisco
São Paulo · Sydney · Tokyo · Toronto

ACADEMIC PRESS INC. (LONDON) LTD.
24/28 Oval Road
London NW1

United States Edition published by
ACADEMIC PRESS INC.
111 Fifth Avenue
New York, New York 10003

British Library Cataloguing in Publication Data
Serra, J.
    Image analysis and mathematical morphology.
    1. Image processing—Mathematics
    I. Title
    621.380′41   TA1632
    ISBN 0–12–637240–3

LCCCN: 81–66397

Typeset by Macmillan India Ltd
and printed in Great Britain by
Page Bros (Norwich) Ltd

Preface

Mathematical morphology was born in 1964 when G. Matheron was asked to investigate the relationships between the geometry of porous media and their permeabilities, and when at the same time, I was asked to quantify the petrography of iron ores, in order to predict their milling properties. This initial period (1964–1968) has resulted in a first body of theoretical notions (Hit or Miss transformations, openings and closings, Boolean models), and also in the first prototype of the texture analyser. It was also the time of the creation of the Centre de Morphologie Mathématique on the campus of the Paris School of Mines at Fontainebleau (France). Above all, the new group had found its own style, made of a symbiosis between theoretical research, applications and design of devices.

These have the following idea in common: the notion of a geometrical structure, or texture, is not purely objective. It does not exist in the phenomenon itself, nor in the observer, but somewhere in between the two. Mathematical morphology quantifies this intuition by introducing the concept of structuring elements. Chosen by the morphologist, they interact with the object under study, modifying its shape and reducing it to a sort of caricature which is more expressive than the actual initial phenomenon. The power of the approach, but also its difficulty, lies in this structural sorting. Indeed, the need for a general theory for the rules of deformations appeared soon. The method progressed as a result of an interchange between intellectual intuitions and practical demands coming from the applications. This finally lead to the content of this book. On the way, several researchers joined the initial team and constituted what is now called the "Fontainebleau School". Among them, we can quote J. C. Klein, P. Delfiner, H. Digabel, M. Gauthier, D. Jeulin, E. Kolomenski, Y. Sylvestre, Ch. Lantuéjoul, F. Meyer and S. Beucher.

A new theory never appears by spontaneous generation. It starts from some initial knowledge and grows in a certain context. The family tree of

v

mathematical morphology essentially comprises the two branches of integral geometry and geometrical probabilities, plus a few collateral ancestors (harmonic analysis, stochastic processes, algebraic topology). Apart from mathematical morphology, three other parallel branches may be considered as current descendants of the same tree. They are stereology, point processes and stochastic geometry as developed by D. G. Kendall's School at Cambridge. Stereology, unlike the other two, is oriented towards applications. The stereologists have succeeded in putting the major theorems of integral geometry into practice. Indeed their society regroups biologists and specialists of the material sciences whose mutual interest lies in the quantitative description of structures, principally at the microscopic scale.

The other and more recent branch of "picture processing" appeared in the United States at the beginning of the 1960s as a consequence of the N.A.S.A. activities. Today, its scope has extended to domains other than satellite imagery, its audience has become more international and is now represented by scientific societies such as that of pattern recognition. These scientists are regrouped more by a common class of problems (enhancement and segmentation of pictures, feature extraction, remote sensing) than by a specific methodology. Here, the theoretical tools often belong to the convolution and filtering methods (Fourier, Karhuhen-Loeve etc.); they also use some algorithms of mathematical morphology without connecting them with the general underlying notions. Finally, to a lesser extent, they borrow their techniques from syntactic and relaxation methods.

The last tree to which mathematical morphology belongs is that of image analysers. It began in 1951, when J. Von Neumann proposed an automatic process which compared each pixel with its immediate neighbours. Over the past twenty years, about thirty prototypes of devices have been built for digital image analysis. Among the few of them which were commercialized, we can quote the quantimets, based on classical stereology, and the Leitz texture analyser, which was the best-selling image analyser at the end of the 1970s.

The necessity to construct a device which could easily perform morphological operations on actual specimens appeared very early in my work. I wanted to design an equivalent of the computer for geometry, where the Hit or Miss transformation and its derivatives would replace the basic algebraic operations cabled in the usual computers. I called this invention the texture analyser, and with an increasing participation of J. C. Klein, I built four prototypes between 1965 and 1975. The Wild-Leitz Company bought the licence in 1970 and currently produces this instrument, after having remodelled it for commercial promotion. Although the quasi-totality of the examples and of the photographs of this book come from texture analysers, I have systematically avoided describing technology and hardware.

Anyway, this particular system, like all computer components, will soon become obsolete. The technologies will change, certain specific routines will be pre-programmed by microprocessors, the internal structure will be made more parallel, or more pipelined, but the concept of the texture analyser will remain the same. This is because, via the basic morphological operations, it links the actual experiments to very fundamental prerequisites of the experimentation of geometrical structures.

Schematically, 40 % of the material of this book has not been published before, another 40 % comes from works of the Fontainebleau school but is not always well known, and the rest comes from other sources. I have set the level of the book at the interface of applications and theory, and have emphasized the links between modes of operation and general underlying concepts. A comprehensive set of pure mathematical results can be found in G. Matheron (1975). I quote, without proofs, his most important theorems.

This book is directed to the triple audience of the *users* of the method (biologists, metallographers, geologists, geographers of aerial imagery . . .), the specialists of *picture processing* and the *theoreticians* (probabilists, statisticians). These different categories of readers are not used to the same formalism, neither do they formulate their thoughts in a common language. Therefore, I had to find a common ground for the three, which is not an easy task. I hope the reader will excuse me if I have not completely succeeded. For example, I have avoided using programming languages and even flow charts, preferring to present the morphological notions within the more general framework of set geometry. Strange terms such as "idempotence" or "anti-extensivity" are likely to shock many naturalists; I know this, but their geometrical interpretations are so intuitive that every experimenter will soon grasp their meaning.

Apart from the language of mathematical morphology, its methodological deductions may bewilder the reader used to another conceptual background. Perhaps this difficulty will be lessened if I make a few comments on the subject.

Mathematical morphology deals with sets in Euclidean or digital spaces, and considers the functions defined in an n-dimensional space as particular sets of dimension $n + 1$ (classically, in picture processing, the function is the primary notion and the set is a particular case). To invert the priority between sets and functions leads us to emphasize the non-linear operations of sup and inf to the detriment of addition and subtraction.

Basically, the objects under study are considered as being embedded in the usual Euclidean space; afterwards they are digitalized (in contrast to this, picture processing is essentially digital). Suitable topologies then allow the robustness of the morphological operations to be studied. This would be impossible in a pure digital framework, where the reference of the Euclidean space is missing. I know that general topology is not familiar to the majority of the readers, but it is the price we have to pay for analysing the stability, the quality, i.e. finally the meaning of all the practical algorithms.

As we have seen, the main objective of morphology is to reveal the structure of the objects by transforming the sets which model them (such a purpose generalizes that of integral geometry and of stereology, which consists of transforming bounded sets into significant numbers). However, all set transformations are not at the same level. Algorithms are governed by more general criteria, which in turn satisfy a few universal constraints. Anyone wishing to master mathematical morphology must assimilate this vertical hierarchy (picture processing is incomparably more "horizontal"). As a consequence, we did not design the book according to various problems such as image enhancement, filtering, or segmentation, but on a classification based on criteria and related questions. Every morphological criterion may help to segment an image, depending upon the type of image and the initial knowledge of it that we possess. It is precisely this information which orients us towards one or another type of criteria.

A strong counterpoint interlaces criteria to models. It generally brings set models into play, but the introduction of probabilistic notions opens the door to the more specific class of random set models. This gives rise to the four main parts of the book: theoretical tools, partial knowledge, criteria, random models.

This book can be read in several different ways, depending upon the field of interest of the reader. The text itself is constructed in a logical order where every new idea is introduced with reference to the preceding notions and not to the following ones. Such a rule, compulsory when one writes a book, does not need to be respected by the readers. A person mainly interested in the algorithms and their experimental implementation may start with the first two chapters and jump directly to the four chapters on criteria. He can also leave out the theoretical sections included in these six chapters if he wishes (i.e. Chapter II, Section E; Chapter IX, Section D; Chapter X, Section F; Chapter XI, Sections B, C, D; Chapter XII, Sections B, G, H). The first reading will possibly inspire him to go further. He can then pursue with the more theoretical chapters, either those centred on probability or alternatively those on geometry. Should he decide to follow the way proposed for statisticians and probabilists, then the second level comprises Chapters V (parameters), VIII (sampling) and XIII (random sets). If he feels more tempted by geometrical methods, then we suggest a further step consisting of Chapters IV (convexity), VI and VII (digital morphology) and the theoretical complements left out in the first reading. Finally, Chapter III is a review of topological results needed practically everywhere in the book, and Chapter XIV, a non-technical conclusion.

I am deeply indebted to G. Matheron for 15 years of scientific help and friendship. In a sense, all that follows is a tribute to his own work. I am also most grateful to the many colleagues from Fontainebleau and from other institutes who have worked with me. Particular thanks are due to S. Beucher, Professor J. L. Binet, Dr L. Bubenicek, Dr Ph. Cauwe, J. C. Celeski, F. Conrad, Dr N. Cressie, Dr P. Delfiner, M. Gauthier, Dr D. Jeulin, Dr J. C. Klein, Dr E. N. Kolomenski, Dr Ch. Lantuéjoul, Dr F. Meyer, Professor R. E. Miles, W. Müller, Dr R. Nawrath, Professor J. S. Ploem, J. Rzeznik, Dr S. R. Sternberg, Y. Sylvestre, E. Wasmund, Professor G. S. Watson.

The English version of the manuscript was read and corrected by Dr N. Cressie, who not only improved the style but also helped me with much constructive advice. Dr M. Armstrong and Dr J. J. Piret also participated in the tedious task of "debugging" the English sentences produced by a French writer. Sections of the manuscript were read and criticized by Dr G. Laslett, Dr H. Maitre and Dr A. Reid; thanks are due to Dr S. Sternberg whose very pertinent and valuable critical comments considerably improved the original manuscript. However, the errors and omissions are all my own. Typing and the preparation of figures were major tasks. I am very grateful to Mrs Kreyberg and Mrs Pipault, helped by Miss Poirier, for having transformed unreadable handwritten documents into a clear manuscript. The figures were drawn by Mr Laurent and Mr Waroquier. I also thank Academic Press, for their encouragement in the process of publication.

J. Serra
Beaurieux, October 1981

(The English version of this book was revised by N. Cressie.)

Am Anfang war die Tat!

W. Goethe

L'action commence par conférer aux objets des caractères qu'ils ne possédaient pas par eux mêmes . . . et l'expérience porte sur la liaison entre les caractères introduits par l'action (et non pas sur les propriétés antérieures de celui-ci).

J. Piaget

Contents

Notation

1. SETS

λ, ρ	scalars (i.e. positive numbers)
α, ω	directions in $\mathbb{R}^2$ and $\mathbb{R}^3$ respectively
α, β, γ	basic directions of the hexagonal lattice
θ	trihedron of directions associated with X in $\mathbb{R}^3$
x, y, b, h etc. . .	(latin lower case letters) points in $\mathbb{R}^n$ or $\mathbb{Z}^n$, and *also vectors* $O_x, O_y \ldots$ when one wants to specify that x is a point (geometric figure) and not a vector, one writes $\{x\}$
$\{x:*\}$	set of points x satisfying property*
X, Y, Z	(latin capital letters) Euclidean or digital sets under study
H	digital hexagon of size one
B, T	structuring elements
$T = (T_1, T_2);$ $T^* = (T_2, T_1)$	mixed structuring element and its dual
$\{X_i\}; \{T^i\}; \{T\}$	sequence of sets; of structuring elements; standard sequence
(X, D)	planar graph, or digital graph with set of vertices X and set of edges D
$\Pi; \Delta$	test plane (resp. test line) generating cross sections on X
Ω	set of the directions ω, i.e. the unit sphere
$\varnothing; \mathbb{R}^n; \mathbb{Z}^n; \vec{\mathbb{Z}}^n$	empty set; Euclidean space; module and affine module of dimension n
$\mathscr{H}$	set of bounded hexagonal convex sets (in $\mathbb{Z}^2$)
$\mathscr{T}; \mathscr{L}$	grid; lattice
$\mathscr{P}(\mathbb{R}^n); \mathscr{P}(\mathbb{Z}^n)$	set of all subsets of $\mathbb{R}^n$; of $\mathbb{Z}^n$

$\mathscr{F}, \mathscr{G}, \mathscr{K}, \mathscr{K}'$	set of all closed, open, compact and non empty compact subsets of $\mathbb{R}^n$ (resp. of $\mathbb{Z}^n$)
$\mathscr{C}(\mathscr{F}); \mathscr{U}(\mathscr{F})$	sets of convex hulls (resp. of umbrae) of the closed sets in $\mathbb{R}^n$
$\mathscr{S}(\mathscr{K}), \overline{\mathscr{S}}(\mathscr{K}), \mathscr{R}(\mathscr{K})$	convex ring; its extension and regular class in $\mathbb{R}^n$
$\mathscr{G}'$	set of non empty open sets in $\mathbb{R}^n$, whose the convex hull of the complement is $\mathbb{R}^n$
$\mathscr{F}_u, \mathscr{P}_i$	set of u.s.c. functions and pictures in $\mathbb{R}^n \times \mathbb{R}$
$\mathscr{K}_u, \mathscr{K}_i$	set of u.s.c. functions and pictures with a bounded support in $\mathbb{R}^n \times \mathbb{R}$

2. LOGIC AND SET TRANSFORMATIONS

$\exists x$	there exists x such that
$\forall x$	for all x
w.r. to	with respect to
$\Rightarrow$	implies
$\Leftrightarrow$, iff	if and only if
Q.E.D.	quod erat demonstrandum (which was to be proved)
$x \in X$; $x \notin X$	point x belongs to set X; point x does not belong to set X
$X = Y$; $X \neq Y$	sets X and Y coincide; set X is different from set Y
$X \subset Y$; $Z \supset B$	X is included in Y; Z contains B
$X \pitchfork Y$	set X hits set Y, i.e. $X \cap Y \neq \varnothing$
$\check{X}$	transposed of set X, i.e. set of points such that $-\dot{x} \in X$
X^c	complement of X, i.e. set of point x such that $x \notin X$
λX	similar of X with scaling factor λ; $\lambda X = \{x : \frac{x}{\lambda} \in X\}$
X_h	translate of X by vector h; $X_h = \{x : x - h \in X\}$
$\Psi(X)$	set transformed of X w.r. to set transformation Ψ
$\Psi^*(X)$	dual transformation (w.r. to the complementation), i.e. $\Psi^*(X) = [\Psi(X^c)]^c$
$X \cup Y$	set union, i.e. set of points belonging to X or to Y
$X \cap Y$	set intersection, i.e. set of points belonging to both X and Y
X/Y	set difference, i.e. set of points belonging to X and not to Y
$\bigcup_{b \in B} X_b$; $\bigcap_{b \in B} X_b$	union (intersection) of all the translates X_b, with $b \in B$
$X \oplus \check{B}$	dilate of X by B, i.e. set of points x such that $B_x \pitchfork X$

$X \ominus \check{B}$ erosion of X by B, i.e. set of points x such that
                                 $B_x \subset X$

X_B opening of X by B, i.e. $(X \ominus \check{B}) \oplus B$

X^B closing of X by B, i.e. $(X \oplus \check{B}) \ominus B$

$X \circledast T$ Hit or Miss transform of X by T, i.e.
                            $(X \ominus \check{T}_1) \cap (X^c \ominus \check{T}_2)$

$X \bigcirc T$ thinning of X by T, i.e. $X / X \circledast T$

$X \odot T$ thickening of X by T, i.e. $X \cup (X \circledast T)$

$X \bigcirc \{T^i\}$ sequential thinning of X by $\{T^i\}$, i.e.
                            $[(X \bigcirc T^1) \bigcirc T^2] \bigcirc T^3 \ldots$
                          (similar notation with $\odot$, $\ominus$ and $\oplus$)

$X \bigcirc T; Y$ conditional thinning, i.e. $(X \bigcirc T) \cup Y$

$X \ominus T; Y$ conditional erosion, i.e. $(X \ominus T) \cup Y$

$X \odot T; Y$ conditional thickening, i.e. $(X \odot T) \cap Y$

$X \oplus T; Y$ conditional dilation, i.e. $(X \oplus T) \cap Y$

$X \bigcirc \{T^i\}; Y$ conditional sequential thinning i.e. $[(X \bigcirc T^1;$
                          $Y) \bigcirc T^2; Y] \ldots$ (similar notation with erosions
                          thickenings and dilations)

$C(X); U(X)$ convex hull of X; umbra of X

3. TOPOLOGY

$\overline{X}; \mathring{X}$ closure; interior of set X

∂X boundary of set X

$X_i \to X$ X_i tends towards X (for the Hit or Miss topology)
                          in $\mathscr{F}$

$\overline{\text{Lim}}, \underline{\text{Lim}}$ upper limit, lower limit

$\underset{t \to t_0}{\text{Lim}} (*)$ limit of $*$ when t tends towards t_0

inf; sup   lower; upper bound

$X_i \downarrow X; X_i \uparrow X$ monotonic sequential convergence of $\{X_i\}$ to X, by
                          upper (resp. lower) values

$d(X, Y); d(x, y)$ distance between X and Y (resp. between x and y)

u.s.c.; l.s.c.   upper semi continuous; lower semi continuous

E; LCS space   topological space; locally compact Hausdorff and
                          separable space

4. PARAMETERS, MEASURES AND PROBABILITIES

$\mu^X(dx)$ measure generated by set X acting on the open set
                          dx

$W_i^{(x)}(X)$	ith Minkowski functional of X in $\mathbb{R}^n$
$N^{(n)}(X)$	Euler Poincaré characteristic in $\mathbb{R}^n$ (also called connectivity number)
$V(X), S(X), M(X)$	volume, surface area and norm of X in $\mathbb{R}^3$ or in $\mathbb{Z}^3$
$A(X), U(X)$	area and perimeter of X in $\mathbb{R}^2$ or in $\mathbb{Z}^2$
$L(X)$	length of X in $\mathbb{R}^1$ or in $\mathbb{Z}^1$
$D(X\|\Pi\omega)$	Diametral variation, i.e. measure of the projection of X on plane $\Pi\omega$
$D^{(2)}(X, \alpha); D^{(3)}(X, \omega);$ $A^{(3)}(X, \omega)$	total diameters in $\mathbb{R}^2$; in $\mathbb{R}^3$; total projection area in $\mathbb{R}^3$
$*V, *A, *L$	where $*$ is a parameter, means "per unit volume", "per unit area" and "per unit length" respectively (notation of the int. soc. for stereology)
$\mathrm{Mes}(X)$	Lebesgue measure of X (volume, area, or length, according to the dimension)
d_X	support function of $X \in \mathscr{C}(\mathscr{K})$
$N^+(X), N^+(X, \alpha)$	convexity member of X, of X in direction α
$n_X^+(dz)$	convexity measure generated by X, acting on dz
$u(d\alpha)$	rose of directions in $\mathbb{R}^2$
$\delta_x; dx$	Dirac measure located at point x; Lebesgue measure
$Pr\{*\}$	probability of the event $*$
$E(X)$	Mathematical expectation of X, i.e. number-weighted moment
$\mathscr{M}(X)$	measure-weighted moment of X
$\mathscr{N}(*)$	number of configuration $*$ (digital)
$P_\alpha(l)$	Prob. that the segment of length l and direction α is included in X
$C(h)$	set covariance of X, stationary case
$K(h)$	set covariance of X, global case
$Q(B)$	functional moment of a random set X, i.e. $Pr\{B \subset X^c\}$
$f(x), g(x)$	numerical function of x
$f(\lambda), g(\lambda)$	density functions associated with the distribution function $F(\lambda)$ and $G(\lambda)$
$k_X(x), l_X(x)$	indicator function of set X

PART I. THE TOOLS

I. Principles—Criteria—Models

A. HOW TO QUANTIFY THE DESCRIPTION OF A STRUCTURE

Among the very vague notions with which we live, that of an "image" opens up a world which is particularly rich with meanings, suggestions . . . and with *ambiguities*. However, the scientific world has taken this notion and made use of it; we have heard for some time now, the term "Image Analysis". This expression does not refer to a class of particular phenomena, for example in the way that the botanists study and classify the vegetable kingdom, but shows more an interest in the *spatial* aspects of the objects under study (whatever they may be).

We need not doubt then, that this analysis, which aims to measure the features of the objects, their sizes, their orientations, their overlapping, may interest many scientists from many different areas. In fact, the methods described in this book have had, at one time or another, the opportunity to be used for the study of arrangements of petrographic phases, of milling of rocks, of histology of the brain, of the dynamics of cloud movements, of automatic reconstruction of relief maps, of computer reading of handwriting, of the detection of cancer cells from a smear, of the study of the competition between trees in a forest, of estimation of mineral orebodies, of estimation of microscopic stringers in steel . . . .

Perhaps it appears a little bold to mix all these uses. Certainly the various kinds of anatomists of the animal, vegetal, and mineral kingdoms all want to handle the spatial distribution of their objects, and sometimes also the temporal. Certainly the vocabularies of specialists contain words such as "texture", "cluster", etc. However, are they talking about the same things? And, what are they looking for from Image Analysis?

Often the final goal is quite clear. The cytologist wants to decide if the actual specimen is pathological or not. The petrologist wants to know if two thin sections taken from two different places come from the same geological horizon, the metallographist wants to know if a given mechanical property of his alloy is represented by a particular microscopic structure, the astronomer wants to know if a certain detail of his photograph is just white noise, or is really an anomaly.

If the ultimate goals can be expressed simply, it is because the purpose is in fact something more than the known geometry of the objects. But these purposes do not imply, by themselves, a method of study. The lack of a direct link between aims and method leads to a new question: how should we describe an image? what kind of help can mathematics give us?

A.1.   We see only what we want to look at

We all have an obvious intuitive feeling about the spatial *structure* of an object. We know that the structure of the Rocky Mountains and of the Swedish plateau is completely different, but if someone asks us to be precise and to exactly define this word, we would meet with real difficulties. The structure of the mountains is perhaps the successive arrangement of their chains, but it is also their typical shapes (plateaus, hanging valleys, ridges), in brief it is that which constitutes the whole. But this is not at all precise. "The successive arrangements . . . ", "that which constitutes . . . " does not tell us how to fit the various parts together, to make up the whole. Moreover, a description is never objective. If you are a mountain climber, you look at the Swedish Plateau as hills without high summits, but if you are fond of cross-country ski- ing, the shape of the peaks means little.

The situation of the scientist confronted with the analysis of a complicated object is still more difficult. Not only is it difficult to give an objective description, but as a scientist, he has a second problem: his description *must* be *quantitative*. An example will demonstrate these difficulties. Figure I.1 shows two polished sections of rocks taken from the Russian Caucasus. Several successive volcanic eruptions have transformed these materials with various conditions of cooling, generating the microstructures that we see in the figure. These two specimens belong to a group of around a hundred taken for the purposes of engineering geology (Ch. X, C. 3). Engineers wanted to build a dam in the region, and were looking for a place where the foundations were solid, i.e. show a resistance to crushing. How can we read from the sections, that we have indeed such a resistance? Does this show itself by the planes of cleavage between solid phases? Is the lining up of pores responsible for the fragility of the rocks, or the big empty spaces? Are these holes, separated in the sections, actually connected in space? Finally, and above all, why would the resistance to crushing be linked to any visible geometrical arrangement? There exists no theory of the mechanical behaviour of this type of completely heterogeneous media, and statistical correlations between mineral grades and

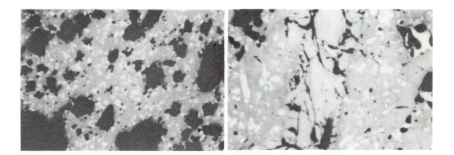

Figure I.1. Polished sections of Caucasian dolerites. Only one of these two rocks strongly resists crushing; a dam is to be built on the corresponding rock mass. Why? The proportions of the four phases which partition the space are practically the same in both cases (magnetite in white, pyroxenes in light grey, plagioclase in dark grey, pores in black). The difference lies in their structures, but how can it be described?

physical properties are in general disappointing. Our engineer must think that the skiers, who know exactly what they want from the morphology of the Rockies, do not know how lucky they are.

Since a description is never neutral, and always emphasizes certain aspects of the phenomenon, why not try to base our morphological approach on this very necessity? Why not exploit the geometrical prerequisites, as they appear in each particular question to be answered, and interpret these constraints as *criteria* for focusing our morphological investigations? (The Caucasian geologist needs an *isotropic* description, since he does not know the orientation of the polished sections, in the horizontal plane. Moreover, grain and pore play *a priori*, a symmetrical role.)

Of course, by approaching the problem this way, we can scarcely hope for a specific answer to each question. Obviously the complexity of the physical laws involved in the crushing of a composite rock greatly exceed the capabilities of the simple assumptions of isotropy, and symmetry between pattern and background. (Although the simple solution which we reach, gives only a first selection among the possible parameters, in practice it often offers sufficient detail.) By deciding that the key question of morphological analysis is the study of such criteria, we reject the specific aspects of the problems, and start to build a general theory.

A.2.   To perceive an image, is to transform it

If we eliminate, even momentarily, the specific aspects of each question, what remains? Nothing, if we want to find a common denominator between two images, unless with a lot of good will, we can find several common features between a lymphocyte swollen by an oedema, and an aerial photograph of a Canadian forest. From the point of view of the experimenter, when we consider the conditions under which he can work, we are immediately surprised by the generality of some of these conditions. For any type of perception, the mind remodels the stimulus, in order to assimilate it to its own patterns. In the area of art, recall for a moment what still-life means for Braque or Picasso; the preceding sentence then becomes obvious. In the field of psychology the work of the gestaltists, and the decisive experiments of W. Kohler have systematically revealed the structuring activity of the mind in even the simplest perceptive phenomena. For the mind, to perceive an image is to transform it.

One might believe that this attitude is linked more or less with the study of humanities. It seems that the necessity to transform the object that one analyses arises also in the exact sciences. In chemistry, when one measures a specific surface, two distinct operations are performed. Firstly, one reduces the initial body to just its boundary (by marking it for example); secondly, the quantity of marker which has been absorbed is weighted. In sieving analysis, when we want to know the size distribution of say the result of the milling of rocks, we generate a sequence of subsets of the milled product, i.e. for each sieve size the subset is the corresponding oversize (that which is left after sieving); then we weigh the oversizes. There are an infinity of such examples, we could present; they will always have two steps: geometrical transformation and then measurement. Anyone who wants to correctly quantify morphology must take into account the existence of these two distinct steps. This point might seem trivial, however the school of stereology (see for example, the proceedings of the first two ISS Congresses 1963, 1967) has not distinguished between the concept of transformation and that of measurement. Consequently, the school has restricted itself to only certain particular problems, well treated we might add, such as the inference from the plane to the space (Weibel (1980), De Hoff (1968), Underwood (1970)), instead of systematically scanning the spectrum of interesting transformations.

B. THE FOUR PRINCIPLES OF QUANTIFICATION

Let us imagine two geologists each independently taking from the same geological horizon, a drill sample from the same region. From this, they extract polished sections, the microscopic texture of which they want to analyse quantitatively. In order that these analyses lead to the same result (at least statistically), they must not depend at all on the particularities of the observer.

Indeed in order to reconstruct their object, the geologists have four constraints within which their methodology must fit:

— They cannot know the location of the drill in the field to within one micron, but at best to within approximately ten centimetres.
— They have to magnify the section in order to see it.
— The field of vision of the magnifying instruments, as well as the restrictions of size of the polished section, extract arbitrarily small zones in the space, imposing only a piecemeal knowledge.
— The various powers of resolution of these instruments divide the scale of the dimensions into three parts: that which is too fine to be seen (optically, everything smaller than 0.2μ), that which is big enough to appear sharp (details bigger than 2μ), and finally that which is between these two limits, a zone where the objects are more or less fuzzy, and have small deformations with respect to a representation with a finer resolution. These are for example the threadlike pores on the photographs of Figure I.1. We can see them, but we cannot delineate them properly. Whatever be the resolution, there will always exist such details.

We shall call *quantitative*, any morphological description which clearly frees itself from these four constraints. In order to arrive at such a description, we shall form concepts in set theoretical terms, which will lead us to numerical results; these results are the *consequences*, and not the starting point of this quantification.

Let us represent the material under study as one or several point sets (in the mathematical sense), without worrying too much for the moment about problems which might arise due to this mathematical transcription.

If there are only two phases, e.g. grains and pores, we will denote by X and X^c ($X^c \equiv$ complement of X) the associated sets. If we split the object into more phases, X will depend upon a discrete valued or continuous parameter. In cytology, X_1, X_2, X_3 can represent the background, the cytoplasma and the cell nuclei, and $X(t)$ is that part of the chromatine which is darker (in the Foelgen staining technique) than the threshold of level t. Any morphological operation is, by definition, the composition of first a mapping or a transformation Ψ, of one set into another, followed by a measure μ. $\Psi(X)$ can be the boundary, the oversize in sieving, etc. . . . ; in each case $\Psi(X)$ is a *new set*. But note that $\mu(\Psi(X))$ is a *number* (weight, volume, surface area).

To the four constraints given above, four principles correspond, which *must* be satisfied by every morphological transformation Ψ, and by every measure μ. In order to make the approach a little easier, we restrict ourselves, for the moment, to consideration of just the transformations Ψ. The axiomatics which follow will be transposed to the measures μ in Chapter IV, Section F.

B.1. First principle: compatibility under translation

Let X_h denote the translate of the set X by vector h; and let λX denote the homothetic of the set X after magnifying by the scalar λ (see Fig. I.2). In order to completely specify a magnification we need to fix an origin; if however, as in the present case, we have compatibility under translation, this point becomes superfluous.

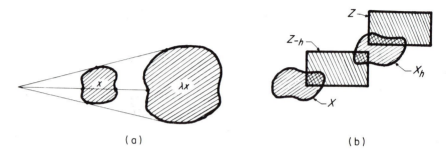

$$(a) \qquad\qquad\qquad (b)$$

Figure I.2. (a) Homothetic λX of set X. (b) Sets X and Z and their translates X_h and Z_{-h}.

There are two types of transformations to distinguish between: those that do not depend on the particular location of the axes of co-ordinates, and those that do. Transformations such as erosion or opening (to be seen in the next chapter) or, more simply, the transformation which consists in eliminating all particles having a surface that can be circumscribed by a given body (in $\mathbb{R}^2$), are independent of the position of the origin of the co-ordinates. In all these cases, we say that Ψ is invariant under translation, and we write symbolically:

$$\Psi(X_h) = [\Psi(X)]_h \qquad\qquad (\text{I-0})$$

In words then, Ψ is invariant under translation when it is equivalent to translate X by h and then apply Ψ, or to begin by using Ψ first, and then shift the result by vector h (rel. I-0). (Clearly, these transformations are insensitive to the origin of the co-ordinates, and consequently, free of the first constraint. We prefer not to pursue invariance under displacements any further: to demand of a Ψ, invariance under rotation would be sometimes injurious; it is not universal: useful in studies on dust for example but not for measurements in histology.)

Suppose now that the transformation Ψ depends on the origin O, and so denote it Ψ^O. Thus Ψ^O is the application to each point x of the space, of a criterion which is a function of the co-ordinates of x (among other things). Let us denote by Ψ^{-h}, the transformation which applies to the point $x - h$, the criterion formerly defined at point x. We meet this type of transformation in the individual analyses of particles (properties connected with the centre of gravity), or more simply, when we intersect the set X with a measuring mask Z. Obviously the transformation Ψ^O

$$\Psi^O(X) = X \cap Z \quad \text{(where } \Psi^h = X \cap Z_h)$$

does not satisfy the invariance-under-translation property, since

$$Z \cap X_h \neq (Z \cap X)_h$$

But in this case, we see from Figure 1.2b that it is the same (up to a translation) to shift the set X by the vector $+h$, as it is to shift Z by $-h$. More precisely,

$$(X_h \cap Z) = (X \cap Z_{-h})_{+h}$$

The same property, written for the general transformation Ψ^O, becomes:

$$\Psi^O(X_h) = [\Psi^{-h}(X)]_{+h} \qquad \text{(I-1)}$$

To apply Ψ^O to X is equivalent to applying Ψ^{-h} to X and shifting the result by the vector h. Note that when Ψ^O does not depend on the origin O, then $\Psi^O = \Psi^{-h}$ and relation (I-1) reduces to relation (I-0). We propose then, to use relation (I-1) as the first quantitative principle of mathematical morphology.

It is easy to construct examples which are not compatible under translation, i.e. which do not satisfy (I-1). For example, take for Ψ the intersection between the set X and a measuring mask $|h|Z_h$ whose size is proportional to the distance $|h|$ to the origin. Another example, this time not involving a measuring mask, but involving digitalization of the images, is presented below. The set X, represented in Figure. I.3, is cut by the lines numbered $1, 2, \ldots, i, \ldots$ of the grid. The transformation consists of adding sequentially one or two picture points to the right of those picture points given by the ith intersection (one when i is odd, two when i is even), except when the intersection is empty. Clearly relation (I-1) does not hold.

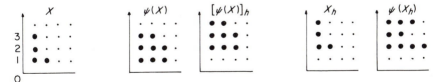

Figure I.3. Example of a transformation which is not compatible under translation: the two operations Ψ and translation by h do not commute.

B.2 Second principle: compatibility under change of scale

There is no guarantee that the two geologists spoken of at the beginning of the section, will look at their polished sections at the same magnification. If they want their approach to be truly quantitative, they have to find a way to be independent of an arbitrary magnification or reduction of their objects. Here again we have to distinguish between two different situations. Let us assume firstly, that the transformations performed, do not depend on any *scale parameter*. The transformations which take the set X into its boundary ∂X, or one of its projections Ψ_1, all belong to this first type (Fig. I.4).

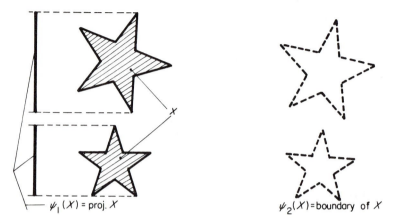

$\psi_1(X) = \text{proj. } X$ $\psi_2(X) = \text{boundary of } X$

Figure I.4. Example of transformations Ψ_1 and Ψ_2 not depending on a scale parameter, and invariant under homothetics.

Independence with respect to image magnification then implies that these transformations Ψ be invariant under homothetics (change of scale), and so:

$$\Psi(\lambda X) = \lambda\Psi(X), \qquad \lambda > 0 \qquad\qquad (I\text{-}2)$$

where λX denotes the homothetic X with the multiplier λ (see Fig. I.2b). Often the transformation Ψ is not of this type, and depends at least, on one scale parameter. Suppose we want to eliminate all the particles whose width is smaller than $10\,\mu$; after magnification 10μ becomes, say, 4 cm. Obviously we use the 4 cm as the cut-off in the magnified image. By so doing, we generate a family of transformations Ψ_λ, depending on the positive parameter λ, from the relation:

$$\Psi_\lambda(X) = \lambda\Psi\left(\frac{X}{\lambda}\right) \qquad\qquad (I\text{-}3)$$

However, it is not obvious that (I-3) is satisfied by any family of transformations $\{\Psi_\lambda\}$. As a counter example consider the isolated particles X_i, and the transformation which eliminates those particles for which the quantity "area + perimeter" is larger than λ. This procedure defines a family $\{\Psi_\lambda\}$ which obviously does not satisfy equation (I-3), since for a homothetic with multiplier λ, the area is proportional to λ^2, and the perimeter to λ. (In what units could we express the quantity "area + perimeter"?) To a physicist, this counter example may seem somewhat unrealistic. However, we could equally have eliminated particles for which the ratio, area/(perimeter)^2 is smaller than λ. Then the transformation for λ considered fixed, is independent of magnifications, and hence satisfies relation (I-2). However, as a transformation depending on the parameter λ, it does not satisfy relation (I-3). So, we are only willing to entertain those families $\{\Psi_\lambda\}$, for which the condition (I-3) holds; i.e. *compatibility* under change of scale.

As a consequence of the compatibility principle of (I-3), we often represent the measurements associated with a family of transformations as functions of a positive parameter. This is the case in particular, for histograms of size (lengths of intercepts, areas, etc. . . .). Here it is the size distribution *curve*, taken as a whole, which is invariant with respect to change of scale (modulo a choice of units on the abscissa of the graph), and is thus intrinsic to the polished section under consideration; the proportion of intercepts bigger than 60μ say, has by itself, no real interest.

B.3 Third principle: the local knowledge

As a general rule, we do not see entirely the set X, but only that part of it which is within a measuring mask, Z say. Think of, for example, aerial photographs of forests, or of clouds, metallographic micro-structures, etc. . . . If X is made up of bounded isolated particles, each of them can be completely surrounded by a mask (statistical problems remain when we can only collect a sample of *all* the particles), but when we look at a connected medium, the edges of the mask limit the object itself. We do not know X, but only $X \cap Z$ (and of course, the set Z itself). Under such conditions, we can only entertain those transformations Ψ for which the resulting $\Psi(X)$, taken in a bounded mask Z', comes from a bounded zone Z in which we know X. The mask Z' must depend only on the known quantities $X \cap Z$ and Z (more often we try to make Z' depend only on Z). This leads us to the third principle of mathematical morphology, which we call the *local knowledge*:

The transformation Ψ satisfies the local approach principle if, for any bounded set Z' in which we want to know $\Psi(X)$, we can find a bounded set Z in which the knowledge of X is sufficient to locally perform (i.e. within Z') the transformation.

In symbolic terms:

$$\forall \text{ bounded } Z', \ \exists \text{ bounded } Z:$$

$$[\Psi(X \cap Z)] \cap Z' = \Psi(X) \cap Z' \qquad \text{(I-4)}$$

Condition (I-4) eliminates, among others, many transformations concerning maxima of X (the longest secant for example). Note that (I-4) becomes vacuous when X is bounded and small enough to be included in the mask Z (an isolated cell in cytology, for example). We can easily construct a counterexample to (I-4) as follows: the set X is the connected background of Figure I.5; here X^c is the union of separated small particles. The transformation $\Psi(X)$ is defined as the union of all points $x \in X$, such that the largest chord passing through x has a positive slope. Therefore we really want to check all chords which pass through x, and hence it is quite possible that we have to do this checking out to infinity, and perhaps even an infinite number of times . . . .

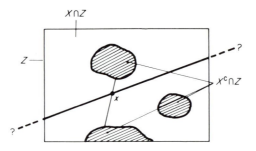

Figure I.5. Example of a transformation which does not satisfy the local approach.

Remark: the sense we have given to the word "local", is not the usual one used in mathematics. However we preferred it to "piecewise" because of its simplicity, and because it is already used by a number of practitioners.

B.4. Semi-continuity

The physical limits of the image analysers, confront us with a tough problem, requiring mathematical tools more powerful than those used above (we hope the reader will forgive us, but such is life!). We know how mathematicians conceptualize the notion of boundary between two complementary sets X and X^c (Ch. III, B. 2). A point is on the boundary ∂X if we can surround it by an arbitrary small ball, the interior of which contains points of X and of X^c. Without for a minute wanting to dampen the interest in this definition, it is possible to find some small contradiction between this and what the physicists call a boundary, namely the contact between two phases. "To reduce the scale arbitrarily?" Every physicist has dreamed of doing this for the past hundred and fifty years . . . .

In fact when we are studying a medium with two phases, the practical detection of a boundary is performed in quite a different manner. A given set

(No. 1) of experimental conditions, including magnification, resolution, etc. . . . , allows us to define two experimental sets X_1 and X'_1, each corresponding to the zones where we are certain to be in one phase or the other. These two sets are said to be "complementary", although they do not completely cover the space. More precisely, the complement of the union $X_1 \cup X'_1$ is a zone Δ_1, about which we know nothing except that it contains the boundary. To improve the sharpness of the contour simply means that we have taken a new set (No. 2) of experimental conditions, which delineates a second unknown zone Δ_2, included in Δ_1, but with a non-zero thickness. As for the actual but experimentally *inaccessible* boundary (the contact between the two phases), it is just the intersection of all the possible Δ_i, for Δ_i's becoming finer and finer.

Let us assume now that this medium is modified by an increasing (image) transformation Ψ. By "increasing", we mean that if A is included in B, then $\Psi(A)$ is included in $\Psi(B)$. The transformation Ψ is acceptable with respect to the fourth principle of mathematical morphology, if, in our case, $\Psi(\Delta_i)$ tends towards the boundary of the transformed image (Fig. I.6). Now the physical problem of the boundary is well posed; it no longer concerns the exact knowledge of the set X and X^c (an impossibility) in order to determine exactly $\Psi(X)$ and $[\Psi(X)]^c$ (another impossibility). It is sufficient to know X to a precision which is modulo a Δ_i (this is always the case), in order to have a transform $\Psi(X)$, which will be known to a certain precision modulo $\Psi(\Delta_i)$. The reader will note a change of point of view; we started from properties of sets, i.e. of objects, out of our own choice, and have finished with properties of transformations Ψ, i.e. of the tools with which we act on the outside world.

We will now translate the above property illustrated in Figure I.6, into a mathematical language. Firstly, we must choose for the Δ_i, those types of sets

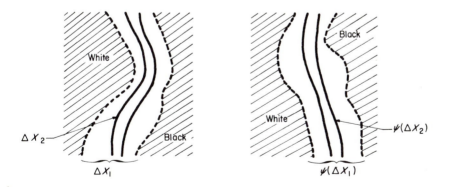

Figure I.6. Semi-continuity: the boundary of the transform is equal to the transform of the boundaries.

which when infinitely nested, do not tend towards the empty set, but towards a limit Δ (possibly having a zero thickness). To do this, we will model the Δ_i to be *topologically closed sets* (Ch. III, B.2); i.e. in the mathematical sense, sets which contain their boundaries. Then, the fourth principle of mathematical morphology, which can be expressed as follows:

For any decreasing sequence of closed sets tending towards a limit Δ, and every increasing transformation Ψ, there must correspond a sequence tending towards the transform of Δ.

We have discussed this principle in such a way that it affects *only* those transformations which are increasing. Moreover, it seems that we favour *decreasing* sequences of sets; however, in the example of Figure I.6, we could equally have studied the complements of the Δ_i, which are increasing. Such a principle can be handled only if we build a topology which is adapted to morphology, as G. Matheron has done (1975) with the Hit or Miss topology (Ch. III). In this framework, we can remove the two restrictions of an increasing transformation, and a decreasing sequence, and give the above principle in the following terms (Theorem III-5):

The only transformations that are relevant to mathematical morphology, are those which are semi-continuous w.r. to the Hit or Miss topology.

As a counter example, suppose we wish to measure the degree of "circularity" of a particle, and we decide to take for $\Psi(X)$ the union of the maximal inscribable circles in X. We will apply this transformation to the decreasing sequence of figures obtained by starting with a rectangle X, and stretching the two longest sides to a height $1/i$ as is shown in Figure I.7. As X_i tends to its rectangular limit, the transform $Y_i = \Psi(Y_i)$, i.e. the maximal circle

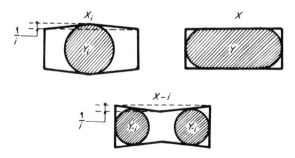

Figure I.7. A discontinuous transformation: X_i and X_{-i} tend towards the rectangular limit, but the transforms Y_i and Y_{-i} do not tend to Y.

centred in the middle of X gets smaller, and at the moment of the passage to the limit, suddenly becomes a long strip with circular extremities (i.e. $\Psi(X)$). A similar discontinuity can be seen for the increasing sequence $\{X_{-i}\}$, obtained by compressing the two longest sides. Hence transformation Ψ is not semi-continuous. One could make the objection that the example is too theoretical, and that in practice one never meets a perfect rectangle. This is true, but an experimental "rectangle" has many small bumps on its boundary, and each generates a maximal circle. Now these elementary distortions (due to experimental "noise") can vary from one experiment to the other and, greatly amplified by the chosen transformation, they lead to transformed images which are totally different from each other. But much more importantly, it is the discontinuity of the transformation Ψ and not properties of the rectangle, which makes the procedure physically unacceptable.

C. MATHEMATICAL MORPHOLOGY

C.1. Its purpose

By deciding that our observations have to satisfy the four prerequisites, we define a method that we shall call *mathematical morphology*. What does this theory address itself to: to the actual experiment, or to the mathematical model? In fact it addresses itself to both; this appears clearly from the four prerequisites. They do not come from a given experiment, nor are they deduced from basic axioms. On the contrary, they precede both the actual object and the theoretical model, being only consequences of constraints imposed by the working conditions of the experimenter in order to make him, as J. Piaget put it "objectif et décentré" (1967). They help us to choose from the large number of possible measurements we could take, and since they are not the results of experiments, we can at the same time formalize them in logical terms. We must then go and find in the maze of mathematical theories, those for which the four principles are satisfied. This makes mathematical morphology a physical theory and not just a mathematical one (as the name suggests). This is all the more true since the mathematical tools that it has called for, already exist independently, and have done so for a long time (except for the recent developments of G. Matheron (1975)). They are, amongst others, integral geometry, geometrical probability, second order random functions, graph theory, and topologies on a locally compact separable space. Mathematical morphology then, simultaneously induces a collection or universe of morphological bodies that can be easily handled by the method, and a mathematical formalism that is adapted to the four methodological decisions. We will now turn to an examination of these two implications.

C.2. The universe of morphological bodies

By duality with the four principles, there appears a collection (or universe) of scientific objects for which these principles can be useful. This "universe" is not really that extensive; we can easily imagine that certain phenomena do not require the four principles, and are even better studied without them. A graphologist knows very well that he must look at a piece of handwriting without changing the scale, because it was at this scale that the document was written. He also knows that whether the address was written towards the left or towards the right of an envelope, reveals rather different psychological characteristics of the writer. More generally a graphologic document has to be considered in its entirety; a mask which eliminates the margins would alter it. Finally, certain very small details, such as for example an isolated dot after each word, often provides first order information for the diagnosis. In brief, all these combine to expel graphology from the domain of quantification that we wish to discuss in this book. This exclusion is not a dogmatic one; the graphologist can always try to decode and differentiate handwriting by the methods of Mathematical Morphology. His result will not be totally meaningless, but he will quickly realize that the comprehensive battery of tools can do little to handle his object. They give him a few global trivialities: the handwriting is big or small, more vertical or more horizontal, etc. . . . It is never possible for him to sift through these average-like results, in order to automatically extract the significant details. Graphology then, seems to be an extreme example, and is probably non-representative; no one would seriously dream of using mathematical morphology for interpreting handwriting.

There is another phenomenon which is relevant to an automatic image analysis treatment, but it does not allow us to impose the principle of invariance under translation. Usually it is by aerial photographs that one studies the distribution of trees in a forest, however one can in fact do exactly the opposite, i.e. put the camera at ground level, and aim it towards the sky. Similar to the view seen by a weary hiker, taking a few minutes rest on his back, the photographic film would see a small window of sky, limited by the tops of the trees surrounding it. The foresters use these kinds of photographs to measure the amount of sunshine at ground level in the woods (Fig. I.8a). In order to do this, they have to know both the position of the zenith on the photograph, and the spherical position of the sky which is limited by the angle of the lens. Then the amount of sunshine is deduced from the projection (i.e. the photograph) of the actual image. The proportion of sky in each circular crown of small width, centred on a point of the zenith line, is weighted by the cosine of the angle φ made by the zenith line and the normal to the unit sphere at the circular crown (see Fig. I.8b). The internal structure virtually demands the use of polar co-ordinates, with an intrinsic origin (at the projection point of

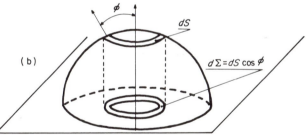

Figure I.8. (a) Photograph of the amount of sunshine in a forest (from the Institute of Water and Forests, France). A transformation invariant under translation is not recommended for an analysis of an image of this type. (b) A hemisphere and its plane projections.

the zenith) and consequently no transformation invariant under translation would respect this internal structure. Similar inadequacies appear in many studies, but fortunately they are usually away from the scale of interest. This is an important point; weather satellites see a vast portion of the northern hemisphere. From two points of view, the image that they transmit is not compatible with invariance under translation, firstly because at their original scale, the curvature of the earth is far from negligible, and secondly because the North Pole plays an important role in the movement of cloud masses.

However, certain methods of dynamic meteorology consider the variation in time of cloud structures at the scale of $10 \, km^2$; under these circumstances, the two preceding objections lose their force.

We could just as easily find other counter-examples (to any degree of objection) to the three other principles. We will leave the reader to think through these ideas. In conclusion, we will simply say that almost all the objects whose observation needs a microscope, and also geographic images (aerial photographs of forests, or urban areas and of clouds, diverse maps, geological profiles, etc. . . .), are in the domain of attraction of the four prerequisites of quantification, provided that we know approximately the upper and lower limiting scale.

We will *not* include in this domain of attraction the sound track of films for example, or more generally any curves $y = f(x)$. Of course these curves are planar in the sense that one needs two dimensions for plotting them, but they are not intrinsically two-dimensional objects (principles 1 and 2). So, we will not in this book, answer questions on speech recognition from plotted data, nor on the reading of electro-encephalograms, etc. We will also leave out the studies on deconvolutions for scanners which reconstitute from thin X-ray beams, horizontal sections of the brain or body. In such studies the purpose is to construct the image, and not—or at least not yet—to analyse it.

In summary then, by using the four principles, we have delineated certain scientific objects in the real world; by scientific we mean reproducible, that is, under the same conditions the same result occurs (J. Ullmo, 1967). In themselves, these phenomena have an existence which we can never know; the principles, and later the instruments (which are their extensions), filter them and basically *define* for us their structure.

C.3. Formalization of the principles

Any theoretical formulation which exists, can only take into account a part of all the aspects of the physical phenomenon. When we change the formalism, the theory sees a completely different object. For example, in Chapter IX, among the applications of the covariance we will have the opportunity of tackling a question of morphogenesis in animal embryology. According to our geometrical point of view, it is fundamental to take into account the spatial distribution of the cells as actually seen on sections. The temporal evolution of the embryo only indirectly intervenes, when we compare average-like results obtained at different times of the foetal life (the requirements of experimentation explain this priority of space over time: it is difficult to cut an embryo into microscopic slices, while still leaving it alive). There is nothing to assure us that this geometrical point of view is the most pertinent in the present case. Anyway, it is not the only one; other theories such as that of R. Thom

(1973) place a discontinuous space-time framework at the centre of the theoretical formulation. The two points of view are not mutually exclusive, but simply different; ours is humbler but closer to experimental reality, and the other provides more synthesis and is bolder, but difficult to confirm or falsify.

The pertinence of a mathematical formalism appears to be real, so it merits some discussion. Not only should natural objects not be thought of as "closed sets", because we do not see very easily how to define their boundaries; but more fundamentally, they are not even "sets", i.e. a collection of points. The physical meaning of the notion of point depends on the scale and is not absolute. At infinitely small scales, our cells, our dendrites, etc. . . . cease to appear as such, and become a mass of atoms, and, in the limit, a wave function. If we could extend our morphological procedures to this scale, they would become completely irrelevant and inoperative.

If the assimilation of natural objects to sets is arbitrary, another arbitrary option arises in the technology. When a cytologist chooses certain selective stainings to isolate the nucleus from the cytoplasm in the lymphocytes, in fact he substitutes a simplified image for a physical reality which is too sophisticated. This image only makes sense in a given scale of magnification, and for certain precise goals. If the scale of observation or the intentions change, he will introduce into the description, other new structures (ribosomes, mitochondriae, etc. . . .), or on the contrary he will ignore completely the cytoplasm, etc. Whatever he does, he always builds a new image different from the actual biological reality. When considered in this light then, the supplementary step by which we decide that his images may be assimilated to mathematical sets, is not the biggest step taken.

However, this so "obvious" formalization leads to new difficulties. The concept of a set is too rich for representing an image. Consider for example the points of the plane which have irrational co-ordinates, and recall the discussion on the mathematical concept of boundary given in Section 2. What experiment can decide whether a point belongs to a boundary, or whether a point has irrational coordinates? So we need to restrict the class of mathematical beings to the ones which are just general enough. A good compromise seems to have been found by G. Matheron (1975), who started from the idea that two sets with the same boundary and the same interior would be experimentally identical; this enables us to define equivalence classes of sets where two sets are identified to a single equivalence class when they have the same boundary and interior (i.e. the same closure). D. Kendall (1975) also proposed the same type of approach in a more general way. The common idea of these theories will serve as a basis for this book, but we lean towards that of Matheron because his is more developed in the direction of applications (size distribution axiomatics, random models, etc.).

Undoubtedly, the fourth principle takes an important part in such a theory, since we introduced it specifically to overcome trouble with the boundaries. But remarkably, the three other principles are not really necessary for a theory of random sets and their main properties. Here, the mathematician's point of view and our own (more that of a physicist's) diverge a little.

A physicist will probably find the fourth principle a little too sophisticated, since he knows he will only ever deal with digitalized images (i.e. *finite* sets of points). In fact, he is not wrong: as soon as one is able to define the translations on a grid of points, it is legitimate to base a digitalized morphology on only the first two principles. However, certain operations, although experimental, are not digitalized (the choice of direction of cut for a section, continuous variation of the magnification, etc. . . .); because of this it is more suitable to start from the object in the Euclidean space, and then study how the digitalized image can approximate it. Assuredly, the theoretical and experimental approach do not give the same resulting structures. They are neighbouring in certain aspects, but their points of view and conclusions made from the structures, do not always coincide.

D. CRITERIA AND ALGORITHMS

D.1. Degrees of generality of the principles

In order to understand how the notions of principles, criteria and algorithms are linked together, we can begin by making some remarks about the level of generality of the four basic prerequisites. Their implications are somewhat different. The last two, i.e. local knowledge and semi-continuity, correspond to quite general operational constraints, which specify a desire for robustness and stability of the transform. Any morphological concept not respecting them could scarcely yield a workable operational procedure. On the contrary, the first two principles, i.e. compatibility under translation and magnification, are more specific. We could quite easily build a morphology for objects known only after distortion (e.g. planar projections of spherical structures as in Figure 1.8, or images taken from a deforming mirror) for which the first two principles no longer hold. By incorporating them into the group of basic axioms, we make a *methodological choice*, a choice which is indeed justified by the variety of applications in which we have been working for the last fifteen years (material sciences, histo-cytology, etc. . . .). The positive side of this choice (or limitation), is the Hit or Miss transformation, and subsequently the design of the Texture Analyser. They are both simple and useful tools, mainly because they are limited in their ambitions.

The two levels of generality we have just described are just the beginning of a sequence of constraints, that becomes more and more specific to the type of question under study. We could even go as far as to say that to correctly formulate a problem is to determine the sequence of constraints it requires. Experience suggests that we classify these constraints into three groups, namely *principles*, *criteria* and *algorithms*. We have already seen that the first group is (for us) obligatory. Then come the more specific criteria, and finally the algorithms are the most specific.

In order to analyse these last two categories, we firstly propose a negative approach, where we discuss notions that are out of the scope of the principles. Then, using the problem of cell extraction from a biological smear as an example, we shall take the more direct approach of trying to isolate the logical steps involved.

D.2. Questions upon which the principles have no bearing

(a) *Invariance under rotation*
As we have already seen, there is absolutely no condition in the principles relevant to rotation. With respect to anisotropy then, mathematical morphology must give us the largest amount of flexibility. The amazing complexity of anisotropy in animal anatomy is as yet virtually unexplored. In the vegetal and mineral worlds, things are slightly simpler; consider dendritic growths (Fig. XII.2), or microscopic sections of wood (Fig. IX.17); however even here the mixtures of order and disorder in their orientations, are extremely complicated.

For a mathematician, translations and relations appear often in parallel, and together define the group of displacements in the Euclidean space. For morphology the main difference between translations and rotations comes from the fact that the circle is a compact space, but not the plane (in everyday words, we can travel around a circle, but not around a straight line). On any specimen we can measure in all the directions of $\mathbb{R}^2$, and they are only limited by the angular accuracy. However, due to the lack of compactness of the plane, we often have two (or more) scales of phenomena; if we fix the origin for studying the larger scale, this origin may have no meaning for the smaller one. Therefore translation invariant transformations are a natural choice.

(b) *Stereology*
Stereological problems, and amongst these the three-dimensional recon-stitution from two-dimensional sections and transformations, have a basic importance in image analysis. Nevertheless, they cannot be considered as un-avoidable experimental constraints, firstly because there are cases where they do not in fact arise (in aerial photographs, or in most biological smears), and secondly because we often have a direct way for measuring in $\mathbb{R}^3$ by taking many thin serial sections (such techniques are time-consuming but not

impossible). But, we have decided to consider as principles only those constraints that the experiment can never directly overcome. In stereology, it is quite possible that although an object is defined in $\mathbb{R}^3$, the problems we ask about might not pertain to three dimensions. For example, when we want to delineate the successive layers of the cerebral cortex by comparing the shape and density of the neurones in the grey matter, the *problem* posed then is to define successive segments on any (*one-dimensional*) probe line penetrating the cortex. This is because we are dealing with *parallel* layers of cerebral cortex.

We can just as easily find intrinsic three-dimensional problems: the growth rate of a dendrite is proportional to the integral of its mean curvature in $\mathbb{R}^3$ (i.e. the higher the growth rate, the sharper the dendrite); how can we measure this growth rate from sections? (A positive answer to this problem is available.) Fluids flowing through a porous medium do so through three-dimensional connections; how can we estimate its permeability from sections? (This is a problem of connectivity which we have already seen is impossible to solve without more model specifications.)

D.3. Example and discussion

The image transformations satisfying the four principles of quantification are numerous. The way for finding the most appropriate transformation in each case consists of picking out constraints specific to the particular case and designating them to be additional "principles"; we will call these, the *criteria*. They correspond to the morphological description that we saw was needed above. Just as for the four basic principles, the criteria cannot be experimentally checked. For example, we will see in Chapter X that the concept of size distribution is expressed by three additional constraints, but we will never be able to check if a given object is "sizeable" or not. In contrast to the principles, the criteria are not compulsory conditions, and (sometimes) can be chosen in very different ways, although still with the same purpose in mind. We will now present an example taken from pattern recognition.

Some current studies taking place in the Centre of Mathematical Morphology, Fontainebleau France (F. Meyer, 1977) have as an aim, to build a device able to automatically perform cytological examinations. The cells have been spread out on a smear, which inevitably damages some nuclei and leaves small scattered fragments. Then, selective staining techniques using fluorescence (due to Dr. Ploem, 1976) isolate separately the nucleus and the cytoplasm by two successive filters. After thresholding they give two sets which can be superimposed as in Figure I.9a. Before describing the cells, we have to find them. How can we program the device for filtering the actual cells from the rest? Here, three constraints can serve as a guide for determining our criteria: firstly, each cell has to contain a nucleus, and so we can eliminate any "cytoplasm" which does not have one. Secondly, if the nucleus has not been

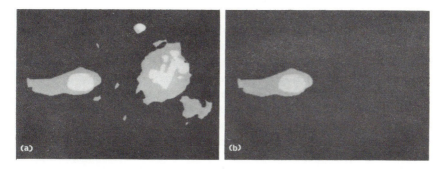

Figure I.9. (a) Nuclei and cytoplasm appear on the screen of the texture analyser as two superimposed sets; they correspond to selective colorations, but some of them may not actually be cells. (b) Actual cell extracted from Figure I.9(a).

crushed when spreading the smear, it is effectively convex. Finally, the very small nuclei come from artefacts or from cells which do not interest us. We decide to keep only those which admit an inscribable circle with a radius bigger then $5\,\mu$.

When the criteria are applied to Figure I.9 and "the dust has settled", only one cell remains from the initial image. The image is now ready for study. (For the exact operations associated with each of the three criteria, see the end of Chapter XI, G.2.)

These criteria are not the only ones possible, and one could imagine other collections leading to practically the same elimination of artefacts. As a general rule, the more concise and easy to put into practice such a collection is, the better it is. Moreover the algorithms which actually apply to the criteria, should satisfy mathematical properties which closely reflect the actual situation. In this example, the four principles have first of all been satisfied, but in the analysis, we have in fact satisfied two other conditions:

(a) All the criteria chosen are happily invariant under rotation, since the smear technique does not respect the orientation of the cell in the human body; the algorithms chosen should respect this.

(b) In spite of appearances, the algorithms corresponding to the three criteria are all independent of the notion of connected particle, and are expressed by a sequence of transformations acting on the whole image.

This second condition, which is more subtle than the first, comes from the fact that initially, the device does not know exactly what a cell is. For example, there is no guarantee that a damaged nucleus is connected, and hence an approach based on the notion of connected particles, might mistake splinters of the same nucleus, to be different nuclei. Furthermore, at the beginning, figure and background are interchangeable phases; the action of the recognition process provokes the emergence of a shape, which finally sits clearly on a neutral background, free from artefacts. Criteria which consider the image as a whole, have two considerable technical advantages. They avoid reiteration of the same algorithm for each particle of the image, and more importantly, they lead to the design of general image analysers. The theory of mathematical morphology shows that we can construct a very extensive set of transformations by combining a small number of basic ones. The power of the Texture Analyser lies precisely in the "cabling" of these basic transformations (i.e. incorporating them into the hardware).

From what we have seen above, we note the following hierarchy of what are notions of a same type; namely at the general level we have the "principles" then the "criteria", and finally, more specific than the criteria, the "algorithms". An algorithm is the mathematical equivalent to a cook's recipe; it gives very precise instructions for handling numerical date, but says nothing about other means that might lead to the same result. For example, the same convex hull criterion that was used in the above example on cell nuclei, can be numerically performed by at least three different algorithms. Firstly we could take the intersection of all half planes containing the object X, secondly we could take the union of all segments whose extremities belong to X, and finally we could perform a closing, by a disk of a very large radius.

We see therefore, an hierarchical structuring of morphological transformations, in a direction leading to more and more particularization. In the table below, we have distinguished between general and specific criteria; the former lead to general notions such as the size distribution axioms, and the latter refer to a particular transformation (e.g. erosion, opening).

Hierarchy of "criteria"

Notions	*Examples*
Principles	Invariance under translation
General criteria	Transformations which are increasing
Specific criteria	Set covariance, convex hull
Algorithms	Union of all segments with both extremities in the object

E. MODELS

E.1. Structural description and laws

If one wanted to explain the difference between natural sciences and physics, the main point of departure would probably be that for the natural sciences, morphological description *precedes* the determination of laws, whereas for physics it *follows*. When Galileo, with the help of an inclined plane, studied the force of gravity, he focused all his attention on two parameters, namely time and a displacement function of time. The main interest of his law comes precisely from the elimination of the morphological characteristics (shape, volume, etc) particular to any body falling in a vacuum. On the contrary, when a sedimentologist wants to know the activity of a river flow (past or present), he must build morphological criteria of smoothness and sphericity to formulate his thoughts.

In the great majority of areas where mathematical morphology can be used, both the structural description and the determination of laws co-exist, and we must try to bring them together. It is not always an easy task, and if we take only one of these two points of view, our ship is in danger of sinking. For example, a pure physicist might consider that the flow in a porous medium is completely described as soon as the partial differential equation is written down, i.e. when he knows the Navier–Stokes law, plus the initial and boundary conditions. However the boundary conditions are so complex and locally unpredictable, that finally the physicist changes his approach (and the observation scale) and uses the Darcy law (flow is proportional to pressure with the coefficient of proportionality being the permeability). At this scale the Navier–Stokes law is virtually meaningless (Matheron, 1967).

Now, from a natural scientist's point of view, porous media can be classified according to their petrographic family: sandstone with or without clay, quartzites, etc. . . . Their permeabilities depend partially on this classification, but the relations are too fuzzy to become operational tools in the hands of the hydrogeologist.

Thus, two opposite points of view arise: the descriptive point of view of the naturalist, and the functional point of view of the physicist. What kind of relationships exist between the two? Up to now, we have only investigated the "naturalist" side of the mathematical morphology (criteria). We have yet to discover its "physicist" side; it is the notion of *model* which gives us access to this aspect.

A model is a set of hypotheses which can be formalized in a mathematical way, and from which we can deduce verifiable consequences. For example, the consequence of a criterion (transformation) on the realizations of the model can be predicted from the mathematical formalism. The above example deals with a morphological model, however we can imagine as consequences of a model, theoretical relationships between physical and morphological parameters (for example, Petch's law in metallography linking mean grain size with physical resistance). We note in passing that the usage of the word model in the literature, is sometimes ambiguous. The concept is often confused with its realization; for example, there is a tendency to consider as a model, the result of a simulation on a computer. Indeed when we say that an object can be represented by a finite union of convex sets, the underlying model is the family of all possible finite unions of convex sets, and one realization is the particular observed shape (for example, the union of an ellipse and a segment). This leads to a temptation in experimental science not to distinguish between cause and behaviour of the phenomena. (This is not the last time we will encounter the behaviourist's point of view.)

With models, we are for the first time able to verify some hypotheses. According to the degree of modelling that we choose, the methods of verification (and the meaning of this verification), can be considerably different. We will now present two examples that will help us to return later to the discussion of this point.

E.2. Examples of modelling

(a) *First example: estimation of a specific surface area*
Since the work of E. Weibel (1963), it has become common practice to estimate the specific surface area of lung alveolae from histological thin sections (Fig. I.10). (This method is used for example to study the pathology of certain emphysema.) Let us examine, step by step, the morphological problems which arise here. Experimentally speaking, we start by sampling one (or several) small block B, in the lung (the size of B is around one cubic centimetre); then a series of thin sections are taken from B. The basic problem is stereologic; we want to find a parameter which can be measured on the sections, and which is related to the surface area per unit volume of the alveolae (denoted S_V). The development of this problem, leads to four questions:

  (i) We have to take a set model, which can represent the structure of the tissue, so that the specific surface area can be related to certain planar parameters. To do this, we will choose the "convex ring" (Ch. V); hence we assume that any small sample of alveolae taken from the lung, can be represented as a finite union of convex sets. In this mathematical framework we will consider a random straight line Δ, with a uniform orientation in the three-dimensional space and which traverses the lung alveolae of the sample B. Let $N_L(\Delta)$ be the number of intercepts of alveolae per unit length on the straight line Δ. Then the mean value N_L of the number of intercepts when Δ takes all possible positions and orientations is equal to:

$$N_L = S_V/4$$

 (ii) We have to construct an estimator N_L^*, of the mean value N_L. All the Δ's would have to be obtained by looking at an infinity of sections oriented in all the directions, and this is obviously not possible. Moreover, due to the digitalization, we can only use a rather small number of directions of Δ. So at best, we will know an approximate value for the mean N_L.

(iii) With respect to the whole lung, the size and the position of the small piece B has some importance. How does the specific surface area S_V defined in this small block, vary through the lung?

(iv) We have to eliminate those artefacts that are in every section, since they can upset the value of the estimator N_L^*. For example, Figure I.10a represents a typical threshold image of a lung thin section. It has in its centre, a small capillary whose walls could be mistakenly taken for boundary of alveolae. A small image transformation by a morphological closing (in the sense of Ch. II) gives a better image, however we must note that this image is the result of a new modelling (where we assume that *everything* eliminated by the closure is an artefact). As can be seen from the numbers in the upper right hand corner (which are in fact proportional to N_L) the cleaning of the image has reduced the estimator by 40 per cent (from 7734 to 4518).

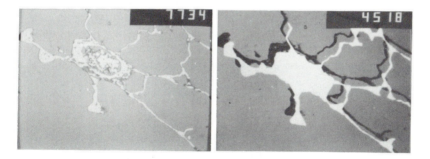

Figure I.10. Thin section of human lung. Alveolae appear in white. How can the artefact be eliminated? Here the perimeter of the white phase (shown in the upper right hand corner) is reduced to 40% of its original value after elimination of the central capillary.

It is not surprising that these four very different questions will need four different theoretical approaches. The first uses integral geometry, the second geometrical probability, the third second order statistical methods, and the fourth pattern recognition techniques. Consequently, the verification of hypothetical answers to the questions will surely take four different paths.

(b) *Second example: Cauwe's Law for the milling of rocks*

Let X be a collection of rocks in a crusher, milled for a certain time t. Is it possible to predict, from the operational conditions (e.g. type of crusher) and the initial state of the rocks, the size distribution (i.e. granulometry) of the milled product obtained after time t? To look at this question, Ph. Cauwe

(1973) studied two very different types of rocks: firstly, a Cu–Pb–Zn ore with a complex multiphase texture, and secondly a concentrate of very homogeneous magnetite. He milled them under thirty different operational conditions, and analysed on polished sections the morphology of the resulting fragments. These experiments led to a surprising result. The size distribution of the milled products was virtually independent of the initial state, i.e. of the texture of the initial rocks (crystal size, pores, contacts between phases, directions of cleavage), but depended mainly on the working conditions of the crusher. From these results, Cauwe interpreted milling by using random sets generated from a model generalizing the Poisson plane process in $\mathbb{R}^3$ (Ch. XIII, E.1). Imagine that an impact in the crusher slices up the mineral space by two very close parallel cuts. This splits up the grain into three parts, namely the intermediary slice and the two parts on either side. Any further impact acts on all three of the descendant particles in the same way, independently of the preceding impact; the direction of each impact is assumed uniformly distributed on the unit sphere.

We can see in Figure I.11 a photograph of a simulation of the model. It is a very nice image, but not realistic because in practice, each of the three particles after being split, do not stay together but move around in the crusher. Therefore we have to imagine all the polygons of Figure I.11 separated from each other and displayed in an arbitrary order. Only then do we have an idea of what the human eye sees on a polished section. With respect to this model,

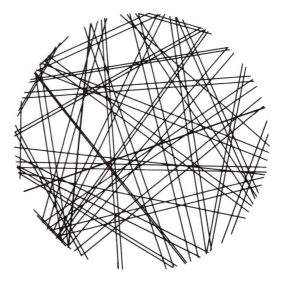

Figure I.11. Section of a realization of the process generated by Poisson doublets.

G. Matheron (1973) derived among other criteria, linear and circular erosions (the measures of which can be interpreted as the probability that a segment of length l, resp. a disc of radius r, is contained in the particle covering the origin). These two criteria provide two theoretical curves $P_1(l)$ and $P_2(r)$, and so we can fit numerical estimates $P_1^*(l)$ and $P_2^*(r)$ from the polished sections. Note that the fitted curve has in fact stereological implications, since it gives us estimates that characterize the (three-dimensional) polyhedra from measurements on plane sections.

This example gives answers to three questions which can be generalized in an obvious way:

(i) How can we find a geometric description of the milling process?
(ii) Having found the model, can we derive formulae that allow fitting of this model from plane sections?
(iii) Does the model which we have fitted, reflect the reality of the actual milled rocks we are studying?

The first question depends upon the creative imagination of the researcher, the second upon the calculus of geometric probabilities, and the third upon sampling problems and statistical estimation.

E.3. The level of the probabilistic approach

From these two examples, we might ask ourselves about the use made of the notion of *randomness*. They show that we can distinguish between three types of models: where the randomness does not appear at all, where it is introduced (artificially) as a *mode of operation* in order to access the object, and where it describes the *genesis* of the object.

When we replace Figure I.10a by Figure I.10b, we assume that all the boundary elements of the initial set removed by the small closure are artefacts, and that only the alveolae walls remain. For a biologist who controls the image analysis by a direct microscopic examination, this is immediately checkable. We can now proceed and apply the same filtering algorithm (i.e. the small closure) automatically to many images of lung, however not check each one. Then the above assumption about the artefacts becomes a little stronger, however remains completely deterministic, since it only holds on these images that we know *in totality* (even when we decide not to examine every one). Therefore, this type of modelling is purely geometric; among the subareas of image analysis we often find these models in pattern recognition. In practice, the deterministic models do not require an additional mathematical formalism since when we work with them, we are working immediately with the consequences of criteria (e.g. after the small closure, the image of the lung is analysed for itself). Their basic formalism is contained in Part I of this book where we develop mathematical morphology in an Euclidean and deterministic framework. This development then leads to a collection of criteria and algorithms which are studied individually and in more detail, in Part III.

The random aspect appears for the first time with questions of sampling. Although the lung under study remains perfectly determined and the local surface area of the alveolae has a unique numerical value, the *means of access* to the alveolae generate sufficient variability to allow the introduction of randomness into the formalism. The fact that we measure only from sections, the choice of the samples, the substitution of a grid of points for a connected object, etc., are all procedures which throw away information and only give access to a partial knowledge of the object (the lung in the first example). But this loss is perhaps not serious. Anyone who has looked at approximately ten images sampled from the same homogeneous object (thin section of lung, of rock, photos of clouds), knows very well the impression of sameness that they give: often if at a small scale we see many fluctuations in the detail, then at a larger scale the images achieve a certain uniformity. This repeatability basically justifies having only a partial knowledge of the object, and as a consequence, also justifies the probabilistic approach. Here we do not want to know the result of application of each criterion to *every* point of the space (e.g. whether or not a disk, centred at a given point x, is included in the alveolae). We are only interested in the *frequency* of the result of each criterion, called (in probabilistic terms) "events". Hence, we do not want to have an illusory complete knowledge of the object; on the contrary we are in a better position to take advantage of the redundancies of the images. Since these random aspects do not refer to any deterministic or random status of the object (but hold only on the results of technological operations performed on it), the simplest way for studying this randomization is by assuming that the object is *fixed* and *bounded* (even if it is experimentally very large). For the most part, it is what we will do in the sampling theory presented in Chapter VIII.

Suppose now that we consider random set models; here the emphasis is very much on the *process* which generates the final object. For example, in the Cauwe model presented above, the grains are no longer assumed to be deterministic but rather to be the realizations of a certain genetic process; i.e. doublets of Poisson planes delineating polyhedra in the space. In this approach the basic characteristics of the model (density of the doublets, statistical distribution of the distance between the two planes) carry the information more efficiently than any functional moments (they are in fact the probabilistic version of the criteria). They also suggest some causal explanation of the phenomena; of course it may not be the actual one, but anyway, it can always be (partially) verified, as we shall see in the fourth part.

F. PLAN OF THE BOOK

The distinctions uncovered by the above discussions determine the plan of the book. We devote the first part to construction of the geometrical formalism necessary for studying mathematical morphology. Beginning with the theory of the image transformations, we are led to first study the algebraic and then the topological properties of the Hit or Miss transformation (Chs II and III). In the next two chapters we move our emphasis from transformation to morphological measurements. Using the simplest model of a shape, i.e. that of a compact convex set (Ch. IV) as a starting point, we look for a class of basic parameters which can be associated with objects modelled by more and more sophisticated sets (Ch. V).

In the second part, the first two aspects of partial knowledge considered above in the example of the lung, are successively treated: digitalization of the images (Chs VI and VII) different random modes of sampling (Ch. VIII). Notions already studied in the first part via purely geometric means in the Euclidean space, must now take on a probabilistic and digital colouring. In the third part, we make a systematic study of the main criteria, namely the covariance (Ch. IX), the size distribution (Ch. X), and the iterative algorithms for sets and gray tone images (Chs XI and XII). These last two chapters are illustrated by a number of pattern recognition examples. Throughout this third part, deterministic and probabilistic (from an operational point of view) versions of the criteria coexist.

In the fourth and last part, we present a study of various models. The probabilistic tool is refined somewhat, in order that it can be used to construct sets that are realizations of genetic processes. In Chapter XIII, we review three main groups of models: those associated with the Boolean processes, those that partition the space, and the point processes. The book ends with a discussion about the use and the meaning of models in mathematical morphology (Ch. XIV).

G. "PRAVDA" OR "ISTINA"?

The "truth" is accessible in just two ways: either we conform to things or they conform to us. The Russian language uses the word "istina" for the first case (when we adapt our mental structures to the outside world) and "pravda" for the adaptation in the reverse sense. We use "truth" as a rather poor translation of both words.

The morphological approach, as we have outlined it, oscillates between these two viewpoints. Its first movement is directed from the mind towards the physical, geometrical world; our method has in fact been especially conceived for the observation of this world. The principles, the criteria and the algorithms all come from this first movement. Their role is to "grip" the object by a series of operational constraints that become more and more restrictive. By construction, they are not verifiable. Their use is analogous to the use the human body makes of its five senses, or to the use a tradesman makes of his panoply of tools. The decision to use the variogram in preference to the opening for the description of wood anatomy, is similar to the choice of pliers instead of the head of a hammer for extracting a nail.

The second movement happens when the pieces of information extracted by the criteria induce a certain interpretation of the object, opening the door to a universe of canonical patterns, models, each with their own logical structure. A small broken pebble might suggest a model of random doublets of planes. The point of view is reversed: it no longer matters whether we, the experimenters, are well adapted to the peculiarities of the object; on the contrary, we now wish to determine how much the object conforms to this creation of our mind (namely the model). The criteria then appear as simple intermediate steps to be taken, to test whether the contingent object resembles the concrete reality of the model (in the whole datum of a model there is in general more than the few properties we test by applying criteria. On the contrary, in the first approach priority is given to the criteria).

It is not due to mere chance that the two viewpoints of mathematical morphology also appear in the Russian concept of truth. They reflect two deep irreducible psychological attitudes. In his famous work on the unconscious, C. G. Jung showed that a large number of the intellectual controversies of history were based on the contrast between two types of minds, namely introversion and extroversion. Probably the most important one was the opposition between Plato and the Megare School, which involved the notions of nominalism and realism. More recently, the two psychological theories of the unconscious, due to Freud and Adler, have provided a new version of the same controversy. For Adler, the subject, who feels belittled, wants to assure an illusory superiority by protestations and other appropriate artifices, which he uses when confronted with any sort of obstacle. In such an approach, the subject is extraordinately emphasized, while the object itself completely fades out. On the other hand, Freud conceives his patients as being completely dependent upon the outside world. What is coming from the subject is essentially a blind thirst for pleasure, thirst which is shaped only by the influence of specific objects.

By these psychological references, we do not mean that theoreticians are all introverts, and experimenters all extroverts, but simply wish to emphasize the duality between criteria and models, which has its source far from scientific thinking. This book attemps to develop a new avatar in the history of the relationships between "Pravda" and "Istina". The last two parts of the book are a result of the two viewpoints. But, more generally, right from the first part, the dialectic between the two points of view enlivens every page: in one place we build up the convex set model in order to obtain stereological criteria, elsewhere we start from an erosion criterion in order to clearly define the plane doublets model, etc., etc. And we shall never finally reach the stage of a true synthesis, nor shall we ever see the definite domination of one viewpoint over the other . . . .

Which one of these two attitudes appeals the most to you? The choice is left to you, dear reader.

II. Hit or Miss Transformation, Erosion and Opening

A. WHY THIS CHAPTER HEADING?

In the construction of mathematical morphology, the Hit or Miss transformation occupies a central place with different paths leading from it; we can get to (or from) the hit or miss transformation in at least five different ways. First we notice that the transformation extends the notion of spatial law of a random function when the latter can only take values one or zero. We will take this direction in part IV only, since it requires clarification of some topological points, and also because it would not be wise at the outset, to link a purely geometric transformation with a (optional) probabilistic interpretation. The hit or miss transformation is an attempt at mathematical formalization of certain ideas of the gestalt psychology. This has historical interest, since it was in fact the study of gestalt psychology fourteen years ago that suggested to us the idea of this transformation. We will not proceed further with this intriguing aspect, since it would deserve a more thorough treatment than could be given in the limited space here (Serra, 1964).

The third direction is the most travelled; the hit or miss transformation comes from (or finishes at) experimentation and the texture analyser. In physics it is not sufficient to construct designs of experiments, to establish subtle distinctions between randomness in the process and randomness in the results, nor to fix the hierarchy of principles, criteria, algorithms. Without a device for performing experiments, the good intentions of Chapter I are wasted. But how can a device be built for transforming images that function with the same generality as the four principles? A very complex computer can perform any desired transformation, but to ask this of our device is to ask too much. However, systems which count points, parameters and particles are forever repeating the same algorithms, and so these may be wired or cabled once and for all. We need a system based on fundamental geometric operations, just as the computer is based on the binary Boolean algebra, and flexible enough for the elaboration of many diverse criteria within the framework of the principles of quantification. But how can we find a road leading to the planning of such a device? Probably it can be found in the fourth direction, guided by a few general theorems which relate certain logical prerequisites to erosions and dilations. If our exposé is to have a strictly deductive order, it is with these theorems that we would have to start. But the simplicity of the presentation suggests a more historical direction. A rapid scan of the literature of the last few years, shows that in almost all the morphological algorithms used in practice, there appears in the form of a skeleton, iterations of very simple transformations (numbering one, two, three, . . . infinity).

B. "ONE, TWO, THREE . . . INFINITY"

Obviously, the atomic physicist Gamow had not foreseen the ideas behind this section when he used the above as a title for his book (1957), although it fits our situation so perfectly that we have adopted it. It describes—in just four words—the whole recent evolution of the methods in Texture Analysis. We will start the contemporary history of these methods and techniques at the beginning of the 1960s, when a crystallization of various events occurred. An interest in quantification from microbiology, petrography and metallography, the Society for Stereology, and the first quantitative television microscope, all date from this time. It was thought that characterizing quantitatively a structure meant assigning to it, a few representative parameters. Three of them are considered as basic in the plane, namely the area, the number of intercepts, and the number of particles; the reason is that they allow reconstruction of respectively the volume, the surface area and the sum of the mean curvature in the three-dimensional space.

B.1. Step "One"

Here, "one" is not an arbitrary label; it means that behind each basic parameter, *one* elementary image transformation is used. The simplest case is that of the area measurement in a digitalized image. It may be interpreted as the number of times that a test point (structuring element B on Fig. II.1), hits the image under study when it is translated on the grid, from vertex to vertex. In the case of Figure II.1, the area of X equals 39. Although it seems trivial, this procedure is actually a transformation performed on the digitalized image. Likewise, the number of intercepts, which is proportional to the length

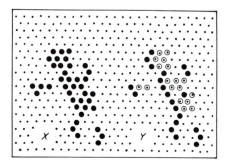

Figure II.1. Initial set X, and its transform Y involved in the intercepts measurements. From Fig. II.1 to II.7 the symbols concerning the structuring elements are as follows:
● points of the structuring element which must belong to X
○ points of the structuring element which must belong to X^c
· points of the structuring element with no condition
↓ location of the origin associated with the structuring element.

parameter has behind it a transformation involving a pair of consecutive points (i.e. the structuring element $B' = b_1 b_2$ on Fig. II.1). The transformation is the following: if $x + b_1$ belongs to the background X^c and $x + b_2$ belongs to X, then the transform gives a "one" at $x + b_2$. When x is allowed to sweep the whole field, a new image Y is generated. The number of intercepts of X, is equivalently the number of picture points of Y, and equals 20 in the case of Figure II.1. The connectivity number (i.e. the number of particles minus their holes) can be calculated by the following similar approach. The associated structuring elements are now triangular configurations, involving adjacent points of the grid. More precisely, one can prove (rel. VI-22) that the connectivity number N equals the following difference:

$$N = \mathcal{N} \begin{bmatrix} & 1 & \\ 0 & & 0 \end{bmatrix} - \mathcal{N} \begin{bmatrix} 1 & & 1 \\ & 0 & \end{bmatrix}$$

$(\mathcal{N}(*) =$ number of configurations of the type $*)$ (II-1)

In other words, if:
Y is the set of points $x + b_2$ such that
$$x + b_1, \qquad x + b_2 \in X \quad \text{and} \quad x + b_3 \in X^c$$
Y' is the set of points $x + b'_3$ such that
$$x + b'_1, \qquad x + b'_2 \in X^c \quad \text{and} \quad x + b'_3 \in X$$

then, the connectivity number N is the difference of areas between Y' and Y (Fig. II.2). The results illustrated on Figures II.1 and II.2 are well known. For simplicity, we have presented them in the digital case, however, they can be transposed to the continuous space $\mathbb{R}^2$ (doublets and triplets of points become supports of distributions, and the sets X, Y have to belong to a fairly regular

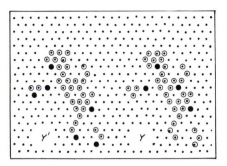

Figure II.2. Transforms Y and Y' involved in the connectivity number computation.

class, such as the convex ring). These figures should show clearly how the set transformations are performed. In particular, it is instructive to notice that the smallest structuring element of zero dimension (i.e. the point), one dimension (the doublet), and two dimensions (the triplet) respectively open the way to the area, length, and number measurements on the set X. Thus if the dimension of the structuring element is added to the dimension of the parameter measured, then we obtain a constant, namely the number of dimensions of the space of the object. This theorem turns out to be a consequence of the kinematic formula due to Van Blaschke (1937) and Santalo (1933), which of course may be generalized to the n-dimensional space. From the stereological point of view, its main interest lies in that it limits the knowledge of X accessible by sections of subspaces of a given dimension. It shows for example that the number of particles in $\mathbb{R}^3$ cannot be estimated from measurements performed on plane sections (except of course, if we specify via a model, the structure under study).

B.2. Step "two"

The structuring elements B that we just saw were infinitesimal ones, limited to the size of the grid spacing. If we now try the above procedure, but for larger B, we are then able to produce well known measures, as well as some new ones. Suppose we are interested in the distribution of the intercepts (linear size distribution). For experimentally determining one value of the histogram, say the percentage of the chords having digital length l, we must successively carry out three steps.

(i) to transform the initial set X according to the above approach, by taking now as the structuring element, the following figure B:

$$B = \underbrace{0 \quad 1 \quad 1 \quad 1 \quad 0}_{l}$$

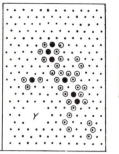
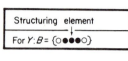

Figure II.3. Chord distribution transformation (size four).

On digital images, l equals the number of intervals (here, $l = 4$). If we start from the set X drawn in Figure II.1, the transformation leads to the set Y of Figure II.3.

(ii) to apply to the set Y the elementary transformation giving the area measurement (here the area of Y equals 7).

(iii) to divide this result by the total number of chords of X, or in other words, by the total number of intercepts determined in Figure II.1. Then the final result is expressed in terms of proportions, and can be interpreted as probabilities in a probability mass function.

The third operation requires no more comment since it simply concerns a calibration of the results. In the two other operations, the process of image iteration appears: a first modification of the set has been performed in step (i) followed by the elementary image transformation of step (ii). Such a two-step approach is extremely frequent. Consider for example, covariance analysis. The structuring element becomes a pair of points distance h apart, in the horizontal direction (for example):

$$B = \underbrace{1 - - - 1}_{h} \quad \text{(here } h = 3)$$

We can see on Figure II.4 the result of the set covariance of size 3 when starting from the same initial set X. Notice the difference between a size distribution transformation and a set covariance. In the latter case two disconnected components of X can be hit by the structuring element, giving the transform Y its discontinuous character.

We could easily find similar examples. The convexity number, which will be presented in Chapter V, the determination of the radii of curvature in $\mathbb{R}^2$ (Ex. X-14), etc. . . . all depend upon this approach: an initial image transformation always precedes the one leading to the measurement; hence the number *two*.

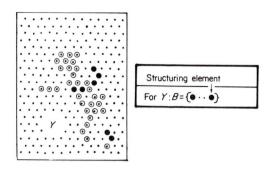

Figure II.4. Set covariance transformation (size three).

Before going further, let us remark on the coherence of the method. It regroups under a single concept all the well known theoretical tools of quantitative microscopy (as were existing for example in 1965, year of the first publications in mathematical morphology). Their origins were so diverse however, that around 1965 no researcher was actually using them all in a coherent fashion. The covariance descends from the theory of random functions of order two, the basic stereological parameters come from H. Minkowski (1903), the use of the laws of intercepts probably began with Crofton's work, and the dilations of convex sets date back from J. Steiner (1840). (For a more systematic review of our "ancestors" of the seventeenth and nineteenth centuries, see the paper of R. Miles and J. Serra (1978).)

We shall now look at this idea of synthesising seemingly different procedures, from a more general point of view. Define the Hit or Miss transformation (Serra 1964, 1969) as the point by point transformation of a set X, working as follows: we choose, and fix, a structuring element B; i.e., the datum being two sets B^1 and B^2. Suppose B is centred at the point x; denote this by (B_x^1, B_x^2). A point x belongs to the Hit or Miss transform $X \circledast B$ of X, if and only if B_x^1 is included in X and B_x^2 is included in the complement X^c of X:

$$X \circledast B = \{x : B_x^1 \subset X; B_x^2 \subset X^c\} \qquad \text{(II-2)}$$

This transformation has been studied in detail in G. Matheron (1967) (where it is called "Serra's transformation") and in J. Serra (1969). A particular but important case is that of the *erosion*. Take for B^2 the empty set. The condition $B_x^2 \subset X^c$ is always fulfilled, and the eroded set Y is the locus of the points x, such that B_x is included in X:

$$Y = \{x : B_x \subset X\} \qquad \text{(II-3)}$$

By intersecting erosions holding on X and on X^c, we are able to construct the Hit or Miss transformation defined by (II-2). Therefore this basic morphological transformation is worthy of more serious study (Section C below). The definition (II-2) can be immediately generalized to the case where several component sets are defined in Euclidean space. Then every point x has to fulfil as many conditions of the type $B_x^i \subset X_i$ as there are set components X^i. Although it is mathematically trivial, this generalization has practical importance in at least three situations:

(a) The "petrographic" situation: p components (often called "phases") exactly partition the space, without overlapping each other, and without leaving empty zones (Ex.: in the basalt seen in the first chapter, Figure I.1, magnetite, pyroxene, plagioclase and pores obey this structure). In set theoretical terms:

$$X_i \cap X_j = \varnothing \quad \text{if} \quad i \neq j \quad \text{and} \quad \bigcup_{i=1}^{p} X_i = \mathbb{R}^n$$

(b) The "geographic" situation: the p components can overlap, and may leave holes between each other. Such an event happens when the X_i represent the zones of extension of different kinds of trees in a forest. At each point x of the space, is defined a vector with p binary components, denoting presence or absence of the various types of trees.

(c) The "grey-tone" situation: the p components are nested in such a way that $X_{i+1} \subset X_i$. This especially arises when the object under study, because of lack of optical (or physical) contrast, does not allow assimilation to a set (i.e. to a 0–1 function). However, what we do is to systematically vary the threshold of the light intensity t. In the limit, the index t can be considered as a continuous parameter; this is followed up in Chapter XII.

For the moment, we restrict ourselves to the morphological study of a set X and its complement X^c. The similarity between the various morphological operations already emphasized above, suggests an examination of the general characteristics of the Hit or Mis transformation. What are its algebraic properties? What are its topological properties? How do the edges of the mask affect it? How can we formalize the concept that a family of transforms tends towards a limit set? By what rules can we iterate successive Hit or Miss transformations? Can a transformation using a multidimensional structuring element be simplified as the "product" of uni-dimensional ones? etc. . . . , etc. . . .

B.3. Step "three"

Now two Hit or Miss transformations precede the measurement of parameters, however, if this new step only involved adding one more iteration, it would not be worth speaking about. In fact, step "three" opens the door to bi-dimensional transformations, and allows the birth of new criteria (it is nice that after *unifying* previous notions, our approach has indeed blossomed new notions). As an example, we will use (see Fig. II.5) the now well-known operation of "closing", dating from the end of the 1960s (Matheron, 1967). It is the product, first, of an erosion of the background X^c by the elementary hexagon H of the grid (i.e. the dilation $X \oplus H$ of X by H), followed by an erosion of the transform $X \oplus H$ by H (Fig. II.5).

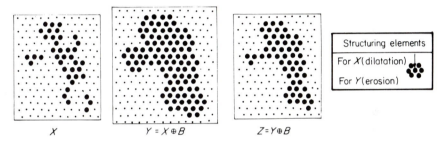

Figure II.5. Closing of X by H.

B.4. Step ". . . Infinity"

From step "three", the question of whether there is a step *"infinity"* is unavoidable. The number of potential transformations is unbounded, compared to the number of possible parameters to measure. By "infinity", we mean here either an infinite iteration of the same algorithm, or systematic modifications of a threshold on an image, leading to series of sets included in each other, or more generally, complex flow sheets of transformations, with logical loops, etc. . . . Chapter XI reviews the major recent contribution of the Centre of Morphology in this field. For the moment we will illustrate with two pedagogical examples, how such infinite image modifications work.

(a) *Connected component extraction*
Let X be the initial set, and x_0 a given point belonging to X. One wants to extract the connected component Y which contains x_0. If H_1 is the smallest regular hexagon of the grid (see Fig. II.6), Y obviously appears as the limit of the induction system:

$$X_1 = x_0; \qquad X_{i+1} = (X_i \oplus H_1) \cap X; \qquad Y = X_\infty$$

since by construction, each point of Y belongs to the set X and is connected by a path to x_0.

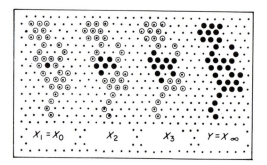

Figure II.6. Connected component extraction. The successive elementary dilations invade the connected particle Y which contains x_0, but cannot leave Y.

(b) Convex hull

We now want to construct the convex hull Y of a given set X. Let B^j be the following elementary triangles, deduced from each other by turning clockwise:

$$\left\{\begin{matrix} & 1 & \\ 1 & & 0 \\ & 1 & \end{matrix}\right\} = B^1 \qquad \left\{\begin{matrix} & 1 & 1 \\ 1 & & 0 \end{matrix}\right\} = B^2 \ldots \left\{\begin{matrix} 1 & & 0 \\ & 1 & 1 \end{matrix}\right\} = B^6$$

The dilations are performed by three points and the result given to the fourth point (the empty circle) which plays the role of the "centre" of the structuring element. The method consists of filling up X, direction by direction, as follows (Fig. II.7):

$$X_1^1 = X; \qquad X_{i+1}^1 = (X_i^1 \ominus B^1) \cup X; \qquad X^1 = X_\infty^1; \qquad Y = \bigcup_{j=1}^{6} X^j$$

Some other iteration procedures can be found in Golay (1967) and K. Preston (1970). Their papers are instructive, giving an approach to the

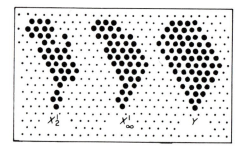

Figure II.7. Hexagonal convex hull of X (X is the same set as in Fig. II.6).

"infinity" step. However, they limit themselves to iterations of infinitesimal Hit or Miss transformations. Most of the macroscopic transforms cannot be considered as iterations of infinitesimal ones (Ex.: openings, covariance . . .). Finally, notice that "infinity", in terms of digitalization, really means fifty, or sometimes five or six . . . .

C. THE HIT OR MISS TRANSFORMATION: AN ALGEBRAIC STUDY

C.1. Definition of the dilation and the erosion

We define the eroded set Y of X as the locus of centres x of B_x included in the set X (rel. II-3). This transformation looks like the classical Minkowski substraction $X \ominus B$ of set X by set B used in integral geometry (Hadwiger, 1957) and defined as the (possibly infinite) intersection:

$$X \ominus B = \bigcap_{b \in B} X_b$$

Indeed, when y sweeps B_0, the point $x + y$ lies in X iff x belongs to the translate X_{-y} of X; i.e.:

$$Y = \{x : B_x \subset X\} = \bigcap_{y \in B_0} X_{-y} = \bigcap_{-y \in B_0} X_y = X \ominus \check{B} \qquad \text{(II-4)}$$

where $\check{B} = \bigcup_{y \in B} \{-y\}$ is the transposed set of B; i.e.: the *symmetrical set* of B with respect to the origin. The analytical expression for the erosion of X by B, given by (II-4), allows us to study the algebraic properties of this operation, and also suggests a way to compute it in practice. Consider now what happens to the complementary set X^c (i.e. "the pores") when eroding set X (the "grains"). Obviously, the pores have necessarily been enlarged, since erosion has shrunk the grains. We shall call this operation *dilation*, and denote it by the symbol $\oplus$. We have:

$$X^c \oplus B = (X \ominus B)^c \qquad \text{(II-5)}$$

Dilating the grains is equivalent to eroding the pores, and conversely. The two operations are said to be dual to each other w.r. to the complementation (see appendix). Just as with erosion, dilation has a local interpretation, since the complement of the proposition "B_x is included in the pores X^c", is the proposition "B_x hits the grains X". Thus the dilate $X \oplus \check{B}$ is the locus of the centres of the B_x which hit the set X:

$$X \oplus \check{B} = \{x : B_x \cap X \neq \varnothing\} = \{x : B_x \pitchfork X\} \qquad \text{(II-6)}$$

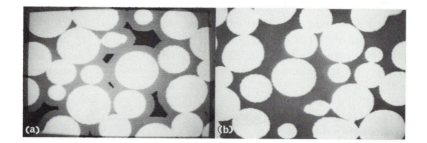

Figure II.8. Dilation by the disk of radius r. The boundary $\partial(X \oplus B)$ is the curve parallel to the boundary ∂X at the distance r. Notice the eroded mask in (b).

where the vertical arrow "⇑", means "hits". By applying (II-4) and (II-5) to the pores X^c, we find an analytical expression of the dilation of X by B.

$$X \oplus \check{B} = \bigcup_{y \in \check{B}} X_y \qquad (II\text{-}7)$$

By noting that X is identical to the union of all its points $(X = \bigcup_{y \in X}\{x\})$, we derive from (II-7) a second more symmetrical expression,

$$X \oplus \check{B} = \bigcup_{\substack{x \in X \\ y \in \check{B}}} \{x + y\} = \bigcup_{x \in X} \check{B}_x \qquad (II\text{-}8)$$

From the first equality of (II-8), we recognize the Minkowski addition, classical in integral geometry (Minkowski, 1903; Hadwiger, 1957). The second equality is obtained by changing the order of the unions in the Minkowski sum: point x remains fixed and point y runs over B. We see also why it was necessary to use the transposed set $\check{B}$ in the definition. Due to this transposition, the roles of X and $\check{B}$ in (II-8) are made symmetrical. Note that the erosion $X \ominus \check{B}$ does not admit a similar symmetrical expression.

A brief example will clearly show the difference between $X \oplus \check{B}$, and $X \oplus B$. Take for X as well as for B the same equilateral triangle of side a. Then $X \oplus B$ is the equilateral triangle of side $2a$, although $X \oplus \check{B}$ turns out to be regular hexagon of side a (Fig. II.9).

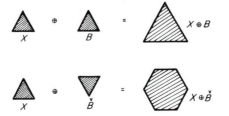

Figure II.9. Difference between the dilations by B and by $\check{B}$.

C.2. Algebraic properties

The relation (II-8) shows that dilation is an associative and commutative operation, so the set $\mathscr{P}(\mathbb{R}^2)$ of all subsets of $\mathbb{R}^n$ is an abelian semi-group; dilation does not admit an inverse. In other words, if we start from the dilated set $X \oplus B$, we cannot recover the original set X, since fine details have been smoothed away by taking unions of the X_b (a fortunate weakness, since it leads to construction of the interesting new set $(X \oplus \check{B}) \ominus B$, which will differ in a significant way from the original set X . . . .). The algebraic properties that we will see now are due largely to H. Hadwiger (1957). Our contribution has consisted of clearing up the questions of local knowledge and edge corrections.

(a) *Dilation and erosion by a point*
The origin $\{o\}$ is the unit element of the semi-group:

$$X \oplus \{o\} = X \ominus \{o\} = X$$

and

$$X \oplus \{x\} = X \ominus \{x\} = X_x$$

For $x \in \mathbb{R}^n$, $X \oplus \{x\}$ and $X \ominus \{x\}$ are the translates of X by x. Consequently, we can take for the origin (i.e. the "centre") any point inside or outside the structuring element B. The results of the two basic morphological operations will be the same modulo a shift; i.e. dilation and erosion are *invariant under translation*.

(b) *Distributivity*
Dilation distributes the union. We can write:

$$X \oplus (B \cup B') = \bigcup_{x \in X} (B_x \cup B'_x) = \left(\bigcup_{x \in X} B_x \right) \cup \left(\bigcup_{x \in X} B'_x \right)$$

which yields relation (II-9), and by duality (II-10) and (II-11):

$$X \oplus (B \cup B') = (X \oplus B) \cup (X \oplus B') \qquad \text{(II-9)}$$
$$X \ominus (B \cup B') = (X \ominus B) \cap (X \ominus B') \qquad \text{(II-10)}$$
$$(X \cap Z) \ominus B = (X \ominus B) \cap (Z \ominus B) \qquad \text{(II-11)}$$

The technological implications of the three relations are enormous. From the first two, we can dilate, or erode X by taking B piece by piece, and then combining the intermediary results by union or intersection respectively. The third formula is more basic again, and is treated below in the subsection on *local knowledge*. We can compare these three relations to the three following set inequalities:

$$X \oplus (B \cap B') \subset (X \oplus B) \cap (X \oplus B')$$
$$X \ominus (B \cap B') \supset (X \ominus B) \cup (X \ominus B')$$
$$(X \cup Z) \ominus B \supset (X \ominus B) \cap (Z \ominus B)$$

We can easily characterize the family Ψ_λ indexed by a positive parameter λ, generated by dilation and compatible with changes of scale (second principle). According to relation (I-3), this transformation Ψ_λ may be written as:

$$\Psi_\lambda(X) = \lambda \Psi_1 \left(\frac{X}{\lambda} \right) = \lambda \left(\frac{X}{\lambda} \oplus B \right) = X \oplus \lambda B \qquad \text{(II-12)}$$

Therefore the family of transformations Ψ_λ is the family of dilations *by the homothetic* λB of B. By duality, the same is true for erosion. This theoretically obvious result in fact limits us in the choice of possible structuring elements. As a counter-example, consider the family B_λ represented in Figure II.10. The structuring element B_λ is defined as the union of two disks, with variable radius λ, and with centres a *fixed* distance d apart. Its size distribution properties (see Section 5) make it a possibly interesting family of structuring elements, however, it does not satisfy the relation $\Psi_\lambda(X) = \lambda \Psi_1(X/\lambda)$, and hence is not compatible under changes of scale.

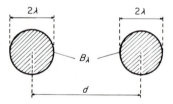

Figure II.10. An example of a family B_λ, not compatible under change of scale.

(c) *Iterativity*
From relations (II-4) and (II-7) we can derive that:

$$\left. \begin{array}{l} (X \oplus Y) \ominus Z = \bigcap_{z \in Z} \bigcup_{y \in Y} X_{y+z} \\[2mm] (X \ominus Z) \oplus Y = \bigcup_{y \in Y} \bigcap_{z \in Z} X_{y+z} \end{array} \right\} \qquad \text{(II-13)}$$

and thus we obtain the (generally strict) set inequality:

$$(X \ominus Z) \oplus Y \subset (X \oplus Y) \ominus Z \qquad \text{(II-14)}$$

In words then it is more severe to erode before dilating than to do the reverse. On the other hand, the iteration of the erosions presents no problem, and we may write:

$$(X \ominus B) \ominus B' = \bigcap_{b' \in B'} (X \ominus B)_{b'} = \bigcap_{b' \in B'} \bigcap_{b \in B} X_{b + b'} = \bigcap_{z \in B \oplus B'} X_z$$

Then:

$$\left.\begin{array}{l}(X \ominus B) \ominus B' = X \ominus (B \oplus B') \\[2mm] (X \oplus B) \oplus B' = X \oplus (B \oplus B')\end{array}\right\} \qquad \text{(II-15)}$$

(The second equation is obtained from the first by taking complements.) Once again it is worthwhile looking behind the dry formalism of relationships such as (II-15) for technological implications. Here we are provided with a powerful morphological tool, which enables us to decompose a given B into the Minkowski sum of several much simpler B_i's. A large number of structuring elements can be composed in this way. For example, a pair of points, added to another pair of points, aligned perpendicularly to the first pair, becomes the four vertices of a rectangle (Fig. II.11). Such decompositions are particularly useful, theoretically and practically speaking, when dealing with *convex* structuring elements. The erosion by the regular hexagon with side length a is the result of three successive erosions by the three segments of length a, which are parallel to the edges of the hexagon (Fig. II.12). In other words, *bi-dimensional* erosions can become iterations of *uni-dimensional* erosions (this will occur when structuring elements are for example the square, the hexagon, the dodecagon, the octogon, but not the triangle or the pentagon!). Similarly, it

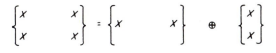

Figure II.11. An example of decomposition of a Minkowski sum into simpler sets.

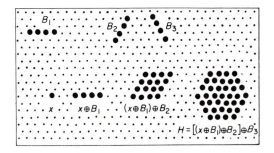

Figure II.12. The hexagon considered as the Minkowski sum of the segments.

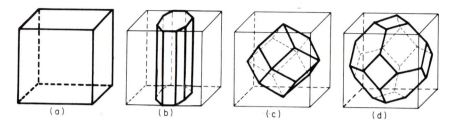

Figure II.13. The four simplest classes of polyhedra able to be represented as the Minkowski sum of segments.

is possible to go from the straight line, or the plane, to three-dimensional space. Imagine we have successive serial sections at our disposal. Then we can erode a 3-D set X (see Ex. VI, 17) with respect to a polyhedron B, by combining 2-D erosions. Four classes of polyhedra particularly lend themselves to this operation: the cube, the hexagonal prism, the rhombododecahedron, and the tetrakaidecahedron (Fig. II.13). In the case of the cube, the three Minkowski terms are three orthogonal vectors with the same length; for the rhombododecahedron, the terms are four vectors normal to the faces of a regular tetrahedron; for the tetrakaidecahedron they are the edges of the same tetrahedron. The hexagonal prism (hexagonal basis, square lateral faces) is obtained by the dilation of a hexagon by a vector perpendicular to its plane. In passing, note the following curious point: these four prototypes are the four Ferodov parallelohedra (Flint, 1957) which are regular compact polyhedra that, when stacked, are able to completely fill the space.

(d) *Increasing and inclusion properties*
Erosion and dilation are *increasing* operations:

$$X \subset X' \Rightarrow \begin{cases} X \ominus B \subset X' \ominus B \\ X \oplus B \subset X' \oplus B \end{cases} \quad \forall B \tag{II-16}$$

and, by duality:

$$B \subset B' \Rightarrow X \ominus B \supset X \ominus B' \quad \forall X$$

These inclusions are due to the fact that union and intersection are increasing transformations. However, the general Hit or Miss transformation is *not* increasing, since the complementary mapping, $X \to X^c$ is decreasing.

The inclusions that we now present are less obvious. The inclusion $\{o\} \in B \ominus \check{B}$ is statisfied by any $B \in \mathscr{P}(\mathbb{R}^n)$, but the equality is not necessarily true. However, if a point $x \neq 0$ belongs to $B \ominus \check{B}$, then $B_x \subset B$ and thus also $B_{ix} \subset B$ for every integer $i \geq 0$; hence the unbounded sequence $\{ix\}$ is included in B. It follows that for any *bounded* set B we can write the equality:

$$B \ominus \check{B} = \{o\}$$

Similarly, if $B \subset C$, the eroded set $B \ominus \check{C}$ is not always empty, but if not (same proof as just above) then B and C are necessarily unbounded sets. (As an example, take for B and C two half planes (in $\mathbb{R}^2$) with parallel boundaries, and such that B is included in C.) It is easy to show that if B is bounded and the inclusion $B \subset C$ is strict, then $B \ominus \check{C}$ is empty.

These last properties show how crucial the notions of *bounded* sets, *closed* sets, *isolated* points etc. . . . are. The necessity of a topological study receives more endorsement from such properties.

(e) *Local knowledge*

We saw that the Hit or Miss transformations, and hence their finite iterations, are invariant under translation and compatible with a change of scale. For demonstrating that local knowledge is available, we need to find a corresponding relation to (II-11), but for dilations. Recall that the set X is assumed to be (partially) known only through the mask, or the frame, Z. This means that an experimenter can only work with the sets $X \cap Z, Z$, and their complements. So it is with this in mind that we try for a relation on dilation. Consider the quantity $X \cup Z^c = (X \cap Z) \cup Z^c$. Then:

$$(X \cup Z^c) \oplus B = (X \oplus B) \cup (Z^c \oplus B) = (X \oplus B) \cup (Z \ominus B)^c$$

The largest mask whose intersection with $(Z \ominus B)^c$ is empty is $Z \ominus B$ itself; thus:

$$[((X \cap Z) \cup Z^c) \oplus B] \cap (Z \ominus B) = (X \oplus B) \cap (Z \ominus B) \qquad \text{(II-17)}$$

Relation (II-17) solves the local knowledge problem for dilation. The generalization to the Hit or Miss transformation acting on k components $(1 < k < \infty)$ is obvious. Consider k sets $X^1 \dots X^k$ defined in the same space, and known inside a same mask Z; they are respectively eroded by $B^1, \dots, B^k$. The final mask in which all the transforms are simultaneously known is:

$$Z \ominus (B^1 \cup B^2 \dots \cup B^k) \qquad \text{(II-18)}$$

If now we focus attention on component $n^0 \cdot i$ only, which is successively eroded or dilated by $B_1^i, B_2^i \dots B_p^i$, then the final transform of this component is known inside the mask:

$$Z \ominus (B_1^i \oplus B_2^i \oplus \dots \oplus B_p^i) \qquad \text{(II-19)}$$

The two relations (II-18) and (II-19) solve the local knowledge problem (or edge effect problem) for finite iterations of the Hit or Miss transformations. The edge effect corrections have caused a large proliferation of papers, and more than a controversy (see for example the "Proceedings of the Successive Symposia of Stereology since 1970"), we would like to make a few comments on our results. First of all, note that the eroded masks $Z \ominus B$ are not the largest possible in general. In the case of dilation for example, it may happen that some points x, located out of Z, are centres of a B_x which hit a point of $X \cap Z$, and thus surely belong to the dilate (similarly for erosion). Nevertheless the mask $Z \ominus B$ is the largest that is *independent of* X and in which the transform is known for each point. This independence turns out to be a very important requirement from a technological point of view. It implies that we can build into the Texture Analyser, a logical unit which automatically erodes the mask Z according to the structuring elements programmed for the analysis of X, independent of the geometric characteristics of the image X.

D. OPENING AND CLOSING

Recall from before that erosion and dilation only satisfy the algebraic structure of a semi-group. After having eroded X by B, it is not possible, in general, to recover the initial set by dilating the eroded set $X \ominus B$ by the same B. This dilate reconstitutes only a part of X, which is simpler and has less details, but may be considered as that part which is the most essential. It turns out that the new set (called the "opening" of X) actually filters out a subset of X which is extremely rich in morphological and size distribution properties. We will now examine this point more precisely.

We call the *opening* of X with respect to B the following set, denoted X_B:

$$X_B = (X \ominus \check{B}) \oplus B \qquad \text{(II-20)}$$

Similarly, we call the *closing* of X with respect to B (denoted by X^B) the following set:

$$X^B = (X \oplus \check{B}) \ominus B \qquad \text{(II-21)}$$

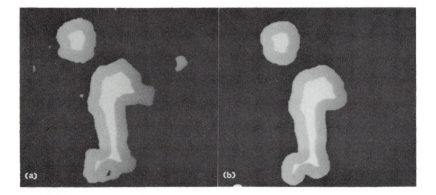

Figure II.14. The two steps of erosion and dilation leading to the opening.

These two concepts are due to G. Matheron (1967, 1975) who studied them within the framework of an axiomatisation of the concept of size. We see in Figure II.14 the two steps leading to the opening of a set X by a regular dodecagon. When the boundary of X is sufficiently smooth, with large radii of curvature, the edges of the opened set X_B coincide with the edges of X; otherwise X_B is inside X and more regular. Now look at the same initial and transformed picture, but invert the roles of figure and background. Then the transformation is seen as a closure acting on the new figure. So, opening and closing are dual with respect to taking complements; opening of the complement X^c of X is precisely the complement of the closing of X, and conversely:

$$(X^c)_B = (X^B)^c \quad \text{and} \quad (X_B)^c = (X^c)^B \tag{II-22}$$

These two relations are derivable directly from the definitions (II-20) and (II-21) and relation (II-5). For example:

$$(X^B)^c = [(X \oplus \check{B}) \ominus B]^c = (X \oplus \check{B})^c \oplus B = (X^c \ominus \check{B}) \oplus B = (X^c)_B$$

Briefly, to open (resp. to close) the grains and to close (resp. to open) the pores are identical operations (see Fig. II.15a and b). Owing to the duality, it is sufficient to study only one of the two operations, and to easily transpose to the other, such properties that we obtain.

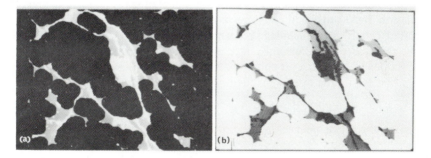

Figure II.15. Duality between opening and closing (the texture analyser operates on the *white* phase).

D.1. Geometrical interpretation of the opening

A very strong—and very intuitive—morphological significance is associated with the opening and closing operations, and is the basis of their use as size and shape descriptors. A point z belongs to the open set X_B if and only if $\check{B}_z$ hits $X \ominus \check{B}$ (Definition II-20):

$$z \in X_B \Leftrightarrow \check{B}_z \pitchfork (X \ominus \check{B})$$

Let y be a point of this intersection. As a point of $\check{B}_z$ it satisfies the following equivalence:

$$y \in \check{B}_z \Leftrightarrow z \in B_y$$

As a point of $X \ominus \check{B}$, it is the centre of a B included in X:

$$y \in X \ominus \check{B} = B_y \subset X$$

Thus, the points of the open set X_B coincide with the points of all the B_y which are completely contained in X. More simply: *the opening is the domain swept out by all the translates of B which are included in the grains X*. This operation smooths the contours of X, cuts the narrow *isthmuses*, suppresses the small *islands* and the sharp *capes* of X (see Fig. II.14 for example). By duality, there is also an interpretation for closing. A point z belongs to the closing X^B if and only if all the translates B_y of B, containing z, hit X. The closure blocks up the narrow *channels*, the small *lakes* and the long, thin *gulfs* of X.

D.2. Algebraic properties

In algebra, an application is said to be an opening (in the sense of Moore (Dubreil, 1964) if it is anti-extensive ($X_B \subset X$), increasing ($X \subset X' \Rightarrow X_B \subset X'_B$) and idempotent $(X_B)_B = X_B$). The morphological opening defined by relation II-20 satisfies these three axioms (thus explaining the origin of its name). Firstly, X_B is included in X since every point $x \in X_B$ belongs to at least one B included in X. Secondly, if X is included in X', then the open set X_B is contained in X'_B, since every point $x \in X_B$ belongs to one B contained in X, so in X'. And finally the opening is idempotent, for we have on one hand:

$$(X_B)_B \subset X_B \quad \text{(anti-extensivity)}$$

and on the other hand:

$$(X_B)_B = \{ [(X \ominus \check{B}) \oplus B] \ominus \check{B} \} \oplus B$$

In this last relation, the expression between braces is just $(X \ominus \check{B})^{\check{B}}$, which thus contains $X \ominus \check{B}$ (extensivity of the closure). Therefore

$$(X_B)_B \supset (X \ominus \check{B}) \oplus B = X_B$$

which establishes idempotence.

By duality *the closing X^B is extensive, increasing and idempotent*.

$$
\left.
\begin{aligned}
&X^B \supset X && \text{(extensive)}\\
&X' \subset X \Rightarrow X'^B \subset X^B && \text{(increasing)}\\
&(X^B)^B = X^B && \text{(idempotent)}
\end{aligned}
\right\} \quad \text{(II-23)}
$$

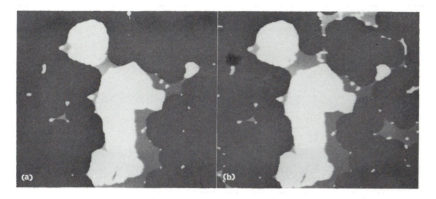

Figure II.16. The two photographs represent closings of X and X' ($\supset X$) by the same dodecagon B. The initial sets X and X' are white, the closures are white and grey. We can check that the three relations (II-23) are satisfied.

D.3. Characterization of the families of open sets

Among the three properties just described, the most novel is the one of idempotence. Just as for closing, dilation also was an increasing and extensive operation. The difference between the two lies in the fact that when we iterate the same dilation once more, we do not stay with the first dilation but get a larger dilated set. In a certain sense, closing and opening may be considered as more stable than the intermediary steps of dilation and erosion. We will attempt a thorough investigation of this question now by analysing the laws of composition of successive closings and openings. The approach will consist of extending the property of idempotence to more general two-step iterations. Firstly, we need to characterize the families of sets that are open and closed with respect to a given B. The following proposition does just this.

The family of sets open (resp. closed) with respect to B is exactly the family of sets of the type $C \oplus B$ (resp. $C \ominus B$), where C ranges over $\mathscr{P}(\mathbb{R}^n)$, the set of all subsets of $\mathbb{R}^n$.

Proof: If X is open with respect to B, then by defintion, we have $X = C \oplus B$ with $C = X \ominus \check{B}$. Conversely, $C \oplus B$ is necessarily open with respect to B, since we have, on one hand:

$$(C \oplus B)_B = [(C \oplus B) \ominus \check{B}] \oplus B = (C^{\check{B}}) \oplus B \supset C \oplus B$$

and on the other hand, because the opening is anti-extensive:

$$(C \oplus B)_B \subset C \oplus B$$

Thus:

$$C \oplus B = (C \oplus B)_B \quad \forall B, \forall C \quad \text{Q.E.D.} \tag{II-24}$$

As a particular case, choose $C = \{o\}$; then B is always open with respect to itself, since $B = \{o\} \oplus B$; i.e.

$$B = (B \ominus \check{B}) \oplus B$$

This characterization of the opening leads to a further property which will allow us to order the closings and openings:

If C is open with respect to B, then the openings X_B and X_C and the closings X^B and X^C of any set $X \in \mathcal{P}(\mathbb{R}^n)$ satisfy the following inclusions:

$$X_C \subset X_B \subset X \subset X^B \subset X^C \tag{II-25}$$

Proof: According to the characterization property, there exists a B' such that $\check{C} = \check{B} \oplus \check{B}'$. Let z be a point belonging to X^B, that is:

$$\check{B}_z \subset X \oplus \check{B}$$

or equivalently (inclusion rule II-16):

$$\check{B}_z \oplus (\check{C} \ominus B) \subset X \oplus \check{B} \oplus (\check{C} \ominus B)$$

On the right-hand side of the set inequality, $\check{B} \oplus (\check{C} \ominus B) \, (= (\check{C} \ominus B) \oplus \check{B}$ by commutativity), is the opening of $\check{C}$ by B, i.e. $\check{C}$ itself. The above inclusion may be simplified into:

$$\check{C}_z \subset X \oplus \check{C}$$

which means that point z belongs to the closure X^c. The inclusion $X^B \subset X^c$ follows, and by duality, the complete formula (II-25) follows. Consider now a third property, which generalizes the idempotence property:

If C is open with respect to B, then for every $X \in \mathcal{P}(\mathbb{R}^n)$:

$$(X_B)_C = X_C$$

If B is open with respect to C, then, on the contrary:

$$(X_B)_C = X_B \tag{II-26}$$

We shall prove the second equality only and leave the first for the reader to verify. Every point z of the open set X_B belongs to a B_y included in X. But by hypothesis, z belongs also to a C_y included in the above B_y, and so $(X_B) \subset (X_B)_C$. But the anti-extensivity implies the inverse set inequality; thus:

$$(X_B)_C = X_B$$

The two properties presented in (II-25), (II-26) allow a mathematical description of the sieving process, and in turn of particle size distribution.

Here, the initial population of separated grains (set X) goes through a sequence of finer and finer sieves (think of the sieves as being a family of structuring elements B_λ, indexed by a positive parameter λ). Each grain is stopped by a certain size of the mesh λ. If we compare two different meshes λ and μ, such that $\lambda \geq \mu$ (i.e. μ is finer than λ), the μ-oversize X_{B_μ} is larger than the λ-oversize X_{B_λ}; i.e. $X_{B_\lambda} \subset X_{B_\mu}$. Moreover, by applying the larger mesh λ to the μ-oversize, we obtain again the λ-oversize X_{B_λ}; but the oversize X_{B_λ}, sieved once again with mesh size μ remains unchanged. These sieving properties should be compared to the relations of (II-26). However, unlike sieving, openings can be carried out on a connected medium. In fact, for us, connectivity has nothing to do with the size criteria; under the same general approach, we will treat size measurements for particles and porous media alike.

This size distribution aspect of the openings has been approached in exactly the same way as putting together a jig-saw puzzle. However, there is one key piece missing at the moment. Can we find families B_λ such that $(B_\lambda)_{B_\mu} = B_\lambda$ for $\lambda \geq \mu$, and which satisfy the four principles of quantification? If so, does there exist a simple geometric characterization for the shapes of such families? At this stage of the presentation, we cannot go beyond just asking the question. Curiously the answer will be a consequence of certain properties of convex sets thus allowing completion of the puzzle.

We will conclude this section with an example of a B_λ family satisfying relation (II-26) but not the second principle of mathematical morphology (i.e. compatibility under change of scale). Take for the family B_λ, the sets defined in Figure II.10, where each B_λ is a pair of disks of radius λ, a fixed distance d apart. Obviously, every point x belonging to one of the two disks of B_λ belongs also to a pair of circles B_μ included in B_λ (Fig. II.17).

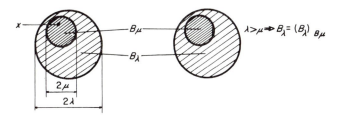

Figure II.17. An example of a family B_λ satisfying the size distribution conditions, but which is not compatible under change of scale.

E. EROSIONS AND INCREASING MAPPINGS, MORPHOLOGICAL AND ALGEBRAIC OPENINGS

We have just met two transformations, opening and closing, which can be represented respectively as the union of erosions and the intersection of dilations; moreover, these transformations are translation invariant and increasing (as are the dilation and erosion). Dare we suspect a more general result: namely if a transformation Ψ is translation invariant and increasing, can we express it as the intersection of dilations, or as the union of erosions? If so, these two basic operations of mathematical morphology will allow access to a broad variety of transformations . . . .

Similarly, we saw that morphological openings satisfy the three axioms which define algebraic openings. Is it only by chance, or on the contrary, for some deep logical reason? G. Matheron tackled these questions (1975) and established several theorems which put the morphological power of erosions and openings into perspective. These will be stated here without proof.

We will only consider mappings $\Psi(X)$ defined on $\mathscr{P}(\mathbb{R}^n)$ and invariant under translation (i.e. mappings satisfying $\Psi(X_h) = [\Psi(X)]_h$, $h \in \mathbb{R}^n$, $X \in \mathscr{P}(\mathbb{R}^n)$). The family:

$$\mathscr{V} = \{X : X \in \mathscr{P}(\mathbb{R}^n), 0 \in \Psi(X)\}$$

is called the *kernel* of the mapping Ψ (i.e. the family of sets whose transforms contain the origin). The kernel $\mathscr{V}$ contains the same "information" as the mapping itself; it follows that:

Any increasing mapping Ψ on $\mathscr{P}(\mathbb{R}^n)$ which is invariatnt under translation is both a union of erosions, and an intersection of dilations. More precisely, for any $X \in \mathscr{P}(\mathbb{R}^n)$, we have:

$$\Psi(X) = \bigcup_{B \in \mathscr{V}} (X \ominus \check{B}) = \bigcap_{B \in \mathscr{V}^*} (X \oplus \check{B})$$

where $\mathscr{V}^*$ denotes the kernel of the dual mapping Ψ^* (i.e. Ψ^* is defined by $\Psi^*(X) = [\Psi(X^c)]^c$).

A second theorem characterizes the algebraic openings and closings:

A mapping $\Psi : \mathscr{P}(\mathbb{R}^n) \to \mathscr{P}(\mathbb{R}^n)$ is an algebraic opening (i.e. an increasing, anti-extensive and idempotent operation) and invariant under translation if and only if it admits the representation $\Psi(X) = \bigcup\{X_B; B \in \mathscr{B}_0\}$ for a class $\mathscr{B}_0 \subset \mathscr{P}$. Then the class of sets invariant under Ψ is generated by unions and translations of B's, and the dual closing Ψ^* has the representation $\Psi^*(X) = \bigcap\{X^B, B \in \mathscr{B}_0\}$.

These two theorems are strictly algebraic. One can improve them by adding the constraints of semi-continuity and of local knowledge (see G. Matheron, 1975 Chs 7-1 and 8-2) but basically the heuristic meaning of these two fundamental results remains unchanged: all increasing morphological mappings come from the erosion, and all the morphological size distributions come from the openings (see Ch. X).

F. A CLASSIFICATION OF THE STRUCTURING ELEMENTS

The universe of all possible object shapes is vast, even when it is reduced to equivalence classes (multiplication by a scale factor does not change the shape). There is therefore a huge offering of potential structuring elements. How can they be classified? Which are best able to give information about the object?

The reply to these questions very much depends on what we the experimenter, want to see in the object. Suppose we take for the set X, the three particles of Figure II.18 and choose for the structuring element, firstly a disk (B), and then a couple of points (B') which are the two extremities of the vector h. After the erosion with the structuring element B, X is reduced to only a part of its largest initial grain; the two other grains, not being able to contain B, have been eliminated. This first erosion has then unveiled some characteristics relative to the *size* of X. Let us now consider the erosion of X with the structuring element B', where the right hand point of B' is placed at the origin. We see easily that $X \ominus B'$ is now made up of only the two small particles on the right, the largest having disappeared. This second transformation only allows a point x of X to remain if the point $x - h$ is also contained in X (the two points x, $x - h$ do not necessarily have to belong to the same particle). Hence the structuring element B' emphasizes more spatial distribution type characteristics, than size characteristics.

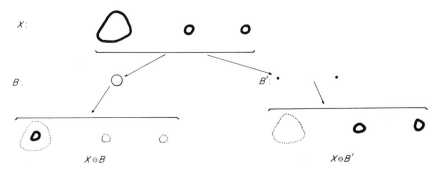

Figure II.18. Two different structuring elements B and B' lead to two different feature discriminations.

Thus analysing the same object X by two dissimilar structuring elements, has resulted in two profoundly different pieces of information on its geometric structure. It is as if the discontinuous aspect of the second element B' has filtered out spatial characteristics of X, and the connected, even convex aspect of the first element B has allowed us to extract only size characteristics. In reflection then, it is probably not wise to say that "X *possesses* a shape". In reality, X by itself possesses nothing at all; only the interaction of X with the structuring element B, has an objective meaning. This point is extremely interesting because although X is forced on us as the object of study, on the contrary we the experimenter can choose B, its size, shape, and orientation. Consequently we are free to take from X, those structural aspects which are the most interesting for us. At this stage, we see the necessity to construct a theory of structuring elements that will associate in a precise way, the geometric characteristics of each B, with the morphological significance of the corresponding transformations.

The theory, acquired experience, and also the bibliographic survey that we skimmed over in the beginning of this chapter, indicates some sort of path through the mass of possible structuring elements. Firstly they must be geometrically much simpler than X; we should not be deceived by the roles of X and B in the formula for dilation (equation (II-8)). We cannot handle, either technologically or even less, theoretically, structuring shapes that are too complicated. In particular, the structuring element is necessarily bounded. A second a priori heuristic suggests that when we have to choose among several shapes having the same property, but to different degrees, we should retain the most extreme of the choices. For example, we have just spoken of isotropy; in the plane the most isotropic shapes are the unions of circular annuli all having the same centre, and in particular the disk. The least isotropic shapes are the subsets of the straight line formed by unions of (aligned) segments and isolated points. We will choose from the second or first group of these structuring elements, according to whether or not we want to analyse a property of X depending on its orientation. Intuitively, the method of taking as structuring elements elongated ellipses in different directions and comparing the resulting erosions, is less powerful than the corresponding method using segments (which are so perfect for studies of orientation).

A second principle of classification was announced by the examples at the beginning of the chapter, and considered again in the case study of Figure II.18. In these examples, we see an inkling of the importance of the convexity of the structuring element. Among the sets which are not convex, the set containing isolated particles of points is even "less convex" than the others, because the segments joining the points two by two, are almost all totally outside the set. So, it is not surprising that the most fruitful morphological concepts deal with either convex structuring elements (size distribution), or with clusters of isolated points (set covariance, infinitesimal iterations).

The two-way classification table of isotropic or not, convex or not, organizes the structuring elements in the plane into the four following classes, shown on the top of next page.

Effectively, these are the four main shapes which we will meet in further chapters; the two convex shapes generate the size distributions (Ch. X), the boundary of the circle (more correctly, its equivalent on a digital grid)

	Isotropic	*Anisotropic*
Convex	The disk	The segment
Non-convex	The circle, i.e. the boundary of the disk.	Aligned points; in particular the couple of points, i.e. the boundary of the segment in $\mathbb{R}^1$.

underlies the majority of the connectivity algorithms (Ch. XI), and the couple of points, in the form of set covariance, describes the superposition of scales, periodicities, and statistical fluctuations (Ch. IX).

A last a priori heuristic contrasts the study of properties, defined very locally around a point, to that which looks at the whole space. Experience shows that we do not know how to handle a structure very well; we are inclined to look at it from a distance which is either very close, or as far away as possible. Stereology and pattern recognition, independently from each other, give primary importance to the infinitesimal structuring element. At the other extreme, we see for example that the convex hull (Ch. IV) is the limit of the closure by a ball with radius $\rho \to \infty$. Also, what do we understand by the notion of say, mean size of grains in a porous medium? One assumes implicitly that the investigated zone becomes infinite and that the average grain size tends towards a limit (the probabilistic tool is introduced, essentially to deal with these types of questions). A contrast between the two types of infinities (the small and the large) then appears via the set of Hit or Miss transformations (including for example, erosions, closures, etc. . . .). In fact this contrast is very general in mathematics, and is usually unavoidable. The Fourier transformations, which invert the properties at the origin and at infinity and the Maclaurin expansions, also support the idea of contrasting the role of the two infinities. The procedures of mathematical morphology have some analogies with their approaches. Concepts such as set covariance and linear erosion, will be considered to be well studied when we know how to interpret their behaviour at the origin (through specific parameters, angularities) and at infinity (range, stars).

G. EXERCISES

G.1. The practice of erosion and dilation

Erode and dilate X by B (X and B compact sets) when:

(a) X is a disk and B a pair of points, and conversely.
(b) X and B are identical (up to a translation).
(c) B is a disk (try various sizes) and X the set in Figure II.19.

Figure II.19

G.2. Openings and closings

(a) Open Figure II.20, a and b, by H_1 (i.e. the smallest regular hexagon of the grid). Interpret [semi-continuity $\Rightarrow$ the two open sets are identical].

(b) Close Figure II.20, c by an arbitrary segment and by an arbitrary hexagon. Interpret [Convex sets are invariant under closure, see Ch. IV.]

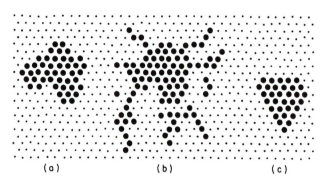

(a)   (b)   (c)

Figure II.20

G.3. Edge correction

The reader knows the classical rule for edge correction, by which one must eliminate the particles cutting the top or the left edge (for example) of the mask, and keep the others. What happens when we take for X a circular crown randomly implanted in the mask? Is the connectivity number of the crown (i.e. zero) biased by the rule? What about the edge correction by mask erosion, as proposed in Section C-3?

G.4. Ordering relations

A binary relation $\prec$ in a set E is an ordering relation when it is:

(a) reflexive: $x \prec x, \quad \forall x \in E$
(b) antisymmetric: $y \prec x$ and $x \prec y \Rightarrow x = y$
(c) transitive: $x \prec y$ and $y \prec z \Rightarrow x \prec z$

Show that "X contains Y", "X is eroded (resp. dilated, open, closed) by Y", define ordering relations.

G.5. One-to-one and onto mappings

A mapping $y = \Psi(x)$ from E to F is one-to-one when all the y are different (the equation $y = \Psi(x)$ has at most one solution), it is onto when its image is the whole space F (the equation $y = \Psi(x)$ has at least one solution) (Fig. II. 21).

When E and F are the space $\mathscr{P}(\mathbb{R}^n)$ of the sets of $\mathbb{R}^n$, and the space $\mathscr{C}(\mathscr{K}')$ of the non-empty compact convex sets of $\mathbb{R}^n$, are the erosion, dilation, and boundary mappings, one-to-one and/or onto?

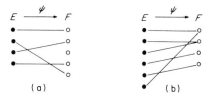

Figure II.21

G.6. Local knowledge and erosions (Lantuéjoul)

(a) Is it possible to commute the roles of masks Z and Z' in the definition of the principle of local knowledge (rel. I-4) and to write:

$$\forall Z, \exists Z' : \Psi(X \cap Z) \cap Z' = \Psi(X) \cap Z'? \quad (Z, Z' \text{ bounded sets})$$

[No! If we accept the empty set for Z', then the above relation becomes trivial; if we impose $Z' \neq \varnothing$, then the relation is not fulfilled for any erosion by a $B \supset Z$.]

(b) Relation (II-11) does not exactly prove local knowledge for the erosions, in the sense of (I-4), since it provides a mask $Z' = Z \ominus B$ when Z is known, but it gives no way of determining Z when Z' is known. Prove that $Z' = Z \oplus \check{B}$ when the transformation Ψ is erosion with respect to B.

[The right-hand side $(X \ominus B) \cap Z'$ is the set of points x belonging to Z', such that B centred in x is included in the set X:

$$(X \ominus B) \cap Z' = \{x : \check{B}_x \subset X; x \in Z'\}$$

Thus, the mask Z must contain, at least, the union $\bigcup_{x \in Z'} \check{B}_x = Z' \oplus \check{B}$. Conversely, write the local knowledge algorithm for the erosion, namely:

$$\{(X \ominus B) \cap (Z \ominus B)] \cap Z' = (X \ominus B) \cap Z' \qquad (i)$$

and replace Z by $Z' \oplus \check{B}$. $(Z' \oplus \check{B}) \ominus B = Z'^B$ contains Z', and (i) is satisfied.]

G.7. Local knowledge algorithms

(a) The transformation $\Psi(X)$ is assumed to satisfy the local knowledge principle (I-4), and Z is an input mask corresponding to the output mask Z'. Show that any set $Z^* \supset Z$ is also an input mask associated with Z'.
   [By applying algorithm (I-4) to set $X \cap Z^*$ and by taking into account the inclusion $Z^* \supset Z$, it becomes:

$$\Psi(X \cap Z^*) \cap Z' = \Psi(X \cap Z) \cap Z' = \Psi(X) \cap Z'.]$$

(b) Sets Z_1 and Z_2 are input masks for the output mask Z'. Prove that $Z_1 \cap Z_2$ is also an input mask.
   [Apply rel. (I-4) to set $X \cap Z_2$.]
(c) Suppose the set transformations Φ and Ψ satisfy the local knowledge axiom (I-4). Prove that the union $\Phi \cup \Psi$ and the composition product $\Psi[\Phi]$ also satisfy relations (I-4). Find examples where these two properties are applied.

G.8. Algebraic properties of erosions

(a) Prove the following equivalence:

$$X \oplus B \subset Y \Leftrightarrow X \subset Y \ominus \check{B} \tag{1}$$

Show that the relation obtained by reversing $\oplus$ and $\ominus$ is false. Deduce that we cannot replace inclusions with equalities in (1).
(b) Prove that the erosion by B is anti-extensive if the origin O belongs to B i.e.

$$O \in B \Rightarrow X \ominus B \subset X$$

III. Hit or Miss Topology

A. WHAT GOOD IS A TOPOLOGY?

This chapter is relatively abstract, however we are led to it via the fourth principle of mathematical morphology, which involves us with certain classes of sets and a very particular topology. The general notions of topology are not usually familiar to biologists and physicists who use mathematical morphology. This is because these notions are rarely useful in the actual application of the model, although they do help in the building of it. Hence this chapter can be omitted on a first reading, by those who are only interested in practical morphology.

For the readers who are ignorant of the use and the language of topology, but want nevertheless to understand this aspect of morphology, we will devote the second section to a simple exposé of some classical results, necessary for what follows. The order of presentation adopted is the same as in the book of G. Choquet (1966), to which we refer the reader for more information. Sections D and F review, with many simplifying restrictions, various chapters of G. Matheron (1975). In particular, we will restrict the presentation to the case of closed sets in the Euclidean space.

A physicist can use the tool of topology for at least three purposes. Firstly, it allows him to restrict the class of sets of $\mathbb{R}^n$ to those which possess a real physical meaning. In nature we will never find a set formed only by the points of the plane with irrational coordinates, nor will we find a disc with two thirds of the boundary belonging to the same phase as the inside, and the other third the same phase as the outside. This is because the notion of an irrational number or the boundary of a set, however useful, are inventions of the mind. In other words, to have a sorting device for irrational points is not possible, since it would assume an infinite power of resolution. Because of this impossibility in experimentation, the subsets of $\mathbb{R}^n$ are grouped into equivalence classes which, for example, do not always distinguish between a closed and open disc.

What are the useful equivalence classes and what constraints do they impose?

Secondly, topology is a mathematical tool constructed to give a meaning to the notions of convergence and continuity. Let us consider for example, the set X given in Figure III.1 and its erosion by the circle $B(r)$ of radius r. We see intuitively that when $r \to 0$, the eroded set $X \ominus B(r)$ will fill up completely the internal zone of X, but will never reconstitute the four threads of zero thickness. On the other hand, if we deal with the dilate $X \oplus B(r)$, when $r \to 0$, it tends to X "from above" and recovers the whole of X, inclusive of threads. The perimeters of $X \ominus B(r)$ and $X \oplus B(r)$ tend to a completely different value when $r \to 0$. Why is dilation here continuous, but not erosions? Consider the following surprising result: if we now invert the roles of X and the background X^c, the same intuitive reasoning shows that the above property is also inverted; i.e. it is now the erosion of X^c which is continuous and not the dilation. What kind of asymmetry between X and X^c has been introduced, leading to a result of this type?

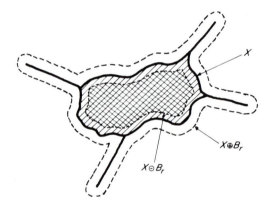

Figure III.1. Set X, $X \oplus B(r)$ and $X \ominus B(r)$. Why is dilation continuous, and not erosion?

The reader should now realize that research on sets that possess a real physical significance (and which can be digitalized), and the questions of continuity, are both consequences of the same type of analysis. He should now understand that these concerns specify (in another way) and illustrate, the two aspects of the fourth principle of mathematical morphology.

But there are other questions which indirectly require that we have already chosen a topology: those which are related to the probabilistic interpretation of the sets under study. If we want to express an idea about the structure of the sets, even such a simple idea as the following: "the object that I am looking at is made up of randomly centred particles", to recourse to probabilistic terms and concepts is inevitable. Now the probabilistic language requires a certain internal coherence; we must define some elementary events (can we take "B hits the boundary of the random set X", as an elementary event?). Moreover, the complement of such an event has to itself be an elementary event. Finally for any family of elementary events, the logical operations "and" and "or", repeated a finite or denumerable number of times, give sets which we call events. Any class of sets, stable with respect to these two properties, is called a σ-algebra. Only these can be used in the logical framework of a probability theory. Right from the start, it seems that in morphology we are in a position that is beyond the capabilities of a σ-algebra. The simplest of the operations, erosion, brings into play in general, a non-denumerable number of intersections. Does that mean that the erosion of a random set by a disk is not a random set? Let us hope not. Anyway, only an adequate topology will throw light on this point.

Faced with the three reasons given above, a strong counter-argument comes quickly to mind; that of digitalization. Figure III.1 was drawn in a continuous space; i.e. the plane, which is a construct of the mathematician. There exists no experiment able to access such a figure; all the actual objects that we can analyse will only be able to be assimilated with finite mosaics of small spots (the finiteness of the power of resolution now becomes a counter-argument to using topology!). So it seems that the structure of the plane is too rich as a framework for representing the physical objects, and that a finite topology based for example on graphs, would better describe reality. This point of view is perfectly defendable, and is even more appropriate for certain models; see Pavel (1969). In the first volume of his famous book, Feller (1960) systematically studied probabilities from a finite point of view. This said, we could also argue that the construction of random sets on graphs seems to be more artificial than on the Euclidean space: regular grids are not isotropic, the definition of a homothetic (change of scale) is quite restrictive, and a scale parameter is introduced (the grid spacing) which is cumbersome and must be fixed, initially at least.

The argument on the finiteness of the power of resolution, is not really a convincing one. In fact, every physical phenomenon is associated with a certain scale of study, and away from this scale, the nature of the phenomenon is changed. This ambiguity in morphology is due to the fact that at the *same scale* technologically discontinuous (digitalization of the image) and continuous (rotations, magnifications) processes, coexist. Therfore there exists no other alternative than to construct a topology for the richer space, i.e. the Euclidean one, and then, to bridge the gap between digital and Euclidean spaces via this topology. The first step is described in this chapter, the second one in Chapter VII.

B. REVIEW OF TOPOLOGY

B.1. Open sets, neighbourhoods

The two prime concepts of topology, namely the notions of an open set and a neighbourhood, allow us to quantify the "order of smallness" of a set about each point. The notions of convergence and continuity follow from there.

Definition III-1. A *topological space* is a pair consisting of a set E and a collection $\mathscr{G}(E)$ of subsets of E called the *open* sets, satisfying the three following properties:

> O_1—Every union (finite or otherwise) of open sets is open.
> O_2—Every *finite* intersection of open sets is open.
> O_3—The set E and the empty set $\varnothing$ are open.

On every set E, several topologies can be defined. In the Euclidean space $\mathbb{R}^n$, the most usual is the one generated by the Euclidean metric (see below).

Definition III-2. A subset X of E is said to be closed when its complement X^c is open.

Remark: Some subsets X of E may be both closed and open. For example, suppose we take for E the set of points of $\mathbb{R}^n$ with integer co-ordinates and associate with it the coarse topology, i.e. the one for which $\mathscr{G}$ is the collection of all subsets of E. Each point is then simultaneously open and closed. In the same way, the introductory example of a set X containing some of its boundary, corresponds to a set neither open nor closed (similarly for the set of irrational co-ordinate points).

Definition III-3. A *neighbourhood V of a point x* of E is a subset of E containing an open set which in turns contains x. Denote $\mathscr{V}(x)$ as the set of all the neighbourhoods V of the point x.

A set is open iff it is a neighbourhood of each of its points. It follows then that the open sets of a space are known whenever the neighbourhoods of x are known for every x. In other words, two topologies on the same set are identical when they have the same neighbourhoods.

We now present four properties which allow us to characterize a topology from the datum of its neighbourhoods:

i_1: every neighbourhood of the point x contains x, and every x has at least one neighbourhood;

i_2: every set containing a neighbourhood of x is a neighbourhood of x;
i_3: the intersection of two neighbourhoods of x is a neighbourhood of x;
i_4: if V is a neighbourhood of x, there exists a sub-neighbourhood W of V (i.e. such that $W \subset V$) such that V is a neighbourhood of every point of W.

The meaning just given to the word "neighbourhood" appears different from the meaning it has in ordinary usage, since for us a point $x \in E$ has many neighbourhoods, one of them even being the space E itself. However, we have in fact enriched a notion which has up to now been imprecise. We can now specify that a point y belongs to a particular neighbourhood of x, and this neighbourhood V makes the degree of proximity of y to x more precise. To be given the family $\mathscr{V}(x)$, it is enough in fact to be given only a subset of the elements of $\mathscr{V}(x)$. More precisely,

Definition III-4. We say that a subset $\mathscr{B}$ of $\mathscr{V}(x)$ forms a (neighbourhood) *base* of $\mathscr{V}(x)$, if every $V \in \mathscr{V}(x)$ contains an element $W \in \mathscr{B}$.

More generally, one defines the (open set) base of a topological space as every family $\mathscr{G}_i$ (here i is thought of as belonging to some index set I) of open sets of E such that the two following equivalent properties are satisfied:

(1) Every $x \in E$ has a neighbourhood base made up of a sub-family of the $\mathscr{G}_i$.
(2) Every open set of E is the union of a sub-family of the $\mathscr{G}_i$.

We will now give some examples of open sets and of bases. If one chooses for E the real line $\mathbb{R}^1$, the open sets are the open intervals (a, b), and the empty set. Then, the neighbourhoods of every point x are all the open intervals containing x. And every point x has a neighbourhood base made up of the open intervals with centre x and half length $1/n$ (n an integer > 0). Therefore the line $\mathbb{R}^1$ has a *countable* (*open set*) *base*. All the bases that we shall have the opportunity to meet here will be countable.

One can give a topological structure to the Euclidean space $\mathbb{R}^n$ by a very simple generalization of the space $\mathbb{R}^1$. The space $\mathbb{R}^n$ is the product of n spaces indentical $\mathbb{R}^1$, or in other words, the set of ordered sequences $(x_1, x_2, \ldots x_n)$ of n real numbers. If $Y_i (i = 1, 2, \ldots n)$ are open sets in $\mathbb{R}^1$, the subset $Y_1 \times Y_2 \times \ldots \times Y_n$ of $\mathbb{R}^n$ is called an *open interval* of $\mathbb{R}^n$ with bases Y_i (it is empty if one of its bases is empty). Every union of open intervals in $\mathbb{R}^n$ is called an open set in $\mathbb{R}^n$. It is clear that the open intervals in $\mathbb{R}^n$ form a countable base in $\mathbb{R}^n$.

B.2. Closure, interior, boundary

When we construct conditional bisectors from morphological openings (Ch. XI) it will be of prime importance to filter the isolated points of X from the non isolated ones. This distinction is made mathematically by the two following topological definitions.

Definition III-5. We call x an *accumulation point* of X if every neighbourhood of x contains at least one point of X other than x (x itself need not belong to X). Also, x is an *isolated point* of X if it belongs to X but is not an accumulation point of X.

This concept is closely linked with the notion of topological closure of a set. A set X is closed if and only if its complement X^c is open.

Definition III-6. The smallest closed set containing the set X is called the *closure* of X and is denoted by $\overline{X}$.

For example, the closure of the set of points $1/n$ ($n = 1, 2 \ldots$) is the original set of points with the addition of the point O. The relation $X = \overline{X}$ defines the closed sets. It can also be easily proved that X is a closed set if and only if it contains all its accumulation points. By duality, one derives the notion of the interior from that of the closure.

Definition III-7. The (possibly empty) union of all the open sets contained in a subset X of E is called the *interior* of X. It is therefore the largest open set contained in X. We denote it by $\mathring{X}$.

Similarly, the relation $X = \mathring{X}$ characterizes the open sets.

The concepts of the *topological* closure and interior of a set must now be compared with the operations of morphological closing and opening. The similarity of the vocabulary is not an accident. In algebra, one calls closure, any operation satisfying the three properties 1, 2, 3 presented just below. In this sense, topological closures and morphological closings, appear as two particular cases of the algebraic closure. (In fact the resemblance goes further, since it is possible to construct a topology such that the interior of every set X is identical with its morphological opening, the opening taken with a given open ball.)

The following properties of the topological closure can be easily proved:

$$X \subset \overline{X} \qquad (extensivity) \qquad \text{(III-1)}$$

$$X \subset Y \Rightarrow \overline{X} \subset \overline{Y} \quad (monotonicity) \qquad \text{(III-2)}$$

$$\overline{\overline{X}} = \overline{X} \qquad (idempotence) \qquad \text{(III-3)}$$

$$\overline{X \cup Y} = \overline{X} \cup \overline{Y} \qquad \text{(III-4)}$$

And for the interior, four equivalent properties can be deduced by using the duality relation:

$$(\mathring{X})^c = \overline{(X^c)} \qquad \text{(III-5)}$$

Definition III-8. The boundary ∂X of a subset X of E is the set of those points x, such that every neighbourhood of x contains at least one point of X and one point of the complement X^c. Thus:

$$\partial X = \overline{X} \cap \overline{X^c} \qquad \text{(III-6)}$$

We can see that the boundary of X and that of the complement X^c are identical, independent of whether X is open, closed, or of a more sophisticated type. More precisely, we can write:

$$\partial X = \overline{X}/\overset{\circ}{X} \qquad \text{(III-7)}$$

(/ symbolizes the difference between sets), i.e. the boundary of a set equals its closure minus its interior.

The concept of boundary allows us to be precise about the notion of a "set without thickness" (see Fig. III.1). Indeed, we have the equivalences:

$$X = \partial X \Leftrightarrow \overset{\circ}{X} = \varnothing \Leftrightarrow \overline{X^c} = E \qquad \text{(III-8)}$$

B.3. Continuity, convergence, and limits

In order to be able to talk about the continuity of a mapping f from a set E into a set E', it is necessary that a notion of neighbouring points is defined on E and E'; in other words, that E and E' be topological spaces.

Definition III-9. A mapping f from a topological space E into another topological space E' is said to be continuous on E, if for every open set X' (resp. closed set) in E', $X \equiv f^{-1}(X')$ is an open set (resp. closed set) in E.

This definition is a general one. We shall meet other, more restrictive definitions for the cases treated below. Notice that it makes use of inverse images under f, and not direct images. The notion of continuity leads to the basic concept of *homeomorphism*. Two topological spaces are homeomorphic if there exists a bijection f (i.e. f is one to one and onto) which interchanges their open sets (i.e. for every open set X in E, $f(X)$ is an open set in E', and conversely). This notion is fundamental in topology, because a homeomorphism is nothing more than an isomorphism of topological structures; it gives the basic equivalence relation in topology. When two spaces are homeomorphic, every property which is true for one is true for the other; they can be considered as two representations of the same geometric entity (from the topological point of view, of course: here a ring and a tea cup with a handle are topologically identical!). We can see in Figure III.2 a number of examples intended to convey the intuitive content of homeomorphism (features on the same line are homeomorphic to one another).

Figure III.2. Examples of homeomorphic figures.

Definition III-10. *Notion of a limit.* Let $\{x_i\}$ $(i = 1, 2, \ldots n, \ldots)$ be a sequence of points in a space E. We say that this sequence converges to, or tends towards, a point x of E, or that x is the limit of this sequence and we write $x_i \to x$, if for every neighbourhood V of x we have $x_i \in V$ for all except at most finitely many values of i.

This definition is of course the most general. Now if E is the usual Euclidean space $\mathbb{R}^3$, then the x_i are points of $\mathbb{R}^3$ in the classical sense of the word. But, we are also interested in the space $\mathscr{F}(\mathbb{R}^3)$, this being the space of all the closed sets of $\mathbb{R}^3$ (see the next section). The "points" in this space are then the *closed sets* X_i of $\mathbb{R}^3$.

This leads us to define a second, very simple kind of convergence, which we shall frequently use in the following sections: *monotonic sequential convergence.* Let us consider a sequence of sets X_i (not necessarily closed, or open) such that $X_{i+1} \subset X_i$ (resp. $X_{i+1} \supset X_i$) and satisfying the relation (III-9)

$$X = \bigcap_i X_i \quad (\text{resp. } X = \bigcup_i X_i) \tag{III-9}$$

Then the set X is said to be the monotonic sequential limit of the X_i, and we denote the convergence by "$X_i \downarrow X$" (resp. "$X_i \uparrow X$"). The implicit topology being used here is that of the order topology on the positive integers.

The notions of convergence and continuity are related to one another. If Ψ is a mapping from one topological space onto another, then Ψ is said to be continuous at the point x iff $\Psi(x_i)$ tends towards $\Psi(x)$ when $x_i \to x$. In what follows, we shall use this formulation to prove the continuity of the operations of union and dilation of closed sets.

We will now consider the notion of adherent points of a sequence. The sequence of real numbers $a_i = (-1)^i$ is not convergent; nevertheless the numbers 1 and -1 do appear as limits in a broader sense. The analysis of this notion leads to the following definition.

Definition III-11. Let $\{a_i\}$ $(i = 1, 2 \ldots)$ be a sequence of points of a topological space E. We say that a point a of E is an *adherent point* of this sequence if for any neighbourhood V of a, there exists arbitrarily large indices i such that $a_i \in V$.

All the spaces under study in this book are separated; i.e. each point of every pair of distinct points possesses a neighbourhood distinct from a neighbourhood of the other point. In such spaces, if a sequence $\{a_i\}$ converges to a, this point a is the only adherent point of the sequence. However, it is false in general, even in a separated space, that if a sequence has a single adherent point, then it converges to this point. For example, the sequence $\{1/2, 2, 1/3, 3, \ldots 1/n, n \ldots \}$ in $\mathbb{R}$ has a single adherent point, namely zero; although it does not converge to zero.

B.4. Compact space and sets

One of the most important properties which a topological space can satisfy is that of compactness. Having this property enables us to guarantee the existence of random closed sets in $\mathbb{R}^n$.

Definition III-12. A space E is said to be *compact* if it is separated and if from every open covering of E one can select a finite subcovering of E.

A ball with a finite radius is a compact set, but not the plane $\mathbb{R}^2$, nor the general Euclidean space. It can be shown that every infinite subset X of a compact space E has at least one accumulation point in E. Moreover, a separated subset X of a space E is said to be compact if every family of open sets in E which covers X contains a finite subfamily which covers X.

For any compact space E:

— every point x admits a base of closed neighbourhoods.
— the class of closed subsets and the class of compact subsets are identical.

In $\mathbb{R}^n$ (which is not a compact space), the compact subspaces are the subsets of $\mathbb{R}^n$ that are closed and bounded. In every separated space, the union of two compact sets is compact, and all intersections of compact sets are compact (clearly the union of an infinite family of compact sets is not in general a compact set).

There exist many spaces which, without being compact, behave *locally* like a compact space. (This is the case for $\mathbb{R}^n$, for example.) They are separated and each of their points has at least one compact neighbourhood. Such spaces are said to be *locally compact*.

B.5. Metric spaces

We will now change our point of view slightly. The proximity of two points will not be defined by reference to neighbourhoods but now by a positive real number depending only on these two points.

Definition III-13. A *metric space* is a pair consisting of a set E and a mapping $(x, y) \rightarrow d(x, y)$ of $E \times E$ into R_+, having the following properties:

$$M_1 \quad x = y \Leftrightarrow d(x, y) = 0$$

$$M_2 \quad d(x, y) = d(y, x) \qquad \text{(symmetry)}$$

$$M_3 \quad d(x, y) \leq d(x, z) + d(z, y) \qquad \text{(triangle inequality)}$$

The function d is called a *metric* and $d(x, y)$ is called the *distance* between the points x and y. It is possible to define *several* distances in the same space; we will for the most part be in $\mathbb{R}^n$, and only use the classical Euclidean distance.

Definition III-14. In a metric space E, the set of points x of E such that $d(a, x) < \rho (\leq \rho)$ is called the open (closed) ball with centre a and radius ρ. The set of points x of E such that $d(a, x) = \rho$ is called the sphere with centre a and radius $\rho \geq 0$.

Among all the topologies which can be defined on a set E having a metric space structure, there is one which is *directly* related to the metric; this is known as the topology of the metric space E. It is defined as follows:

Definition III-15. A subset X of a metric space E is said to be open if it is empty or if for every $x \in X$ there exists an open ball with centre x and non zero radius which is contained in X. It is immediate that the collection of open sets in E satisfies the three axioms of a topological space. More precisely, in the case of $\mathbb{R}^n$, if one chooses the Euclidean metric, the topology generated coincides with the topology of Section B-1, defined by reference to intervals. Notice that all these spaces are Hausdorff and separable. Finally, the properties of limits and convergence are extremely easy to express within the framework of metric spaces. (See exercises below.)

C. THE HAUSDORFF METRIC

After this brief exposition, let us now come to the very essence of the chapter. What topology is the most suitable for the sets which model the objects

investigated by mathematical morphology? The simplest way to proceed
would be to define a metric, so that the more two sets are neighbours, the
smaller is their distance. Of course, such a metric would act in an abstract
space, in a space where a "point" is actually a set of the usual space. In general,
this procedure is not possible, save in one special, but extremely useful case. Its
simplicity, and its restrictions, suggest a presentation before we present the
general case. The analysis goes back to Hausdorff (beginning of twentieth
century; Hausdorff is famous for having found the simplest axiomatic system
(Def. III-1) from a multitude of possibilities). Hausdorff's metric is defined on
the space $\mathscr{K}'$ where each point is a non empty compact set of $\mathbb{R}^n$.

Proposition III-16. *The space $\mathscr{K}'$ is a metric space.*

If K_1 and K_2 denote two non empty compact sets in $\mathbb{R}^n$ (or equivalently, two
points of $\mathscr{K}'$), and $B(\varepsilon)$ the *closed* ball with a radius ε, then the quantity:

$$\rho(K_1, K_2) \equiv \inf\{\varepsilon : K_1 \subset K_2 \oplus B(\varepsilon), K_2 \subset K_1 \oplus B(\varepsilon)\} \quad \text{(III-10)}$$

defines a metric on $\mathscr{K}'$, known as the *Hausdorff metric*. From (III-10), ρ is the
radius of the smallest closed ball B such that both K_1 is contained in the dilated
set $K_2 \oplus B$ and K_2 is contained in the dilated set $K_1 \oplus B$. The quantity ρ is
indeed a distance. According to relation (III-10), the identity $K_1 \equiv K_2$ implies
$\rho = 0$; conversely if $\rho = 0$, then by definition $K_1 \subset K_2$ and $K_2 \subset K_1$, thus
$K_1 = K_2$ (incidentally, notice that if K_1 and K_2 were not closed the converse
part of the proof would not hold). The second axiom is a trivial consequence of
relation (III-10). To prove the triangle inequality axiom, consider three
compact sets K_1, K_2 and K_3, and the associated distances:

$$\rho_1 = d(K_2, K_3); \qquad \rho_2 = d(K_3, K_1); \qquad \rho_3 = d(K_1, K_2)$$

From $K_3 \subset K_1 \oplus B(\rho_2)$ and $K_3 \oplus B(\rho_1) \supset K_2$, we have that K_2
$\subset [K_1 \oplus B(\rho_2)] \oplus B(\rho_1) = K_1 \oplus B(\rho_1 + \rho_2)$. Similarly:

$$K_1 \subset K_2 \oplus B(\rho_1 + \rho_2)$$

Thus $\rho_3 \leq \rho_1 + \rho_2$.

Figure III.3 illustrates the notion of the Hausdorff distance between two
compact sets. In the particular case when K_1 and K_2 are reduced to two points,
the Hausdorff distance $\rho(K_1, K_2)$ coincides with the Euclidean distance.

In integral geometry, study of the convex ring (cf. Ch. V) uses only the
topology generated by the Hausdorff metric. Thanks to the Hausdorff metric,
we can for example, define the surface of a three-dimensional body. However it
does demand that three restrictive hypotheses be fulfilled.

Firstly, the Hausdorff metric ignores the empty set $\varnothing$. In fact, relation
(III-10) makes no sense for K_1 or $K_2 = \varnothing$.

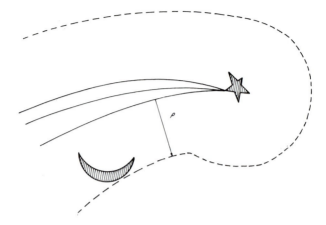

Figure III.3. An example of the Hausdorff distance.

Secondly, it is concerned only with closed sets. Imagine, for example (III-10) applied to the two sets K_1 (closed) and $K_2 = \mathring{K}_1$. Their Hausdorff distance is zero but the two sets are *not* identical. Hence the first axiom for metrics is not satisfied.

Thirdly, we have assumed K_1 and K_2 to be bounded. Here too, the restriction is essential. Let us give a counter example: K_1 and K_2 are two lines in a plane, forming the angle θ (and consequently, two non-bounded closed sets). For every finite ρ, the relation (III-10) is impossible to satisfy, and so $\rho = \infty$. Now, if we imagine the angle θ decreasing, the lines come closer together, but ρ remains infinite. Then, when the limit $\theta = 0$ is attained, ρ suddenly drops to zero! Hence the Hausdorff distance is very clumsy in describing how a straight line (or more generally a non bounded set in $\mathbb{R}^n$) tends towards another. Unfortunately, the stationariness of a random function implies that its realizations are non-bounded. Hence this topology cannot be used to elaborate models most frequently under consideration, namely stationary ones.

D. TOPOLOGICAL STRUCTURE FOR THE CLOSED SETS OF $\mathbb{R}^n$

From now on, we will focus attention on $\mathscr{F}$ the space of all closed sets of $\mathbb{R}^n$ (elements of $\mathscr{F}$ are not necessarily bounded; $\varnothing \in \mathscr{F}$). How can we construct a topology on $\mathscr{F}$? G. Matheron (1969) had the idea of defining the degree of proximity between two sets from a Hit or Miss transformation point of view. In his approach, the more two sets hit the same sets and the more they are disjoint from other sets, the closer neighbours they are said to be.

This intuitive approach does in fact yield a topology which we call "the Hit or Miss topology". G. Matheron studied its main properties when the starting space is locally compact, Hausdorff, and separable (denote below by L.C.S.), i.e. each point in E admits a compact neighbourhood, and the topology of E admits a countable base (1969; 1975). At around the same time, D. G. Kendall (1974) independently constructed the same structure on more general abstract spaces. Mathematically speaking, Kendall's description is more extensive, thus richer. But Matheron's, although more severe in its prerequisites, fits in better with phenomena under study. Euclidean spaces, the sphere, the circle, etc. . . . and the finite product of such spaces, are all locally compact Hausdorff and separable spaces. Moreover, Matheron's topology is directly adapted to the probabilistic version of morphology, in which the σ-algebra of events is generated by denumerable combinations of unions and intersections of the open sets (see Ch. XIII, G.1).

D.1. The Hit or Miss topology

The discussion in Section C showed us the limitations of an approach which uses a metric. Hence we shall now define a topology on $\mathscr{F}$ directly from its neighbourhoods. A collection of closed sets in $\mathbb{R}^n$(i.e. of *points* in $\mathscr{F}(\mathbb{R}^n)$) has to be sufficiently "packed" to constitute a neighbourhood. More precisely,

Definition III-17. Given the two finite sequences $G_1 \ldots G_i \ldots G_m$ of open sets and $K_1 \ldots K_j \ldots K_p$ of compact sets in R^n, the class of all the closed sets which hit every G_i and miss every K_j defines an open neighbourhood in $\mathscr{F}(\mathbb{R}^n)$ (Fig. III.4).

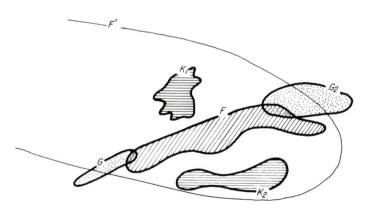

Figure III.4. The two sets F and F' belong to the same neighbourhood defined by $K_1 K_2$ and $G_1 G_2$.

It is easy to check that the four characteristic properties of the neighbourhoods (III-3) are satisfied. In this way a topology can be generated on the space $\mathscr{F}(\mathbb{R}^n)$, namely the *Hit or Miss topology*.

Let $\mathscr{V}$ be the family of sets stable under finite intersection and generated by the $\mathscr{F}_G$ and the $\mathscr{F}^K$; its elements are written as:

$$\mathscr{F}^{K_1} \cap \ldots \mathscr{F}^{K_p} \cap \mathscr{F}_{G_1} \ldots \cap \mathscr{F}_{G_m} = \mathscr{F} \frac{K_1 \cup \ldots \cup K_p}{G_1, \ldots, G_m}$$

We will now give the following fundamental result:

THEOREM III-18. *The space $\mathscr{F}$ of closed sets of $\mathbb{R}^n$, equipped with the Hit or Miss topology, is compact, and has a countable base.*

Notice that $\mathscr{F}$ is compact but not $\mathbb{R}^n$. However, the reason why $\mathscr{F}$ has a countable base is because the same occurs for $\mathbb{R}^n$. In what follows, Theorem III-18 will be particularly useful in formulating the probabilistic version of the theory (Ch. XIII), and for studying questions of convergence.

D.2. Convergence, continuity

Because $\mathscr{F}$ admits a countable base, we can restrict ourselves to the convergence of sequences: $i \to F_i$ in $\mathscr{F}$ (see Def. III-10). Using the definition of the Hit or Miss topology on $\mathscr{F}$, we see that a sequence $\{F_i\}$ converges in $\mathscr{F}$ towards a limit F, if after a certain time the F_i hit the same open sets G and miss the same compact sets K, as F itself. More precisely

Definition III-19. A sequence $\{F_i\}$ converges towards F in $\mathscr{F}$ if and only if it satisfies two conditions:

1. If an open set G hits F, then G hits all the F_i except at most a finite number.
2. If a compact K is disjoint from F, then it is disjoint from all the F_i except at most a finite number.

We shall write "$\lim F_i = F$", or equivalently "$F_i \to F$".

A simpler, but equivalent version of the convergence in $\mathscr{F}$, involving only relationships between $\{F_i\}$ and the points of R^n, is given in Exercise III-2.

From Definition III-19, several immediate consequences can be deduced. Firstly, the union operation is continuous:

PROPOSITION III-20. *The mapping $(F, F') \to F \cup F'$ is continuous from the product space $\mathscr{F} \times \mathscr{F}$ onto $\mathscr{F}$.*

In other words, if two sequences $\{F_i\}$ and $\{F_i'\}$ respectively converge in $\mathcal{F}$ towards the limits F and F', then the sequence $\{F_i \cup F_i'\}$ converges towards $F \cup F'$. Notice that the two sequences do not necessarily converge in a monotone way. Nevertheless the majority of basic transformations Ψ used in mathematical morphology are monotone (i.e. $Y \subset X$ implies either $\Psi(Y) \subset \Psi(X)$ or $\Psi(Y) \supset \Psi(X)$, depending on whether Ψ is increasing or decreasing). The operations of erosion, $X \ominus B$, and dilation, $X \oplus B$ are monotonic (either acting on X with B fixed, or acting on B with X fixed). A similar situation occurs for openings X_B and for closings X^B, considered as mappings acting on X. Thus operations which are monotone appear frequently in morphology. The topological tool adapted for their study is that of monotonic sequential convergence (rel. (III-9)). With respect to this convergence Proposition III-21 gives us an important collection of properties:

PROPOSITION III-21. *Let* $\{F_i\}$ *and* $\{F_i'\}$ *be two sequences in* $\mathcal{F}(\mathbb{R}^n)$; *then:*

(a) $F_i \downarrow F$ *implies* $Lim \, F_i = F$ *in* $\mathcal{F}$

(b) $F_i \uparrow X$ *implies* $Lim \, F_i = \overline{X}$ *in* $\mathcal{F}$ $\qquad$ (III-11)

(c) $F_i \downarrow F$ *and* $F_i' \downarrow F'$ *imply* $Lim \, (F_i \cap F_i') = F \cap F'$

$\quad$ *and* $Lim \, (F_i \cup F_i) = F \cup F'$ *in* $\mathcal{F}$.

(d) $F_i \uparrow X$ *and* $F_i' \uparrow X'$ *imply* $Lim \, (F_i \cup F_i') = \overline{X \cup X'} = \overline{X} \cup \overline{X}'$

$\quad$ *and* $Lim \, (F_i \cap F_i') = \overline{X \cap X'} \subset \overline{X} \cap \overline{X}'$ *in* $\mathcal{F}$ $\qquad$ (III-12)

(e) $Lim \, F_i = F$, $Lim \, F_i' = F'$ *and* $F_i \subset F_i'$ *for every* i, *imply* $F \subset F'$

All these results are useful and will be used in what follows. One of the most curious results among them, is the non-continuity of the intersection (property (d)). It is worth spending a small amount of time on this. As an example, consider the two families of squares C_i and C_i' shown in Figure III-5. The

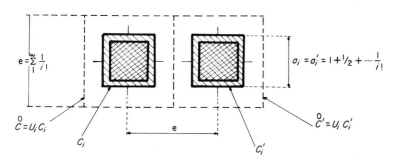

Figure III.5. Example of discontinuity of the intersection.

distance between their two centres is e (Euler's constant), and their sides a_i and a_i' are given by:

$$a_i = a_i' = 1 + \frac{1}{2} + \ldots \frac{1}{i!}$$

By hypothesis, the two squares are closed, and respectively tend towards the two limit squares $\overset{\circ}{C}$ and $\overset{\circ}{C}{}'$, with sides $a = a' = e$. $\overset{\circ}{C}$ and $\overset{\circ}{C}{}'$ are open sets. We are thus in the situation described in case (d) of Proposition III-21: the limit of the union $C_i \cup C_i'$ is the union of the two *closed* squares C and C'; but the limit of the intersection $C_i \cap C_i'$ is the empty set (since $C_i \cap C_i' = \varnothing$ no matter how large the index i is!) although the intersection of the limits C and C' in $\mathscr{F}$ is the lateral edge common to these two squares. Clearly then, the two operations of intersection and union are to be treated differently when it comes to questions of convergence. The latter is continuous, the former is not. It then becomes clear what happened in the introductory example (Fig. III-1), where the dilation (based on unions) seemed intuitively continuous but not the erosion (based on intersections).

The example also illustrates points (a) and (b) of Proposition III-21; it shows that the limit of the C_i in $\mathscr{F}$ is the closed square C, while the C_i sequentially converges towards the open square $\overset{\circ}{C}$. This seemingly paradoxical result is simply a result of using two different modes of convergence (due to two different topologies); it is important therefore to know the links between them. The convergence $C_i \uparrow C$ means that there exist closed sets C_i such that $j > i$ implies $C_j \supset C_i$ and $C = \bigcup_i C_i$ (but C is not necessarily closed). However, convergence in $\mathscr{F}$ requires that the corresponding limit (Lim C_i) is closed. This convergence is based on the two conditions of Definition III-19, which have nothing at all to do with monotonicity; there is no requirement that the C_i are included in one another, but only whether they hit or miss the same reference sets. Hence the implications (a) and (b) of Proposition III-21 are not at all trivial! It seems unavoidable then that a large proportion of morphological operations will not be "completely" continuous (those based on intersections), but only partially, as indicated in points (c) and (d) of Proposition III-21. We shall now be more precise on this point by introducing the notions of upper and lower semi-continuity.

D.3. Upper and lower limits, semi-continuity

Definition III-22. Let $\{F_i\}$ be a seqence in $\mathscr{F}$. We denote by $\underline{\text{Lim}}\ F_i$ the intersection of the adherent points of the sequence in $\mathscr{F}$, and by $\overline{\text{Lim}}\ F_i$ the union of the adherent points. ($\underline{\text{Lim}}\ F_i$ and $\overline{\text{Lim}}\ F_i$ are respectively said to be "the lower and upper limits of the F_i".)

Clearly: $\underline{\text{Lim}}\ F_i \subset \overline{\text{Lim}}\ F_i$. It can also be shown that:

$$\overline{\text{Lim}}\ F_i = \bigcap_{i > 0}\ \overline{\bigcup_{j > i}\ F_j} \tag{III-14}$$

An example will easily illustrate the notions of lower and upper limits. For F_i we take the boundary of the portion of cheese (Fig. III.6). When i is even, a slice at the top (of angle $\theta_i = \pi/i$) is taken away from the circle, and when i is odd, a slice at the bottom (of angle $\theta = (2\pi)/(i-1)$) is taken away. When i increases the corners of F_i become sharper and sharper, alternating between two operations. There is clearly a place for two types of limits; the points which are adherent to all the F_i are classed separately from the union of adherent points of subsequences of $\{F_i\}$. The first is exactly the lower limit, here the circle plus its centre, and the second is the upper limit, namely the circle plus its vertical diameter.

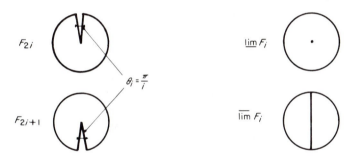

Figure III.6. An example of a sequence F_i and of its upper and lower limits.

The notion of lower and upper limits are related very simply to the notions of continuity, due to the following proposition:

PROPOSITION III-23. *If $\{F_i\}$ is a sequence in $\mathscr{F}$, then:*

(a) $\underline{\text{Lim}}\ F_i$ *is the largest closed set such that every open set which hits* $\underline{\text{Lim}}\ F_i$, *hits all the* F_i *(except at most a finite number). In other words,* $x \in \underline{\text{Lim}}\ F_i$ *iff for any integer i large enough there exists* $x_i \in F_i$ *and* $\text{Lim}\ x_i = x$ *in* $\mathbb{R}^n$.

(b) $\overline{\text{Lim}}\ F_i$ *is the smallest closed set such that every compact set which misses* $\overline{\text{Lim}}\ F_i$, *misses all the* F_i *(except at most a finite number). In other words,* $x \in \overline{\text{Lim}}\ F_i$ *iff there exists a subsequence* $\{F_{i_k}\}$ *and a sequence* $i_k \rightarrow x_{i_k} \in F_{i_k}$ *such that* $x = \text{Lim}\ x_{i_k}$.

Therefore, the definition of convergence seen above (Def. III-19), which involved conditions on the G and the K, can now be split into separate

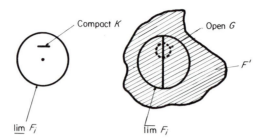

Figure III.7. Geometrical interpretation of the upper and lower limits in the intersection topology.

parts. Intuitively at least, we can now understand why the global notion of convergence is not rich enough to describe what happens to the sequence of Figure III.6. If we take for the limit, the circle plus its centre ($\underline{\text{Lim}}\ F_i$), we see (Fig. III.7), that a small horizontal segment, placed just above the centre, misses $\underline{\text{Lim}}\ F_i$, although it hits all (but a finite number) of the sets F_{2i}. Similarly, $\overline{\text{Lim}}\ F_i$ cannot serve as the limit: the small open circle G (Fig. III.7) hits $\overline{\text{Lim}}\ F_i$, but is disjoint from all (but a finite number) of the sets F_{2i+1}! Then the general definition of a limit (Def. III-19) combines two notions with the two subdefinitions providing balance one against the other. If we suppress one condition (the first for example), then we can find an arbitrarily large number of limits F' fulfilling the second condition; any closed set F' which covers all the F_i will be suitable. If we want to recover uniqueness of the limit, we must replace the counterbalance of the first condition of Definition III-19 by another constraint, namely: "the smallest closed sets such that . . ." (Prop. III-23). The notion of semi-convergence for sequences leads to the study of semi-continuity for mappings. The following definition makes a start in this direction:

Definition III-24. Let E be a separable space, and Ψ a mapping from E onto $\mathscr{F}$. Then Ψ is said to be *upper semi-continuous* (u.s.c.) iff:

$$\Psi(x) \supset \overline{\text{Lim}}\ \Psi(x_i) \tag{III-15}$$

for any $x \in E$ and any sequence $\{x_i\}$ converging towards x in E. Similarly, it is said to be *lower semi-continuous* (l.s.c.) iff under the same conditions:

$$\Psi(x) \subset \underline{\text{Lim}}\ \Psi(x_i) \tag{III-16}$$

In Definition III-24, the topological space E (i.e. the domain of the mapping) has less structure on it than the space $\mathscr{F}$ of closed sets of $\mathbb{R}^n$. Frequently used mappings, such as erosion, intersection, and many others, start from two closed sets, or a closed set and a compact set, and generate a new closed set. In such cases, we map the product spaces $\mathscr{F} \times \mathscr{F}$, or $\mathscr{F} \times \mathscr{K}$, onto $\mathscr{F}$. We therefore need to leave a certain degree of generality with the starting space in Definition III-24.

Clearly, a mapping Ψ is continuous, if and only if it is both upper and lower semi-continuous. Among the elementary set mappings, intersection and the transformation which take a closed set into its boundary are respectively upper and lower semi-continuous. We have already seen an example illustrating semi-continuity of the intersection:

$$(F, F') \to F \cap F' \quad \text{from } (\mathscr{F} \times \mathscr{F}) \text{ onto } \mathscr{F}$$

See Figure III.5. There the quantity $\overline{\text{Lim}} \ C_i \cap C_i'$ is the empty set $\varnothing$, whereas the intersection $C \cap C'$ of the two limits is the common edge of the two rectangles C and C', thus illustrating relation (III-15). Now the boundary mapping:

$$F \to \partial F \quad \text{from } \mathscr{F} \text{ onto itself}$$

is lower semi-continuous. To illustrate this, return to Figure III.6, but now take for F_i not only the outside edge of the cut cheese, but also its interior. When θ_i tends towards zero, the F_i tend towards the full disk, whose boundary ∂F (the circle), is only a subset of $\underline{\text{Lim}} \ \partial F_i$ (the centre is lacking). In this case, the inclusion (III-16) is in fact strict. In Proposition III-21, we tried to give some link between convergence in $\mathscr{F}$ and monotonic sequential convergence. The more powerful mathematical tool of *semi*-continuity allows us to explore this matter further, with the following proposition:

PROPOSITION III-25. *Let E and E' be two LCS spaces and Ψ an increasing mapping from $\mathscr{F}(E)$ onto $\mathscr{F}(E')$. Then Ψ is upper semi-continuous iff $F_i \downarrow F$ in $\mathscr{F}(E)$ implies $\Psi(F_i) \downarrow \Psi(F)$ in $\mathscr{F}(E')$.*

The proof of this theorem is quite simple. We hope it will give the reader an idea of the mathematical style used in topology (which is very different from a probabilistic proof, and different again from the inductive proofs).

Proof: Suppose that Ψ is u.s.c. If $F_i \downarrow F$, we have $F = \text{Lim} \, F_i$ in $\mathscr{F}(E)$ (Prop. III-21a), and thus $\overline{\text{Lim}} \ \Psi(F_i) \subset \Psi(F)$. On the other hand, $F_i \supset F$ implies $\underline{\text{Lim}} \ \Psi(F_i) \supset \Psi(F)$. Hence $\Psi(F) = \text{Lim} \, \Psi(F_i)$ in $\mathscr{F}(E')$. But Ψ is increasing, and we have also $\Psi(F_i) \downarrow \bigcap_i \Psi(F_i)$. Thus $\bigcap \Psi(F_i) = \Psi(F)$ (by Prop. III-21c); i.e. $\Psi(F_i) \downarrow \Psi(F)$.

Conversely, suppose that $\Psi(F_i) \downarrow \Psi(F)$ if $F_i \downarrow F$. Let $\{F_i\}$ be a sequence in $\mathscr{F}(E)$ converging towards a limit F. By relation (III-14) we have $F = \bigcap_{i > 0} A_i$, with $A_i = \overline{\bigcup_{j > i} F_j}$. But the sequence $\{A_i\}$ is decreasing, and thus $\Psi(A_i) \downarrow \Psi(F)$. By $F_i \subset A_i$, we have also $\overline{\text{Lim}} \ \Psi(F_i) \subset \overline{\text{Lim}} \ \Psi(A_i) \subset \Psi(F)$, and so Ψ is u.s.c. Q.E.D.

Using Proposition III-25 then, the property of semi-continuity in $\mathcal{F}$ is equivalent to the fourth principle of morphological quantification, for increasing mappings. But semi-continuity in $\mathcal{F}$, which from now on we shall assume holds for every morphological transformation we consider, nowhere uses the notion of increasing mappings: it is by far the most sensible property we could demand of our mappings. We will now give, without proof, a final theorem, which will play an important role in a probabilistic version of the theory.

PROPOSITION III-26.   *Let E be a LCS space, and Ψ be a semi-continuous mapping from E onto $\mathcal{F}$. Then Ψ is measurable with respect to the σ-algebra associated with $\mathcal{F}$.*

D.4. Hausdorff's metric and the myopic topology

It remains then to compare the Hit or Miss topology (on closed sets) with the topology generated by the Hausdorff metric (on non-empty compact sets). The simplest method of comparison is via an intermediary step, that of the "myopic topology" of G. Matheron. The myopic topology is generated by the base of neighbourhoods that is made of sets of the form: all those compacts sets K which hit a given open set, and miss a given closed set. The topological space $\mathcal{K}$ ($\mathbb{R}^n$) of the compact sets of $\mathbb{R}^n$ equipped with the myopic topology is locally compact, Hausdorff, and separable (LCS). Note that unlike $\mathcal{F}$ ($\mathbb{R}^n$), it is only *locally* compact (it would have been compact if we had started from a compact initial space, the sphere for example, instead of the Euclidean space $\mathbb{R}^n$). However, the two topologies have the same convergence criteria for those sequences K_i that:

1. converge under the Hit or Miss topology (i.e. converge as closed sets);
2. are contained in a fixed compact set K_0.

Under the above two conditions the sequence converges in $\mathcal{K}$ ($\mathbb{R}^n$) under the myopic topology. This result becomes clearer when we "compactify" the space $\mathcal{K}$ by adding to it the element $\mathbb{R}^n$ itself. Then a sequence of sets K_i which satisfies only the first condition, converges in $\mathcal{K}$ compactified, but uniquely towards $\mathbb{R}^n$. In other words, the myopic topology is not able to distinguish between two different closed sets which extend out to infinity (hence its name!).

In the myopic topology, the empty set $\varnothing$ plays a special role, since no sequence of non-empty compact sets can converge to it. It is an isolated point in $\mathcal{K}(\mathbb{R}^n)$, and can be removed from $\mathcal{K}(\mathbb{R}^n)$ without destroying the topological relationships between the other elements. Then it can be shown that the restriction of the myopic topology to $\mathcal{K}' \equiv \mathcal{K}/\varnothing$ is identical to the topology generated by the Hausdorff metric. These series of equivalences, in spite of the limitations placed on them, imply that the main properties of semi-continuity (particularly Props. III-25 and III-26) can be extended to compact sets, and can be given in the language of the Hausdorff metric.

E. PHYSICAL CONSEQUENCES OF THE HIT OR MISS TOPOLOGY

The results presented in this section are all fundamental. Consider the following phenomenon: certain heuristics of physics, having been taken and formalized by mathematicians, are returned to the physicists in a clearer and more coherent framework (for those who take the trouble to overcome the difficulties of vocabulary). Their simplicity resembles a little the simplicity of Mozart's symphonies, whose immediate charm is due to his perfect mastery of the rules of harmony, and the writing of music. This analogy should signal to the reader that in this section we will be more intuitive than rigorous.

The Hit or Miss topology draws its techniques from the Hit or Miss transformation. In this topology, two closed sets become more neighbouring as there exist more open sets that hit them both, and more compact sets that miss them both. Obviously this topology has been constructed in such a way, that the topological aspects of the Hit or Miss transformations are made easier to study. Underlying it, there is a certain idea of the notion of shape, note the following remark of G. Matheron (1975, Introduction): "In general, the structure of an object is defined as the set of relationships existing between elements or parts of the object. In order to experimentally determine this structure, we must try, one after the other, each of the possible relationships and examine whether or not it is verified. Of course, the image constructed by such a process will depend to the greatest extent on the choice made for the system $\mathscr{R}$ of relationships considered as possible. Hence this choice plays a priori a constitutive role (in the Kantian meaning) and determines the relative worth of the concept of structure at which we will arrive".

Often the media under study have no recognizable pattern. A porous medium, a histological section taken from a bone, etc. . . . suggest nothing immediately. This lack of meaning (for us) leads us to the position of probing them systematically, using the simplest relations that one can imagine. From this comes the idea to choose independently of the medium X, a shape B; i.e. the structuring element; and to check if B falls inside $(B \subset X)$ or outside $(B \cap X = \emptyset)$ X.

These basic binary pieces of information are then combined together by iteration, in a manner described in the preceding chapter. If we want only a denumerable number of such logical operations (we never require anything more complex for modelling), the family of all possible relations is called the

σ-algebra $\mathcal{R}$ generated by the two types of relations, $B \subset X$ and $B \cap X = \emptyset$, where B runs over some defining class $\mathcal{B}$. From the deterministic point of view, we consider the structure of the set X as known, if for every possible relation we know whether it is true or false. In the same way, from the probabilistic point of view, we consider the structure of the *random* set X as known, if for every possible relation, we know the *probability* of it being true. In other words, a random set X can be (classically) defined by specifying a probability P on the σ-algebra $\mathcal{R}$.

What are the morphological consequences of the above approach? It is not obvious that a transformation from X into X' does not modify the object so that the mode of structural description through a σ-algebra, is no longer applicable. We want to be able to use the same approach for studying both X and X'. We require the set of relations which define X', to come only from those relations used in defining X; if this is so, then the mapping from X to X' is said to be *measurable*.

It is here that we start using our topological tool; it guarantees that any semi-continuous mapping (upper or lower) on closed sets, is measurable (Theorem III-6). The fourth principle of mathematical morphology, developed in this topological framework, has already selected out semi-continuous mappings (Theorem III-5). The circle now closes, since we are sure that any set obtained by semi-continuous mappings will always be measurable.

In the previous sections we emphasized the study of closed sets with both the open and the compact structuring elements (which generate a σ-algebra, σ_f). We can in fact, check that the complete set of relations is generated by only one of these two families, either by the open sets that hit X, or by the compact sets which miss X. Moreover, if we equip the set $\mathcal{P}(\mathbb{R}^n)$, the set of *all* subsets of $\mathbb{R}^n$, with the σ-algebra generated by the relations "the open set G hits X, where $X \in \mathcal{P}(\mathbb{R}^n)$", then the equivalence

$$X \cap G \neq \emptyset \Leftrightarrow \bar{X} \cap G \neq \emptyset$$

shows that these relations have the same meaning in $\mathcal{F}$ and $\mathcal{P}(\mathbb{R}^n)$. So in order to study any set by the use of the σ-algebra given immediately above, it is equivalent to study its closure by the use of σ_f. By using the information of whether X is hit by any open set, we cannot "perceive" any difference between it and another set X' having the same closure.

In response to the requirements of physicists, the Hit or Miss topology limits our choice of mathematical sets. To take X to be all the irrational points of the plane, means now to take X to be the whole plane. In the same way, it is as equivalent to erode X by a disk B (a non-denumerable number of intersections) as it is to erode X by the rational points of B (a denumerable number of intersections). This topology then, resolves all the paradoxes of the introductory section.

To the experimenter, the question of whether a boundary belongs to the grains or the pores is vacuous, and by the same reasoning, he can arbitrarily take the compact sets as structuring elements (he could equally have taken the open sets). To the theoretician is left the freedom to choose his models amongst closed sets. This allows him to build models with points, lines, and surfaces without thickness, it being understood that the morphological properties discovered (on which statistical inferences will take place) can never allow verification of the closure structure of the said models.

This use of the Hit or Miss topology is not a trick, nor is it mathematical wizardry, but it simply reflects the adequacy of a logical framework, (chosen from other possibilities) with one a priori condition of morphology; i.e. the fourth principle.

F. TOPOLOGICAL PROPERTIES OF THE HIT OR MISS TRANSFORMATION

F.1. Properties of convergence

Henceforth, the sets X are to be interpreted as closed or compact sets (so their complements as open sets), and the structuring elements as non-empty compact sets. When we want to specifically insist on the topological type of a set under study, we will denote it F, G or K, according as it is a member of the closed, open or compact sets. For every structuring element $K \in \mathcal{K}'$ ($\mathcal{K}' = \mathcal{K}/\varnothing$) the following rules are satisfied:

PROPOSITION III-27

$$F \in \mathcal{F} \Rightarrow F \oplus K, F \ominus K, F_K \text{ and } F^K \text{ are closed sets}$$

$$G \in \mathcal{G} \Rightarrow G \oplus K, G \ominus K, G_K \text{ and } G^K \text{ are open sets} \qquad \text{(III-17)}$$

$$K' \in \mathcal{K} \Rightarrow K' \oplus K, K' \ominus K, K'_K \text{ and } K'^K \text{ are compact sets}$$

Some of these relations are obvious. For example:

$$G \oplus K = \bigcup_{x \in K} G \oplus \{x\}$$

is an open set since it is the union of open sets. Also, by duality, we see that $F \ominus K$ is a closed set. The other relations need a slightly more complicated proof. The properties (III-17) show that the three classes of sets (closed, open and compact) are each stable under finite iterations, unions and intersections of the four basic morphological operations. We will now consider continuity properties of the most common transformations:

PROPOSITION III-28. *The following mappings are continuous: the dilation of a closed set:* $(F, K) \to F \oplus K$ *from* $\mathscr{F} \times \mathscr{K}$ *onto* $\mathscr{F}$. *The dilation of a compact set:* $(K', K) \to K' \oplus K$ *from* $\mathscr{K} \times \mathscr{K}$ *onto* $\mathscr{K}$. *Transposition:* $F \to \check{F}$ *from* $\mathscr{F}$ *onto itself,* *and* $K \to \check{K}$ *from* $\mathscr{K}$ *onto itself. Magnification (change of scale):*

$(\lambda, F) \to \lambda F$ *from* $\mathbb{R}_+ \times \mathscr{F}$ *onto* $\mathscr{F}$, *and* $(\lambda K) \to \lambda K$ *from* $\mathbb{R}_+ \times \mathscr{K}$ *onto* $\mathscr{K}$.

PROPOSITION III-29. *The following mappings:*

$$— erosion \quad (X, K) \to X \ominus K$$
$$— opening \quad (X, K) \to X_K$$
$$— closing \quad (X, K) \to X^K$$

from $\mathscr{F} \times \mathscr{K}'$ (resp. $\mathscr{K} \times \mathscr{K}'$) onto $\mathscr{F}$ (resp. onto $\mathscr{K}$) are upper semi-continuous.

We will illustrate these properties with two simple examples. The first deals with the semi-continuity of erosion. Consider the sequence of compact disks X_i of radius $\rho_i = r_0 - 1/i$, which converges (from below) towards the compact disk $X = \overline{\bigcup_i X_i}$, of radius r_0. Consider also the sequence $\{B_i\}$ made up of the compact disks of radius $r_i = r_0 + 1/i$, which tend towards the same limit X, but from above. The inclusion $X_i \subset B_i$ (no matter how large the index i is) means that: $X_i \ominus B_i \equiv \varnothing$ for every finite index i. Nevertheless, if individual limits are taken, then the erosion $X \ominus X$ equals the set $\{0\}$, which is different from the empty set (in fact, $\varnothing \subset \{0\}$). The same pair of sequences may be used for illustrating the semi-continuity of the opening. For every index i, the open set $(X_i)_{B_i}$ is empty, although when individual limits are taken, $(X)_X$ is the disk of radius r_0. The second example is also very simple, and differs from the first one, in that it does not involve monotonicity. The set X_i is the segment X with fixed length $|X|$ and orientation θ (Fig. III.8b). The B_i all have the same length $|B|$ $< |X|$ but their orientation α varies; $\alpha = \theta + 1/i$. While i remains finite, the eroded set $X_i \ominus B_i$, and the opening $(X_i)_{B_i}$ are empty. But in the limit, they respectively become the segment $(|X| - |B|, \theta)$ and the segment X itself. "Au contraire", notice that the dilation $X_i \oplus B_i$ is continuous (in the present case, so also is the closure $(X_i)^{B_i}$). Therefore, it is not necessary to make an increasing or decreasing type assumption for the X_i or the B_i in order to demonstrate upper semi-continuity. The only requirement is that the transform of the limit set contains the upper limit of the transforms (rel. (III-15)). In more colloquial terms we could say that in the limit, the transform makes an *upward* jump.

In practical morphology, the set X under study is considered fixed, whereas the structuring element K often varies in a monotonic way. The following proposition addresses itself to the modes of convergence in this situation:

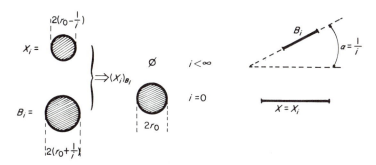

Figure III.8. Semi-continuity of some morphological operations. (a) the opening of the disk X_i by the disk B_i is always empty, except in the limit, when it becomes the disk of radius r; (b) the erosion, the opening and the closure of the segment X by the segment B_i with orientation $\alpha = 1/i$, are empty for every finite value of i, but not in the limit.

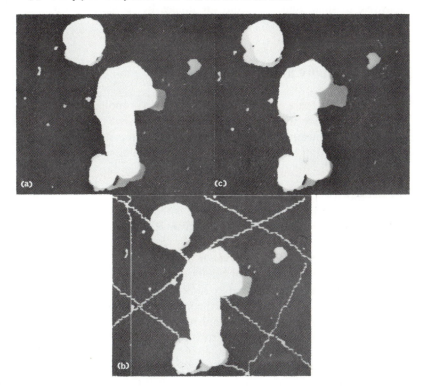

Figure III.9. Semi-insensitivity of the opening to noise. The initial set X (union of the white and grey parts of (a)) becomes the white part of the same photograph, after being opened by a small dodecagon. If we add points and lines to X, then provided they are fine enough, the opening remains unchanged (b). But remove just three points from X, and its opening is completely upset (c).

PROPOSITION III-30. *Let X be a closed (resp. a compact) set, and $\{K_i\}$ an increasing sequence in $\mathscr{K}'$. If $\overline{\bigcup K_i} = K \in \mathscr{K}$, we have*

$$(X \ominus K_i) \downarrow (X \ominus K) \qquad \qquad \text{(III-18)}$$

and $Lim(X \ominus K_i) = X \ominus K$ in $\mathscr{F}$ (resp. in $\mathscr{K}$).

For the experimenter, semi-continuity as stated by Propositions III-29 and III-30, appears as a *semi*-insensitivity to noise. The three photographs of Figure III.9, taken from the T.V. monitor of the Texture Analyser show this clearly. This (semi) insensitivity, together with the idempotence property, make the opening and closing operations excellent image cleaning devices in the practice of image analysis. In fact, it would be more appropriate to speak of semi-cleaning, since by duality to what occurred in Figure III.9, closing is insensitive to small removals from the image, but completely upset by the adding of any supplementary noise.

F.2. Connectivity and homotopy

We now leave the space $\mathscr{F}(\mathbb{R}^n)$, and the Hit or Miss topology, and go back to the Euclidean space $\mathbb{R}^n$ itself, equipped with its usual metrics. Indeed the properties of connectivity and homotopy concern the classes of points of a given set X, and not at all collections or sequences of sets in $\mathscr{F}(\mathbb{R}^n)$.

(a) *Connectivity*

A curve, or a path, in a topological space is the image of the closed interval $[0, 1]$ under a continuous transformation from $\mathbb{R}$ onto E. The images of 0 and 1 are the initial and terminal points of the curve respectively. When they coincide, the curve is a closed curve, or a loop.

A set X is arcwise connected, or briefly, connected, when every pair x, $y \in X$ may be joined by a curve included in X. Set X is simply connected when it is homotopic to the unit hall. A classical theorem due to jordan states is that in $\mathbb{R}^2$ any loop divides the plane into two connected regions X and Y, one of which is bounded, and that every path going from $x \in X$ to $y \in Y$ hits the loop.

PROPOSITION III-31. *If X and B are connected, then $X \oplus B$ is also connected.*

Proof: Let x_1 and x_2 be two points of X, and y_1 and y_2 two points of B. We have to find a path going from $x_1 + y_1$ to $x_2 + y_2$ which is contained in $X \oplus B$. The two points $x_1 + y_2$, $x_2 + y_2$ belong to X_{y_2}, so are connected;

similarly for the pair of points $x_1 + y_1$, $x_1 + y_2$ since they belong to B_{x_1}. Thus $X \oplus B$ is connected. Notice that $X \ominus B$ is not always connected, even if both X and B are.   Q.E.D.

PROPOSITION III-32. *Let B be a connected set, and X be a set with connected components X_i, then it is equivalent to erode X considered as a whole, or to erode each X_i separately, and take the union of the results:*

$$X \ominus B = \bigcup_i (X_i \ominus B) \tag{III-19}$$

Proof: Firstly, $X_i \subset X$ implies $X_i \ominus B \subset X \ominus B$, thus $\bigcup (X_i \ominus B) \subset X \ominus B$. Conversely, let $z \in X \ominus B$. Then $\check{B}_z \subset X$. Now $\check{B}_z$, a connected set, is necessarily included in one of the connected components of X, say X_i. Thus $z \in X_i \ominus B$, and hence $X \ominus B \subset \bigcup_i (X_i \ominus B)$.   Q.E.D.

PROPOSITION III-33. *If B is a connected set, it is equivalent to open X by B or to take the union of the openings by B, of the connected components X_i of X.*

$$X_B = \bigcup_i (X_i)_B \tag{III-20}$$

Proof: It is immediate from the two preceding propositions.   Q.E.D.

(b) *Homotopy*

Two transformations Ψ_1 and Ψ_2 of a topological space E_1 onto a topological space E_2 are homotomic if there is a continuous mapping $\Phi(x, t)$ from $E_1 \times [0, 1]$ onto E_2 (i.e. continuous w.r. to both x and t), for which $\Phi(x, 0) = \Psi_1(x)$ and $\Phi(x, 1) = \Psi_2(x)$, $\forall x \in E_1$. Moreover, for each value of t, $\Phi(x, t)$ is a one to one mapping, i.e. does not bring points together. Such transformations shrink, twist, etc. . . . in any way without tearing. When the two spaces E_1 and E_2 are the plane $\mathbb{R}^2$ minus its point at the infinity, a simpler definition of the homotopy can be given. With each compact set K of the plane associate its *homotopy tree*, whose trunk corresponds to the back ground K_0 (i.e. the infinite connected component of K^c), the first branches correspond to the connected components K_1 of K adjacent to K_0, the second branches to the pores of K_1 adjacent to K_1, etc. . . . (see Fig. III.10).

A transformation Φ is homotopic when it does not modify the tree associated with every compact set of the plane. None of the transformations met in Chapter II are homotopic. (Note that the definition based on the homotopic tree is longer valid in $\mathbb{R}^3$, since it considers a ball and a torus as homotopically equivalent.)

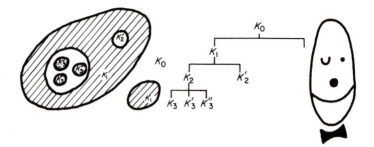

Figure III.10. Homotopic figures and their tree.

G. EXERCISES

G.1. Equivalent topologies

(a) Let d and d' be metrics on a space E. Prove that if d and d' tend simultaneously towards zero (i.e. for every $\varepsilon > 0$ there exists $\eta > 0$ such that $\{d(x, y) < \eta\} \Rightarrow \{d'(x, y) < \varepsilon\}$ and vice versa), then the topologies associated with d and d' are identical.

[the mapping $x \to x$ of E with the d-topology, onto E with the d'-topology is bi-continuous, hence it is a homeomorphism].

(b) Derive from a, that in $\mathbb{R}^n$ the distances

$$\left[\sum (x_i - y_i)^2 \right]^{1/2}; \quad \sup |x_i - y_i|; \qquad \sum |x_i - y_i|$$

lead to the same topology.

(c) Consider the Hausdorff distance III-10, and take for $B(\varepsilon)$ the homothetics εB of a compact convex set B, arbitrarily chosen, but having a non empty interior. Prove that the topology induced by the corresponding Hausdorff metric is independent of the choice of B. In $\mathbb{R}^2$ for example, the disk, the square and the hexagon generate the same topology (see an application of this result in Ch. VI, D.3b).

G.2. Pseudo-distance from a point to a set

Let d be a metric on the space $\mathbb{R}^n$, $x \in \mathbb{R}^n$ and $X \in \mathscr{P}(\mathbb{R}^n)$. Consider the quantity $d(x, X)$ depending on the pair (x, X) via:

$$d(x, X) = \inf_{y \in X} d(x, y) \qquad x \in \mathbb{R}^n, \, X \in \mathscr{P}(\mathbb{R}^n)$$

(a) Prove that d is not a distance, but satisfies the following Lipschitz property:

$$|d(x, X) - d(y, X)| \le d(x, y) \qquad (1)$$

(b) Prove that the two sets X and $\overline{X}$ are identical for the pseudo-distance d, i.e.:

$$d(x, X) = d(x, \overline{X}) \qquad \forall x \in \mathbb{R}^n$$

(c) Prove that a sequence $\{F_i\}$ converges in $\mathscr{F}$, w.r. to the Hit or Miss topology, iff for any $x \in \mathbb{R}^n$ the sequence $\{d(x, F_i)\}$ converges in $\overline{\mathbb{R}}_+$.

[The "only if" part is an immediate consequence of (1). Conversely, suppose that for any $x \in \mathbb{R}^n$, the sequence $\{d(x, F_i)\}$ has a limit $f(x) \in [0, \infty]$. Since $\mathscr{F}$ is a compact space, there exists a subsequence $\{F_{i_k}\}$ converging towards a limit $F \in \mathscr{F}$, and $d(x, F_{i_k}) \to d(x, F)$, from the "only if" part of the proof. Hence $f(x) = d(x, F)$ and $\{F_n\} \to F$.]

G.3. Topological properties of the dilations and the erosions (G. Matheron, 1967)

(a) Show that if the set $\mathring{B}$ is topologically open, $X \oplus \mathring{B}$ is open and $X \ominus \mathring{B}$ is closed for every set X. If $\overline{C}$ is topologically closed $\overline{C} \ominus X$ is closed.

(b) Similarly, show that:

$$X \ominus \mathring{B} = \mathring{X} \ominus \mathring{B}; \qquad X \oplus \mathring{B} = \overline{X} \oplus \mathring{B}; \qquad \overline{B} \ominus X = \overline{B} \ominus \overline{X}$$

(c) If B is bounded, show that:

$$\overline{X \oplus B} = \overline{X} \oplus \overline{B}; \qquad \overset{\circ}{X \ominus B} = \mathring{X} \ominus \overline{B};$$

$$\overset{\circ}{B^c \ominus X} = \overline{B^c \ominus \overline{X}}$$

(d) $\rho \mathring{B}$ and $\rho \overline{B}$ respectively denote the open and the closed ball of radius ρ. Prove that:

$$\mathring{X} = \bigcup_{\rho > 0} X \ominus \rho \mathring{B} = \bigcup_{\rho > 0} X \ominus \rho \overline{B}$$

$$\overline{X} = \bigcap_{\rho > 0} X \oplus \rho \mathring{B} = \bigcap_{\rho > 0} X \oplus \rho \overline{B}$$

(e) The set X is assumed to be open ($X = \mathring{X}$). For every $\rho > 0$, define the two decreasing families $\{G_\rho\}$ and $\{F_\rho\}$ with:

$$G_\rho = X \ominus \rho \overline{B} \quad \text{and} \quad F_\rho = X \ominus \rho \mathring{B}$$

Prove that G_ρ is open and F_ρ is closed and that $\mathring{F}_\rho = G_\rho$ but $F_\rho \supset \overline{G}_\rho$.

(f) Establish the two following formulae:

$$G_\rho = \overset{\circ}{F}_\rho = \bigcup_{\varepsilon > 0} F_{\rho+\varepsilon} = \bigcup_{\varepsilon > 0} G_{\rho+\varepsilon}$$

$$F_\rho = \bigcap_{\rho' < \rho} F_{\rho'} = \bigcap_{\rho' < \rho} G_{\rho'}$$

Derive that the family $\{G_\rho\}$ (resp. $\{F_\rho\}$) is right (resp. left) continuous.

IV. The Convex Set Model

A. A BASIC MODEL

It is impossible to continue with the elaboration of a morphology, without making use of the notion of a convex shape, since it gives us the best chance of obtaining a mathematical description of a set. The reader might find this reason a little theoretical; indeed our aim is not to write down a lot of mathematical equations, nor do we want to force convexity on natural objects (they are often too complicated), but quite simply, *we just want to arrange our thoughts*. The mind imagines concepts such as surface area or width of a body, only by more or less implicit reference to convex figures. In the next chapter we will see that all the numerical parameters for general shapes, are generalizations of those for the convex set model.

Considered as a structuring element, a compact convex set is the key to the operations involving size distribution: when we speak of the size of a body, we always have in mind underlying convex structuring elements. Considered as an object, its "good" properties, with respect to projections, thick sections, dilations, etc, makes us choose it as a descriptor for isolated particles in cytology or in the milling of rocks (sometimes using the particles' convex hulls as an intermediary step). But in fact, it is mainly *finite unions of convex sets* which play a primary role—first as a general model for sets, i.e. the convex ring (Ch. V), second as a prototype of almost all the random sets that we find in this book; e.g. Boolean schemes, Poisson and Voronoï polyhedra, etc.

When we see the ramifications of convexity, it is clear that the theory of the geometry of convex sets could itself be the contents of a whole book. Therefore, we have decided to present only those parts of the theory which will pertain at least once, to a practical study. In any model, we can split the analysis into three categories:

(a) What are the set theoretical properties of convex sets (in particular with respect to the basic morphological operations)?

(b) What numerical parameters can we measure?
(c) How do these change when we deform the convex set by basic morphological operations?

The section following this one is devoted to question (a). Any response to the question in (b) must quite clearly make the distinction between global and local representations. The two quite different methods of integral geometry and measure theory respectively refer to global and local representations. We will tackle them successively, and in the process give a brief summary of the prerequisite measure theory. The study of the last question, (c), is more scattered through the various sections of this chapter. A first answer is given by Steiner's formula. Another is given in the last two sections, where we collect together various properties related to anisotropy. The theoretical tool is the support function of a convex set; it is first used for the Steiner class of compact sets, then for the convex sets in the plane (but not necessarily Steiner). The latter are very rich in properties due to the space being only of dimension *two*.

The first important work in integral geometry dates from the middle of the last century (Cauchy, 1840; Steiner, 1840). With Crofton (1860), and particularly Minkowski (1903), we see the first attempt to make the mathematical structure of the objects under study more precise. (For Cauchy, Barbier, etc. . . . , the hypotheses of the regular model of Chapter V were so obvious that they never discussed them.) This new rigour led to focusing research onto the simplest shape, using compact convex sets. The concepts which they elaborated upon, although geometric, surprisingly have a probabilistic interpretation (Crofton). We owe to Minkowski, not only the study of the functionals which bear his name, but the systematic use of the supporting hyperplane, the isoperimetric inequalities, and the first results on surface measures. In the twentieth century, Deltheil (1926) extended the work of Crofton on geometrical probability, and Fenchel and Bonnensen enlarged the geometrical theory of the convex set. S. A. Santalo (1933) and W. Von Blaschke (1936) extended the basic stereologic theorems and Steiner's formula, to the finite union of compact convex sets. In 1957, H. Hadwiger, in a book which synthesized much of integral geometry, characterized the Minkowski functionals by certain properties related to the four principles of morphology (we shall discuss them later). Recently, G. Matheron (1975) defined and studied the class of Steiner compact sets, looking again at, and developing the study of, surface measures and perimeter measures (1977). In doing so, he interprets the basic results of integral geometry in a *local* way, which leads him to propose a new generalization of the Minkowski functional to the convex ring, which is different from the generalization of Santalo and Von Blaschke, and based on the notion of convexity numbers. In 1975, R. E. Miles, and in 1976, G. Matheron extended Crofton's formula to thick sections of convex sets, the first by probabilistic proofs, the second by a geometrical approach. There are many other references we could cite (Giger, Streit, . . .). See the bibliographies in Hadwiger's, Matheron's and Santalo's books.

In this book, we split the study of convex sets into two parts, depending on whether we are dealing with a model or with criteria. The first part is developed in this chapter, while the second part, contained mainly in Chapter X, enables us to develop the size distribution criteria. In fact, convexity can be found in many other places: in digitalization where we have to think of convexity on a grid of points; in the Boolean model, where the choice of a primary convex grain enriches the model with a Markovian property, etc. . . .

B. SET PROPERTIES OF THE CONVEX SETS

B.1. Definition

In this section, we emphasize the most important properties by presenting them as "propositions". All the corresponding proofs can be found in G. Matheron (1975). The symbols $\mathscr{C}$, $\mathscr{C}(\mathscr{K})$ and $\mathscr{C}(\mathscr{F})$ respectively denote the convex sets, the compact convex sets and the closed convex sets in $\mathbb{R}^n$. There are at least two ways of defining a convex set. First, $X \in \mathscr{P}(\mathbb{R}^n)$ is said to be a convex set if for every pair of points x_1, x_2 belonging to X, the whole segment $[x_1, x_2]$ belongs to X. This definition is analytic in nature and has no counterpart when we replace the space $\mathbb{R}^n$ by a lattice of points. However, the second definition (given below) although identical to the first for closed sets $\mathscr{F}$ in $\mathbb{R}^n$ can be extended to digital spaces. The closed set X is said to be convex if and only if it is the intersection of all the half spaces which contain it. Com-

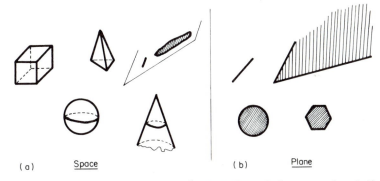

(a)   Space   (b)   Plane

Figure IV.1. Examples of convex sets. (a) in $\mathbb{R}^3$: cube, ball, tetrahedron, cone, plane, half space, disk, segment. (b) in $\mathbb{R}^2$: disk, hexagon, segment, internal portion of an angle.

pare this definition with the invariance property of convex sets under morphological closing, Proposition IV-4. The two definitions of convexity are respectively formalized by relationships (IV-1) and (IV-2):

$$X \in \mathscr{C} \Leftrightarrow x_1, x_2 \in X \Rightarrow [x_1, x_2] \in X \tag{IV-1}$$

$$X \in \mathscr{C}(\mathscr{F}) \Leftrightarrow X = \bigcap_i \Pi_i; \ \Pi_i \supset X \tag{IV-2}$$

Figure IV.1 shows a few examples of convex sets.

B.2. Basic morphological operations

If X is a convex set, then its transpose $\check{X}$, its projection onto any subspace (onto a plane or a straight line if $X \in \mathbb{R}^3$, for example), and its intersection with any subspace, are convex sets. Moreover, if X' is also a convex set, then the intersection $X \cap X'$ is convex. If $X \mid P$ denotes the projection of X onto the subspace P, we can thus write:

$$X, X' \in \mathscr{C} \Rightarrow \check{X}; \ X \cap P; \ X \mid P; \ X \cap X' \in \mathscr{C} \tag{IV-3}$$

We will now review the properties relating to dilation and erosion. Notice first the asymmetrical role played by these two operations with respect to convexity. If X and B are convex sets, then dilation, erosion, closing and opening of each one by the other (in both senses) are also convex:

$$X, B \in \mathscr{C} \Rightarrow X \oplus B, X \ominus B, B \ominus X, X_B, X^B, B_X, B^X \in \mathscr{C} \tag{IV-4}$$

In the case of dilation, one proves relation (IV-4) as follows: if $y_1 = x_1 + b_1$ and $y_2 = x_2 + b_2$, with $x_1, x_2 \in X$ and $b_1, b_2 \in B$, then $\lambda y_1 + (1 - \lambda) y_2 = \lambda x_1 + (1 - \lambda) x_2 + \lambda b_1 + (1 - \lambda) b_2 \ (0 \le \lambda \le 1)$ is the result of an element of X added to an element of B, and so belongs to $X \oplus B$. In the case of erosion, the property is true since $X \ominus B$ is a product of intersections of convex sets (namely the $X_h, h \in B$). Note that the proof shows the property to hold for every B, even a non-convex set, i.e.

$$X \in \mathscr{C} \Rightarrow X \ominus B \in \mathscr{C}, \quad \forall B \tag{IV-5}$$

B.3. Convexity and magnifications

The basic questions posed in this section are more concerned with the study of structuring elements than the sets themselves. The second principle, namely compatibility with magnifications, limits the choice of families $B(\lambda)$ of structuring elements depending upon a positive parameter λ, to the homothetics λB, of a reference figure B. If we now take B to be convex, then three rather profound properties emerge (Props. IV-1, 2, 3). The first one is closely linked with the iteration of dilations:

PROPOSITION IV-1. *A compact set B is infinitely divisible with respect to dilation, if and only if it is convex.*

This proposition means that if one takes an arbitrary positive integer k, and dilates the homothetic $1/k \cdot B$ of the compact set B with itself k times, then the result is identical to B if and only if the compact set B is convex.

$$B = \left(\frac{1}{k}B\right) \oplus \left(\frac{1}{k}B\right) \oplus \dots \underbrace{\left(\frac{1}{k}B\right)}_{k \text{ times}} = \left(\frac{1}{k}B\right)^{\oplus k} \qquad \text{(IV-6)}$$

The proof one way is rather obvious (start from $(1/k)B$ and check relation (IV-6)). The converse, namely relation (IV-6) implies the convexity of B, and is far less easy. Notice the technological consequence of relation (IV-6). To carry out dilations (and by duality erosions: thus openings and closings) using homothetic convex sets, it is enough to build a module into the computer, that will dilate by the smallest B_0 of the chosen shape (i.e. smallest square, hexagon, segment, etc. . . . defined on the grid). By iterating this dilation and allowing the possibility of taking complements, we are able to reconstitute any basic morphological operation involving the homothetics kB_0 of B_0 (see Ex. VI.15). The method is relatively time-consuming but simple. Unfortunately the "only if" part of Proposition IV-1 indicates that such an iterative process does not hold as soon as B is non convex (which is the case for covariance, contouring logic, etc. . . .). The second proposition is a more quantitative expression of the previous one:

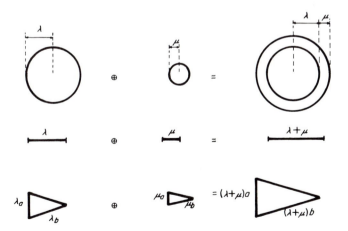

Figure IV.2. The dilates of a disk, of a segment, and of a triangle by their respective homothetic sets. The same shape always results.

PROPOSITION IV-2. *A family* B_λ $(\lambda \geq 0)$ *of non-empty compact sets is a one-parameter continuous semi-group (i.e.* $B_\lambda \oplus B_\mu = B_{\lambda+\mu}$, λ, $\mu \geq 0$*) if and only if* $B_\lambda = \lambda B$ *where* B *is a convex compact set.*

Thus the dilates of a ball (a cube, a segment, etc. . . .) with radius (side, length) λ, by a ball (a cube, a segment) with corresponding parameter μ, is again a ball (a cube, a segment) and with parameter $\lambda + \mu$ (Fig. IV.2).

The third proposition, derived from the first two, completes the analysis of the way convexity and "similarity" interact:

PROPOSITION IV-3. *Let* B *be a compact set in* $\mathbb{R}^n$. *Then the homothetic* λB *is open with respect to* B *(i.e.* $(\lambda B)_B = \lambda B$) *for every* $\lambda \geq 1$, *if and only if* B *is convex.*

Once more, proof of the "if" part is obvious (note that if B is convex, every point $x \in \lambda B$ belongs to a B_y included in λB), and proof of the converse is extremely delicate. From a practical point of view, Proposition IV-3 means that when we open a disk (triangle, segment) with another disk (triangle, segment), smaller than the original, we leave it unchanged. This proposition is one of the strongest of mathematical morphology. It is one of the links of the logical chain leading to the morphological concept of *size*. In Chapter II, Section E, we saw that a family of openings of X by $B(\lambda)$ was monotonic (and thus had a size distribution meaning) when each $B(\lambda)$ was open with respect to all $B(\mu)$, $\lambda > \mu$. This result, combined with Proposition IV-3 allows us to exchange the monotonicity property of $(X)_{B(\lambda)}$ with the convexity property of the structuring element. The whole of Chapter X will be devoted to the consequences of this substitution.

B.4. Convex hull, invariance under closure

A large number of properties that are satisfied by the set X when convex, are more generally satisfied by its convex hull when X is not convex. There are two ways of dealing with the convex hull, corresponding to the two definitions of convexity given above. First, we can say that for every set $X \in \mathscr{P}(\mathbb{R}^n)$, the convex hull $C(X)$ is the smallest convex set which contains X. When X is a topologically closed set, the convex hull is also the intersection of all the half spaces which contain X (Fig. IV.3). For both definitions, the convex hull is identical to X iff X itself is convex.

With respect to dilation, we have the following relation:

$$C(X) \oplus C(X') = C(X \oplus X') \quad \forall X, X' \in \mathscr{P}(\mathbb{R}^n) \tag{IV-7}$$

We will now consider the relationship between the convex hull $C(F)$ and the morphological closing F^B of a (topologically) closed set F.

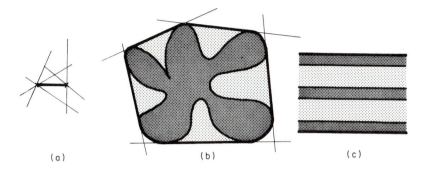

Figure IV.3. Convex hulls for a set X which is: (a) two points; (b) a compact set; (c) a closed unbounded set.

PROPOSITION IV-4. *If* $F \in \mathscr{F}$ *and if* B *is an arbitrary bounded set, then* F^B $\subset C(F)$. *Moreover, if* F *is convex, then it is identical to its closure*:

$$F \in \mathscr{C}(\mathscr{F}), \ B \text{ bounded} \Rightarrow F^B = F \qquad \text{(IV-8)}$$

Furthermore, if F and $F' \in \mathscr{F}(\mathbb{R}^n)$ are convex, then the equation $F \oplus B$ $= F' \oplus B$ allows "cancellation" of the B from both sides:

$$F, F' \in \mathscr{C}(\mathscr{F}), \ B \text{ bounded set}, \ F \oplus B = F' \oplus B \Rightarrow F = F' \qquad \text{(IV-9)}$$

The two results of Proposition IV-4 are fundamental. From a heuristic point of view, we have already noticed that dilations (or erosions) "simplify" or "smooth" the objects, and in general, a given X cannot be recovered from its dilate $X \oplus B$. Proposition IV-4 presents the convex shape as exactly poor enough in features to escape this restriction: neither a dilation nor an opening can rob it of its features. Relationships (IV-8) and (IV-9) express this formally. (A warning: this last relation does not mean that the data of $F \oplus B$ and B, automatically determine F; we have to assume *a priori* that F is a convex set: a disk and an annulus with inner ring of radius r, give the same resulting set after dilation by any circle with a radius $> r$.)

The simplification $F^B = F$ when F is convex, is the basis of several experimental techniques for testing the convexity of a particle. Such a utilization suggests further pursuance and we might wonder about what happens when F is no longer convex. A partial answer already appears in the inclusion, $F^B \subset C(F)$ (Prop. IV-4). The following three propositions set out the results:

PROPOSITION IV-5. *The mapping* $K \rightarrow C(K)$, *which associates every compact set* K *with its convex hull* $C(K)$, *is continuous on* $\mathscr{K}$. *Similarly,* $F \rightarrow C(F)$ *is lower semi-continuous on* $\mathscr{F}$.

Moreover, the mapping $X \to C(X), X \in \mathscr{P}(\mathbb{R}^n)$, is invariant under translation and magnification, but compatible with a local knowledge only for *bounded sets*. This restriction, together with Proposition IV-5, means that use of the convex hull as a quantitative morphological tool is confined exclusively to compact sets. The next proposition is the analogue of Proposition IV-1, for non-convex sets:

PROPOSITION IV-6. *For every non-empty compact set K, the sequence $\{(1/i \cdot K)^{\oplus i}\}$ converges towards the convex hull $C(K)$.*

Convergence is from below, since $\{(1/i)K^{\oplus i}\}$ is always included in $C(K)$. In practice, Propositions IV-1 and IV-6 do not lead to good criteria; the magnifications realizable on a lattice must have an *integer* ratio. So sets such as $(1/i)K$, that are more reduced than the original, can often only be approximately known through interpolation. The next proposition provides an algorithm which is easier to handle.

PROPOSITION IV-7. *If K is a compact convex set with a non-empty interior, and if K admits a finite curvature at each point of its boundary, then for every closed set F, we have:*

$$C(F) = \operatorname*{Lim}_{\lambda \to \infty} F^{\lambda K} \qquad (IV\text{-}10)$$

The limit is taken in $\mathscr{F}$ or in $\mathscr{K}$ if F is compact. (The proposition is given in Matheron (1975) without any statement about curvatures. Nevertheless, in the proof which follows, the adherent point G of the sequence $(\rho_n + \varepsilon)K$ is a half space only if the radii of curvature of $(\rho_n + \varepsilon)K$ all tend to infinity, which is implied by finite curvatures of K.) In practical morphology, relation (IV-10) is particularly interesting when F is itself a compact set.

The limit $C(F)$ is approached monotonically from below, since $F^{\lambda K}$ is always embedded in $C(F)$, and the closure operation is monotonic. Figure IV.4 clearly shows the evolution of $F^{\lambda K}$; when K admits some angularity (case (b)) or has an empty interior (case (a)) then there is no approach towards $C(F)$. The reader is invited to compare these results, for $\mathbb{R}^n$, with the equivalent notions in a digital analysis (Ch. VI, D.4, and Ex. XI-1). In digital image processing, two methods may be used for convex hull computation: either algorithm IV-10, or an iterative process due to Golay–Preston, specifically designed for, and well adapted to digitalization, but more time-consuming than algorithm IV-10. The monotonicity of this latter algorithm as a function of λ, suggests its use as a test for "progressive" convexity in pattern recognition techniques. A family of bounded pictures (nuclei of cells for example) are ordered according to the amount of area incremented when a given closing acts

Figure IV.4. Closings and convex hulls. (a) K is a segment, its interior is empty: $C(F)$ is not approached. (b) K is angular (infinite curvature for some points of ∂K): $C(F)$ is not approached. (c) K is sufficiently regular: the convex hull $C(F)$ is reached.

on them. Finally, note that the results stated in this section are all independent of the number of dimensions of the space but, however, do not provide a way of computing, for example, the convex hull of $X \in \mathscr{P}(\mathbb{R}^n)$ from its plane section. In fact, Proposition IV-7 emphatically denies any possibility of stereology, since all 2-D structuring elements have an empty interior in $\mathbb{R}^3$.

B.5. Set union and convexity

In general the union of two convex sets is not a convex set. This is a fortunate circumstance, since it leads to the very fruitful idea of the convex ring. Under what conditions then is the union of two compact convex sets, still convex? To answer this question, we first need to study how two sets might be separated by a third set (Matheron, 1975). Let K, K' and C be three compact sets in $\mathbb{R}^n$. Then we say that K and K' are separated by C if, for any $x \in K$ and $x' \in K'$, there exists a real number λ such that $0 \le \lambda \le 1$ and $\lambda x + (1 - \lambda)x' \in C$ (Fig. IV.5)

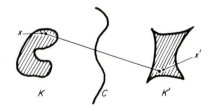

Figure IV.5. The set C separates the compact sets K and K'.

PROPOSITION IV-8. *Let K and K' be two compact sets, the union $K \cup K'$ of which is convex. Then K and K' are separated by $K \cap K'$.*

PROPOSITION IV-9. Matheron (1975): *Let K and K' be two compact convex sets. Then $K \cup K'$ is also convex if and only if:*

$$K \oplus K' = (K \cup K') \oplus (K \cap K') \qquad \text{(IV-11)}$$

Finally, we will indicate a formula that is essential in the establishment of Steiner's formula (Sect. C.2). We saw that dilation distributes the union (i.e. for every K, K' and X: $(K \cup K') \oplus X = (K \oplus X) \cup (K' \oplus X)$). Similar distributivity also holds for the intersection, provided the compact set $K \cup K'$ and the set X are convex:

$$(K \cap K') \oplus X = (K \oplus X) \cap (K' \oplus X) \quad K \cup K', X \in \mathscr{C}(\mathscr{K}) \quad \text{(IV-12)}$$

C. THE GLOBAL POINT OF VIEW: INTEGRAL GEOMETRY FOR COMPACT CONVEX SETS

Extraordinarily enough, integral geometry succeeds in answering, at once, three very different but equally crucial questions. First, it concerns itself with numerical parameters associated with compact convex sets, parameters which satisfy a few very strong properties of invariance more severe than the four principles of morphology. A powerful theorem due to Hadwiger (1957) exhausts the subject, since he showed that these parameters are all linear combinations of a small subset of them, called the Minkowski functionals.

Moreover, integral geometry shows how the projections of the compact convex set K, or its sections (defined by translates of subspaces of smaller dimension) allow us to reform its Minkowski functionals. These very attractive properties, due to Cauchy for projections and to Crofton for sections, open up a fruitful path for the study of stereology. (In fact the parameters engendering a stereological interpretation do not all belong to this category. The notions of "star", and of "measure weighted moments", give us another way of approaching the stereological problems, but more often than not, experimenters draw from the battery of Minkowski functionals.)

The third success of integral geometry appears in the work of Von Blaschke, Santalo and Hadwiger. Their search was for the family of sets having compact, convex sets as a sub-family (called a convex ring), which continue to satisfy the properties of invariance and stereology (e.g. reconstitution via Minkowski functionals) mentioned above. We will pursue this further in the next chapter.

C.1. The Minkowski functionals and their characterization

(a) *Definition*

Points, straight lines, planes and directions are parametrized as follows. The symbol x represents a point with co-ordinates (x) in $\mathbb{R}^1$, (x_1, x_2) in $\mathbb{R}^2$, and

(x_1, x_2, x_3) in $\mathbb{R}^3$. We use the co-ordinates in algorithms only when it becomes strictly necessary to do so. The symbols α and ω denote directions respectively in the plane and the space. So ω is a unit vector centred at the origin, whose extremity describes Ω, the boundary of the unit ball. On the unit circle and the unit sphere, the equivalent of Lebesgue measure is a unit mass spread uniformly, resulting in (uniform) densities, $d\alpha/2\pi$ in $\mathbb{R}^2$ and $d\omega/4\pi$ in $\mathbb{R}^3$. Capital greek letters Δ and Π respectively denote the test lines and test planes for sectioning the set X. The symbol $\{x\}$ represents a test point (or more generally, a set reduced to a point). In the plane, the straight line passing through x with a direction α, is written $\Delta(x, \alpha)$. In the space, the straight line with direction ω and the plane normal to ω, both passing through the point x, are respectively $\Delta(x, \omega)$ and $\Pi(x, \omega)$ with $\Pi_\omega = \Pi(x, \omega)$, $\Delta_\omega = \Delta(x, \omega)$, $\Delta_\alpha = \Delta(x, \alpha)$. (The sense of the vectors α and ω has no importance in the stereological theorem, but does intervene in the convexity number algorithm). Symbol "$|$" means "projection onto", and the expression "$X|\Delta$" is read: "projection of X onto Δ".

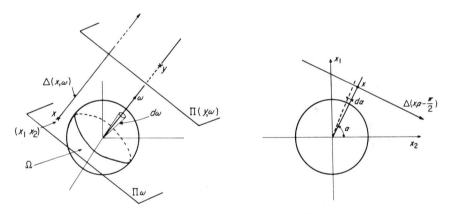

Figure IV.6. Parametrization of sections of the set X in $\mathbb{R}^3$ and $\mathbb{R}^2$ respectively.

A functional is, by definition, a global parameter associated with a set, i.e. a mapping from $\mathscr{P}(\mathbb{R}^n)$ onto $\mathbb{R}$ (Fig. IV.6). For every $X \in \mathscr{C}(\mathscr{K})$ in $\mathbb{R}^n$, there exist $n+1$ Minkowski functionals. For the usual spaces ($\mathbb{R}^0$ is the space reduced to the origin), they are given in the table in Figure IV.7.

For $X \in \mathscr{C}(\mathscr{K})$ in $\mathbb{R}^n$, the connectivity number $N^{(n)}(X)$ is the functional equal to one when $X \neq \varnothing$, and to zero if not. When the boundary ∂X is regular enough to have $n-1$ finite radii of curvature R_j at each point, then the $W_i^{(n)}$ ($i = 1, \ldots, n$) can be represented as surface integrals of the basic symmetric functions of the curvatures (Ex. VI-3: Bonnensen–Fenchel, 1934). In $\mathbb{R}^1$ in particular, the norm $M(X) = \int_{\partial X} \frac{1}{2}(1/R_1 + 1/R_2)\, ds$ is the integral of

Space of dim. k	Minkowski functionals of number i			
	$i = 0$	$i = 1$	$i = 2$	$i = 3$
$k = 3$	Volume V	$\frac{1}{3}$.. Surface area S	$\frac{1}{3}$. Norm M	$\frac{4}{3}\pi$. Con. Number $N^{(3)}$
$k = 2$	Area A	$\frac{1}{2}$. Perimeter U	π. Con. Number $N^{(2)}$	
$k = 1$	Length L	2. Con. Number $N^{(1)}$		
$k = 0$	Con. Number $N^{(0)}$			

Figure IV.7. Table of the Minkowski Functionals $W_i^{(k)}$.

the mean curvature. However, such a regularity of the boundaries is not essential and the curvatures may be generalized by measure (G. Matheron, 1975; see below Sect. D).

In $\mathbb{R}^n$, the Minkowski functionals are defined by a recurrence relation on sub-dimensions of the space as follows:

$$\left.\begin{aligned}
&\text{for } n = 1 \quad W_0^{(1)} = L(X); \qquad W_1^{(1)}(X) = 2 \\[2mm]
&\text{for } n > 1 \quad W_0^{(n)} = V^{(n)}(X); \\[2mm]
&W_k^{(n)}(X) = \frac{1}{nb_{n-1}} \int_{\Omega_n} W_{k-1}^{(n-1)}[X|\Pi_\omega^{(n-1)}]d\omega \qquad 1 \le k \le n
\end{aligned}\right\} \quad \text{(IV-13)}$$

($V^{(n)}(X)$, $b_n = n$-volume of X and of the unit ball (Ex. IV-1); ω and Ω_n, direction and set of directions (i.e. unit sphere) in $\mathbb{R}^n$).

(b) *Cauchy and Crofton formulae*
For the sake of simplicity, in this section we restrict ourselves to the cases of the usual spaces. By taking n for 3 and 2 in relation (IV-13), we find:

$$\frac{1}{4}S(X) = \frac{1}{4\pi}\int_{4\pi} A(X|\Pi_\omega)d\omega \qquad \text{(IV-14)}$$

The surface area of X equals four times the rotation average of the projected area of X. Similarly:

$$M(X) = \frac{1}{2}\int_{4\pi} L(X|\Delta_\omega) = \frac{1}{\pi}\int_{4\pi} U(X|\Pi_\omega)d\omega \qquad \text{(IV-15)}$$

The norm of X equals 2π times the average width of X, and also four times the average perimeter of its 2-D projections. In $\mathbb{R}^2$, we have:

$$\frac{1}{\pi} U(X) = \frac{1}{2\pi} \int_0^{2\pi} L(X|\Delta_\alpha) \, d\alpha \qquad \text{(IV-16)}$$

The perimeter of X equals π times its average projection. Formulae (IV-14) to (IV-16) constitute the Cauchy relations and concern the projections of X. (Indeed, Cauchy found only the first one.) Equivalently, they can be interpreted in terms of sections.. Consider, for example, Figure IV.8a: the straight line $\Delta(x, \omega)$ normal to Π_ω at point $x \in \Pi_\omega$ hits X if and only if $x \in X|\Pi_\omega$, then relation (IV-14) implies:

$$\frac{1}{4} S(X) = \frac{1}{4\pi} \int_{4\pi} d\omega \int_{\Pi_\omega} N^{(1)} [X \cap \Delta(x, \omega)] \, dx \qquad \text{(IV-17)}$$

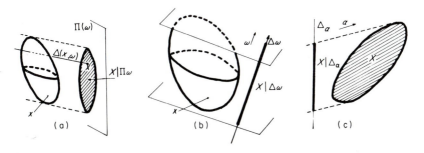

Figure IV.8. Projection of the compact convex set X. For $\mathbb{R}^3$: (a) on the plane Π_ω with a normal ω, i.e. $X|\Pi_\omega$; (b) on the straight line Δ_ω with a direction ω, i.e. $X|\Delta_\omega$. For $\mathbb{R}^2$: (c) on the straight line Δ_α with a normal α, i.e. $X|\Delta_\alpha$.

A similar interpretation of relations (IV-15) and (IV-16) leads to:

$$\frac{1}{2\pi} M(X) = \frac{1}{4\pi} \int_{4\pi} d\omega \int_{-\infty}^{+\infty} N^{(2)} [X \cap \Pi(x, \omega)] \, dx \qquad \text{(IV-18)}$$

and to

$$\frac{1}{\pi} U(X) = \frac{1}{\pi} \int_0^{\pi} d\alpha \int_{-\infty}^{+\infty} N^{(1)} [X \cap \Delta(x, \alpha)] \, dx \qquad \text{(IV-19)}$$

These three relations are classically called Crofton's formulae for the usual spaces (a direct proof of rel. (IV-18) is given in Ex. V-3).

(c) *Crofton formulae for thick sections (in $\mathbb{R}^3$)*
The notion of a planar intercept with zero thickness is an idealization of the "thin section". In practice, such an approximation is not always acceptable

(imagine, for example, a section of thickness 10μ that cuts a neuron of width 15μ . . .). Fortunately, this handicap can be overcome, at least for compact convex sets. R. Miles (1976) derived the basic formulae (IV-24) from certain properties of the Boolean process. In fact, relations (IV-24) and (IV-25) are purely geometrical and do not depend on any probabilistic hypothesis on the compact convex sets. G. Matheron (1976) proved them, using methods of integral geometry in the more general setting of n dimensions. His proof, however, is rather complex, so we will restrict ourselves to $\mathbb{R}^3$ and proceed in a direct and simple manner.

Consider the section of X taken by a slice with thickness e and the projection of this section on one of the faces of the slice, namely the plane $\Pi(x, \omega)$. We see in Figure IV.9 that studying the projection of the thick section is equivalent to first dilating X with a vector $e \cdot \omega$ normal to the face $\Pi(x, \omega)$ and of length e, and then studying the section of the dilate $X \oplus e \cdot \omega$ by plane $\Pi(x, \omega)$. Symbolically:

$$Y = [X \cap (\Pi(x, \omega) \oplus e \cdot \omega)] | \Pi(x, \omega) = (X \oplus e \cdot \omega) \cap \Pi(x, \omega) \quad \text{(IV-20)}$$

Now, V, S and M are respectively volume, surface and norm of X. Define by V', S' and M' similar quantities for $X \oplus e \cdot \omega$ by using, as definition algorithms, Crofton's formulae (V-17) to (V-18) i.e.

$$\left. \begin{aligned} V'(X) &= \frac{1}{4\pi} \int_{4\pi} d\omega \int_{-\infty}^{+\infty} A\left[(X \oplus e \cdot \omega) \cap \Pi(x, \omega)\right] dx \\ \frac{\pi}{4} S'(X) &= \frac{1}{4\pi} \int_{4\pi} d\omega \int_{-\infty}^{+\infty} U\left[(X \oplus e \cdot \omega) \cap \Pi(x, \omega)\right] dx \\ \frac{1}{2\pi} M'(X) &= \frac{1}{4\pi} \int_{4\pi} d\omega \int_{-\infty}^{+\infty} N^{(2)}\left[(X \oplus e \cdot \omega) \cap \Pi(x, \omega)\right] dx \end{aligned} \right\} \quad \text{(IV-21)}$$

Notice these integrals hold on the only experimental sets we have at our disposal. In other words, V', S' and M' are the parameters directly deduced from the actual measurements on the sections (area, perimeter, connectivity

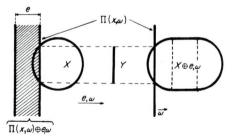

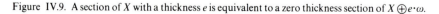

Figure IV.9. A section of X with a thickness e is equivalent to a zero thickness section of $X \oplus e \cdot \omega$.

number) by averaging according to Crofton's formula. With direction ω is associated on ∂X a (not necessarily plane) equatorial line, i.e. the locus of the points which are simultaneously input and output of intercepts having the direction ω. It separates ∂X into two parts (two "hemispheres"). To dilate X by $e \cdot \omega$ is equivalent to considering the set obtained by inserting, between the two hemispheres, a cylinder of direction ω, of height e and of generator the equatorial line. This geometrical remark allows us to calculate the three integrals in dx in (IV-21). More precisely, we find

$$
\left.
\begin{aligned}
\int_{-\infty}^{+\infty} A[(X \oplus e \cdot \omega) \cap \Pi(x, \omega)]\,dx &= V(X \oplus e \cdot \omega) \\
&= V(X) + e \cdot A[X \,|\, \Pi(x, \omega)] \\
\int_{-\infty}^{+\infty} U[(X \oplus e \cdot \omega) \cap \Pi(x, \omega)]\,dx &= \int_{-\infty}^{+\infty} U[X \cap \Pi(x, \omega)]\,dx \\
&\quad + e \cdot U[X \,|\, \Pi(x, \omega)] \\
\int_{-\infty}^{+\infty} N^{(2)}[(X \oplus e \cdot \omega) \cap \Pi(x, \omega)]\,dx &= D(\omega) + e
\end{aligned}
\right\} \quad \text{(IV-22)}
$$

(where $D(\omega)$ is the diameter of X in direction ω), and, by averaging in ω, and solving the system of equations, we get:

$$
\left.
\begin{aligned}
V' &= V + \frac{1}{4}eS \\
S' &= S + \frac{2}{\pi}e \cdot M \\
M' &= M + 2\pi e
\end{aligned}
\right\} \text{(IV-23)} \iff
\left.
\begin{aligned}
V(X) &= V' - \frac{1}{4}S'e + \frac{1}{2\pi}M'e^2 - e^3 \\
S(X) &= S' - \frac{2}{\pi}M'e + \pi e^2 \\
M(X) &= M' - 2\pi e
\end{aligned}
\right\} \quad \text{(IV-24)}
$$

These results can easily be extended to the case of v separate compact convex sets X_i, $(i = 1 \ldots v)$, where the number v is unknown. We must assume that every thick section projection is made up of non-overlapping particles. Then, system (IV-23) has to be applied to set $X = \bigcup_{i=1}^{v} X_i$, after substituting

$$
M' = M + 2\pi e \cdot v
$$

for the last equation (the first two remain unchanged). Solving this new system, we see that v multiplies the last correction terms in each equation of (IV-24) $(-ve^3, \pi ve^2,$ and $-2\pi ve$ instead of $-e^3, \pi e^2$ and $-2\pi e$ respectively). Consequently, if e is small enough, we can neglect these last three terms and recover V, S and M with rather good accuracy. Anyway, these three parameters concern only the sum of all the particles X_i, and give no mean information about the particles.

However, if we have at our disposal two sets of measurements, performed on two kinds of thick sections having two different thicknesses e_1 and e_2, then

$$M'_1 = M + 2\pi e_1 v \quad \text{and} \quad M'_2 = M + 2\pi e_2 v$$

and by difference

$$v = \frac{1}{2\pi} \cdot \frac{M'_1 - M'_2}{e_1 - e_2}$$

In practice, this estimate of v will be helpful only if the two thicknesses e_1 and e_2 are really different, i.e. $e_2 = 3e_1$ for example.

In some areas of application, such as drill holes in geology, the object is studied only via linear probes. However the test "line" is in fact a *cylinder* of radius r. A similar proof to the above leads to the relations:

$$
\left.
\begin{aligned}
V(X') &= V(X) + \frac{\pi r}{4} S(X) + \frac{r^2}{2} M(X) \\[2mm]
S(X') &= S(X) + 2rM(X) + 4\pi r^2
\end{aligned}
\right\}
\qquad \text{(IV-25)}
$$

Apart from assuming some sort of hypothesis on M, we clearly cannot recover the initial volume and surface area of X from the cylindrical intercepts.

(d) *The polyhedra (in* $\mathbb{R}^3$*)*
Volume and surface area have no *special* interpretation for polyhedra. On the other hand, the norm $M(X)$ is intimately connected with the edge lengths L_i (edges are labelled with the index i) and with the dihedron angles θ_i between the two faces adjacent to the same edge i. We have

$$M(X) = \tfrac{1}{2} \sum_i L_i \theta_i \qquad \text{(IV-26)}$$

where the summation is over all the edges. The simplest way to prove (IV-26) consists of opening the polyhedron X with the ball $B(r)$, considering $M(X)$ as the integral of the mean curvature, and then letting $r \to 0$. The method implicitly introduces a number of measures, and shows how the "mean curvature" measure is concentrated on the various edges, each of them contributing a weight proportional to the corresponding dihedron angle. The same proof shows also that the "total curvature" measure is concentrated on the vertices S_i, with a weight equal to $4\pi - \Omega_j$, where Ω_j is the solid angle subtended by vertex S_j. It turns out that the sum of these solid angles is given by:

$$\sum_j \Omega_j = \pi [s - 4] \quad (s = \text{number of vertices}) \qquad \text{(IV-27)}$$

a result honoured as the earliest of integral geometry. Stated by Descartes, its first known proof goes back to Euler (1750). In $\mathbb{R}^2$, it is expressed for polygons as:

$$\sum_j \alpha_j = \pi[s-2] \quad (\alpha_j = \text{internal angles})$$

Formulae (IV-26) and (IV-27) become inexact as soon as the faces of the polyhedra have some curvature (as would be the case for soap lather in a bottle, or for metallic grains).

(e) *Compact convex sets with an empty interior* (*in* $\mathbb{R}^3$ *and* $\mathbb{R}^2$)
We now consider planar compact convex sets X and segments Y treated as 3-D bodies, as well as segments Z, treated as 2-D bodies. We can study these sets by dilating them with a small ball $B(r)$ (resp. a small disc), and then letting $r \to 0$. Cauchy relations imply that in the limit:

$$\begin{cases} V(X) = 0 \\ S(X) = 2A(X) \\ M(X) = (\pi/2)U(X) \end{cases} \begin{cases} V(Y) = 0 \\ S(Y) = 0 \\ M(Y) = \pi L(Y) \end{cases} \begin{cases} A(Z) = 0 \\ U(Z) = 2L(Z) \end{cases} \quad \text{(IV-28)}$$

(f) *Hadwiger's characterization theorem* (*in* $\mathbb{R}^n$)
The Minkowski functional W_k possesses the following properties:

(a) Increasing: $X \subset Y \Rightarrow W_k^{(n)}(X) \le W_k^{(n)}(Y), \quad \forall n, k$
(b) Continuous: $d(X_i, X) \to 0 \Rightarrow W_k^{(n)}(X_i) \to W_k^{(n)}(X), \forall n, k$ (*d*: Hausdorff's distance)
(c) Homogeneous of degree $n-k$: $W_k^{(n)}(\lambda X) = \lambda^{n-k} W_k^{(n)}(X)$
(d) Invariant under displacements: translations *and rotations*
(e) Additive in the following sense:

If X, Y and $X \cup Y \in \mathscr{C}(\mathscr{K})$, then:

$$W_k^{(n)}(X \cup Y) + W_k^{(n)}(X \cap Y) = W_k^{(n)}(X) + W_k^{(n)}(Y) \quad \text{(IV-29)}$$

We shall call this last property *C-additivity*, in order to clearly distinguish it from the additivity of measures of the next section. Note that formula (IV-29) has no local meaning at all, since it associates numbers with sets considered globally.

The five properties given above, are extremely strong, but even more significant is the converse proposition, which characterizes the Minkowski functionals via these properties. This has been proved by Hadwiger (1957):

(a) *Any functional φ on $\mathscr{C}(\mathscr{K})$, which is continuous, C-additive and invariant under displacements may be written as follows:*

$$\varphi(X) = \sum_{k=0}^{n} a_k W_k^{(n)}(X) \qquad \text{(IV-30)}$$

with suitable constants $a_k(-\infty < a_k < +\infty)$.

(b) *Any functional φ on $\mathscr{C}(\mathscr{K})$, which is increasing, C-additive and invariant under displacements admits a representation of the form* (IV-30) *with suitable positive constants $a_k(0 < a_k < \infty)$.*

The preceding properties are common to all the $W_k^{(n)}$ associated with X. In addition, the volume and the norm satisfy two supplementary characteristics:

(a) The volume $W_0^{(n)}$ vanishes on the class of all those compact convex sets contained in a subspace of $\mathbb{R}^n$. Conversely, any functional φ satisfying (IV-30) and vanishing on this class, admits a representation of the form $\rho(X) = C \cdot W_0^{(n)}$.

(b) If X and Y belong to $\mathscr{C}(\mathscr{K})$, then the norm satisfies the following property:

$$M(\lambda_1 X \oplus \lambda_2 Y) = \lambda_1 M(X) + \lambda_2 M(Y), \qquad \lambda_1, \lambda_2 > 0 \quad \text{(IV-31)}$$

Thus the norm is additive with respect to dilation. Intuitively, the reason for relation (IV-31) is clear. In every direction of the space, the width of the dilate equals the sum of the width of both sets, and the property remains true under isotropic averaging. Conversely, any functional $\varphi(X)$ satisfying relations (IV-30) and (IV-31) may be written as:

$$\varphi(X) = aM(X) + b$$

Hadwiger's Characterization Theorem is a fundamental result. However if we compare it to the four principles of mathematical morphology, we find it at the same time too demanding, and not demanding enough. It is too demanding, because it imposes an invariance under rotations that we would have preferred to avoid, and which looks rather artificial (in formulae (IV-17) to (IV-19) isotropic averaging is imposed onto a basically anisotropic functional). It is not demanding enough because of the global aspect of the approach: we must look hard for local parameters that lie behind the Minkowski functionals.

Note that C-additivity *is not* a condition on the same level as the morphological principles. If so, the experimenter would have to reject a product or a ratio of two functionals: for example; the excellent shape descriptor (for convex sets) formed by the ratio "Area (perimeter)²" is not C-additive! The C-additivity property should be thought of as the logical link that gets us to the convex ring (see the next chapter) and not a *sine qua non* condition for experiments.

C.2. Steiner's formula

Stated in 1840, within an interval of less than one year, the formulae of Cauchy and of Steiner are the two earliest themes of integral geometry. Since then there have been many variations of Steiner's formula. In all the models of this and the next chapter, we see it reappearing, sometimes under several different guises. The first version (Steiner's original) connects the functionals of a compact convex X to those of its dilate by a ball in $\mathbb{R}^3$. The next step is to replace the ball with a compact convex set, then to replace dilation with erosion, then to make the global approach a local one, then to replace the set X with a Minkowski sum of segments .... We will now present some of these variations.

(a) *Steiner's formula for dilations*
B is an arbitrary compact convex set in $\mathbb{R}^n$, and let θ denote the orientation of its coordinate system. To define a uniform orientation on θ, in $\mathbb{R}^3$ for example, means to uniformly take a first vector ω as the unit sphere, then a second one, say α, on the unit circle normal to ω, and finally the third must be in one of the two remaining orthogonal directions, the choice being made by say, tossing a coin. The approach is easily generalized to $\mathbb{R}^n$, and, because of the compactness of the set of all directions, it leads to a unique probability law $\varpi(d\theta)$ which is invariant under rotation (bounded Haar measure). Then the Steiner formula for dilation is given by relation (IV-32), where $\overline{W}_i^{(n)}$ denotes the rotation average:

$$\int W_i^{(n)}[X \oplus \lambda B(\theta)]\varpi(d\theta) = \frac{1}{b_n} \sum_{k=0}^{n-i} \binom{n-i}{k} \lambda^k W_{k+i}^{(n)}(X) \cdot W_{n-k}^{(n)}(B)$$

$$= \overline{W}_i^{(n)}[X \oplus \lambda B] \qquad \text{(IV-32)}$$

The average value of the ith functional of $X \oplus B$ is a bilinear function of the functionals of both X and B. We have in particular in $\mathbb{R}^3$:

$$\overline{V}(X \oplus B) = V(X) + \frac{M(X)\cdot S(B) + S(X)\cdot M(B)}{4\pi} + V(B) \qquad \text{(IV-33)}$$

and in $\mathbb{R}^2$

$$\overline{A}(X \oplus B) = A(X) + \frac{U(X)\cdot U(B)}{2\pi} + A(B) \qquad \text{(IV-34)}$$

For stereological applications, relation (IV-22) is used when a 3-D set X is known from its sections (which implies $B \in \mathbb{R}^2$ or $\mathbb{R}^1$). By taking into account relation (IV-38), we may write Steiner's formula as in Figure IV.10.

Apart from some changes due to digitalization, the formulae of Figure IV.10

Minkowski	Structuring element B	
Functional	Segment l	Disk r
$\overline{V}(X \oplus B) =$	$V(X) + \dfrac{l}{4}S(X)$	$V(X) + \dfrac{\pi}{4}rS(X) + \dfrac{r^2}{2}M(X)$
$\overline{S}(X \oplus B) =$	$S(X) + \dfrac{l}{2}M(X)$	$S(X) + \dfrac{\pi r}{2}M(X) + \pi r^2$
$\overline{M}(X \oplus B) =$	$M(X) + \pi l$	$M(X) + \pi^2 r$

Figure IV.10. Steiner formulae in $\mathbb{R}^3$.

are the basis for computing V, S and M in image analysers (when X is a convex set or even simply a regular set).

As an example of a typical integral geometric argument, we will present the proof of Steiner's formula (IV-32), with the Characterization Theorem being used in the usual way. Let $\Phi(X, B)$ be the left-hand side of equation (IV-32), which we consider as a function of X only. The dilation distributes the union, as well as the intersection in the case of C-additivity (Prop. IV-9). Therefore we have:

$$W_i^{(n)}(X \oplus B) + W_i^{(n)}(X' \oplus B) = W_i^{(n)}[(X \cup X') \oplus B]$$
$$+ W_i^{(n)}[(X \cap X') \oplus B] \qquad \text{(IV-35)}$$

The property (IV-36) remains true after rotation averaging. Thus $\Phi(X)$ is C-additive. It is also invariant under displacements (by construction), and monotonic (integral of monotonic functions). Similarly, $\Phi(X, B)$ considered as a function of B satisfies these three properties. Then, according to the Characterization Theorem of Hadwiger, $\Phi(X, B)$ is a bilinear form of X and B. In this form, only the terms with degree $n - i$ are different from zero, since Φ is of degree $n - i$, and so:

$$\Phi(X, B) = \sum_{k=0}^{n-i} a(k, i)\, W_{k+i}^{(n)}(X) \cdot W_{n-k}^{(n)}(B) \qquad \text{(IV-36)}$$

We go from (IV-36) to the final result (IV-32) by taking for B the ball of radius r, and identifying the coefficients $a(k, i)$ with the corresponding coefficients (see Ex. IV-2).

(b) *Steiner's formula for erosions*

We continue the variations of Steiner's formula by studying what happens when dilations are replaced by erosions. It is curious that this question and its partial answer, go back only four years, whereas the case of dilation has been solved for one hundred and forty years! In spite of their duality, the two concepts of erosion and dilation did not play the same role in the evolution of the ideas of set geometry (in particular, the notion of erosion is totally absent from the work of Minkowski). Erosion made its first humble appearance in a theorem due to Fejes–Toth (1953), which very surprisingly links Steiner's formula with the skeleton of a polygon. The gap between dilation and erosion analysis is perhaps due to the requirement of a supplementary hypothesis in the latter case; i.e. we must assume X to be invariant under openings by balls sufficiently small in radius (this is the same hypothesis as for the regular model of Chapter V). R. Miles (1975) proved that in $\mathbb{R}^2$, such an assumption is a necessary condition for Steiner's formula; G. Matheron (1977) improved this result by extending it to $\mathbb{R}^n$, and proving that in $\mathbb{R}^2$, it was also sufficient. Steiner's theorem as applied to erosions may be stated as follows:

Let X and $B(\theta)$ be two compact convex sets in $\mathbb{R}^n$. Then provided X is open with respect to $B(\theta)$, irrespective of the orientation θ, they satisfy the formulae deduced from (IV-22) by changing dilation into erosion (replace the symbol $\oplus$ by $\ominus$ in the left-hand side) and by changing λ into $-\lambda$ in the right-hand side of the formulae; λ must be < 1.

Proof: Assume that X is open with respect to B irrespective of its orientation. Put $X = X_1 \oplus B$ with $X_1 = X \ominus \check{B}$. For $0 \le \lambda \le 1$ we have:

$$X \ominus \lambda \check{B} = [(X \ominus \lambda \check{B}) \ominus (1-\lambda)\check{B}] \oplus (1-\lambda)B = X_1 \oplus (1-\lambda)B \qquad (IV\text{-}37)$$

From (IV-32) we deduce that the average volume:

$$\int V^{(n)}[X \ominus \lambda \check{B}(\theta)]\varpi\,(d\theta) = \int V^{(n)}[X_1 \oplus (1-\lambda)B(\theta)]\varpi\,(d\theta) = Q(\lambda)$$

$$(IV\text{-}38)$$

is a polynomial in λ, namely $Q(\lambda); 0 \le \lambda \le 1$. The equality $X \oplus \lambda B = X_1 \oplus (1+\lambda)B(\lambda > 0)$ implies:

$$\int V^{(n)}[X \oplus \lambda B(\theta)]\varpi\,(d\theta) = \int V^{(n)}[X_1 \oplus (1+\lambda)B(\theta)]\varpi\,(d\theta) = P(\lambda)$$

Thus $P(\lambda) = Q(-\lambda)$ for $0 \le \lambda \le 1$. But P and Q are polynomials, i.e. finite linear combinations of coefficients; so we can write $Q(\lambda) = P(-\lambda)$ for $0 \le \lambda \le 1$.   Q.E.D.

The identical proof goes through for the other $W_i^{(n)}$. Conversely, it can be proved that in $\mathbb{R}^2$, the relationship:

$$\int A[X \ominus \lambda B(\theta)]\varpi\,(d\theta) = A(X) - \frac{\lambda L(B) \cdot L(X)}{2\pi} + \lambda^2 A(B) \qquad (IV\text{-}39)$$

implies that X is open with respect to all $B(\theta)$ (the generalization of this converse to $\mathbb{R}^n$, $n > 2$, is not proved).

The condition which says that X is open with respect to $B(\theta)$ and all its rotations, has a simple geometrical interpretation in $\mathbb{R}^3$. Let M be an arbitrary point of the boundary of X, in which the curvatures are finite. We fix the orientation θ of B. By hypothesis, there exists a $B(\theta)$ passing through M and contained in X. Now intersect the figure with a plane Π going through M; $B(\theta) \cap \Pi$ is included in $X \cap \Pi$, so the radius of curvatures of $X \cap \Pi$ of M is larger than that of $B(\theta) \cap \Pi$. The same remark is true for all possible sections Π, and hence by a theorem of Meunier (see in James/James, 1976) the inequality can be extended to the radii of curvature in $\mathbb{R}^3$ at M. In other words, the sup of the radii of curvature of B is bounded above by the inf of the radii of curvature of X. Or equivalently: there exists a ball $C(r)$ with a radius $r \geq 0$ such that X is open with respect to $C(r)$ and $C(r)$ is open with respect to the $B(\theta)$. We see in particular that B cannot be a segment, or a square, etc. . . . or any figure having an infinite curvature, even at only one point. What then constitutes a "good" X? The theorem plays on the fact that it is equivalent to erode X or to dilate X_1 to obtain $X \ominus \lambda B$, which implicitly implies that X itself is the dilate of X_1 by a ball with a radius larger than those of all the λB.

D. MINKOWSKI MEASURES

D.1. Measures

In $\mathbb{R}^1$, the closed interval $[a \ \ b]$ is the compact convex set where a and b are the end points. More generally, in $\mathbb{R}^n$ a closed interval is the set of those points x whose coordinates satisfy the inequalities $a_i \leq x_i \leq b_i \ (1 \leq i \leq n)$, for some fixed numbers $a_1 < b_1; a_n < b_n$. In other words, the closed intervals in $\mathbb{R}^n$ are the compact rectangular parallelepipeds. Consider now the family of sets generated by the countable union of the closed intervals in $\mathbb{R}^n$, and the complements of these unions (i.e. the σ-algebra generated by the intervals). It defines a classical class of sets called Borel sets $\mathscr{B}(\mathbb{R}^n)$; from now on, the measuring masks (or measuring fields) Z used in morphology are assumed to belong to the Borel class.

A set measure is a parameter $\mu(Z)$ associated with each mask $Z \in \mathscr{B}(\mathbb{R}^n)$ such that $\mu(\varnothing) = 0$ and $\mu(\bigcup_i Z_i) = \sum_i \mu(Z_i)$ for any countable sequence $\{Z_i\}$ for which $i \neq j \Rightarrow Z_i \cap Z_j \neq \varnothing$. This condition, called σ-additivity, allows us to calculate the measure μ over elementary disjoint masks dz which partition Z, and to sum up the results in order to obtain $\mu(Z)$. Let us quote two particularly important measures:

(a) *Lebesgue measure.* There exists only one measure equal to $a_1 \times a_2 \times \ldots a_n$ for each rectangular parallelepiped of sides $a_1 \ldots a_n$. It is called Lebesgue measure and is denoted dz. Any measure which is translation-invariant and positive is proportional to dz.

(b) *Dirac measure*. This measure, denoted δ_x, is generated by the set reduced to point $\{x\}$, as follows: $\delta_x(Z) = 1$ when $x \in Z$ and $\delta_z(Z) = 0$ when not, $\forall Z \in \mathscr{B}(\mathbb{R}^n)$

We have just presented the measures as local parameters acting on sets. Their action may be extended to functions in the following way. A real function $\varphi(x)$, $x \in \mathbb{R}^n$, is measurable if for any real number a the set of all the x for which $\varphi(x) > a$ is a mask in $\mathbb{R}^n$. Consider a "step" function, a *simple function*, i.e. a simple function taking only a finite number of values, and defined as follows:

$$\varphi(x) = \sum_{i=1}^{p} \lambda_i k_{Z_i}(x); \qquad \lambda_i > 0, \qquad Z_i \in \mathscr{B}(\mathbb{R}^n);$$

$$p < \infty \qquad k_Z = \text{indicator function of } Z.$$

With such a function, we associate the measure:

$$\mu(\varphi) = \sum_{i=1}^{p} \lambda_i \mu(Z_i)$$

It can be shown (Neveu, 1964) that any positive measurable function φ is the monotonic limit of at least one increasing sequence of simple functions $\varphi_j < \varphi$. The measure $\mu(\varphi)$ is then defined over the class of the measurable functions as the sup of the $\mu(\varphi_j)$. In brief, the role of the simple functions is that of a hinge between the notion of a measure on the valued measurable functions, and of a set measure $\mu(Z)$.

The support of a function φ is the domain in which φ takes defined values. In the same way, the support of the measure μ is the smallest closed set $Y \in \mathscr{F}(\mathbb{R}^n)$ such that $\mu(\varphi)$ is identically zero for all φ whose support is contained in Y^c. The support of the Dirac measure δ_x is reduced to point $\{x\}$, while that of the Lebesgue measure is the whole space (the support of a function and the support function, as defined in Section G below, are two radically different notions).

D.2. Minkowski measures

The simplest way to define a local version of the Minkowski functionals is, as we shall see, to use Steiner's formula of Section 3b as motivation, but consider instead projections of all the points of the space, on the compact convex sets under study (G. Matheron, 1975). Let X be a compact convex set and let x be a point of $\mathbb{R}^n$. For every point x there exists a unique point $x' \in X$ minimizing the distance $|x - y|$ as y runs over X. This point x' is called the projection of x onto

X and is denoted by $x' = x|X$. When x belongs to X, its projection x' is just the point x itself. A projection point x' can come from several very different points x; in fact an entire cone of the space can have the same projection point.

Let $\rho(x)$ be the distance $|x - x'|$. Since both x' and ρ are of interest, we consider the transformation $x \rightarrow (x', \rho)$ which maps $\mathbb{R}^3$ (resp. $\mathbb{R}^2$) onto $X \times \mathbb{R}^+$, the mapping being continuous. (Imagine a haze of very fine iron filings, disseminated homogeneously throughout the space. Now suppose each particle x of the iron filings precipitates onto the nearest point of X. The particles inside X do not move, while those outside fall onto the boundary ∂X; the higher is the curvature, the greater is the number of falling particles condensed on that part of the boundary . . . .) Now the mapping $x \rightarrow (x', \rho)$ transforms the Lebesgue measure in $\mathbb{R}^3$ (resp. in $\mathbb{R}^2$) into a measure which turns out to be the sum of four terms (resp. three terms). Integrated against an arbitrary continuous function $\varphi(x)$, these measures give the following formulae, in $\mathbb{R}^3$ and $\mathbb{R}^2$ respectively:

$$\int_{X \oplus \rho B} \varphi(x|X)\,dx = \int_{\mathbb{R}^3} \varphi(x)[v(dx) + \rho s(dx) + \rho^2 m(dx)$$

$$+ \frac{4}{3}\pi\rho^3 n^{(3)}(dx)] \qquad \text{(IV-40)}$$

$$\int_{X \oplus \rho B} \varphi(x|X)\,dx = \int_{\mathbb{R}^2} \varphi(x)[a(dx) + \rho u(dx) + \pi\rho^2 n^{(2)}(dx)] \quad \text{(IV-41)}$$

where B is the unit ball. When the compact convex set X is sufficiently regular, and ∂X possesses finite curvatures, the two equalities (IV-40) and (IV-41) give the classical Bonnensen–Fenchel formulae (1936). It is less obvious, although intuitive, that the two relations remain true in the general case with angularities, empty interiors, etc. Local description of the Minkowski functionals immediately results: take $\varphi(x)$ to be the constant function equal to one and identify the coefficients of the polynomials (IV-50) and (IV-51) with those of Steiner's formulae (IV-33) and (IV-34). We obtain Figure IV.11.

Minkowski measure W_i^X of order	$\mathbb{R}^3$	$\mathbb{R}^2$
0	$V(X) = \int v(dx) = \int h_x(x)dx$	$A(X) = \int a(x)dx = \int h_X(x)dx$
1	$S(X) = \int s(dx)$	$U(X) = \int u(dx)$
2	$M(X) = \int m(dx)$	$N^{(2)}(X) = \int n^{(2)}dx$
3	$N^{(3)}(X) = \int n^{(3)}(dx)$	

Figure IV.11. Table of the Minkowski measures in $\mathbb{R}^3$ and $\mathbb{R}^2$.

These integrations are taken over the whole space, however the supports of the measures are limited to X or to ∂X. Indeed, the first is the Lebesgue measure dx restricted to the points of X, and the others are concentrated on the boundary ∂X. The second one is the *superficial measure* of ∂X, obtained by uniformly distributing a unit mass per unit of the surface area of ∂X (in $\mathbb{R}^3$) or per unit length (in $\mathbb{R}^2$); the last one associates with each element $d(\partial X)$ of the boundary the proportion of the directions of the space whose projection falls on $d(\partial X)$ (all these results can immediately be generalized to $\mathbb{R}^n$).

The Minkowski measures are local, *quantitative morphological* parameters, in the sense of Chapter V.A. They are compatible with the displacements and the magnifications of X, and satisfy the local knowledge principle. Moreover, for any continuous function φ the mapping $X \to \int \varphi(x) W_i^X(dx)$ from $\mathscr{C}(\mathscr{K}')$ onto $\mathbb{R}^+$ is continuous (Matheron, 1975). Compatibility with displacements is again more than we need since it includes compatibility with rotations. This last constraint is too strong for the needs of morphology, so we shall try now to eliminate it.

E. CONVEX SETS AND ANISOTROPIES

We start with the development of anisotropic properties of the compact convex sets in the general space $\mathbb{R}^n$. Nevertheless, the 2-D space very quickly plays a special role: the characterization of the Steiner compact sets is much easier in the plane than in the space. Therefore the last part of this section will be devoted exclusively to the plane.

E.1. The support function

Up to now, we have ignored that definition of a compact convex set, which says $X \in \mathscr{C}(\mathscr{K})$ if it is the intersection of its supporting half-spaces. Consider, in $\mathbb{R}^n$, the class $\mathscr{C}_0(\mathscr{K})$ of compact convex sets containing the origin, and the set Ω consisting of all unit vectors ω of the space (i.e. the unit sphere). To each direction ω we associate the distance $d_X(\omega)$ from the origin to the supporting hyperplane of X with positive normal ω (Fig. IV.12).

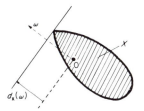

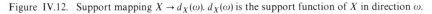

Figure IV.12. Support mapping $X \to d_X(\omega)$. $d_X(\omega)$ is the support function of X in direction ω.

To say that X is defined as the intersection of its supporting half spaces is equivalent to saying that:

$$X = \bigcap_{\omega \in \Omega} \{x : \langle x, \omega \rangle \leq d_X(\omega)\} \qquad \text{(IV-42)}$$

where the scalar product $\langle x, \omega \rangle$ is just the coordinate of the point x on the axis Δ_ω (the axis with orientation ω). If x is a point of X, this coordinate is necessarily smaller than the value of the support function $d_X(\omega)$, and conversely. Thus relation (IV-42) can be read in the following different way:

$$d_X(\omega) = \text{Sup}\{\langle x, \omega \rangle, x \in X\}$$

i.e. the support function in the direction ω, equals the largest projection of a vector $x \in X$ on the axis Δ_ω. One can show that on $\mathscr{C}_0(\mathscr{K})$, $d_X(\omega)$ is a continuous function of ω. Moreover the set $\mathscr{C}_0(\mathscr{K})$ and the set of all support functions (provided with the topology of uniform convergence) are homeomorphic to each other. In other words, the datum of a compact convex set is equivalent to the datum of its support function, modulo a translation.

With respect to dilation, inclusion and magnification, the support function obviously satisfies the following properties:

$$d_{X \oplus X'} = d_X + d_{X'} \qquad \text{(IV-43)}$$

$$\left.\begin{array}{c} X \subset X' \Leftrightarrow d_X \leq d_{X'} \\[1mm] d_{\lambda X} = \lambda d_X \end{array}\right\} \qquad \text{(IV-44)}$$

Thus, the supporting mapping transforms Minkowski sums into simple additions (here lies its major interest). As far as erosions are concerned, we only have the inequality:

$$d_{X \ominus \check{X}'} \leq d_X - d_{X'}$$

More precisely, the eroded set $X \ominus \check{X}'$ is the largest compact convex set Z to satisfy $d_Z \leq d_X - d_{X'}$.

If we arbitrarily choose two sets X and Y belonging to $\mathscr{C}_0(\mathscr{K})$, it is not always true that the difference $f = d_X - d_Y$ of their two support functions is still a support function (for example f might be negative for some directions ω). If however f is a support function, the relation $d_X = d_Y + f$ shows that there exists a set $Z \in \mathscr{C}_0(\mathscr{K})$ such that $X = Y \oplus Z$ (and obviously conversely).

From this equivalence we derive an *order relation*: inequality $d_X \geq d_Y$ (if $d_X - d_Y$ is a support function) is equivalent to X *being open with respect to Y*. This order relation simply means that X can be split up into the Minkowski sum of a Y "smaller" than it, and another term Z. Then we ask whether the set Y can be decomposed. If it cannot, then we say that Y is an extremal element. The segment, the triangle, the tetrahedron, the pyramid, are all extremal elements.

At this point we might ask: is the decomposition of X into its extremal elements a unique one? Obviously the answer is no, since the same regular hexagon, in $\mathbb{R}^2$, is either the Minkowski sum of three segments at angles of $120°$ to each other, or the sum of two equilateral triangles, one the transpose of the other (Ex. no. 4). From this comes the idea to concentrate on that sub-class of $\mathscr{C}_0(\mathscr{K})$, where each set admits a unique decomposition into its extremal elements. We shall class this sub-class the Steiner class; it turns out that the extremal elements will always be segments.

E.2. The Steiner class

Consider the parallelogram obtained by dilating two segments of length L_1 and L_2, directions ω_1 and ω_2, and centred at the origin (Fig. IV.13).

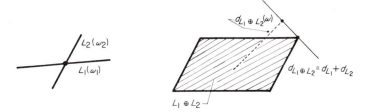

Figure IV.13. Support function of $L_1 \oplus L_2$.

Applied to the dilate $L_1 \oplus L_2$, formula (IV-43) implies that:

$$d_{L_1 \oplus L_2} = L_1 \left|\cos(\omega-\omega_1)\right| + L_2 \left|\cos(\omega-\omega_2)\right|$$

Instead of two directions, consider now an angular distribution of measures $L(d\omega)$, taken on the unit half sphere $\frac{1}{2}\Omega$. An immediate generalization of relation (IV-43) leads to:

$$d_X(\omega) = \int_{1/2\Omega} \left|\cos(\omega-\omega')\right| L(d\omega') \qquad \text{(IV-45)}$$

We shall call a *Steiner compact set* any symmetric (i.e. $X = \check{X}$) set $X \in \mathscr{C}(\mathscr{K})$ whose support function satisfies a decomposition (IV-45). In $\mathbb{R}^2$, the class of Steiner compact sets is in fact all the symmetric compact convex sets. Among other members, the class includes all polygons having an even number of sides (it also includes circles, ellipses, and so forth . . .). In three dimensions, such an equivalence is no longer true, and the Steiner class appears as a proper subset of the symmetric $X \in \mathscr{C}(\mathscr{K})$. To see this, imagine a triangle T, embedded in a plane which does not go through the origin. Take its transposed set $\check{T}$, and the convex hull $C(T \cup \check{T})$, of the union of the two triangles. Although it is symmetric, $C(T \cup \check{T})$ does not belong to the Steiner class, since its faces are triangles (they must be at least parallelograms).

In the second chapter, we met several Steiner polyhedra, namely the cube, the prism with a hexagonal base, the dodecahedron, and the tetrakaidecahedron. We noticed then that all of them could generate partitions of the space into equal polyhedric alveolae. Of course, this is not true in $\mathbb{R}^2$, since the Steiner class also includes the sphere, the ellipsoid, the cylinder symmetric about its "upright", and an infinity of polyhedra.

(a) *Characterization of the Steiner compact sets*
    and of their anisotropies

Let us now return to equation (IV-45), and suppose we have a positive symmetric function $d(\omega)$ of the directions of the space. Under what conditions can the function $d(\omega)$ be the support function of a Steiner compact set X? In other words, does there exist a group of unit vectors ω which, when multiplied by $L(d\omega)$ and (Minkowski) summed, give a Steiner compact set X, $L(d\omega)$ then being connected with $d_X(\omega)$ via relation (IV-45)?

In answer to these questions, Matheron (1975) showed that a given function $d(\omega)$ can be decomposed into a sum of the type (IV-45) if and only if the function $\exp[-|h|d(h/|h|)]$, defined for every $h \in \mathbb{R}^n$, is a continuous covariance. When this is the case, the decomposition is unique, and hence it is equivalent to be given either $d(\omega)$ or the measure $L(d\omega)$. The above criterion is in fact more checkable than it would seem at first glance, since covariances are functions which admit a positive integrable Fourier transform. It is therefore possible to judge whether a given function is or is not a support function of a Steiner compact set. In $\mathbb{R}^2$, the criterion is much simpler, since we merely have to check that d is symmetric.

The Steiner class would be collecting dust in a museum of mathematical curiosities if it were not right in the middle of an a priori completely different question. Suppose X is an arbitrary compact set (not necessarily convex) from the regular class or of the extended convex ring in $\mathbb{R}^3$. Compute the total projection $A^{(2)}(X, \omega)$ of X in direction ω. We are interested in the anisotropy of the medium X and would like to quantify it using the sum of the intercepts as a function of direction. Are we able to simply take any symmetric function $f(\omega)$? The answer is no! The only functions able to play this role are the *support functions of the Steiner compact sets of $\mathbb{R}^3$*; with each medium X is associated a unique Steiner compact set, which is a descriptor of its anisotropy.

In two dimensions, the same approach yields a simpler result. The total diameter $D^2(X, \alpha)$ (in the sense of Chapter V), plotted as a function of α, must describe the boundary of a symmetric compact convex set (a triangle or a pentagon could not describe anisotropy in $\mathbb{R}^2$). The Steiner class appear again in morphology; it plays a central role in describing stochastic Poisson hyperplanes.

(b) *Measures and functionals on the Steiner class*

Since the measure $L(d\omega)$ completely determines a Steiner compact set (modulo a translation), it must be possible to derive all the basic functionals from $L(d\omega)$. In fact, with the help of measures, we can represent Steiner's formula in a local form.

In $\mathbb{R}^2$, the area A and the perimeter U of the Steiner compact set X are given by the relations:

$$\left.\begin{aligned}
A(X) &= \frac{1}{2} \int_0^\pi L(d\alpha_1) \int_0^\pi L(d\alpha_2) A(\alpha_1 \oplus \alpha_2) \\
U(X) &= 2 \int_0^\pi L(d\alpha)
\end{aligned}\right\} \qquad \text{(IV-46)}$$

where $A(\alpha_1 \oplus \alpha_2)$ is the area of the parallelogram obtained by dilating the two unit vectors α_1 and α_2. Figures IV-14a and b illustrate the formulae (IV-46) when the measure $L(d\alpha)$ puts means on only four directions. Then the area becomes the discrete double sum:

$$A(X) = \sum_{i=1}^4 \sum_{j>1}^4 L_i L_j |\sin(\alpha_i - \alpha_j)|$$

The six terms in the sum are the areas of the six polygons given by the dotted lines in Figure IV-14. We can also check that the perimeter of X is twice the sum of all the lengths L_i.

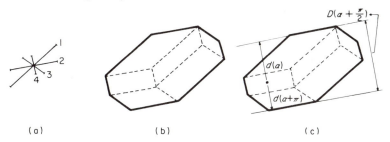

(a)   (b)   (c)

Figure IV.14. (a) and (b): measure $L(d\alpha)$ and corresponding Steiner polygon. (c) relationship between support function and apparent contour.

In $\mathbb{R}^3$, the Minkowski functionals are given by the three relations:

$$\left.\begin{aligned}
V(X) &= \frac{1}{6} \int_{2\pi} L(d\omega_1) \int_{2\pi} L(d\omega_2) \int_{2\pi} L(d\omega_3) V(\omega_1 \oplus \omega_2 \oplus \omega_3) \\
S(X) &= \int_{2\pi} L(d\omega_1) \int_{2\pi} L(d\omega_2) A(\omega_1 \oplus \omega_2) \\
M(X) &= \pi \int_{2\pi} L(d\omega)
\end{aligned}\right\} \qquad \text{(IV-47)}$$

where the domain of integration is the half unit sphere (2π steradiants). Notice that the volume $V(\alpha_1 \oplus \alpha_2 \oplus \alpha_3)$ of the dilation of three unit vectors could be null for certain $\alpha_1, \alpha_2, \alpha_3$ (this occurs say, when X is the prism with a hexagonal base). The last result of (IV-47) about the norm, is more useful than formula (IV-26), since now we have $M(X) = \pi \Sigma L_i$, regardless of the angles between faces. Moreover, for all X compact and convex, $M(X) = 2\pi D^{(3)}(X)$ ($D^{(3)}(X)$ $\equiv$ mean width); thus for any Steiner polyhedron, the mean width equals half the total length of the edges.

The 1970 version of the Texture Analyser pioneered the 2-D analysis with convex structuring elements, by basing the approach on Steiner polygons (J. Serra, 1969, 1970). The method has since become well known, and is used extensively in image analysis. Today it is not very expensive to store a large number of images, such as successive thin sections of a 3-D object (biologists are particularly interested in this facility). In such cases the Steiner geometry for polyhedra just presented, provides the experimenter with the means for making 3-D image transformations such as dilation, by first performing 2-D dilations on the thin sections, and then iterating on the third dimension.

E.3. Perimetric measures in $\mathbb{R}^2$

(a) *Support function and perimetric measure*

We shall confine our comments in this last sub-section to the Euclidean plane $\mathbb{R}^2$. Here the compact convex sets, whether symmetric or not, possess a remarkable property that is particular to $\mathbb{R}^2$. Their support functions $d(d\alpha)$ and their perimetric measures $u(d\alpha)$ are locally and linearly connected by the differential equations:

$$u_X(d\alpha) = d_X(d\alpha) + d_X''(d\alpha) \qquad \text{(IV-48)}$$

where the derivative is taken in the sense of distribution theory. By integrating (IV-48) we find:

$$d_X(\alpha) = a \cos(\alpha - \alpha_0) + \int_0^\alpha \sin(\alpha - \theta) u_X(d\theta)$$

The two arbitrary integration constants a and α_0 mean that knowledge of u_X determines the set X only up to a translation. Notice the local structure of relation (IV-48): when we know the support function on an open segment of the unit circle, we can calculate the perimetric measure on the same segment. For $n \geq 3$, this local character remains, but not the linearity relation (Matheron, 1978). The perimetric measures in $\mathbb{R}^2$ (and in general the surface measures in $\mathbb{R}^n$) are C-additive in the sense of relation (IV-29). We see directly from relations (IV-43) and (IV-48) that in $\mathbb{R}^2$ (but not in $\mathbb{R}^n$ $n \geq 3$):

$$u_{X \oplus Y}(d\alpha) = u_X(d\alpha) + u_Y(d\alpha)$$

and consequently, that X is open with respect to Y if and only if its perimetric measure is always larger than the perimetric measure of Y.

$$u_X \geq u_Y \Leftrightarrow X = (X \ominus \check{Y}) \oplus Y \qquad \text{(IV-49)}$$

This equivalence yields a curious property of the planar compact convex sets. If X and K belong to $\mathscr{C}(\mathscr{K})$ in $\mathbb{R}^2$, then X is open with respect to every non empty eroded set $X \ominus \check{K}$.

Proof: Put:

$$X_1 = X \ominus \check{K} \quad \text{and} \quad K_1 = X \oplus X_1$$

We have $u_{x_1} \leq u_x$ since the mapping $X \rightarrow u_x$ is decreasing on erosions. Thus X is open with respect to X_1, i.e. $X = X_1$. We then deduce that

$$C = X \ominus \check{X}_1 = K_1, \quad \text{and hence} \quad X = X_1 \oplus K_1 \quad \text{Q.E.D.}$$

Similarly, we can easily verify that $K = K_1$ (i.e. K is closed with respect to X^c) if and only if $X = X_K$ (X is open with respect to K). (In $\mathbb{R}^n$, $n \geq 3$, the equality $X = X_K$ still implies that $K = K_1$, but the converse proposition is false).

The support function $d_X(\alpha)$ depends on the choice of the origin of coordinates. On the contrary, the apparent contour $D_X(\alpha)$ is invariant under translation. Since we put the origin inside the set X, the apparent contour $D_X(\alpha)$ equals the sum of the support function for the angles α and $\alpha + \pi$:

$$D_X\left(\alpha + \frac{\pi}{2}\right) = d_X(\alpha) + d_X(\alpha + \pi) \qquad \text{(IV-50)}$$

(see Fig. IV.14c). Modulo a symmetry, one can go back from the apparent contour to the perimetric measure, i.e.

$$u_X(d\alpha) + u_X[d(\alpha + \pi)] = D_X\left[d\left(\alpha + \frac{\pi}{2}\right)\right] + D_X''\left[d\left(\alpha + \frac{\pi}{2}\right)\right] \quad \text{(IV-51)}$$

(The reader will excuse us for the two confusing notations $d(\alpha + \pi)$ (differential element) and $d_X(\alpha + \pi)$ (supporting function)). Relation (IV-50) is instructive, and shows us that the apparent contour of a convex set X, can be a constant even if X is not a circle. Now $D_X(\alpha) = $ constant, implies that the radius of curvature $R_X(\alpha)$ has to be a positive function such that $R(\alpha) + R(\alpha + \pi) = $ constant. For example, take:

$$R(\alpha') = k + \sin(2i + 1)\alpha, \quad \text{where } k > 1 \quad i\text{:positive integer}$$

For $i = 1$, the set X looks like the round triangles of Figure IV.15.

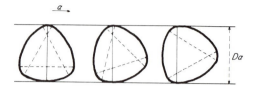

Figure IV.15. 2-D convex sets with a constant apparent contour. Cylinders with such a section could be used for rolling a board parallel to a flat terrain as was done in ancient Egypt. But what about wheels of this shape?

F. EXERCISES

F.1. Unit ball in $\mathbb{R}^n$

Let b_n denote the volume of the unit ball in $\mathbb{R}^n$, i.e.:

$$b_n = \frac{\pi^{n/2}}{\Gamma(1 + n/2)}$$

where Γ is the Euler (or Gamma) function:

$$\Gamma(n + 1) = n!; \qquad \Gamma\left(n + \frac{1}{2}\right) = \sqrt{\pi}\, \frac{(2n)!}{4^n\, n!}$$

(a) Show that the surface area of the unit sphere in R^n is equal to nb_n.
(b) Prove that the least Minkowski functional $W_n^{(n)}(X) = b_n$ for every $X \in \mathscr{C}(\mathscr{K})$ [proof by induction from relation (IV-28)].
(c) Prove that $W_i^{(n)} = b_n$ $(i = 0, 1, \ldots n)$ for the unit ball in $\mathbb{R}^n$ [proof by induction].

F.2. Steiner's formula for balls in $\mathbb{R}^n$ (Hadwiger)

(a) Prove relation (IV-32).
[The method is by induction on the dimension of the space: for $n = 1$, $V[X \oplus B(r)] = V(X) + 2r$. Suppose (1) is satisfied up to $n - 1$, and let $X \in \mathscr{C}(\mathscr{K})$ in $\mathbb{R}^n$. According to (IV-13), and the induction hypothesis, the surface area $S(r)$ of $X \oplus B(r)$ can be written:

$$S(r) = \sum_{\mu=0}^{n-1} n \binom{n-1}{\mu} W_{\mu+1}^{(n)}(X) r^\mu$$

By integration we get

$$V[X \oplus B(r)] = V(X) + \int_0^r S(\rho)\, d\rho \qquad \text{Q.E.D.}]$$

(b) Prove the relationship:

$$W_i^{(n)}[X \oplus B(r)] = \sum_{k=0}^{n-i} \binom{n-i}{k} W_{k+i}^{(n)}(X) r^k; \quad \forall r > 0$$

[Notice that $X \oplus B(\sigma + \rho) = [X \oplus B(\rho)] \oplus B(\sigma)$. Apply (IV-20):

$$V[X \oplus B(\rho + \sigma)] = \sum_{i=0}^{n} \binom{n}{i} W_i^{(n)}(X \oplus \rho B) \sigma^i = \sum_{j=0}^{n} \binom{n}{j} W_j^{(n)}(X)(\sigma + \rho)^j$$

Identify the multipliers of σ^i on both sides of the equation].
(c) Calculate the surface area S of the cube $X(a)$ with side a, dilated by the ball ρ (in $\mathbb{R}^3$):

$$[W_1^{(3)}[X \oplus B(\rho)] = W_1^{(3)}(X) + 2 W_2^{(3)}(X)\rho + W_3^{(3)}(X)\rho^2$$
$$S = 6a^2 + 6\pi a \rho + \pi \rho^2]$$

F.3. Minkowski functionals and radii of curvature

We assume that $X \in \mathscr{C}(\mathscr{K})$ in $\mathbb{R}^n$ admits $(n-1)$ radii of curvature R_i ($i = 1, \ldots, n-1$) at any point $x \in \partial X$. Denote by Φ_i the ith elementary symmetric function of the principal curvatures: $\Phi_1 = \sum (R_i)^{-1}, \Phi_2 = \sum\sum (R_i R_j)^{-1}$, etc. . . . Prove that the Minkowski functional $W_k^{(n)}$ is equal to:

$$W_k^{(n)}(X) = \frac{1}{k\binom{n}{k}} \int_{\partial X} \Phi_{k-1} \, dS \quad (k = 2, \ldots n) \tag{1}$$

[Consider the outer parallel set $X \oplus \rho B$. At $x + \rho n \in \partial(X \oplus \rho B)$, the radii of curvature are $R_i + \rho$, $i = 1, 2, \ldots n-1$; then, from Gauss' formula:

$$\frac{dS(X)}{R_1 \ldots R_{n-1}} = \frac{dS(X \oplus \rho B)}{(R_1 + \rho) \ldots (R_{n-1} + \rho)}$$

and

$$S(X \oplus \rho B) = \int_{\partial X} \frac{(R_1 + \rho) \ldots (R_{n-1} + \rho)}{R_1 \ldots R_{n-1}} dS(X)$$

$$= S(X) + \sum_{k=2}^{n} \rho^{k-2} \int_{\partial X} \Phi_{k-1} \, dS(X)$$

By identification with Steiner's formula (IV-20), we obtain (1).]

F.4. Decompositions of polygons

Show that the same hexagon can be considered as the dilation of two triangles, or of three segments; the same octogon, as the dilation of two squares, or four segments, or a triangle and a pentagon.

Is the octahedron a Steiner polyhedron? Decompose the cube, the rhombohedron and the tetrakaidecahedron into Minkowski sums of 3, 4 and 6 segments respectively.

F.5. Convex sets and symmetry (Matheron, 1969)

1. Let B and $B' \in \mathscr{C}(\mathscr{K})$ in $\mathbb{R}^n$ be such that

$$B \stackrel{\mathscr{T}}{=} \lambda B' \oplus (1-\lambda)\check{B}' \qquad (\lambda \in [0, 1])$$

Where $\stackrel{\mathscr{T}}{=}$ means "equal modulo a translation". Denote the relation between B and B' as $B \succ B'$. Prove that the relation $\succ$ is reflexive and transitive, and so is a pre-ordering relation.

2. Let $\equiv$ denote the equivalence associated with the pre-ordering relation $\succ$; we say that B and B' have the same symmetry when $B \equiv B'$.

Prove that B and B' have the same symmetry in $\mathscr{C}(\mathscr{K})$ if and only if they are equal, modulo a translation and eventually a transposition:

$$B \equiv B' \Leftrightarrow B \stackrel{\mathscr{T}}{=} B' \quad \text{or} \quad B \stackrel{\mathscr{T}}{=} \check{B}'$$

[Firstly, prove that for $B \in \mathscr{K}$, $B' \in \mathscr{C}(\mathscr{K})$, and $\lambda \in [0, 1]$ the relation $B = \lambda B \oplus (1-\lambda)B'$ implies $\lambda = 1$ or $B = B'$.]

F.6. Convex sets, ordering and pre-ordering relations

Let us review the ordering and pre-ordering relations that appeared in the chapter, in order to compare them with each other. If X and Y belong to $\mathscr{C}(\mathscr{K})$ we can have:

(a) the inclusion $X \subset Y$, or equivalently $d_X \leq d_Y$ (rel. IV-56)
(b) the inclusion modulo a translation: $X_h \subset Y$ or $Y \ominus \check{X} \neq \varnothing$
(c) Y is open with respect to X, or equivalently $d_Y - d_X$ is a support function
(d) X is closed with respect to Y^c
(e) (in $\mathbb{R}^2$ only) $u_X(d\alpha) \leq u_Y(d\alpha)$ (u: perimetric measure)

Prove that (b), (c), (d) and (e) are ordering relations in the quotient space $\mathscr{C}_{\mathscr{T}}(\mathscr{K})$ of the compact convex sets considered up to a translation. Prove that (a) is an ordering relation for the families of nested sets. Check the following implications:

$$\begin{array}{cc} a \Rightarrow b & a \Rightarrow b \\ \Uparrow & \Uparrow \\ c \Rightarrow d & c \Leftrightarrow d \Leftrightarrow e \\ \text{in } \mathbb{R}^n (n \geq 3) & \text{in } \mathbb{R}^2 \end{array}$$

V. Morphological Parameters and Set Models

A. DEGREES OF IRREGULARITY

A.1. Quantitative morphological parameters

A morphological parameter is a numerical value associated with a set. In the first chapter we gave a precise treatment in morphological terms, of the notion of a quantitative image transformation. Similarly, we now have to transpose the four basic prerequisites of quantification to the morphological parameters. This time we will distinguish right from the start global measurements (which are easier to handle) from local measurements.

(a) *Global quantitative parameters*
Here, we assume that the compact set X is not only bounded, but also experimentally accessible from a single "swoop" (one T.V. scanning for example). We denote by $\{W(X)\}$ the various parameters associated with the set X. These functionals will be morphologically acceptable if they fulfill the following three conditions:

(i) *Invariance under translation:* $W(X) = W(X_h)$ h vector in R^n
(ii) *Homogeneity:* $W(\lambda X) = \lambda^h W(X)$ $\lambda > 0,$ h real number
(iii) *Semi-continuity:* $\{X_i\} \to X$ in $\mathscr{F}$ implies $\overline{\text{Lim}}\ W(X_i) \subset W(X)$ or $\overline{\text{Lim}}\ W(X_i) \supset W(X)$ (see Fig. V.1).

We shall say that a global parameter W is *quantitative*, in the morphological sense, when the three preceding conditions are satisfied (the fact that all of X is accessible frees us from a fourth, local knowledge type condition). Note that nothing to do with C-additivity has yet been demanded.

Some parameters, but not all, have both local and global definitions. Take for example the perimeter of a 2-D convex set. We saw that it can be defined (in

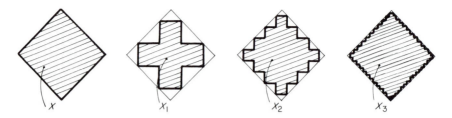

Figure V.1. Although $X_i \to X$, the perimeter $U(X_i) \not\to U(X)$. We have only $\lim U(X_i) \supset U(X)$. The perimeter is a lower semi-continuous parameter.

integral geometry) without any local considerations. But it is possible to proceed differently, by weighting the boundary ∂X with a uniform mass that is equal to one per unit length. Consider now the ratio $[U(X)]^2/A(X)$, of the perimeter squared, divided by the area of X. This classical shape descriptor is defined via the global approach and satisfies the three conditions given above for $X \in \mathscr{C}(\mathscr{K})$ i.e. it is quantitative. However there is no way of attributing to each small open set dx of the plane a numerical value such that by summation we recover $[U(X)]^2/A(X)$: the ratio is not "localizable", (therefore, non accessible from cross sections).

(b) *Local quantitative parameters*
In a local analysis, we no longer assume that the set X is compact, but only that it is closed. We denote by $\mu^x(dx)$ a measure generated by the set X, and we demand that $\mu^x(dx)$ satisfy the following four axioms:

(i) Compatibility with translations: $\mu^{X_h}(G) = \mu^X(G_{-h})$

The measure generated by X acts on the open mask G exactly as the measure generated by the translate X_h acts on the translate G_{-h}. For example, if $\mu^x(G)$ is the portion of volume of X hit by G, it is equivalent to shift X to the right by h (which generates a *new* measure μ^{X_h}) or to shift G to the left by $-h$ (which leaves us with the same measure μ^X). Note that invariance under translation is *stronger* than compatibility.

(ii) Compatibility with magnifications (homogeneity): $\mu^{\lambda X}(\lambda G) = \lambda^k \mu^X(G)$
$\lambda > 0;\ -\infty < k < +\infty$

(iii) Local knowledge: Whatever is the measuring mask Z', there exists a bounded mask Z such that

$$k_{Z'} \cdot \mu^X(dx) = k_{Z'} \cdot \mu^{X \cap Z}(dx) \qquad (k_{Z'}: \text{indicator of } Z') \qquad \text{(V-1)}$$

This relation is exactly analogous to the third principle for the transformations; recall relation (I-4):

$$Z' \cap \psi(X) = Z' \cap \psi(X \cap Z) \qquad \text{(I-4)}$$

Now we are no longer dealing with sets, but with numbers, so Z' is replaced by its indicator function 1_Z; the mapping ψ is replaced by the measure μ, and the set intersection by the simple product $1_{Z \cap Z'} = 1_Z \cdot 1_{Z'}$. Notice that condition (V-1) does not make double use of σ-additivity for measures (Ch. IV, D.1). Here, not only do the small open sets dx divide the space into portions (hence the necessity of σ-additivity), but also the set X itself, is not known everywhere. Relation (V-1) results from this constraint.

(iv) Semi-continuity: For every semi-continuous positive function φ, defined on $\mathbb{R}^n$, the mapping $X \to \mu^X (\varphi)$ must be semi-continuous on $\mathscr{F}(\mathbb{R}^n)$.

(c) *Stereological constraint*

Typical studies of the relationships between physico-chemical properties in microstructures of metallography or petrography for example, associate physical characteristics in $\mathbb{R}^3$ with structural measurements on plane sections. In such cases it would be appropriate to go up from the planar parameters to those having a three-dimensional meaning. Hence we must find planar measures for which this passage to three dimensions is possible: this is called the *stereological constraint*. It does not seem possible to assign a precise once-and-for-all criterion to this constraint; so it will reappear periodically when considering the formulation of models and criteria. Therefore it concerns more a colouring of our point of view, than a definite constraint (in actual fact, the demands of stereology go further than the rather narrow one just proposed: a *tensor* of permeabilities of a porous medium, cannot be correlated with a *scalar* such as porosity, even if the latter is what we have defined to be stereo-logical).

A.2. Counter-examples

The results of the last chapter, corroborated by an abundant literature in applied image analysis teach us that a basic set of quantitative and stereological parameters is provided by the Minkowski functionals. Since both theoreticians and experimenters find them fundamental, what is left to say about morphological parameters? Of course we do need more explanations (why these and not others? Do they satisfy the morphological principles?), but this can be done in a few pages. The study of these parameters becomes much more difficult as soon as we wonder about the domain of validity of the basic relationships such as Crofton's formula. It is not enough to simply apply an algorithm, say, "the number of intercepts" (found in any image analyser) in order for the resulting number to mean something. The following two counter-examples clearly illustrate this point:

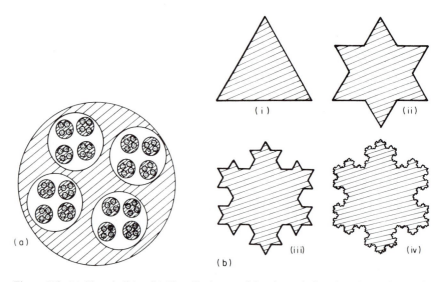

Figure V.2. (a) Nested disks. (b) Von Koch snowflake. At each iteration the perimeter is multiplied by 4/3, but the area and the connectivity are bounded.

(a) *Nested disks* (Fig. V.2a)

The set X is a disk of radius γ_0, in which are placed four circular disjoint pores with a radius γ_1. In each of the pores we implant four circular disjoint grains, each with a radius γ_2, etc. . . . , iterating indefinitely. We can be sure that the area of X is bounded, since all we have done is remove a portion of the initial disk. On the contrary, the fate of the perimeter depends on the way the radii successively decrease; if each is one fifth of the preceding ($\gamma_n = 5^{-n}\gamma_0$) then the infinite sum for L converges to $10\pi\gamma_0$. If the n-th radius is only one quarter of the previous ($\gamma_n = 4^{-n}\gamma_0$) then the perimeter of X becomes infinite. Finally, the connectivity number and the convexity number show a third type of behaviour; independently of the radii, $N^{(2)}$ is never defined and N^+ is always infinite!

Let us return to the set X and take its morphological closure by the disk of radius γ_n (we assume that the pores and grains are implanted far enough from the boundaries of the preceding iterations, as in Chapter V, Section E). This closure is identical to the set we would have obtained, had we stopped the construction of X at the n-th step. Here the four basic parameters are of course finite (although the value $N^{(2)}$ has wild oscillations). Hence it is much easier to define the dilate or the closure of this type of set, than its perimeter or its connectivity number, and the only robust measurement is the area.

(b) *The Von Koch snowflake* (Fig. V.2b)

As the measures move down in dimensionality we might think that if we find one measure which is undefined, then all those after it will also be undefined. In fact this is not the case, as we will see in this second counter-example, that of the famous Von Koch snowflake. The definition of this curve dates from 1904 and since then it has seduced the interests of many people, from pure mathematicians (Cesaro) to popular writers (Northrop, 1945). Recently, Mandelbrot (1977) has taken it as a starting point for the more general and more flexible notion of self-similarity, often interpreted in probabilistic terms.

What is this curve? Starting from an equilateral triangle X_0 (Fig. V.2b) for each side we take out the middle third and replace it with two segments identical in length to the one removed, thus forming a spike. Each side of the resulting polygon is then in turn made spiky in exactly the same way, etc. . . . indefinitely. We see on Figure V.2b, β and δ the polygons X_2 and X_4 associated with respectively the second and fourth iterations. Eventually the evolution of the figure becomes restricted by the fineness of the line that can be drawn. Now the limiting snowflake, as well as all the intermediary polygons, have a connectivity number equal to one. Moreover, its area is bounded by the area of the circle that circumscribes the initial triangle. However, the perimeter U_n diverges, since $U_n = U_0 (4/3)^n$. The boundary of X_0 is continuous, but almost everywhere without derivative; if any arbitrarily small circle is centred at a point of the boundary, then the circle will contain an infinite number of angular points.

Perhaps the reader will find these two examples excessively pathological, and too far removed from reality. If this is the case, we would like to refer him to the end of the present chapter; he will see that a simple mound of clay (the most common material found in the ground!) is much closer to these "jagged" models than to a smooth set with continuous derivatives on its boundary.

A.3. Using Steiner's formula as a guideline for regularity

Fortunately as a general rule we are not interested in extremely fine details. The physico-chemical properties of the bodies under study usually indicate the scale under which the structural irregularities can be neglected (e.g. the capillary laws for studying permeability of porous media). However this information is not always available, so that we should always remember that the sets have to be sufficiently regular. Therefore we must

(i) classify sets according to their degree of regularity;
(ii) find good criteria which enable us to choose the most appropriate class in each case.

We shall look for answers to both questions, within the set of all compact convex sets $\mathscr{C}(\mathscr{K})$. In Chapter IV, it was proved that not only were the compact convex sets invariant under closure with respect to a ball $B(r)$ of arbitrary radius r (a characteristic property of the compact convex sets), but also the sets $X \in \mathscr{C}(\mathscr{K})$ satisfied Steiner's formula (IV-33) that gives the volume of the dilate $X \oplus B(r)$ as a polynomial in r. These two properties are related. By successively weakening them, we can generate increasingly general classes of sets, as we show here in the case of $\mathbb{R}^3$.

First of all, one can define the class of compact sets which are invariant under closure with respect to a ball with *fixed* radius r. For reasons of symmetry, we shall also demand these sets to be invariant under opening by the small ball, and shall call the corresponding family, the regular model $\mathcal{R}(\mathcal{K})$ (Sect. C). The second model, which contains $\mathcal{R}(\mathcal{K})$ and hence is slightly more irregular, is the extended convex ring $\bar{\mathcal{S}}(\mathcal{K})$. The compact sets $X \in \bar{\mathcal{S}}(\mathcal{K})$ are generally not invariant under closure with respect to an arbitrarily small ball $B(r)$. However when $r \to 0$, X and its closure $X^{B(r)}$ are not very different, and the volume of the dilate $X \oplus B(r)$ has an expansion at the origin, with the first two terms of Steiner's formula:

$$V[X \oplus B(r)] = V(X) + rS(X) + \varepsilon(r) \qquad (\varepsilon \to 0) \qquad \text{(V-2)}$$

In both models $\mathcal{R}(\mathcal{K})$ and $\bar{\mathcal{S}}(\mathcal{K})$, the stereological relations (V-1) are satisfied, although the first model rejects sets with angularities whereas the second model accepts them. Moreover, in both cases, one can define Minkowski measures, which allows extension of the models to the closed sets $\mathcal{F}$ (here again, we have $\mathcal{R}(\mathcal{F}) \subset \bar{\mathcal{S}}(\mathcal{F})$). The distinctions between the set models $\mathcal{K}$, $\mathcal{R}(\mathcal{K})$ and $\bar{\mathcal{S}}(\mathcal{K})$ play a fundamental role in the problems of digitalization.

The next and final model considered, is the class $\mathcal{K}$ itself (Sect. D) which includes $\bar{\mathcal{S}}(\mathcal{K})$ and $\mathcal{R}(\mathcal{K})$ as subclasses. The only functional which is still defined on $\mathcal{K}$ is the volume; and Lebesgue measure is only semi-continuous on $\mathcal{K}$. Invariance under closure is obviously not in general satisfied, nor is the limited Steiner expansion (V-2) (the surface area does not always exist). However, Minkowski showed that the class $\mathcal{K}$ admits a generalization of relation (V-2) when one substitutes r^{α} for r ($0 \le \alpha \le 3$). Thus we find boundary sets ∂X of a dimension greater than two in $\mathbb{R}^3$ (or greater than one in $\mathbb{R}^2$)!

B.   THE CONVEX RING AND ITS EXTENSION

The very simple mathematical basis of the convex ring $\mathcal{S}(\mathcal{K})$ is due to H. Hadwiger (1957). It is obtained by constructing the class stable under finite unions of compact convex sets. Thus by definition:

$$X \in \mathcal{S}(\mathcal{K}) \Leftrightarrow X = \bigcup_{i=1}^{N} Y_i \qquad Y_i \in \mathcal{C}(\mathcal{K}) \qquad \text{(V-3)}$$

i.e. the set X belongs to the convex ring if it can be decomposed into a finite union of compact convex sets. There is an immediate advantage in using such a class. The C-additivity condition of relation (IV-29) now appears in a much more natural way than it did for convex sets (where we were obliged to assume that the union of two compact convex sets X and Y also belongs to $\mathscr{C}(\mathscr{K})$!).

B.1. The Euler–Poincaré constant

The quantity known as the Euler–Poncaré constant, or connectivity number, is the most important notion for sets of the convex ring. All the stereological parameters are derived from it by various integrations. Furthermore, in R^3, the way it is computed experimentally from serial sections, is an exact translation of its definition.

The Euler–Poincaré constant $N^{(n)}(X)$ is defined by induction on the dimension of the space. We start in $\mathbb{R}^0$, with the algorithm:

$$N^{(0)}[X \cap \{x\}] = \begin{matrix} 1 & \text{if} & X \cap \{x\} = \{x\} \\ 0 & \text{if} & X \cap \{x\} = \varnothing \end{matrix} \qquad \text{(V-4)}$$

(this is nothing more than counting the points of X, "à la Delesse" (1847)). Suppose now that X is defined in $\mathbb{R}^1$, and is a finite union of disconnected segments (convex ring assumption). Put:

$$h^{(1)}(x) = \operatorname*{Lim}_{\varepsilon \to 0} \{ N^{(0)}(X \cap \{x\}) - N^{(0)}(X \cap \{x + \varepsilon\}) \}, \quad \varepsilon > 0 \qquad \text{(V-5)}$$

Obviously $h^{(1)}(x)$ equals one at the right extremity x_i of the ith segment, and equals zero elsewhere. We shall define the connectivity number $N^{(1)}(X)$ (and see Fig. V.3) as the sum of those $h^{(1)}(x_i) \neq 0$:

$$N^{(1)} = \sum_i h^{(1)}(x_i) \qquad \text{(V-6)}$$

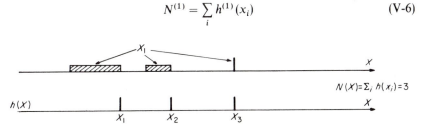

Figure V.3. 1-D connectivity number.

The induction mechanism now becomes clearer. In the plane, we arbitrarily choose the x-axis. Let $\Delta(x)$ be the straight line perpendicular to the x-axis at the point $\{x\}$. As $\{x\}$ moves along the axis, so the test line $\Delta(x)$ sweeps out the plane, and intersects the 2-D set X; then $X \cap \Delta(x)$ has connectivity number $N^{(1)}[X \cap \Delta(x)]$ (Fig. V.4).

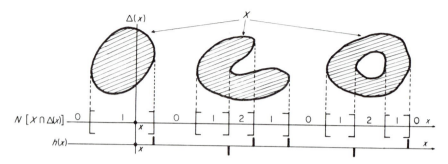

Figure V.4. 2-D connectivity number.

In an analogous way to $\mathbb{R}^1$, we define the increments $h^{(2)}(x)$ now in terms of the $N^{(1)}$'s:

$$h^{(2)}(x) = \lim_{\varepsilon \to 0} \{ N^{(1)}[X \cap \Delta(x)] - N^{(1)}[X \cap \Delta(x+\varepsilon)] \} \qquad \varepsilon > 0$$

The limit $h^{(2)}(x)$ differs from zero at a finite number of points $\{x_i\}$; it is $+1$ when $\Delta(x_i)$ is leaving the grain at a boundary point which is locally convex, and is -1 when $\Delta(x_i)$ is leaving the grain at a point which is locally concave. Consequently, we can define its sum $N^{(2)}(X)$:

$$N^{(2)}(X) = \sum_i h^{(2)}(x_i)$$

as the connectivity number, or Euler–Poincaré constant, of the 2-D set X. It is important to note that $N^{(2)}(X)$ does not depend upon the choice of the x-axis (Hadwiger, 1957), and as we see from Figure V.4, is independent of the particle *shapes* that make up the set X. $N^{(2)}(X)$ is simply the difference between the number of particles and the number of holes contained in them. Therefore, by taking the complement X^c of X, the notions of particles and pores are interchanged, and so by convention, we can put:

$$N^{(2)}(X^c) = 1 - N^{(2)}(X) \tag{V-7}$$

where the constant 1 represents the whole plane. (We say by convention because the unbounded set X^c *cannot* belong to $\mathscr{S}(\mathscr{K})$. In fact, exactly as for the convexity number, the connectivity number $N^{(2)}(X)$ manifests properties of the boundary ∂X. When ∂X possesses curvatures, $N^{(2)}(X)$ equals the total rotation of the normal along ∂X, divided by 2π.)

The topological parameter $N^{(2)}(X)$ is invariant under translation, rotation, magnification, affine transformation, ..., and all other interesting homeomorphisms. Fundamentally, it is however *not* connected with the existence of individual particles, since it acts on the figure and the background in a symmetric way. But, if X is the union of isolated grains without pores, then $N^{(2)}(X)$ is the total number of particles.

In $\mathbb{R}^3$, the same type of induction is carried out. The plane $\Pi(x)$ which is orthogonal to the arbitrary (but fixed) x-axes, sweeps out the space as its common point $\{x\}$ moves along the axis. For each location of $\Pi(x)$, we know how to calculate the Euler–Poincaré constant $N^{(2)}[X \cap \Pi(x)]$ of its intersection with the 3-D set X. By putting

$$N^{(3)}(X) = \sum_i h^{(3)}(x_i) = \sum \lim_{\varepsilon \to 0} \{N^{(2)}[X \cap \Pi(x)] - N^{(2)}[X \cap \Pi(x+\varepsilon)]\}$$

(V-8)

we define the Euler–Poincaré constant in $\mathbb{R}^3$. As in the case of $N^{(2)}(X)$, the connectivity number $N^{(3)}(X)$ is a topological parameter, i.e. a number invariant under homeomorphisms, and does not depend upon the choice of the x-axis. The algorithm (V-8) suggests how we might experimentally compute $N^{(3)}(X)$, when we have serial sections at our disposal. We have only to subtract the connectivity numbers of two successive sections (to give an estimate of $h^{(3)}(x)$) and then sum over all successive pairs. N. Suwa and T. Takahashi (1978) used this method successfully for analysing morphological changes in cirrhotic livers.

The connectivity number $N^{(3)}(X)$ has a very intuitive geometric interpretation which is due to Poincaré, and which has been taken up again by R. De Hoff (1972) (Poincaré's context was different from that of this book. He was looking for solutions to certain differential equations, using the Betti numbers of algebraic topology). Suppose we define the *genus G* of the boundary ∂X to be the maximum number of curves that we can draw on it without disconnecting it into two parts. The value of G is zero for a sphere, one for the boundary of a torus, two if we add a supplementary handle to the torus, etc. . . . Then the Euler–Poincaré constant $N^{(3)}(X)$ is equal to:

$$N^{(3)}(X) = \sum_i [1 - G(\partial X_i)]$$

(V-9)

where the discrete sum is taken over all the connected components of ∂W by distinguishing the inside boundaries from the outside ones. For example, if X is a soap bubble without thickness ($X = \partial X$), then ∂X must be counted twice, each time with a genus equal to zero, thus $N^{(3)}(X) = 2$. Figure V.5 illustrates relation (V.9) with three elementary examples. Just in passing, note that the basic ball with an internal hole has a connectivity number which is two in $\mathbb{R}^1$ (two segments) and two in $\mathbb{R}^3$, but zero in $\mathbb{R}^2$. By induction, we immediately see that the result is general, and only depends upon the evenness of the dimension of the space: a peach without its pit in a 1979 dimensional space has a connectivity number equal to two!

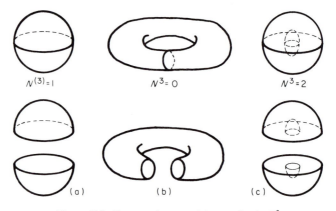

$N^{(3)}=1$ $N^3=0$ $N^3=2$

(a)   (b)   (c)

Figure V.5. Genus and connectivity number in $\mathbb{R}^3$.

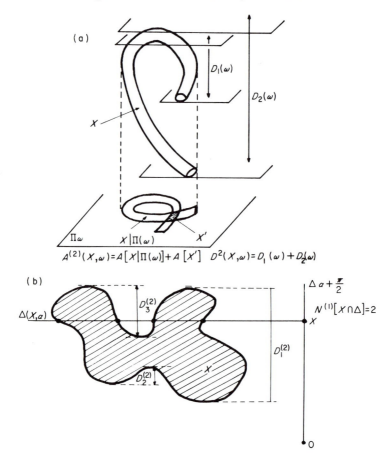

(a)

$D_1(\omega)$

$D_2(\omega)$

X

Π_ω $X|\Pi(\omega)$ X'

$A^{(2)}(X,\omega)=A[X|\Pi(\omega)]+A[X']$ $D^2(X,\omega)=D_1(\omega)+D_2(\omega)$

(b)

$\Delta a + \frac{\pi}{2}$

$\Delta(X,a)$

$D_3^{(2)}$

$N^{(1)}[X\cap\Delta]=2$

X

$D_1^{(2)}$

X

$D_2^{(2)}$

0

B.2. Stereology of the convex ring in $\mathbb{R}^3$ and $\mathbb{R}^2$

We saw that in going from $\mathbb{R}^0$ to $\mathbb{R}^3$, the four connectivity numbers are strongly related to each other. When this inductive geometrical construction is extended n times, one generates the number $N^{(n)}(X)$, $X \in \mathscr{S}(\mathscr{K})$ in $\mathbb{R}^n$. For us, the most important property of the Euler–Poincaré constant is stated by the following theorem, due to Hadwiger (1957):

The only functional defined on the convex ring in $\mathbb{R}^n$, of degree zero, invariant under displacements, C-additive and constant for the compact convex sets, is the Euler–Poincaré constant $N^{(n)}(X)$.

It is often said that the $N^{(i)}(X)$ are topological parameters, as against metric parameters. This is perfectly true, although quite a large number of other topological parameters exist (number of grains in the holes, etc. . .). We have to keep in mind that the $N^{(i)}(X)$ are associated with a given dimension of the space. When we use one of them in another space, its meaning changes and it takes on a metric interpretation. If the polished sections of a piece of coke have a given mean connectivity number $N^{(2)}(X)$ (after rotation averaging for example), the coke itself, which is a 3-D object, does not possess an $N^{(2)}(X)$, but an $N^{(3)}(X)$ which is not accessible from the $N^{(2)}(X)$. In other words, the four connectivity numbers $N^{(0)}$ to $N^{(3)}$ of X and its cross sections (after rotation averaging) provide us with four independent pieces of information, which extend the Minkowski functionals to the convex ring. Indeed, the three basic Crofton's formulae (IV-17) to (IV-19) still remain valid on taking the Euler–Poincaré constant for $N^{(1)}$. One can complete them by the relations:

$$V(X) = \int_{\mathbb{R}^3} N^{(0)}[X \cap \{x\}]\,dx \quad \text{in } \mathbb{R}^3 \qquad \text{(V-10)}$$

and in $\mathbb{R}^2$

$$A(X) = \int_{\mathbb{R}^2} N^{(0)}[X \cap \{x\}]\,dx \qquad \text{(V-11)}$$

Alternatively, we can generalize Cauchy's relations by introducing the notions of *total projection area* $A^{(2)}(X, \omega)$ of X in direction ω, and of *total diameters* $D^{(3)}(X, \omega)$ and $D^{(2)}(X, \alpha)$ of X in direction ω (in $\mathbb{R}^3$) and α (in $\mathbb{R}^2$) respectively. Put, by definition:

$$A^{(3)}(X, \omega) = \int_{\pi_\omega} N^{(1)}[X \cap \Delta(x, \omega)]\,d\omega \qquad \text{(V-12)}$$

Figure V.6.(a): in $\mathbb{R}^3$, total projection $A^{(3)}(X, \omega)$ and total diameter $D^{(3)}(X, \omega)$. (b): in $\mathbb{R}^2$, total projection $D^{(2)}(X, \alpha)$.

$$D^{(3)}(X, \omega) = \int_{-\infty}^{+\infty} N^{(2)}[X \cap \Pi(x, \omega)] dx \qquad \text{(V-13)}$$

$$D^{(2)}(X, \alpha) = \int_{-\infty}^{+\infty} N^{(1)}[X \cap \Pi(x, \alpha)] dx \qquad \text{(V-14)}$$

Figure V.6 shows the geometrical meaning of these quantities. For any $X \in \mathscr{S}(\mathscr{K})$, Cauchy relation (IV-14) becomes

$$\frac{1}{4} S(K) = \frac{1}{4\pi} \int_{4\pi} A^{(3)}(X, \omega) d\omega \qquad \text{(V-15)}$$

(A similar treatment applies to relations (IV-15) and (IV-16) by using $D^{(3)}(X, \omega)$ and $D^{(2)}(X, \omega)$.)

The characterization theorems of Hadwiger (IV-C.1f) carry over to the convex ring, by restricting the conditions of continuity and monotonicity to the subclass of the *convex* compact sets. Similarly, Steiner's formulae (IV-33 and 34) admit a generalization due to W. Von Blaschke and S. A. Santalo. For the usual spaces $\mathbb{R}^3$ and $\mathbb{R}^2$, it looks like:

$$\int \omega(d\theta) \int_{\mathbb{R}^3} N^{(3)}[X \cap B(x, \theta)] dx = V(X) + \frac{M(X) \cdot S(B) + S(X) \cdot M(B)}{4\pi}$$
$$+ V(B) \qquad \text{(V-16)}$$

$$\frac{1}{2\pi} \int_0^{2\pi} d\alpha \int_{\mathbb{R}^2} N^{(2)}[X \cap B(x, \alpha)] dx = A(X) + \frac{U(X) \cdot U(B)}{2\pi} + A(B) \quad \text{(V-17)}$$

where $B(x, \theta)$ is a B centred at point x, and oriented by the system of axes θ. In spite of the formal identity of the right-hand sides of relations (IV-33–34) and relations (V-16) and (V-17), here the left-hand sides are no longer the integration of a volume of a dilation. In Figure V.7 we can compare the different extensions of Steiner's formula proposed by Von Blaschke–Santalo

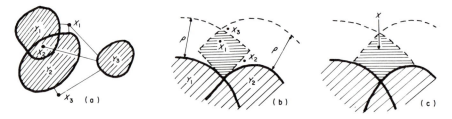

Figure V.7. (a) Projections $x|X$ of points x on the set $X = Y_1 \cup Y_2 \cup Y_3$. (b) and (c) extensions of Steiner's formula according to Matheron (b) and Von Blaschke–Santalo (c). In both cases, the dashed zone (X) is to be counted twice, and the other parts of the dilate only once. The surface area of the dilate is no longer considered.

on one hand (Fig. V.7c) and by Matheron on the other hand (Fig. V.7b). In Figure V.7c, when zones are superimposed onto each other after dilation, they have to be counted twice, or more often, according to the number of overlappings.

B.3. A local study of the convex ring

The approach used in constructing the Minkowski measures on $\mathscr{C}(\mathscr{K})$ is easily extendable to $\mathscr{S}(\mathscr{K})$. The only difference lies in the fact that the projection $x|X$ of the point x on the set $X \in \mathscr{S}(\mathscr{K})$ does not reduce to a single point x, but rather becomes i $(1 \leq i \leq p)$ points: $\{x\}$ has a projection onto each Y_i ($i = 1, \ldots, p$), but the total number of points might be less than p due to Y_i's overlapping when taking the union $\bigcup Y_i$. As an example, look at Figure V.7a: the point x_1 has three distinct projections, the point x_2 has two projections which are confused with x_2 itself and a third which is distinct (although x_2 is an interior point of X!); finally, two of the three projections of x_3 remain, but the third one vanishes in the interior of $Y_1 \cup Y_2$.

In order to give a local version of Steiner's formula, we have to integrate the contributions of all the points belonging to the dilate $X \oplus rB$. Figure V.7b shows that the three points x_1, x_2, x_3 can belong to the same dilate, but at the same time, produce three different types of projections. The point x_1 has only one projection (with distance between x_1 and $x_1|X$ smaller than r), x_2 has two projections (with both distances smaller than r), and x_3 also has two projections (with one distance smaller, and one distance greater than r). We will allow for such a variety of possibilities by associating with each x, the number of projections at a distance from x that is less than or equal to r. More generally, for every continuous function φ and every point x we define the sum Φ by adding up the values of φ taken at each projection point of x:

$$\Phi(X, \varphi, r; x) = \sum_{x' \in (x|X) \cap B_r(x)} \varphi(x') \qquad (\text{V-18})$$

where $I = \{x' : x' \in (x|X) \cap \mathrm{Br}(x)\}$.

Then under these conditions, the two relations (IV-50) and (IV-51) remain true, provided we substitute the left-hand side of the equalities with $\int \Phi(x)\,dx$. The expressions obtained generalize Steiner's formula to the convex ring, expressing it as a scalar product of a sequence of measures concentrated on X and on its boundary. Suppose for example that in $\mathbb{R}^2$ and $\mathbb{R}^3$, we take φ to be the constant function, equal to one. Then Φ is the number $n(X, r, x)$ of projections x' of x, lying within a disk of radius r and centre x. (Here is the departure from the Von Blaschke–Santalo approach; see Fig. V.7.) By integration over the space, we find:

(a) in $\mathbb{R}^2$:

$$\int n(X, r, x)\,dx = \int 1_X\,dx + r \int u(dx) + \pi r^2 \int n^{(2)+}(dx)$$

$$= A(X) + rU(X) + \pi r^2 N^{(2)+}(X) \qquad \text{(V-19)}$$

(b) in $\mathbb{R}^3$:

$$\int n(X, r, x)\,dx = \int 1_X\,dx + r \int s(dx) + r^2 \int m^+(dx) + \frac{4}{3}\pi r^3 \int n^{(3)+}(dx)$$

$$= V(X) + rS(X) + r^2 M^+(X) + \frac{4}{3}\pi r^3 N^{(3)+}(X) \qquad \text{(V-20)}$$

In relations (V-19) and (V-20), the first two terms do not differ from those of Santalo's formulae for B a ball, and involve the first two Minkowski functionals. Besides, the measures $x^{(2)+}$, m^+ and $n^{(3)+}$ generalize the notion of symmetrical functions of positive radii of curvature, and lead to new functionals. The most important ones are the *convexity* numbers $N^{(2)+}$ and $N^{(3)+}$. The seven measures involved in these two formulae are all compatible with displacements and magnifications, and are additive in the sum of relation (IV-43). As far as semi-continuity goes, they satisfy the following rather weak result: the mapping $(X, r) \to \int u(X, r, x)\,dx$ is lower semi-continuous on $\mathscr{S}(\mathscr{K}) \times \mathbb{R}^+$. Then, *only the Steiner formula* is semi-continuous (and of course its first term, $V(X)$, or $A(X)$ which are semi-continuous for general compact sets).

B.4. The classes $\mathscr{S}(\mathscr{F})$ and $\bar{\mathscr{S}}(\mathscr{F})$

While the convex ring $\mathscr{S}(\mathscr{K})$ offers a great deal mathematically, its morphological fecundity seems less certain. The sets X must be bounded, and its concave portions and saddles must necessarily be *angular*, i.e., interior angles in $\mathbb{R}^2$, interior edges or cones in $\mathbb{R}^3$ (Fig. V.8). The notion of "normal bodies" introduced by H. Hadwiger (1959) provides a class of sets with better behaved concavities, but it is a strictly global concept restricted to the compact sets. Therefore we would prefer not to take this approach, but rather start from the class $\mathscr{S}(\mathscr{F})$ already proposed by G. Matheron (1975) in order to construct our model. Indeed, the local version of the Minkowski functionals enables us to overcome the barrier that "X is bounded", allowing extension of the preceding results to a more general class. We denote by $\mathscr{S}(\mathscr{F})$ the class of closed sets $F \in \mathscr{F}$ such that $F = \bigcup F_i$ for a locally finite family $\{F_i, i \in I\}$ of closed convex sets (i.e. each $F_i, i \in I$ is closed and convex, and for any compact mask $Z, F_i \cap Z = \varnothing$ for all but a finite number of $i \in I$). There are no new problems involved in

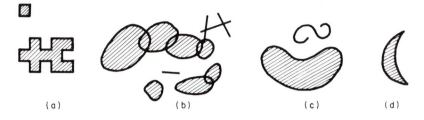

Figure V.8. (a) and (b): elements of the convex ring $\mathscr{S}(\mathscr{K})$. (c) and (d): figures which are not members of $\mathscr{S}(\mathscr{K})$ but belong to $\overline{\mathscr{S}}(\mathscr{F})$.

defining the (finite) set of projections of a point x on the set F, where the projection points are restricted to lie within a ball of centre x and radius r. Then the local approach for $\mathscr{S}(\mathscr{K})$ can be re-formalized in terms of measures and relations (V-19) and (V-20) always hold (Matheron, 1975).

We still have to enlarge the model in order to include such figures as Figure V.8c and Figure V.8d for example. In fact, these types of figures can be represented as limits of sets of $\mathscr{S}(\mathscr{F})$, which suggests that $\mathscr{S}(\mathscr{F})$ should be enlarged to its closure $\overline{\mathscr{S}}(\mathscr{F})$. More precisely, let $\overline{\mathscr{S}}(\mathscr{F})$ denote the class of sets X:

(a) which are limits in $\mathscr{F}$ of elements $X_i \in \mathscr{S}(\mathscr{F})$;
(b) which have Minkowski measures that are defined as the limits of the corresponding Minkowski measures of the X_i, with respect to the weak topology on the measures (i.e. $\mu = \lim \mu_i$ if $\int \mu \varphi = \lim \int \mu_i \varphi$ for any continuous function φ with a compact support).

The mapping defined by relation (V-18) is still measurable (limit of measurable mappings) and relations (V-19) and (V-20) always hold. Moreover, the class $\overline{\mathscr{S}}(\mathscr{F})$ is stable for finite iterations of the Hit or Miss transformations. In terms of applicability, it is the basic set model for 90 per cent of the structures encountered in practical morphology (by replacing in (a) "limit in $\mathscr{F}$" by "limit in $\mathscr{K}$" and in (b) "Minkowski measures" by "Minkowski functionals" we equally define the extended class $\overline{\mathscr{S}}(\mathscr{K})$).

B.5. Convexity numbers

The main new idea introduced by a local approach is that of the *convexity numbers*. In the space $\mathbb{R}^3$, the number $N^{(3)+}$ can be interpreted via the first contact points of sweeping test planes, i.e. planes which "enter" the set: $N^{(3)+}$ is the sum of a measure concentrated on ∂X, whose value in $dx \in \partial X$ is the solid angle carved out by normals of planes having dx as the "entry" set. This measure is identically zero for those $dx \in \partial X$ which are not on a locally convex part of the boundary (holes, saddles, etc. . . . .). The convexity number $N^{(3)+}$ is a generalization of the notion of total curvature and in the same way, it cannot be estimated by stereological formulae, but only from serial sections. It is consequently less often used in practice than the 2-D convexity number $N^{(2)+}$ presented below.

The concept of $N^{(2)+}$ is due to Haas, Matheron and Serra (1967) and independently, to De Hoff (1967). These articles developed the subject in different ways. At the same time, De Hoff (1978) studied the three-dimensional meaning of the 2-D convexity number, called by him the "tangent count", whereas we preferred to stay in two dimensions and concentrate on the connection between convexity numbers and the distribution of the radii of curvature; (see Ch. X, Ex. 14). (We prefer to avoid the nomenclature "tangent count", since the bodies under study do not always possess tangents.)

In Figure V.9 we see a grain progressively becoming two pseudo-grains under attack from a pore. How can such an evolution be described? The connectivity number $N^{(2)}$ is the number of connected components minus their holes (Fig. V.5), which in the present case, is equal to one in (a), (b) and (c), but suddenly jumps to two, or to zero, at the last step (d). We would like to construct a more continuous descriptor of the process.

Consider, in Figure V.9b the test line $\Delta(\alpha)$ with an orientated normal α. When $\Delta(\alpha)$ sweeps across the plane in direction α, it (separately) hits the two pseudo-grains of the set X; thus the straight line "sees" two distinct grains. If we had scanned in the opposite direction $\alpha + \pi$, the same grain would have offered only one first contact point to $\Delta(\alpha + \pi)$. Let $N^+(X, \alpha)$ be the number of first contacts in direction α. We are looking for its mean value over all directions α; i.e., the integral

$$\frac{1}{2\pi} \int_0^{2\pi} N^+(X, \alpha) d\alpha$$

Thus the convexity number $N^+(X)$ is just the average value of the numbers of first contact points. Moreover, if ∂X has radii of curvature $R(x)$, $N^+(X)$ is the sum of their positive values over ∂X:

$$N^+(X) = \frac{1}{2\pi} \int_0^{2\pi} N^+(X, \alpha) \, d\alpha \left(= \frac{1}{2\pi} \int_{R>0} \frac{du}{R}, \quad \text{if } R \text{ exists} \right) \quad \text{(V-21)}$$

Similarly, denote by $N^-(X, \alpha)$ the number of first contact points with the *pores* in direction α, and by $N^-(X)$ the corresponding average value in α (we call $N^-(X)$ the convexity number of the pores $\overline{X}^c$; in fact, N^+ and N^- are both parameters associated with the *boundary* ∂X):

$$N^-(X) = \frac{1}{2\pi} \int_0^{2\pi} N^-(X, \alpha) \, d\alpha \left(= \frac{1}{2\pi} \int_{R<0} \frac{du}{|R|}, \quad \text{if } R \text{ exists} \right)$$

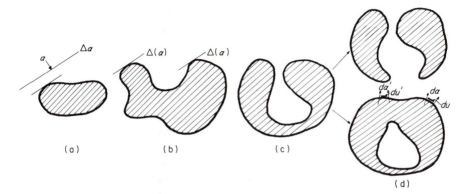

Figure V.9. Progressive individualization of two sub-grains. (a) $N^+ = 1.01$; $N^- = 0.01$; (b) $N^+ = 1.5$; $N^- = 0.5$; (c) $N^+ = 1.8$; $N^- = 0.8$; (d) (top) $N^+ = 2.3$; $N^- = 0.3$; (bottom) $N^+ = 1.1$; $N^- = 1.1$.

Figure V.10. (a) The convexity number N^+ is not C-additive. In direction North-South, sets $X \cup X'$ and $X \cap X'$ respectively have two (x_1 and x_2) and one (x_3) entry points. However $N^+(X) = N^+(X') = 1$. (b) Example of a number of negative convexity $N^{(2)-}$, which equals 1 $- \theta/2\pi$ in z (a) and $- 2\pi$ in z' (b).

Note that the convexity numbers are not C-additive (see Fig. V.10).

More generally, let $X \in \mathcal{S}(\mathcal{K})$ in R^n; let $\Pi(r, \omega)$, $r > 0$, denote the hyperplane orthogonal to direction ω and including the point $r\omega$, and let Y_i denote the connected components of $X \cap \Pi(r, \omega)$. One of those components, say Y_{i_0} is an *entering set* in X, w.r. to direction ω, if there exists an open and connected set $G \supset Y_{i_0}$ such that $G \cap Y_i = \varnothing$ for $i \neq i_0$ and $G \cap X \cap \Pi(r - \varepsilon, \omega)$ $= \varnothing$ for any $\varepsilon > 0$ small enough. The convexity number $N^+(X, \omega)$ is the number of entering sets in direction ω, and $N^+(X)$ is the rotation average of $N^+(X, \omega)$.

In $\mathbb{R}^2$, the gap between the convexity and connectivity numbers is easy to bridge. The latter turns out to be the difference of two integrals with respect to measures of convexity $n^{(2)+}(dx)$ and $n^{(2)-}(dx)$, respectively associated with X and $\bar{X}^c \in \bar{\mathcal{S}}(\mathcal{F})$:

$$N^{(2)}(X) = N^{(2)+}(X) - N^{(2)-}(X) = \int n^{(2)+}(dx) - \int n^{(2)-}(dx) = \int n^{(2)}(dx)$$

Therefore, we can define a connectivity measure as the difference $n^{(2)} = n^{(2)+} - n^{(2)-}$ of the two convexity measures associated with X and $\bar{X}^c$. (Note that if X is made up of disjoint particles possibly with holes, the number of particles is an exclusively global parameter, but $N^{(2)}(X)$ is not).

C. THE REGULAR MODEL

We saw that within the framework of $\mathscr{S}(\mathscr{F})$, some Minkowski functionals could be discontinuous. As a matter of fact, the *angularities* encountered in the sets $X \in \mathscr{S}(\mathscr{F})$ make the digitalization of such sets very difficult. The regular model $\mathscr{R}(\mathscr{K})$ provides an alternative, which is particularly useful when studying connectivity questions (Ch. VI, C.4, Ch. XI, G.1) or radii of curvature (Ch. X, Ex. 14). Although we restrict its presentation to the compact case in $\mathbb{R}^2$ (in order to make a parallel with the integral geometry of the convex ring), it can easily be extended over $\mathscr{F}$ in $\mathbb{R}^n$, by using the curvature measures.

Definition: The regular model $\mathscr{R}(\mathscr{K})$ is the class of compact sets X that are morphologically open and closed with respect to the compact ball rB of radius $r > 0$:

$$X = X_{rB} = (X \ominus rB) \oplus rB = X^{rB} = (X \oplus rB) \ominus rB \qquad \text{(V-22)}$$

Properties: (1) *Normals*. Take an arbitrary point x on the boundary ∂X. The two compact balls λB which contain x and are respectively included in X and $\bar{X}^c$ are necessarily tangential at the point x. Thus ∂X possesses a normal at the point x.

(2) $\mathscr{R}(\mathscr{K})$ *is closed in* $\mathscr{K}$. Firstly, it is easy to see that every $X \in \mathscr{R}(\mathscr{K})$ is a topologically open-closed set, i.e. $X = \overline{\mathring{X}}$ (each $x \in X$ is a limit of interior points of X). Now, let $X_i \to X, \{X_i\} \in \mathscr{R}(\mathscr{K})$. We have (i) $X = \lim X_i = \lim[(X_i)_{rB}]$ $\subset X_{rB}$ (the opening is semi-continuous); (ii) $X_{rB} \subset X$ (anti-extensiveness) thus $X = X_{rB}$. The same proof, applied to $\mathring{X}$, shows that $\mathring{X} = (\mathring{X})^{rB}$, but we also have $X = \overline{\mathring{X}} = \overline{(\mathring{X})^{rB}} = (\overline{\mathring{X}})^{rB} = X^{rB}$.   Q.E.D.

(3) *Homotopy*. Let $X \in \mathscr{R}(\mathscr{K})$ in $\mathbb{R}^2$. The dilation $X \oplus \lambda B (0 < \lambda < r)$ may modify the homotopy of X either by suppressing small holes in X^c, or by creating new bridges in X (Fig. V.11). By definition, each hole X_i^c of X contains at least one disk $r\mathring{B}$. After dilation, the new hole $X_i^c \ominus \lambda B$ contains $(r - \lambda)B$ and thus is non empty. To study possible new bridges, we will firstly define the notion of a *neck* as follows. The set X^c shows a neck in the bounded open zone D if $X \cap D$ contains at least two disjoint connected components X_1 and X_2 and if there exists $r_0 > 0$ such that $A(r_0) = (X_1 \oplus r_0 B) \cap (X_2 \oplus r_0 B)$ is non

(a)   (b)

Figure V.11. (a) a non allowed bridge of $X \oplus \lambda B$. (b) one-to-one correspondence between open zones of the boundaries.

empty and is included in D. In other words, a certain size of dilation generates an overlapping zone, included in D (Fig. V.11). Let r' be the smallest radius of a disk such that $A(r') \neq \varnothing$. The interior of $A(r')$ is empty (if not, we would have $A(r' + \varepsilon) \neq \varnothing$). Therefore for any $y \in A(r')$ the disk $r'B_y$ centred at the point y, is tangent to X_1 at x_1 and to X_2 at x_2. On the other hand, $\overline{X}^c$ is open with respect to rB, and any disk rB containing x_1 and included in $\overline{X}^c$ is necessarily tangent at x_1; this implies $r \leq r'$. Therefore, for any $\lambda < r$, we have $(X_1 \oplus \lambda B) \cap (X_2 \oplus \lambda B) = \varnothing$. Small dilations do not change the homotopy of X, and by duality, neither do small erosions.

(4) *Minkowski measures and stereology.* The regular model is a subset of $\mathscr{F}(\mathscr{K})$. Therefore, its elements possess Minkowski functionals satisfying the stereological relations (IV-17 to IV-19) and also Minkowski measures. But the simplicity of the model enables definition of these measures directly. Consider an outside point x at a distance $d < r$ from ∂X. It almost always has only *one* projection onto ∂X, namely the foot of the normal to ∂X going through x (except for points such as centres of spherical cavities for example, however they are negligible under Lebesgue measure). Then the local approach developed for Minkowski measures of convex sets in Chapter IV, D.4 is still valid. In particular, for any compact convex set C such that rB is open with respect to it (i.e. $rB = (rB)_C$), Steiner's formulae for dilation and for erosion are applicable.

For example, taking for C the ball $\lambda B (0 < \lambda < r)$, we obtain

$$V(X \overset{\oplus}{\underset{\ominus}{}} \lambda B) = V(X) \pm \lambda S(X) + \lambda^2 M(X) \pm \tfrac{4}{3}\pi\lambda^3 N^{(3)}(X) \qquad \text{(V-23)}$$

$$A(X \overset{\oplus}{\underset{\ominus}{}} \lambda B) = A(X) \pm 2\lambda U(X) + \lambda^2 \pi N^{(2)}(X) \qquad \text{(V-24)}$$

and similar relations for the other functionals (note in particular that the Euler–Poincaré constants of $X, X \oplus \lambda B$ and $X \ominus \lambda B$ are identical).

(5) *The Minkowski functionals are continuous.* Consider a sequence $X_i \to X$ in $\mathscr{R}(\mathscr{K}')$. The Hausdorff distance ρ between X and X_i tends towards zero and relation (V-22) implies $X_i \ominus \rho B \subset X \subset X_i \oplus \rho B$, and therefore $V(X_i) \to V(X)$. According to Steiner's relation (V-23), we see that $V(X_i \oplus \lambda B) \to V(X \oplus \lambda B)$ for every λ $(0 < \lambda < r)$, which implies that each coefficient of the polynomial in λ is continuous.   Q.E.D.

(6) *Curvature measures.* Radii of curvature might not exist. Consider for example in $\mathbb{R}^2$, an ellipse X opened by a ball ρB with radius in between the two extremum radii of curvature of the ellipse. Some parts of $\partial(X_{\rho B})$ are arcs of a circle, others are portions of ∂X left unchanged by the opening. At the junction point of two such parts the radius of curvature is not defined. Nevertheless these junction points are negligible with regard to the perimetric measure (resp. the surface measure in $\mathbb{R}^3$). It is precisely for this reason that the measure approach was used above to introduce the Minkowski functionals. However, when the radii of curvature always exist on ∂X, the functionals are integral of their symmetric functions, and we have, in $\mathbb{R}^3$:

$$M(X) = \frac{1}{2} \int_{\partial X} \left(\frac{1}{R} + \frac{1}{R'} \right) dS \qquad N^3(X) = \frac{1}{4\pi} \int_{\partial X} \frac{1}{R \cdot R'} \cdot dS$$

(R, R' are the main radii of curvature at the point $x \in \partial X$) and in $\mathbb{R}^2$:

$$N^{(2)}(X) = \frac{1}{2\pi} \int_{\partial X} \frac{dU}{R}$$

We will intentionally avoid working with curvatures of ∂X, for this model. Not only is it quite useless, as has been illustrated, but it is also non operational. A model is an efficient tool when we have some means at our disposal for testing it from digital images; obviously openings and closings are digitally accessible. It is not possible to say the same about curvatures.

D. THE CLASS $\mathscr{K}$ AND THE FRACTAL DIMENSIONS

In order to orientate ourselves among the confusion of models which confront us, we will again look at their reaction to isotropic closure. In the convex ring $\mathscr{S}(\mathscr{K})$, we encountered sets with interior angles and sufficiently irregular so that a closure by a ball ρB, no matter how small, never leaves the sets invariant. However these sets were not so pathological that they did not possess a surface area $S(X)$; recall that it was defined by first considering the volume of the dilated set $X \oplus \rho B$, and then setting it equal to a certain limit, i.e.:

$$\lim_{\rho \to 0} \left[\frac{V(X \oplus \rho B) - V(X)}{\rho} \right] = S(X) \qquad (V\text{-}25)$$

The next step in the direction of irregularity (and probably the last to possess any physical interpretation) is achieved when we no longer insist that we have a surface area. Here the model can become so chaotic that with the exception of the volume, none of the basic functionals can be defined.

The adjective "fractal" is a new word created by B. Mandelbrot (1977a), in order to denote extraordinary shapes with abrupt and tortuous edges. These shapes appear sometimes in natural phenomena; for example, consider the coast of Brittany with its sharpness and tortuousness, or the galaxies in space where myriads of stars are accumulated in one place and leave between them, large empty zones. In the fractal models, the notion which carries on from surface area, is that of dimension or content. Initially, this seems surprising since this notion may be approached in a number of ways that do not necessarily lead to the same result (Recall that this is not the first time we have seen such multi-possible generalizations: the connectivity number has been extended in two different ways when going from sets in $\mathscr{C}(\mathscr{K})$ to sets in $\mathscr{S}(\mathscr{K})$. See also the various mathematical definitions of the surface area in Federer (1969) which coincide for a convex body.) The reader will find in the book by Mandelbrot (1977) an excellent intuitive presentation of the main concepts of the notion of dimension within mathematics. He has no use for the Minkowski dimension (although he mentions it). However it is of interest to us since it is better adapted for the purposes of Mathematical Morphology. Our aims are modest; we want firstly to give the reader an idea about what is meant by a fractal object, so that he has a better feeling for what assumptions are being implicitly made about the real world, when he measures set boundaries. Secondly we would like to show that it is possible to design simple experiments, resulting in concrete objects, for which such sophisticated models provide good fits. It is interesting that it can indeed be done, thus warranting the subject a good deal of consideration (more than will be given here).

D.1. The Minkowski Dimension

How can the boundary ∂X of a compact set be measured? The original idea of Minkowski (1903) was to reduce the question to a simpler one, involving the measuring of volumes. For every compact convex set Y, symmetrical with respect to the origin (i.e. such that $Y = \check{Y}$) a surface area of the compact set X can be defined according to the formulae:

$$
\left.
\begin{aligned}
S(X,Y) &= \underline{\lim} \; \frac{V(\partial X \oplus \rho Y)}{2\rho} \\
S^*(X,Y) &= \overline{\lim} \; \frac{V(\partial X \oplus \rho Y)}{2\rho}
\end{aligned}
\quad \rho \to +0 \;
\right\}
\qquad \text{(V-26)}
$$

and since $V(\partial X) = 0$, it is essentially just relation (V-25), the approach taken by us up to now. The numbers S and S* are said to be the lower and upper surface areas of X with respect to Y. Notice that it is quite possible that they are both finite but do not coincide. In the sufficiently regular classes $\mathscr{R}(\mathscr{K})$ and $\mathscr{F}(\mathscr{K})$, the surface areas $S(X,Y)$ and $S*(X,Y)$ are identical. But it happens in some cases that both S and $S*$ are infinite (Von Koch snowflake, for example . . .). Minkowski overcame this difficulty by introducing the concept of a *dimension* of the set X, with respect to Y. For any $\rho > 0$ and for $-\infty < \tau < +\infty$, the following ratios are introduced:

$$q_\tau(X,Y,\rho) = \frac{1}{\rho^{n-\tau}} V(\partial X \oplus \rho Y) \qquad \text{(V-27)}$$

Then, for $\rho \to 0$, take the two limits:

$$q_\tau(X,Y) = \underline{\lim} \ q_\tau(X,Y,\rho); \qquad q_\tau^*(X,Y) = \overline{\lim} \ q_\tau(X,Y,\rho)$$

and put:

$$d(X,Y) = \inf\{\tau, q_\tau(X,Y) = 0\}; \ d*(X,Y) = \inf\{\tau, q_\tau^*(X,Y) = 0\}$$

The two numbers d and $d*$ always exist, and satisfy the inequalities:

$$0 \le d(X,Y) \le d*(X,Y) \le n \qquad \text{(V-28)}$$

They are called the lower and upper dimensions of the set X, with respect to Y. They may be distinct from each other. When $d(X,Y) = d*(X,Y) = n-1$, the set ∂X is said to be *superficial* with respect to Y. Then the two surface areas of (V-26) are finite and equal to each other.

The existence of two simultaneous dimensions in the compact set model brings to our attention, the possibility of large measurement fluctuations. Thus, variation in the experimental values of d can be caused either by a sampling error, or by the intrinsic irregularity of the object itself. This basic uncertainty makes statistical inference for fractal objects a difficult task. It is worth saying at this point that for all studies we made within this framework, the estimation procedure found just one d, which seemed to be an extremely robust estimator (perhaps even too robust . . .).

D.2. An application—the nested disks

As a theoretical example, once again consider the nested disks (Fig. V.1), introduced as a counter-example at the beginning of this Chapter. Firstly consider the case when the radii decrease according to:

$$r_i = \left(\frac{1}{4}\right)^i r_0$$

Let us focus our attention on the boundary ∂X of the resulting set. We can find positive numbers ρ such that the dilate $\partial X \oplus \rho B$ (B is the unit disk) is made up of disjoint circular crowns up to order i, and beyond order i, of 4^{i+1} smaller disks surrounded by their crowns (the smaller particles of X are clustered together by the dilation ρB). For example, the above claim holds true if we take $\rho_i = (1.2)r_{i+2}$. The area of the dilate is then:

$$A(\partial X \oplus \rho B) = 4\pi\, i\, \rho\, r_0 + 4^{i+1}[r_0(4)^{-i-1} + \rho]^2 \qquad \text{(V-29)}$$

We see that the characteristic ratio $\rho^{\tau-2} \cdot A(\partial X \oplus \rho B)$ tends to infinity when $\tau \geq 1$, and tends to zero otherwise, resulting in the following paradox:

(a) The Minkowski dimension of ∂X is one
(b) The total length of ∂X is infinite!

Imagine now that the circle radii decrease more slowly, and that instead of the factor 0.25, we use 0.27:

$$r_i = (0.27)^i r_0 = k^i r_0$$

Once again, it is easy to verify that the dilates $\partial X \oplus \rho B$ are made up of finite unions of disjoint crowns and circles. The choice of $\rho_i = (1 - \sqrt{2}/2)r_{i+1} = \theta r_{i+1}$ gives such images. Then an easy calculation shows:

$$\rho_i^{n-\tau} \cdot A(\partial X \oplus \rho B) = \rho_i^{\tau-1} \frac{4\pi r_0[(4k)^{i+1} - 1]}{4k - 1}$$

$$+ \rho_i^{-2} 4^{i+1} \cdot k^{2(i+1)} \cdot \pi r_0^2 \left(1 + \frac{\theta}{k}\right)^2 \qquad \text{(V-30)}$$

The expression (V-30) diverges for $\tau > \log 4/\log(1/k)$, tends to zero for $\tau < \log 4/\log(1/k)$ and tends to a finite non-zero limit, when $\tau = \log 4/\log(1/k)$. Thus the dimension d is:

$$d = \frac{\log 4}{\log(1/k)} = \frac{\log 4}{\log 3.7} = 1.06$$

The Minkowski dimension of ∂X now lies between one and two, and hence its length is infinite!

This example shows the kind of results we can expect from fractals. When dealing with fractals experimentally, the following three steps outline the approach:

(a) Produce a sequence of finer and finer images of the same object by increasing the optical or digital resolution. Each magnification corresponds to a ρ_i (i.e. the power of resolution).

(b) For any ρ_i, compute the "length of the boundary set" defined by the following algorithm

$$L(\partial X, \rho_i) = \frac{A(\partial X \oplus \rho_i B)}{2\rho_i}$$

(Note that the dilation can be carried out digitally; e.g. Coster and Deschanvres (1978); or be directly produced from the photographs themselves as in the example below.)

(c) On log–log paper, plot the quantity $\log L(\partial X, \rho_i)$ against $\log(\rho_i)$.

We saw that there are three types of behaviour of the curve when $\rho_i \to 0$ (Fig. V.12):

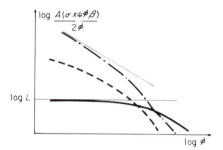

Figure V.12. The three possible behaviours of a dilated boundary.

(a) The boundary ∂X is sufficiently regular; $\log L(\partial X, \rho_i)$ tends towards a horizontal asymptote with ordinate equal to the logarithm of the perimeter of X. The dimension d is equal to $n - 1$ (e.g. for the plane, $d = 1$).

(b) The curve $\log L(\partial X, \rho_i)$ does not asymptote, but shows logarithmic behaviour for negative values of $\log(\rho_i)$ (i.e. ρ_i small). In this case, the perimeter of X is infinite, but the dimension d remains equal to one ($n - 1$ in $\mathbb{R}^n$).

(c) The curve $\log L(\partial X, \rho_i)$ tends towards an oblique asymptote. According to Minkowski's results above, in $\mathbb{R}^2$ the slope c of this asymptote is necessarily between zero and minus one.

The reason is simple:

$$\log\left[\frac{A(\partial X \oplus \rho B)}{2\rho}\right] \to (1 - d)\log \rho + K$$

$$(K = \text{bounded constant}) \tag{V-31}$$

and in the plane, $1 \le d \le 2$. By estimating the slope c from the graph, we determine the dimension d to be $1 - c$. Note that relationship (V-31) is an *experimental* version of relationships (V-26) and (V-28). The symbol "→" here means that the curve, plotted in log–log co-ordinates looks like a straight line as $\rho \to 0$.

D.3. Discussion

Can the Minkowski dimension d, be considered as a morphological parameter? It is invariant under translations, under rotations (when the Y chosen in (V-26) is a ball), and even under magnifications. Note that when we change the scale, we shift the curve (expressed in logarithmic co-ordinates) but we do not change the slope of the asymptote. Therefore, not only is the dimension d compatible with magnifications (in the sense of (IV-E)) but it is independent of them. We will assume that the Minkowski dimension, given above in a global approach and for compact sets, can be extended in a local manner (see Mandelbrot, 1977, and Matheron, 1975, Ch. 3) to random closed sets, which are usually infinitely divisible and of dimension different from $n - 1$. Also, the semi-continuity property completely loses its importance; recall that it was a consequence of the constraint that the boundary of the transform is the transform of the boundary. And from the fractal point of view, the boundary escapes measurement no matter how infinitesimal the scale.

Let us return now to the interesting property of independence with respect to magnifications. For ρ infinitely small, the variation of the log of the Lebesgue measure of $\partial X \oplus \rho B$, no longer depends on the scale: this logarithm increases as much for ρ ranging from ρ_0 to $\rho_0/2$, as for $\rho_0/100$ to $\rho_0/200$, since for ρ small the behaviour of the log–log graph is essentially linear. We can say that in some sort of way, the set ∂X presents a certain self-similarity, and each small part of the boundary possesses the same internal information as any piece of it one tenth or one hundredth of the size (from H. Von Koch to Mandelbrot via P. Lévy, this somewhat mystical property has seduced many minds). One can extrapolate this asymptotic self-similarity and construct self-similar models, starting from a basic boundary and cascading down (the snowflake of H. Von Koch) or up (the Poisson plane process; see Ch. XIII). But these properties are *a lot stronger* than the simple self-similarity observed when we discussed the notion of dimension. They treat the whole space at once, and do not only consider the behaviour of the set with respect to infinitesimal structuring elements. In other respects, some self-similar structures, such as the Poisson plane process, are simple enough to possess a finite specific surface area: self-similarity does not imply fractal structure, nor conversely.

A last remark. Why did we choose the Minkowski dimension over the others? The reason for this choice is mainly that it, and only it, explicitly involves a structuring element and a dilation. The advantage allows us to use two different experimental techniques that both result in the same method of fractal analysis. These techniques are based on the opto-photographic limits of resolution, and on digital dilation. Also, the Minkowski dimension is well adapted for estimation on an automatic device; the other dimensions such as the content dimension, Hausdorff dimension, etc. . . . are based on covering ∂X by balls in $\mathbb{R}^3$ (or discs in $\mathbb{R}^2$) and hence they do not seem as easy to handle experimentally. Finally, we are free to choose the structuring element (among the symmetric compact convex sets) that is in some way "optimal" for a given situation. For example, if in the illustration of the preceding section we took the snowflake of Von Koch instead of the noticed disks, a regular hexagon would have been better adapted to the symmetry of the phenomenon, than a circle. On a square (or hexagonal) grid, the digitalization of the image means that boundaries can only be made up of lines in two (or three) directions. So it is fortunate that the mathematical formalism leaves us with the freedom to take the square (or the hexagon) as a structuring element for estimating the dimension.

The stereological advantage is even more obvious; for the other dimensions such as the Hausdorff dimension for example, when the set X is defined in $\mathbb{R}^3$, we have to find the minimal covering by *balls* of a fixed radius ρ. How can we proceed in practice, when X is only known by random section? With the Minkowski dimension, the problem is solved. Suppose we can only know the three-dimensional set X via random sections. Then to estimate a Minkowski dimension of the boundary of the section, we might for example use a disk as the structuring element. Seen from three dimensions, we are in fact estimating the Minkowski dimension of ∂X with the *flat* (i.e. zero thickness) disc as structuring element. Of course, this dimension will be different from the Minkowski dimension of the boundary of the same X, with the ball as structuring element. (Both are interesting; the problem remains how we might, for a given X, link the two values.)

D.4. The fractal behaviour of clay material

The following example, which is taken from a more comprehensive study of the microstructures of clay rocks (Kolomenski and Serra, 1975), will illustrate how to test for the convex ring model, and, because the model will be rejected, how then to estimate a fractal dimension.

In engineering geology, the relationship between microstructures and physical properties of clay, such as its water absorption, play a crucial role (Sergueev, 1971) (Galadkovskaïa, 1968). The clay specimens are usually observed under an electron microscope scanner. This is due to the opaqueness of the thin sections, and the large range of possible particle sizes of the clay. In the present case, a group of five specimens has been scanned under the five different magnifications $\times 10^2$; $\times 3\cdot10^2$; $\times 10^3$; $\times 3\cdot10^3$; $\times 10^4$. For each of these magnifications, we looked at about twenty fields per specimen. The five specimens exhibit rather similar fractal structures, so we will only present the results of two: series A and B. Mineralogically speaking, the two series are identical; they have been sectioned at different depths, normal to the sedimentation plane, from the same homogeneous geological horizon. By physical methods, the following have been measured: their size distribution (by sedimentation in a liquid phase), their water content W_r (by drying), their porosity P and their water saturation percentage W_f (W_f is the point beyond which the clay becomes liquid). As we see from the table in Figure V.13, the first two physical parameters are different and the last two about the same. Along with this table, Figure V.17 shows two views of specimens A and B, each photographed at magnifications $\times 3,000$ and $\times 10,000$. Obviously, the two materials are different; in each case, the lower magnification smooths out small details that only appear with high resolution.

		Size Distribution (in μ)					W_r	W_f	P
	Depth (in m)	>0.25	0.25 −0.1	0.1–0.05	0.05–0.005	<0.005	(%)	(%)	(− %)
A	3.05–3.75	0	0.3	5.5	13.8	80.4	2.9	55	46.2
B	5.70–6.05	0	0.1	2.0	31.8	66.1	3.4	55	42.9

Figure V.13. Physical properties of the clays under study.

The interplay of light and shadow give the photographs a certain aesthetic appeal, however this leads to inherent experimental problems. The interpretation of the grey tones is ambiguous. A clay particle is lighter or darker according to its depth in the sample, its orientation, and whether or not it is blocked by another particle. Angles, edges, and the small fragments re-emit more electrons and hence look lighter than flat zones. Moreover, when a point is dark, we cannot decide whether it belongs to a dark particle or is in a shadow.

If we were trying to estimate the specific surface area of the clay, the ambiguities of the scanner image would be inhibitory to the results. The apparent number N_L^* of particles intersected by a line randomly drawn on the photograph is certainly a biased estimator of the actual number N_L of intersections on the line; N_L is needed to calculate the specific surface area. Unfortunately $N_L^* = K \cdot N_L$, where K is unknown.

Suppose we want to test the model $\mathscr{S}(\mathscr{F})$. We assume that for a given specimen, the coefficient K is approximately constant (i.e. the proportion of the cast shadows does not change as we increase the magnification), but we do not need to know its numerical value. Now testing the model consists of seeing whether N_L is independent of the magnification, which is equivalent to seeing whether N_L^* is constant. We derive N_L^* from the covariance $\mathrm{Cov}(h)$ of the light intensity (see Ch. IX, and Fig. V.14).

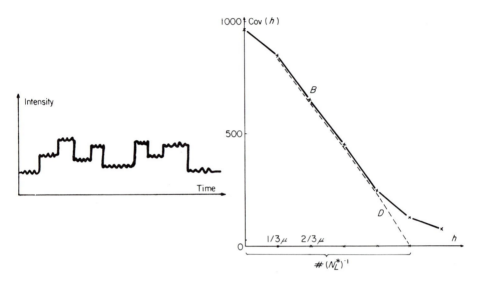

Figure V.14. (a) Example of the grey intensity along a scanning line. Electronic smoothing gives the vertical jumps a less abrupt appearance. (b) mean covariance for the specimen B, at magnification $\times$ 1000.

If the jumps of light intensity were truly vertical, the behaviour of the covariance at the origin would be linear. Here, smoothing of the image makes the value of the covariance at the origin, smaller; the linear behaviour begins at the second point. $(N_L^*)^{-1}$ is the h-intercept of the "tangent" D. Note that the y unit (a product of light intensities) is independent of the value $(N_L^*)^{-1}$. In the present case we find $N_L^* = 606 \ \mathrm{mm}^{-1}$. For the two specimens A and B, the estimated N_L^*, expressed in mm^{-1}, are as in Figure V.15.

Specimen	Magnification rate				
	10^2	3.10^2	10^3	3.10^3	10^4
A	31.5	76	159	330	870
B	25	91	235	606	1200

Figure V.15. Tables of N_L^*.

Obviously N_L^* cannot be considered a constant; It is indeed a function $N_L^*(G)$ of the magnification rate G. The range of variation is five thousand per cent for a given specimen! We therefore have no qualms about rejecting the convex ring model.

What then can be said about the fractal structure of clay? If we could work with an infinite resolution, the resulting signal would be very sharp. In fact each magnification is just a moving average of this ideal signal, which removes particles smaller than the power of resolution, and which smooths the vertical jumps as in Figure V.14a. We can interpret this elimination in a non-rigourous (but adequate) manner by assuming that $N_L^*(G)$ is proportional to the quantity $A(\partial X \oplus \rho B)/2\rho$ which occurs in relation (V-31). Here ∂X is the infinite apparent contour of all the particles of specimen. The quantity $\log \rho$, which also appears in (V-31) is just $-\log G$, modulo an additive factor. Hence, we can re-write the relation in the more accessible form:

$$\log N_L^* = (d-1)\log G + K' \tag{V-31}$$

The experimental data is compatible with the above relation, thus the clay microstructure can be fitted to a fractal model with a Minkowski dimension d (of course, within the limits of the magnifications investigated). Figure V.16 shows the plots of the two tables of N_L^*, in log–log co-ordinates. To a first approximation, $\log N_K^*$ looks like a linear function of $\log G$, particularly for group A. The mean slopes, taken from the middle three points, are:

$$d-1 \simeq 0.62 \quad \text{for} \quad A$$
$$d-1 \simeq 0.80 \quad \text{for} \quad B$$

If we now estimate the same slope from the first and last points, it becomes 0.60 for A and 0.54 for B. The self-similarity hypothesis seems acceptable in the first case, but questionable in the second. In the latter case, the fluctuations of d are statistically significant (the number of photographs guarantees a low variance of the results). Moreover, by comparing Table V-13 and Table V-15, we see immediately that d varies with the size distribution of the clay particles for B but not for A.

What general conclusions can be deduced from this analysis? We choose an example from the field of scanning microscopy because the Minkowski dimension is particularly suitable to this technology: the scanning microscope covers a large range of magnifications, and produces quick images in situations where a direct interpretation is ambiguous. We could push the structural description further and construct a random set model (for example the dead leaves model, Ch. XIII). It would show us that the Minkowski dimension of the particles in $\mathbb{R}^3$, can be obtained by adding 1 to the Minkowski dimension of the projections, and hence should vary between 2.54 and 2.80. The model would also provide us with algorithms connecting observed projections to the

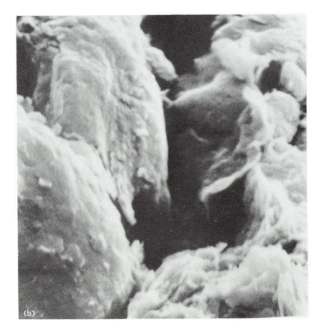

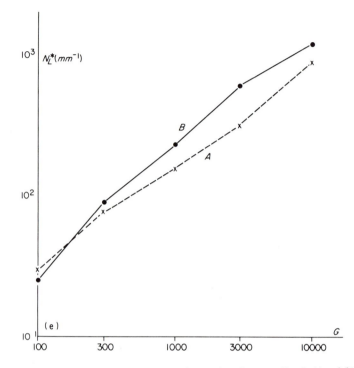

Figure V.16. Fractal dimension of the clays (see also previous 2 pages). Clay A: (a) and (b). Clay B: (c) and (d). (a) G = 3000; (b) G = 10 000; (c) G = 3000; (d) G = 10 000.

size distribution of the 3-D particles. More generally (i.e. with or without a stochastic model), we can always study the fractal objects via convex structuring elements. The area proportion of a dilation, or an opening, always exists and leads to more structural information than just the Minkowski dimension.

E. EXERCISES

1. Stereology on particle counting

Intuitively, one could guess that the more particles there are in the 2-D section of X, the more there are in X itself. In fact, if there are N_V identical particles per unit volume, then relation (V-12), expressed in specific terms, shows that the number $N_A(\omega)$ of particles per unit cross section area satisfies

$$N_A(\omega) = N_V D(\omega)$$

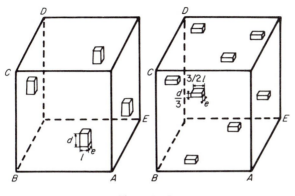

Figure V.17.

where $D(\omega)$ is the width of the elementary 3-D particle in direction ω. As an illustration, consider the following two cubes, each of unit volume (Fig. V-17).

In both cases, compute the mean number N_A for horizontal sections. Interpret the result [the larger is N_V, the smaller is N_A].

2. Uses of the convexity number

(a) Figure V.18a is a typical ferritic structure seen on a polished section of steel. Some edges are rectilinear, others are not. We want to measure the

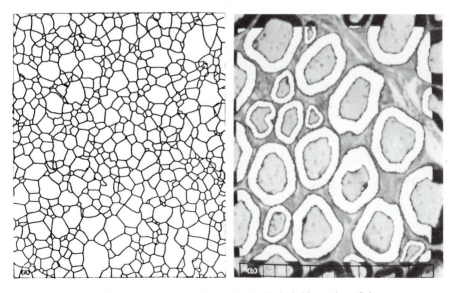

Figure V.18. (a) cross-section of steel. (b) histological thin section of the eye.

mean curvature $\bar{C}$ along the boundary, per unit area. Show that:

$$\bar{C} = \int \frac{du}{|R|} = N_A^{(2)} - N_A^+$$

(b) Some objects (Fig. V.18b) or some techniques, (such as phase contrast in optics) emphasize the edges of the objects and make them annular:

Show that if the particles are roughly convex, the convexity number N_A^+ may serve as an algorithm for particle counting, when the connectivity number is identically zero.

3. The norm and the 2-D connectivity number in the regular model

A set $X \in \mathcal{R}(\mathcal{K})$ is intersected by the test plane $\Pi(x, \omega)$ (∂X is assumed to possess finite curvatures everywhere). Let z be a point of $\partial X \cap \Pi(x, \omega)$; ds and R are the elementary arc and the curvature in z respectively. R_1 and R_2 are the radii of curvature of ∂X in z, β the angle of the tangent in dS with the first curvature direction, and θ the orientation of the outside normal to ∂X in z (see Fig. V.19).

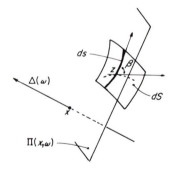

Figure V.19.

Prove the stereological result

$$M(X) = \frac{1}{2} \int_{4\pi} D^{(3)}(X, \omega) d\omega$$

[According to the classical Euler and Meusnier theorems, we have

$$\frac{1}{R} = \left(\frac{\cos^2 \beta}{R_1} + \frac{\sin^2 \beta}{R_2} \right) \frac{1}{|\sin(\theta - \omega)|}$$

Hence:

$$D^3(X, \omega) = \frac{1}{2\pi} \int_{\partial X} \left(\frac{\cos^2 \beta}{R_1} + \frac{\sin^2 \beta}{R_2} \right) dS$$

Therefore $D^3(X, \omega)$, being bounded, always exists. To average over ω, define the orientation of Π by the product of the two independent random vectors:

(a) one vector in the plan e tangent to X in z (uniform orientation in β);
(b) one perpendicular vector (uniform orientation in the plane perpendicular to β).]

PART II.   PARTIAL KNOWLEDGE

VI. Digital Morphology

A. INTRODUCTION

Most image analysis technologies resort, at least partially, to digital means, both for storing images and operating on them. Thus an image under study may be represented by a finite number of values which are regularly displayed over the space. They may correspond to discrete sampling points such as the vertices of a rectangular grid, or contiguous elements of mosaics (pixels (picture elements)). Adopting these experimental approaches poses for us firstly the problem of constructing a digital morphology and then secondly, making the connection with the Euclidean morphology already studied in Part II. This chapter and the next one cover these two considerations, and because it is the most important in practice, the 2-D case receives the most emphasis.

To digitalize, it is not enough merely to replace a Euclidean figure by a mosaic of pixels, as occurred in Chapter II for example (Figs II.1 . . . II.7). We also have to replace Euclidean translations, rotations, convex hulls, etc. by analogous versions and define what a connected particle means, what a distance means, etc. on a grid of points. A number of umbrella notions result. The first is the notion of a *module*, classical in linear algebra, which governs increasing transformations, projections and convexity analyses. Then there is our notion of a *grid*, which is the key to all the rotation and connectivity problems. Finally, by equipping the grid with a metric, we are able to talk about such notions as stereology, skeletons, etc. Someone interested only in applications might find these subtleties rather academic. However it is important to know which underlying *hypothesis* on the digital space we are implicitly choosing; for example, it is important to know that size distributions do not depend directly on metric analyses, although it is sometimes wrongly stated in the literature that they do (size distributions involve only the less demanding module structure). Several pertinent studies on the general

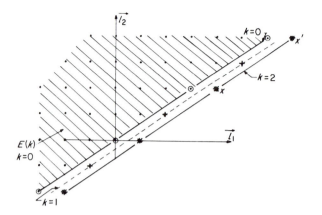

Figure VI.1. Points $x = (4, 2)$ and $x' = (4 + 3, 2 + 2)$ are two consecutive points on the straight line $2x_1 - 3x_2 = 2$. Note the half planes $E(h)$, successively delineated by the lines $2x_1 - 3x_2 = k (k = 0, 1)$.

properties of digital sets have already been written, although they do not use our classification, nor obtain the same results. Among the most representative we can quote that of A. Rosenfeld (1970, 1976) where connectivity and the square grid play a central role, and that of M. L. Minsky and S. Papert (1970).

B. MODULES AND THE ASSOCIATED MORPHOLOGY

Digital image data are generally labelled according to two integer parameters (row number and column number in a matrix). One often speaks of a grid of points, or of pixels to signify this spatially ordered structure. The corresponding mathematical notion is the *module*. The module structure gives us a digital version of such morphological notions as the Hit or Miss transformation, projection, convexity and of such parameters as the area and number of intercepts in a given direction.

B.1. The $\mathbb{Z}^2$ and $\mathbb{Z}^n$ modules

Let $\mathbb{Z}^1$ be the set of all integers (>0; <0 and $= 0$) and $\mathbb{Z}^2$ the set of ordered pairs of elements of $\mathbb{Z}^1$:

$$x \in \mathbb{Z}^2 \Leftrightarrow x = (x_1, x_2); \qquad x_1, x_2 \in \mathbb{Z}^1$$

For every $x = (x_1, x_2)$ and $y = (y_1, y_2)$ belonging to $\mathbb{Z}^2$ we define addition as follows:

$$x + y = (x_1 + y_1, x_2 + y_2)$$

Similarly scalar multiplication of x by λ is given by:

$$\lambda x = \lambda(x_1, x_2) = (\lambda x_1, \lambda x_2) \qquad \forall \lambda \in \mathbb{Z}^1 \qquad \text{(VI-1)}$$
$$\forall x \in \mathbb{Z}^2$$

Equation (VI-1) represents a mapping $\mathbb{Z} \times \mathbb{Z}^2 \to \mathbb{Z}^2$, which satisfies the following conditions:

(α) For all λ, $\mu \in \mathbb{Z}^1$ and x, $y \in \mathbb{Z}^2$, we have

$$(\lambda + \mu)x = \lambda x + \mu x \qquad \lambda(x + y) = \lambda x + \lambda y \qquad \text{(VI-2)}$$

(β) $\qquad\qquad (\lambda\mu)x = \lambda(\mu x)$

(γ) $\qquad\qquad 1x = x \qquad (1 = \text{unit element of } \mathbb{Z})$

Thus $\mathbb{Z}^2$ is a module over the ring $\mathbb{Z}^1$, which we denote by $\vec{Z}^2$; $\vec{Z}^2$ is commutative, and can be generated from any basis made up of two elements of $\mathbb{Z}^2$.

In algebra, the notion of a module generalizes the notion of a vector space over a field. (The module of course is defined over a ring, rather than a field). Geometrically, it has the following interpretation: let the points in the plane be represented as the linear combination of two orthogonal vectors Ox_1 and Ox_2 and suppose we consider all the points x whose coordinates (x_1, x_2) are integers. The $\vec{Z}^2$ is identified with the set of vectors Ox.

Although the above presentation was for the two-dimensional case, the definition can obviously be generalized to the n-dimensional space. The module $\mathbb{Z}^n$ defined over the ring Z, is the set of all n-tuples of integers satisfying the same axioms as previously; it has a similar geometrical interpretation in the Euclidean space $\mathbb{R}^n$.

We now wish to construct groups of displacements (i.e. rotations and translations) in the $\vec{Z}^n$-modules and in the associated affine spaces. In order to do this, it is necessary to first present some theorems on automorphisms, and on subsets of $\vec{Z}^n$. (An automorphism is an isomorphism of a set E, provided with an algebraic structure (S), onto the same set provided with the same structure.) By convention, we will use the same symbol (R for example) for a linear transformation and its associated matrix, i.e. $R(\vec{x}) = R\vec{x}$.

THEOREM VI-1. *R is an automorphism of $\vec{Z}^n$ if and only if $|\text{Det } R| = 1$.*

Proof: For any morphism R, $|\text{Det } R|$ is a *positive* integer r (all the terms of the matrix are integers). If $r > 1$, R does not possess an inverse, since $|\text{Det } R^{-1}| = 1/r$ would be less than 1. Conversely, let us assume $|\text{Det } R| = 1$. Let $\bar{R}$ be the morphism which extends R onto $\mathbb{R}^n$. Since $|\text{Det } R| \neq 0$, $\bar{R}$ is an automorphism. Let R^* be the restriction of $\bar{R}^{-1}$ to $\vec{Z}^n$; R^* is a map of $\vec{Z}^n$ into itself, since all the entries of the matrix R are integers and $|\text{Det } R| = 1$. Every $\vec{x} \in \mathbb{Z}^n$ satisfies the relation:

$$R^* \cdot R \cdot \vec{x} = \bar{R}^{-1} \cdot R \cdot \vec{x} = \vec{x}$$

and so the inverse mapping exists in $\vec{\mathbb{R}}^n$, namely R^*. Hence R is an automorphism.   Q.E.D.

COROLLARY 1. $(\vec{b^1}, \ldots, \vec{b^n})$ is a basis if and only if $\text{Det}(b_j^i) = \pm 1$.

Definition. We call a *primitive vector* or a *direction*, any element $\vec{a} \in \vec{\mathbb{Z}}^n$ whose coordinates, represented in terms of a given basis, are relatively prime. When $\vec{a}$ is fixed, and $k \in \mathbb{Z}^1$ variable, the set generated by the $k\vec{a}$'s is a sub-module of $\vec{\mathbb{Z}}^n$, called an *axis*.

COROLLARY 2. *Let R be an automorphism of $\vec{\mathbb{Z}}^n$. Then R is an invertible mapping from the set of the axes of $\vec{\mathbb{Z}}^n$ onto itself.*

Proof: First, we recall Bezout's identity, according to which n integers $k_1 \ldots k_n$ are relatively prime (i.e. their common denominator equals one) if and only if one can find n integers $a_1 \ldots a_n$ such that $\sum_1^n a_i k_i = 1$.

The set of axes of $\vec{\mathbb{Z}}^n$ isomorphic to the set of primitive vectors. Indeed, Bezout's identity shows that every automorphism R changes a primitive vector into another primitive vector. Conversely, if $\vec{b}$ is a primitive vector of the transformed space, then $R^{-1}\vec{b}$ exists and is a primitive vector of the initial space (Bezout's identity again).   Q.E.D.

Definition. Let $(\vec{a^1}, \ldots, \vec{a^n})$ be a given basis of $\vec{\mathbb{Z}}^n$. The sub-modules generated by $n-1$ of these vectors are called *hyperplanes*. The hyperplane $\bar{P}(\vec{a^1}, \ldots, \vec{a^n})$ is the set of points $\vec{x}$ satisfying the relation:

$$\vec{x} = k_1\vec{a^1} + k_2\vec{a^2} + \ldots k_{n-1}\vec{a^{n-1}}, \qquad \forall k_1, \ldots, k_{n-1} \in \mathbb{Z}^1 \qquad \text{(VI-3)}$$

which is equivalent to solving:

$$\Delta = \det \begin{bmatrix} x_1 & a_1^1 & \ldots & a_1^{n-1} \\ \vdots & & & \\ x_n & a_n^1 & \ldots & a_n^{n-1} \end{bmatrix} = 0 \qquad \text{(VI-4)}$$

By expanding Δ along its first column, one finds a linear form $f(\vec{x})$, defined on $\vec{\mathbb{Z}}^n$, whose kernel is precisely the hyperplane $\vec{P}$. This form characterizes $\vec{P}$, and may be written as follows:

$$f(\vec{x}) = A_1 x_1 + \ldots A_n x_n = 0 \Leftrightarrow \vec{x} \in \vec{P} \qquad \text{(VI-5)}$$

where the A_1 are the corresponding minors of the determinant Δ.

THEOREM VI-2. *A linear form $g(\vec{x})$ represents a hyperplane if and only if its coefficients are relatively prime.*

Proof: Let $g(\vec{x}) = B_1 x_1 + \ldots + B_n x_n$. If g is the equation of the hyperplane $\vec{Q}$, one can find a basis $\vec{b}^1, \ldots, \vec{b}^n$ in $\mathbb{Z}^n$ whose first $n-1$ vectors generate $\vec{Q}$. According to Corollary 1, the determinant of $\vec{b}^1, \ldots, \vec{b}^n$ (expressed on an arbitrary basis) is equal to ± 1. Thus one can write, by appropriate sign changes of the b_1^n's:

$$b_1^n B_1 + b_2^n B_2 + \ldots + b_n^n B_n = 1 \tag{VI-6}$$

According to Bezout's identity, equation (VI-6) shows that the B_1 are relatively prime. The converse argument is immediate by taking each step of the proof in reverse.

Note: The primitive vector $(B_1, \ldots, B_n)$ is termed the normal to the hyperplane.

B.2. Affine modules and translations

In order to construct a group of *displacements* in $\vec{\mathbb{Z}}^n$, we have to define the operation of translation. When starting from the vector space structure, the concept of translation is provided by the affine space associated with the vector space. Here, we shall proceed in a similar way, and construct an "affine module" from the module $\vec{\mathbb{Z}}^n$. This affine module is denoted by $\mathbb{Z}^n$ (without the arrow), and its elements are called "*points*". It is defined by the map $\mathbb{Z}^n \times \mathbb{Z}^n \to \vec{\mathbb{Z}}^n$, which associates with each pair (x, y) in $\mathbb{Z}^n \times \mathbb{Z}^n$ an element $\overrightarrow{x-y}$ of $\vec{\mathbb{Z}}^n$, satisfying the following conditions:

(a) For all $x, y, z \in \mathbb{Z}^n$, the Chasles relation holds:

$$\overrightarrow{y-x} + \overrightarrow{z-y} + \overrightarrow{x-z} = 0 \tag{VI-7}$$

(b) For all $a \in \mathbb{Z}^n$, the map $x \to \overrightarrow{x-a} = \vec{h}$ from $\mathbb{Z}^n$ into $\vec{\mathbb{Z}}^n$ is invertible.

These two axioms are analogous to those that define translations in the usual affine space. We can then conclude that for all $x \in \mathbb{Z}^n$ and all $\vec{h} \in \vec{\mathbb{Z}}^n$, there exists a unique $y \in \mathbb{Z}^n$, such that $\overrightarrow{y-x} = \vec{h}$. The point is called the translate of x by $\vec{h}$, and is denoted by $x + \vec{h}$:

$$y = x + \vec{h} \tag{VI-8}$$

The use of the addition symbol in (VI-8) is no accident; it has the usual properties of addition since, by the Chasles relation:

$$x + (\vec{h} + \vec{k}) = (x + \vec{h}) + \vec{k}$$

Thus the translations make up a commutative group of $1 - 1$ mappings of $\mathbb{Z}^n$ onto itself.

Definition: Let $Z_0 \in \mathbb{Z}_n$ be a point of $\mathbb{Z}^n$ and $\vec{P}$ be a hyperplane in $\vec{\mathbb{Z}}^n$. As point $\vec{x} \in \vec{P}$ spans $\vec{P}$, the set of points $x = x_0 + \vec{x}$ is said to be a hyperplane Q of $\mathbb{Z}^n$, parallel to $\vec{P}$.

THEOREM 3. *Let* $f(\vec{x}) = 0$ *be the equation of a hyperplane* $\vec{P}$ *of* $(\vec{\mathbb{Z}}^n)$. *The hyperplanes* $Q \in \mathbb{Z}^n$ *parallel to* $\vec{P}$ *are characterized by the equation* $f(x) = k$, $x \in \mathbb{Z}^n$, *where* k *is an arbitrary integer.*

Consequently, as k varies from $-\infty$ to $+\infty$, hyperplane $Q(k)$ sweeps out *the whole space,* encountering each point only once. Those points x such that $f(x) \leq k$ generate the half space $E(k)$. We have

$$E(k) \subset E(k+1) \qquad \forall k \in \mathbb{Z}^1 \tag{VI-9}$$

B.3. Basic morphological operations associated with modules

From now on, apart from the final section on $\mathbb{Z}^3$, we will deal only with $\mathbb{Z}^2$. The sets under study, which we continue to denote by X, B, etc., are subsets of $\mathbb{Z}^2$. Elementary notions such as complementation (X^c), inclusion, intersection and translation, are immediate. We obtain:

$$x \in X^c \Leftrightarrow x \in \mathbb{Z}^2, \qquad x \notin X \tag{VI-10}$$
$$X \subset Y \Leftrightarrow \forall x \in X \Rightarrow x \in Y \tag{VI-11}$$
$$X \Updownarrow Y \Leftrightarrow \exists x \in X \cap Y \tag{VI-12}$$
$$X_b = \bigcup_{x \in X} \{x + b\}, \qquad b \in \mathbb{Z}^2 \tag{VI-13}$$

Dilation and erosion immediately result:

$$X \ominus \check{B} = \bigcap_{b \in B} X_b, \quad X \oplus \check{B} = \bigcup_{b \in B} X_b,$$

and their algebraic properties already studied within the Euclidean framework in Chapter II, will carry over. In particular, Matheron's theorem stating that any increasing morphological mapping is a union of erosions (and an intersection of dilations), remains valid in $\mathbb{Z}^2$. Its use of course, is that it offers us a comprehensive range of digital transformations to choose from. The parameter which we call *area* $A(X)$, is defined to be the number of points of X, and thus is formalizable in this module approach.

B.4. Intercepts, projections

In order to describe the notion of an intercept, we first define the notion of a segment. Consider the points a and b. If $\overrightarrow{b-a}$ is a primitive vector, then the segment $[a, b]$ is made up of the two points a and b, which are said to be *consecutive points*. If $\overrightarrow{b-a}$ is not a primitive vector then the segment $[a, b]$ contains all the points between a and b on the straight line joining a to b:

$$[a, b] = \left\{ a + \frac{q}{p}\overrightarrow{(b-a)} \right\} q, p \in \mathbb{Z}; \quad q \le p; \quad \frac{q}{p}\overrightarrow{(b-a)} \in \mathbb{Z}^2 \quad \text{(VI-14)}$$

Let $X \in \mathscr{P}(\mathbb{Z}^2)$ be a set in $\mathbb{Z}^2$, $\alpha = (\alpha_1, \alpha_2)$ a direction in $\mathbb{Z}^2$ and Δ_α a straight line normal to α with equation $\alpha_1 x_1 + \alpha_2 x_2 = k$. The *intercepts* of $X \cap \Delta_\alpha$ are the segments and isolated points induced by X on the line. As the integer runs over Z, Δ_α completely sweeps out $\mathbb{Z}^2$ (Th. 3). The resulting union of intercepts is the digital equivalent of the total projection defined in Chapter V. (Note that although this "projection" parameter is well defined, it is not single-valued in general (Fig. VI.2).)

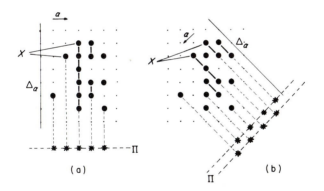

Figure VI.2. Intercepts of the same set X in two different directions α. The corresponding projections on a line Π parallel to α in the first case exhibit one mode, and in the second case two modes.

B.5. Convexity

Let $X \in \mathscr{P}(\mathbb{Z}^2)$. Define its convex hull $C(X)$ as the intersection of the half planes $H(\alpha, k)$ which contain X:

$$C(X) = \bigcap_{\alpha, k} \{ H(\alpha, k) : H(\alpha, k) \supset X \} \quad \text{(VI-15)}$$

We say that X is a *convex set* in the $\mathbb{Z}^2$ sense, when it is identical with its convex hull. A convex set is not necessarily bounded. This definition is not the only possible one, and we could for example be more analytical: if $x_i \in X$ and integers $\lambda_i \geq 0$ are such that $\sum \lambda_i = 1$, then since $x = \sum \lambda_i x_i \in \mathbb{Z}^2$, $x \in X$ if and only if X is a convex set. This barycentric approach is equivalent to the half plane approach if X is bounded.

In order to describe erosion and dilation, first notice two very asymmetrical properties. Since the intersection of two convex sets is convex, then an erosion of a convex set X by an arbitrary set B is again convex. On the contrary, the dilate of a convex set by another is not necessarily convex (Fig. VI.3a). As regards connectivity, we see that there is no special reason why a digital convex set is to be connected (Fig. VI.2b). Also note that the property x, $y \in X \Rightarrow \mathbb{Z}^2 \cap [x, y] \subset X$ follows from the convexity notion in $\mathbb{Z}^2$, however the converse is false, even when X is bounded (Fig. VI.3c). In practice, one often uses a more restrictive convexity, where only a finite number of digital directions α of the $H(\alpha, k)$ are used (VI-15). We define the convexity number of X in direction α as the number of first contact between X and a straight line sweeping the plane normal to α; by averaging over α we obtain the convexity number of X. An instructive use of digital convex hulls may be found in J. Slansky (1972).

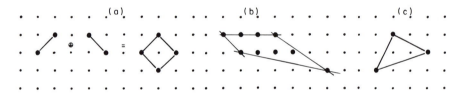

Figure VI.3. Paradoxes of digital convexity. (a) The dilate of a convex set by another is not necessarily convex. (b) Look at this convex set: do you think that it is connected? (c) for every pair $x, y \in X$, the segment $[x, y] \in X$, however, X is not convex!

C. GRIDS AND ROTATIONS

C.1. Definition of a grid

To describe digital equivalents of notions such as connectivity, or rotations, the datum of a module is too poor. As an illustration of the first point, consider the following well-known counter example presented in Figure VI.4 (Haas *et al.*, 1967).

Knowing just the points that make up X does not tell us enough about the connectivities of X and of X^c. As for rotations, start from a set X as defined in

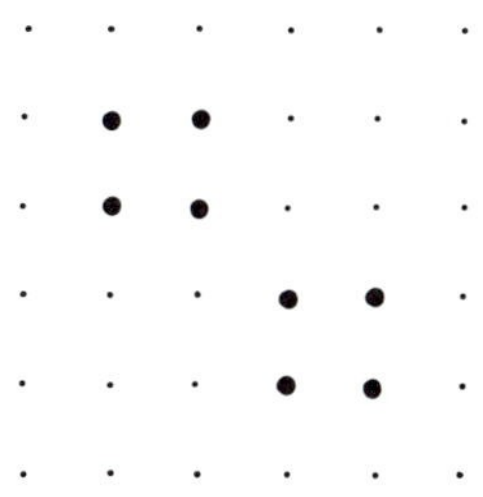

Figure VI.4. How many particles are contained in this set? Is it one or two? How many holes in the background? These questions cannot be answered from just the points.

Figure VI.5a, and try to turn it through an angle of $\pi/4$ around the point O. If we wish for the operation to respect the alignments of points (which seems to be a minimum morphological requirement) then the first point in direction α, namely B, should become the first point in direction $\alpha + \pi/4$, namely B', the line ABC becomes $A'B'C'$, etc. resulting in the set X'. But what about the points M and N in the space of the transform? We must either add them to X', resulting in the rotation algorithm not respecting area, or add them to X'^c, resulting in a hole inside X'. Finally, are we sure that a digital $\pi/4$ rotation is possible?

By enriching the concept of a module into that of a *grid*, we are able to handle the questions raised by these two counter examples.

The grid is the support on which the digital images are drawn up and through which, neighbouring points can be connected. The raw material for constructing such grids is $\mathbb{Z}^1$ or $\mathbb{Z}^2$ itself. In $\mathbb{Z}^1$, a grid $\mathscr{T}$ (Y, D) is defined by two sets:

(i) a subset $Y \subset \mathbb{Z}^1$ whose elements are called *vertices*
(ii) a subset $D \subset Y \times Y$ of pairs of points (y, y') whose elements are called edges.

The sets Y and D have to satisfy two conditions:

(i) there exists a sub-group t of translations in $\mathbb{Z}^1$ for which the grid, i.e. the two sets X and D are invariant (t-invariance condition)

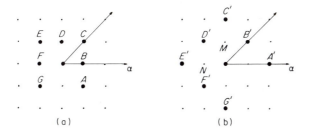

(a)   (b)

Figure VI.5. (a) set X; (b) set X', i.e. the "same", after a "rotation" of 45°.

(ii) if (y^1, y'^1) and (y^2, y'^2) are two arbitrary edges, the points of $\mathbb{Z}^1$ between y^1 and y'^1 must be disjoint from those between y^2 and y'^2 (non-crossing condition), i.e.:

$$(y^i - y^j)(y'^i - y'^j) \geq 0 \quad \text{and} \quad (y^i - y'^j)(y'^i - y'^j) \geq 0$$

$$\text{for} \quad i, j = 1, 2 \tag{VI-16}$$

In $\mathbb{Z}^2$ the translation invariance (condition (i)) remains, but with the supplementary condition that the translation vectors must act in at least two different directions. Moreover, if y_1^i and y_2^i denote the two co-ordinates of y^i, the non-condition (ii) must be satisfied by *at least one* of the set of co-ordinates y_1^i or y_2^i. Figure VI.6 shows several types of grids; others can easily be imagined. Note that the two grids (d) and (e) are identical; it is only their representation in the plane $\mathbb{R}^2$ which gives them two different appearances. Note also that the groups of translations are poorer for the octagonal and the dodecagonal grids than for the square and hexagonal grids.

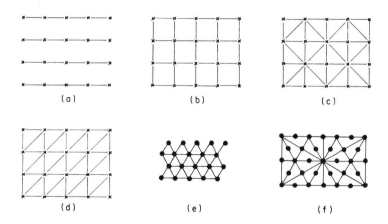

Figure VI.6. Examples of grids. (a) linear; (b) square; (c) octagonal; (d) hexagonal I; (e) hexagonal II; (f) dodecagonal.

The morphological reason for t-invariance is as follows: consider the restriction $\mathscr{T} \cap \Delta_\alpha$ of the grid $\mathscr{T}$ to a straight line Δ_α of $\mathbb{Z}^2$; $\mathscr{T} \cap \Delta_\alpha$ forms a grid in $\mathbb{Z}^1$. Insisting on t-invariance means that when we shift Δ_α anywhere by a t-translation, we get exactly the same pattern of grid on any Δ_α; in other words Δ_α can legitimately be used as a test line on the grid. In order to construct a group of *displacements*, our next task is to resolve how to define groups of rotations of the grids. Before doing this, we first have to analyse the conditions under which two straight lines in $\mathbb{Z}^2$ hit each other.

C.2. Principal directions

Hyperplanes in $\mathbb{Z}^2$ are just the straight lines. Two non-parallel straight lines can have an empty intersection; for example, consider the two lines D_1 and D_2 of Figure VI.7, defined by:

$$D_1 = \{x_1 = x_2\} \qquad D_2 = \{x_1 + x_2 = 1\}$$

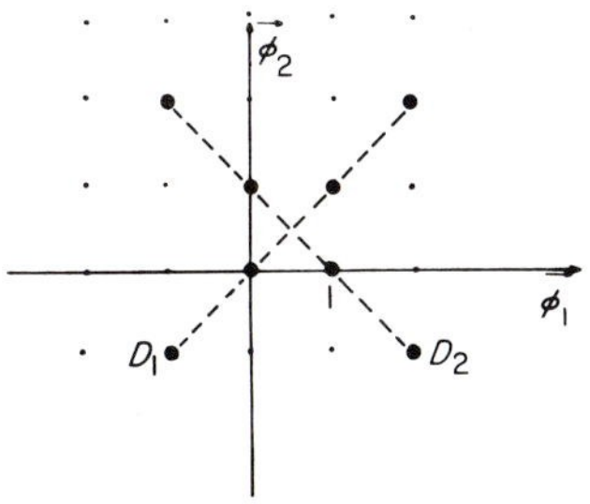

Figure VI.7. Crossing of straight line.

Definition: Let $f_1(x) = \lambda_1$ be the equation of the straight line D_1. Then D_2 is said to *cross* D_1 if there exists consecutive $b, b' \in D_2$, such that

$$f_1(b) < \lambda_1 \quad \text{and} \quad f_1(b') > \lambda_1$$

If D_2 crosses D_1, then D_1 crosses D_2. Also D_1 and D_2 are said to *meet each other* if there exists one and only one $b \in D_2$ such that $f_1(b) = \lambda_1$.

THEOREM VI-4. *Let Δ_1 and Δ_2 be two axes defined by the primitive vectors $\vec{\alpha}$ and $\vec{\beta}$, respectively. Any line D_1 parallel to Δ_1 meets any line D_2 parallel to Δ_2, if and only if $\vec{\alpha}$ and $\vec{\beta}$ define a basis in $\mathbb{Z}^2$.*
(easy proof)
   Thus, one can find an infinity of pairs of directions to build sets of such parallel intersecting straight lines. Their intersection is a property stable under the group of translations of $\mathbb{Z}^2$. In particular, we have the basic pair $\vec{e}_1, \vec{e}_2$.

Figure VI.8. Possible choices for the second principal direction β.

COROLLARY. Let $\vec{\alpha}$, $\vec{\beta}$, $\vec{\gamma}$ be three primitive vectors and u, v, w three arbitrary points of $\mathbb{Z}^2$ at which we draw the parallels to $\vec{\alpha}$, $\vec{\beta}$ and $\vec{\gamma}$ respectively. Each line meets the other two lines if and only if:

(i) $\vec{\alpha}$ and $\vec{\beta}$ define a basis

(ii) $\vec{\alpha} + \varepsilon\vec{\beta} + \varepsilon'\vec{\gamma} = 0$, where $\varepsilon, \varepsilon' \in \{-1, +1\}$ (VI-17)

We say that the triple $(\vec{\alpha}, \vec{\beta}, \vec{\gamma})$ defines a set of *principal directions*.

When $\vec{\alpha}$ and $\vec{\beta}$ are chosen, relation (VI-17) shows that there are four possible solutions for $\vec{\gamma}$, namely the two represented in Figure VI.9, and the two obtained by reversing the direction of $\vec{\gamma}$ in the figure. Notice that more than three principal directions is not possible. If there were a fourth, let us say $\vec{\delta}$, it would also have to satisfy the relation

$$\vec{\alpha} = \varepsilon\vec{\beta} + \varepsilon'\vec{\delta} = 0$$

Figure VI.9. Two systems of principal directions.

By comparing this to relation (VI-18), we see that (modulo the sign)

$$\vec{\gamma} = \vec{\alpha} + \vec{\beta} \qquad \vec{\delta} = \vec{\alpha} - \vec{\beta}$$

But then, $\text{Det}(\vec{\gamma}, \vec{\delta}) = -\text{Det}(\vec{\alpha}, \vec{\beta}) + \text{Det}(\vec{\beta}, \vec{\alpha}) = 2\text{Det}(\vec{\beta}, \vec{\alpha})$. Thus $(\vec{\gamma}, \vec{\delta})$ is not a basis. Hence, according to Theorem VI-4, the parallel lines associated with $\vec{\gamma}$ and with $\vec{\delta}$ do not necessarily hit each other.

C.3. Rotations and grid invariance

In $\mathbb{Z}^2$, a rotation is a $1-1$ mapping from $\mathbb{Z}^2$ onto itself which: (i) transforms any straight line $\Delta\{\alpha_1 x_1 + \alpha_2 x_2 = k\}$ into a straight line Δ' (ii) induces on the slopes α_2/α_1 of the Δ's an *increasing* mapping from the rational numbers onto themselves (this condition excludes the symmetry type mappings) (iii) takes the transformed line back to the initial Δ after a finite number of iterations.

If we now demand that a rotation leave a grid globally invariant, then the straight lines parallel to the principal directions must be transformed into each other. This supplementary condition limits the choice of possibilities. In the case of the square grid, the group of admissible rotations is obviously just the transformations R_1 and its successive powers:

$$R_1 = \begin{pmatrix} 0 & 1 \\ -1 & 0 \end{pmatrix}, \quad R_1^2 = -I, \quad R_1^3 = \begin{pmatrix} 0 & -1 \\ 1 & 0 \end{pmatrix}, \quad R_1^4 = I = \begin{pmatrix} 1 & 0 \\ 0 & 1 \end{pmatrix}$$

i.e. the "90°" rotations are the only permissible ones. The reason why a 45° rotation in the square grid is not possible (Fig. VI.4) is now clear. The matrix of this linear transformation is M:

$$M = \begin{pmatrix} 1 & 1 \\ -1 & 1 \end{pmatrix} \quad \text{and} \quad \mathrm{Det}\,(M) = +2$$

Such a "rotation" is not an automorphism, taking $\mathbb{Z}^2$ onto only a part of itself, and leaving an infinite number of points that cannot be reached by M.

In the hexagonal grid, we have a richer admissible group of rotations, now with six elements. Depending upon whether $\vec{j}$ is $(+1, +1)$ or $(+1, -1)$, the two starting matrices are R_2 and R_2^*:

$$R_2 = \begin{pmatrix} 1 & 1 \\ -1 & 0 \end{pmatrix} \quad R_2^* = \begin{pmatrix} 0 & 1 \\ 1 & 1 \end{pmatrix}$$

whose successive powers describe the group.

The cycle of R_2^* is symmetric with respect to the direction $(0, 1)$ (see Fig. VI.10), and since the representation of the hexagonal grid in R^2 is symmetric with respect to the y-axis, both cycles are equivalent. An example of how they perform is given in Figure VI.11.

It may happen that several sub-cycles of directions can exist in the same grid. For example, in the square grid the sub-group $(0, \pi/2)$ co-exists with $(-\pi/4, +\pi/4)$.

We see that none of these subgroups possess more than three directions. The co-occurrence of several groups is the reason why the eight structuring elements of Figure VI.12 are not equivalent to each other, but to two times four similar sets.

The smallest *isotropic* and *centred* convex sets are generated by turning the vector $\vec{a}$ of Figure VI.8 and taking the union of all the possible positions (Fig. VI.12). This construction results in the hexagon H for the hexagonal rotations and in the square C_1 for the square ones. In the latter case, one also uses the square C_2, defined from the rotations of vectors $\vec{a}$ and $\vec{j}$. Use of these rotation groups not only give us a mechanism for interpreting structuring elements, but they are also indispensable in all problems of local estimation (normal, curvatures, gradient for grey tone functions, etc. . . .) and global estimations (circularity, anisotropy). The richer is the rotation group, the more detailed can be the analyses. Visual perception of a digital image at a given point density seems more "natural" for the "richer" grid, i.e. the hexagonal grid.

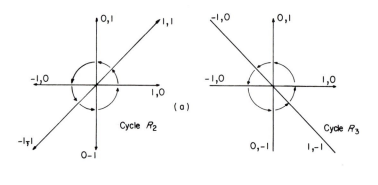

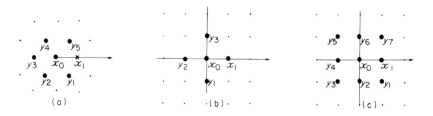

Figure VI.10. (a): the two hexagonal cycles; (b) an example of the group of hexagonal rotations.

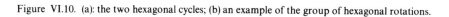

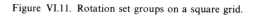

Figure VI.11. Rotation set groups on a square grid.

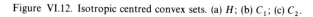

Figure VI.12. Isotropic centred convex sets. (a) H; (b) C_1; (c) C_2.

D. GRIDS AND DIGITAL GRAPHS

D.1. Digital graphs

By removing the translation invariance condition from the definition of a grid, we obtain a *digital graph* (X, D) with vertices X and edges D. The digital graph (X, D) associated with a set $X \in \mathscr{P}(\mathbb{Z}^2)$ is the concept from which we will be able to quantify the notion of the homotopy of X, the notion of its boundary, and also what the new set looks like after rotation. The simplest way to construct a digital graph (X, D) with vertices X is to start from a grid and to modify it either by removing or adding edges. In either case, there is the double effect of also generating a digital graph (X^c, D^*) associated with the complement X^c of X in $\mathbb{Z}^2$. The three most usual modes of construction, which we now present, lead to the square, the 8-connected and the hexagonal graphs.

(a) *Square graphs*
Starting from the square grid, we are only going to remove edges. Two points of X (resp. of X^c) are extremities of an edge D of (X, D) (resp. of (X^c, D^*)), if and only if they are consecutive in one of the two principal directions of the square grid. Applying the method to the sets X and X^c of Figure VI.13, we obtain the two square graphs drawn in Figure VI.13. This technique has the advantage of being an internal computer operation, since in the image banks, the data are generally stored in lines and columns. Secondly, it gives symmetric roles to the sets X and X^c. On the other hand, it gives a strange interpretation to diagonal configurations. In the continuous space, from which digitalization has extracted the configuration $\begin{smallmatrix} 0 & 1 \\ 1 & 0 \end{smallmatrix}$, it is highly likely that either the 1's or the 0's are connected. However, from the square graphs (X, D) and (X^c, D^*), neither the ones nor the zeros are linked. Note also another unfavorable property: the rotation group is not the richest possible.

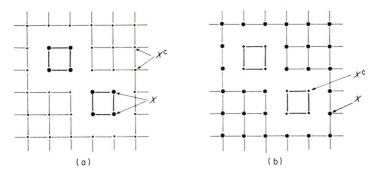

(a)   (b)

Figure VI.13. Square graphs associated with the sets X and X^c of Figure VI.4.

(b) *Eight-connected graphs*

Once again start from the square grid and a set $X \in \mathscr{P}(\mathbb{Z}^2)$. Construct the two graphs (X, D) and (X^c, D^*) by applying the following rules:

$$
\begin{matrix} 1 & 1 \\ & \\ 1 & 1 \end{matrix} \quad \rightarrow \quad \begin{matrix} 1{-}1 \\ |\ \ | \\ 1{-}1 \end{matrix} \qquad \text{(Similarly for } X^c)
$$

$$
\begin{matrix} 1 & 1 \\ & \\ 1 & 0 \end{matrix} \quad \rightarrow \quad \begin{matrix} 1{-}1 \\ |\diagup \\ 1\quad 0 \end{matrix} \qquad \text{(Similarly for } X^c\text{, and for the four directions)}
$$

$$
\begin{matrix} 1 & 1 \\ & \\ 0 & 0 \end{matrix} \quad \rightarrow \quad \begin{matrix} 1{-}1 \\ \\ 0\quad 0 \end{matrix} \qquad \text{(Similarly for } X^c\text{, and for the four directions)}
$$

$$
\begin{matrix} 1 & 0 \\ & \\ 0 & 1 \end{matrix} \rightarrow \begin{matrix} 1\quad 0 \\ \diagdown \\ 0\quad 1 \end{matrix} \ \text{and} \ \begin{matrix} 0 & 1 \\ & \\ 1 & 0 \end{matrix} \rightarrow \begin{matrix} 0\quad 1 \\ \diagup \\ 1\quad 0 \end{matrix} \ \text{(there is no equivalent for } X^c)
$$

Applied to the basic test set of Figure VI.4, eight-connected reconstruction provides the following graphs (Fig. VI.14):

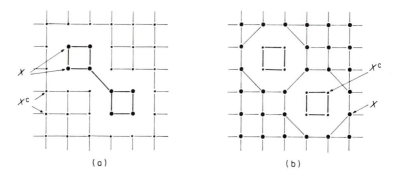

(a)   (b)

Figure VI.14. 8-connected graphs constructed on the sets X and X^c of Figure VI.4.

The properties of the eight-connected graph have been thoroughly studied by A. Rosenfeld (1970, 1972, 1978). The two advantages are that it removes the ambiguities of diagonal configurations, and that it is an internal computer operation. However, it still does not improve the rotation group, and now the X and X^c play asymmetric roles. The problem is fraught with difficulties— both from the morphology and the software points of view. On the figure we see that one part of the set under study will either be isolated from, or connected with the rest of the set, depending upon whether we call the set X or X^c. This asymmetry in the digital representation could be an advantage when the object under study is itself asymmetric with its complement; e.g. convex shapes on a connected background, such as isolated red cells on a blood smear. Then by working with 8-connected graphs on the *background* (and not on the object) the digitalization is adaptive to the structure under study. But even in this case, we must be sure that after a certain number of image transformations, we have not destroyed, or inverted, the figure-background relationship. For example, suppose that both X and Y are 8-connected graphs, then what about $Z = X/Y = Y^c \cap X$? Is Z 8-connected or not? The 8-connectivity turns out to be inconsistent with the set differences.

As for the software involved, we see that all the definitions and algorithms concerning rotations, connectivity, or boundaries, have to be *doubled up*. In each of these cases, we cannot use duality under complementation for performing a morphological calculation. Among the three types of digital groups studied here, the eight-connected graph is undoubtedly the worse.

(c) *Hexagonal graphs*

To construct hexagonal graphs, we start from a hexagonal grid and process as for the square grid case. (i.e. remove all grid edges whose two extremities do not belong to the same set). Figure VI.15 shows an example of the two graphs associated with the two sets X and X^c.

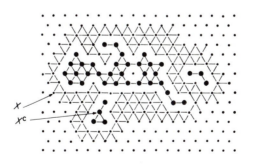

Figure VI.15. Hexagonal graphs constructed from X and X^c.

The hexagonal graph structure deals symmetrically with X and X^c, it does not have ambiguous diagonal configurations, and it is based on the richest possible rotation group on a grid. In addition, the associated algorithms turn out to be very simple. These properties make it the most suitable for the large majority of images.

D.2. Paths, loops, and connectivity

We stated previously that the digital Hit or Miss transformation involved only the module structure $\mathbb{Z}^2$. Nevertheless the same transformation result can be interpreted differently depending on the grid into which we embed the sets (Fig. VI.16). In this section, we will discuss the consequences of going from X to (X, D). In a digital graph (X, D) a *path* is defined to be a finite ordered sequence X' of vertices $(X' \subset X)$, where each pair of successive vertices is adjacent to an edge of D. A *loop* is a path whose initial and terminal point coincide. The smallest loops determined by three (hexagonal, eight-connected graphs) or by four (square graphs) non aligned vertices are called the *elementary cells* of the graph (Fig. VI.17).

Figure VI.16. Dilation of the same graph (X, D) by the same structuring element. Changing the grid type does not modify the vertices of the transform but does change its edges. (a) square grid; (b) hexagonal grid.

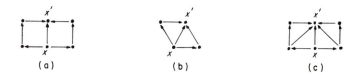

Figure VI.17. Elementary cells in a set X. (a) square; (b) hexagonal; (c) 8-connected graphs.

The elementary path joining two consecutive points x, x' is said to be *equivalent* to any other path joining x, x' which follows the edges of elementary cells of edge $x \, x'$ (Fig. VI.18). Similarly, the *loop* from x to x along an elementary cell which contains x, is equivalent to point x.

Let (X, D) be a digital graph. A graph (X', D') is a subgraph of (X, D) if $X' \subset X$ and $D' \subset D$. A *connected component* of (X, D) is defined to be any subgraph (X', D') such that each pair of points x, $x' \in X'$ can be considered as the initial and terminal points of a path included in (X', D'). Moreover, the connected component (X, D) is said to be *simply connected* when for every pair of points x, $x' \in X$, all the paths having x and x' as initial and terminal points are equivalent.

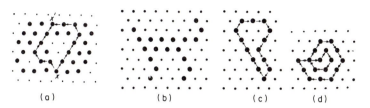

Figure VI.18. (a) two paths going from x to x'; (b) simply connected components; (c) loop surrounding a hole; (d) loop which does not surround a hole.

We saw in the previous section that the technique which associates a graph (X, D) with a digital set X, associates also by duality, a graph (X^c, D^*) with the background X^c. The connected components of (X^c, D^*) are called the *holes* of (X, D). The graph (X, D) is simply connected iff (X, D) and (X^c, D^*) are both connected. (This property specific to $\mathbb{Z}^2$, no longer holds for $\mathbb{Z}^n$, $n > 2$).

D.3. Boundaries and contours

There are different ways of attacking the idea of a digital boundary, depending upon the morphological aim we have in mind; the four main ones are presented below.

First, we might be interested in perimeters; we want to calculate a *digital perimeter* as an estimate of the Euclidean perimeter of the underlying continuous set, of which X is the digital version. In this case, it is *not* necessary to define a boundary of (X, D). Formula (VI-40) defines a digital perimeter from local measurements on (X, D) and relation (VII-14) shows the correspondence between this and the Euclidean perimeter, when the latter exists. Or, we might be interested in a digital perimeter for itself, without reference to $\mathbb{R}^2$ (such as in the estimation of the length of a skeletonized fibre). Suitable algorithms for this type of inquiry are given in Exercise XI, 10.

Then by definition of the erosion, the boundary ∂X is the set difference between X and X eroded by the elementary isotropic convex set of the grid. Such a boundary depends on the choice of a group of rotations on the grid. For example, using the notation of Figure VI.12, we can construct (at least) the following three different boundaries for the sets X and X^c

(a) $\partial X = X/X \ominus H$ $\partial X^c = X^c/X^c \ominus H$
(b) $\partial X = X/X \ominus C_2$ $\partial X^c = X^c/X^c \ominus C_2$
(c) $\partial X = X/X \ominus C_1$ $\partial X^c = X^c/X^c \ominus C_2$

which are illustrated in Figure VI.19a, b, c respectively. It is interesting to notice that in any case:

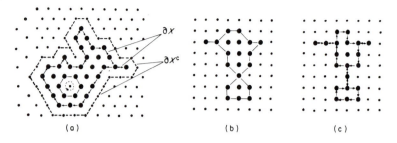

Figure VI.19. Boundary sets and contour loops, for graphs which are (a) hexagonal, (b) square, (c) 8-connected.

— the sets X and X^c have two disjoint boundaries (in contrast to what happens in $\mathbb{R}^2$)
— ∂X and ∂X^c can consist of several components even when (X, D) is connected, but not when (X, D) is simply connected.
— the boundary set can possess isolated points.

The fourth method of defining boundaries arises when we want to follow a contour. The contours of a bounded set in $\mathbb{R}^2$ and of its complement are not sets but loops, whose definition depends on the grid. In the hexagonal case, they are constructed as follows: consider one connected component ∂X_1 of ∂X and two consecutive points x_0 and $x_1 \in \partial X_1$. The next vertex x_2 of the contour is the first point of the sequence y_i of the neighbourhood H which belongs to ∂X_1 (see Fig. VI.12). The same definition holds for X^c. H is replaced by C_1 in case of square graphs, and by C_2 (for X) and C_1 (for X^c) in case of eight-connected graphs. Each connected component of the boundary ∂X provides a distinct loop of the contour, which is eventually reduced to an isolated point. It is easy to prove that the sets of vertices of the contour loops of X and X^c coincide with ∂X and ∂X^c. Similarly, the edges involved in the contour loops of X and X^c are those of the graphs $(\partial X, D)$ and $(\partial X^c, D^*)$ generated by ∂X and ∂X^c respectively.

Although the contour of a simply connected graph consists of a single loop, it is not true in general that, given a loop, one can find a simply connected set with the loop as its contour (Fig. VI.18d). If so, the loop is called a *contour loop*. The interest of the class of the contour loops is due to the following theorem which transposes to digital grids, the classical Euclidean result of Jordan quoted in Chapter III.

THEOREM VI-5. *Every contour loop (L, D) divides the basic grids of $\mathbb{Z}^2$ into two regions Y_1 and Y_2, called the inside and the outside of the loop such that:*
(α) *Only Y_1 is bounded*
(β) *Both $Y_1 \cup L$ and Y_2 are simply connected*

(γ) *If $Y_1 \neq \emptyset$, every path going from $y_1 \in Y_1$ to $y_2 \in Y_2$ hits at least one vertex of the loop.*

Proof: Points (α) and (β) are trivial. If X denotes a set such that $\partial X = L$, it suffices to put $Y_2 = Y^c$ and $Y_1 = X/\partial X$. Suppose now that $Y_0 \neq \emptyset$; to say that a path p going from $y_1 \in Y_1$ to $y_2 \in Y_2$ misses all the vertices of ∂X is equivalent to saying that two consecutive points z_1 and z_2 of the path p belong to Y_1 and Y_2 respectively. When y_2 spans Y_2, the set of its successive points is $Y_2 \oplus H$ (hex. graph) or $Y_2 \oplus C_2$ (square graph) or $Y_2 \oplus C_1$ (eight-connected graphs) which are, in any case, disjoint from Y_1. Therefore, such a path p is impossible.   Q.E.D.

An interesting consequence of Theorem VI-5 follows:

COROLLARY. *Let X be a bounded graph in $\mathbb{Z}^2$ and let x_1, x_2 be two points of the same connected component of ∂X. Join x_1 to x_2 by a path $[x_1, y_1 \ldots y_k, x_2]$ whose intermediary vertices all belong to X^c. Then the adjunction of the set $\{y_i\}$, $i = 1 \ldots k$, to X generates a new hole.*

D.4. Euler's relation

In this section, we limit ourselves to the hexagonal case. Suppose that $X \in \mathcal{P}(\mathbb{Z}^2)$ is finite, and that (X, D) is the associated hexagonal graph. The letters v, t, e, n and h denote the numbers of vertices, of elementary triangles, of edges, of connected components and of holes of (X, D) respectively. Thus, by Euler's famous relation:

$$n - h = v + t - e - 1 \qquad \text{(VI-20)}$$

(The adventures of this relationship are some of the most fascinating in the history of mathematics; see J. C. Pont, 1975. In fact, Euler discovered it only for certain classes of polyhedra (1751), but the seeds sown by the initial work have from Euler to Poincaré, led to a huge harvest of results: planar graph theory (Cauchy, 1809), analysis situs (Von Staudt, 1847) . . . . The reader can find the original version of Euler's work in R. E. Miles and J. Serra, 1978.)

Proof: Generate a sequence of sets $X = X_v \supset X_{v-1} \supset \ldots \supset X_1 \supset X_0$ obtained by omitting none, one, two, etc. v points of the set X, and consider the associated graphs (X_i, D_i), $i = 0, 1 \ldots v$. We will prove relation (VI-20) by induction. For $i = 0$ (empty set) and $i = 1$, it is obviously satisfied. Assume that (VI-20) is also satisfied for (X_1, D_1) and add a neighbouring vertex x_{i+1}; this can happen in any one of the fourteen ways shown in Figure VI.20 (modulo a hexagonal rotation). For configuration numbers 1 to 6, $\Delta h \equiv 0$, and (VI-20)

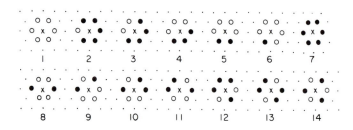

Figure VI.20. The 14 possible ways a neighbouring point x_{i+1} can be added. $\bullet \in x_i$, $\bigcirc \in x_i^c$, $\times$ = point $x_i + 1$.

remains valid. In configuration 7, one hole vanishes, $\Delta h = -1$, and (VI-20) is still satisfied. For each of configurations 8 to 14, two different situations are possible. First, the two components of the neighbourhood are subparts of a same connected component, then $\Delta n = 0$, but according to the corollary of Theorem VI-5, a new hole is generated and $\Delta h = +1$. Second, the two components of the neighbourhood belong to two distinct components of X. Then $\Delta n = -1$ but no new hole is created and $\Delta h = 0$. Similarly, in configuration No. 14, three possible variants occur; we have $\Delta n = -2$ and $\Delta h = 0$; or $\Delta n = -1$ and $\Delta h = -1$, or $\Delta n = 0$ and $\Delta h = +2$, depending on whether three, two or one connected component respectively, already existed in X_i. Q.E.D.

Euler's relation (VI-20) makes for great ease in developing a digital mode of computation. Putting $N(X, D) = n - h + 1$ and denoting $\mathcal{N}(*)$ as the configuration number of type *, we have

$$N(X,D) = \mathcal{N}(1) + \underbrace{\mathcal{N}\begin{pmatrix} 1 & 1 \\ & 1 \end{pmatrix} + \mathcal{N}\begin{pmatrix} 1 \\ 1 & 1 \end{pmatrix}}_{} - \underbrace{\left[\mathcal{N}(1 \;\; 1) + \mathcal{N}\begin{pmatrix} 1 \\ 1 \end{pmatrix} + \mathcal{N}\begin{pmatrix} 1 \\ & 1 \end{pmatrix}\right]}_{}$$

$$v \quad + \qquad t \qquad\qquad - \qquad\qquad\qquad e \qquad\qquad \text{(VI-21)}$$

which can be simplified into

$$N(X, D) = \mathcal{N}\begin{pmatrix} & 1 \\ 0 & & 0 \end{pmatrix} - \mathcal{N}\begin{pmatrix} 1 & 1 \\ & 0 \end{pmatrix} \qquad\qquad \text{(VI-22)}$$

This result is well known (J. Serra, 1969); it is what governs the computation of numbers in many image. Its equivalent for the 8-connected grid is even better known (A. Haas *et al.*, 1967); Exercises VI-4 and VI-11 discuss several variants for the square, 8-connected and cubic versions.

D.5. Homotopy

Definition: Consider the finite parts X's of $\mathbb{Z}^n$ and the graphs induced on them by a grid $\mathcal{T}$. A mapping Ψ from the set of the finite parts of $\mathbb{Z}^n$ into itself is said to be homotopic with respect to the grid $\mathcal{T}$ (or to preserve the homotopy) when it transforms each pair of equivalent paths of (X, D) and also of $(X \stackrel{c}{,} D^*)$ into a pair of equivalent paths of $(\Psi(X), D_1)$ and of $([\Psi(X)]^c, D_1^*)$ respectively. Two finite graphs (X, D) and (X', D') are said to be homotopic, with respect to the grid, when there exists a homotopic transformation Ψ such that $X' = \Psi(X)$.

In $\mathbb{Z}^2$ this general definition has a very simple geometrical interpretation in terms of homotopy, in the sense of Chapter II, F.2b. The two finite sets X and $X' \in \mathcal{P}(\mathbb{Z}^2)$ are homotopic with respect to a grid $\mathcal{T}$, if there exists a one-to-one and onto correspondence between the connected components and the holes of X and of X' (Corollary of Jordan's theorem).

We now give a few results concerning the hexagonal grid. Digital transformations which preserve homotopy have no particular reason to be compatible with translation, nor to satisfy the local knowledge principle. If we now add these two conditions, then we are asking that the transformation be performed point by point, where each point x is transformed according to criteria that depend only on the set X and a bounded structuring element centred at X. Let S_x be the support of the structuring elements associated with a given mapping Ψ, i.e. the set of 1's and 0's brought into play in deciding how to modify the state of the point x. By Jordan's theorem, if the mapping Ψ neither creates nor removes a new class of paths in $X \cap S_x$, for every $x \in X$, then it respects homotopy. Note that the condition is not necessary. One might have a transformation which preserves the homotopy globally, but not locally.

Now suppose that the support S_x is contained in the elementary hexagon H^1 centred at x (i.e. the point x and its 6 neighbours). If we wish to respect the homotopy in $X \cap S_x$, the six neighbours must comprise exactly one connected component of 1's, and one of 0's. Indeed, when the state of x changes from 1 to 0, then a number of things can happen:

— if the 6 neighbours are 1's, we generate a hole in (X, D)
— if the 6 neighbours are 0's, we remove a connected component of (X, D)
— if the 6 neighbours comprise more than one connected component of 1's (thus of 0's), we remove a class of paths of 1's (thus we create a class of paths of 0's).

In all three cases the homotopy is locally modified. Only the five neighbourhoods 2 to 6 of Figure VI-20 remain possible (modulo a complementation and a rotation). They can be reduced to the three structuring elements L, M and D of Figure VI.21 (for the corresponding analysis on an octagonal grid, see A. Rosenfeld (1970) and Castan (1977)). Therefore, by

Figure VI.21. (a), (b), (c) the only 3 types of structuring elements of size one that respect homotopy. (d) Hexagonal neighbourhood of size two.

iterating the Hit or Miss transformations with the two prototype structuring elements of Figure VI.20, together with those derived by rotation and complementation (18 patterns), we *generate* the class Ψ_1 of all possible morphological transformations that are translation invariant and homotopically compatible, and use the elementary neighbourhood. All the algorithms act on X by peeling away its boundary. The class Ψ_1 is central to the current methods of image analysis (Golay (1969), S. Levialdi (1971), M. Duff (1971), V. Goetcharian (1979), S. Sternberg (1978) etc). However it is not the only class possible. Take for example a basic support of size 2 (Fig. VI.21d); new classes will appear, which are not reducible to Ψ_1.

E. DIGITAL METRICS AND THE ASSOCIATED MORPHOLOGY

The third, and final grouping of morphological properties for digital sets is obtained when we put a metric on the bounded parts $\mathscr{K}(\mathbb{Z}^2)$ of $\mathbb{Z}^2$, allowing us to talk about such notions as the conditional bisector, skeletons, ultimate erosions, etc. . . . developed in Chapter XI. It also is the basis of a *stereology* for the bounded graphs (X, D) analysed in detail below for the case of the hexagonal grid.

E.1. Isotropic convex sets

One can define various different metrics on a given grid. In the hexagonal case for example, the role devoted to the regular hexagon below, could alternatively be played by the dodecagons; the results would visually be more similar to those of the Euclidean metric in $\mathbb{R}^2$. Nevertheless, for the sake of simplicity, we restrict ourselves to the simplest metric, based on the hexagon. Either of three equivalent algorithms can be used to define the regular hexagon nH of size n:

(α) nH is the set obtained by $n-1$ successive dilations of the elementary hexagon H:

$$nH = H^{\oplus n} = \underbrace{H \oplus \ldots H}_{n} \qquad \text{(VI-25)}$$

(β) nH is the set of points h such that

$$\overrightarrow{Oh} = n_\alpha \vec{\alpha} + n_\beta \vec{\beta} + n_\gamma \vec{\gamma} \qquad \text{(VI-26)}$$

where integers n_α, n_β, n_γ vary from 0 to n (see Fig. VI.22). The set of points satisfying relation (VI-26) is confined to lie between the two parallels to $\vec{\gamma}$ given by $(n_\alpha = 0;\ n_\beta = n)$ and $(n_\alpha = n;\ n_\beta = 0)$; i.e. the points a and 0 of Figure VI.22. Similarly for the directions $\vec{\alpha}$ and $\vec{\beta}$. Hence nH is limited by the boundary "$a\ b\ c\ d\ e\ f$".

(γ) nH is the result of successive dilations of the three segments $[0\ a]$, $[0\ c]$, $[0\ e]$, of Figure VI.22, of equal length n and parallel to the principal grid directions:

$$nH = [0\ a] \oplus [0\ c] \oplus [0\ e] \qquad \text{(VI-27)}$$

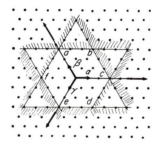

Figure VI.22. nH = regular hexagon of size n, centred at the origin (here $n = 3$).

E.2. The hexagonal Hausdorff metric

Define the digital Hausdorff metric $d(X, X')$ as:

$$d(X, X') = \inf\{\, p, q : X \oplus pH \supset X';\ X' \oplus qH \supset X \,\} \qquad \text{(VI-28)}$$

This is completely analogous to what happens in $\mathbb{R}^2$. Note that the distance between two points x and x' is the number of units in the shortest path between extremities x and x'. Analytically speaking, if $(i_\alpha, i_\beta, i_\gamma)$ and $(i'_\alpha, i'_\beta, i'_\gamma)$ are the co-ordinates of the points x and x', then:

$$d(x, x') = |i_\alpha - i'_\alpha| + |i_\beta - i'_\beta| = |i_\beta - i'_\beta| + |i_\gamma - i'_\gamma|$$
$$= |i_\gamma - i'_\gamma| + |i_\alpha - i'_\alpha| \qquad \text{(VI-29)}$$

If in (VI-22) the set X becomes a point x, then:

$$d(x, X') = \mathrm{Sup}\{d(x, x');\ x' \in X'\} \qquad \text{(VI-30)}$$

Consider the rotations defined above by the two cyclic groups of auto-morphism R_2 and R_3. They turn out to be *isometric* mappings of $\mathcal{K}\,(\mathbb{Z}^2)$ onto itself with respect to the Hausdorff distance (indeed, the regular hexagons nH, centred at the origin, are invariant under the transformations R_2 and R_3).

The metric relationship between a point x and a set $X \in \mathcal{P}\,(\mathbb{Z}^2)$ can be refined further, by using the concept of pseudo distance (Ex. II-2). It does not use the Hausdorff distance; rather it is defined to be the value ρ of the smallest regular hexagon, centred at x, which hits X:

$$\rho\,(x,\,X) = \inf\,\{n{:}x \oplus nH \Uparrow X\} \qquad \text{(VI-31)}$$

The difference between this and the Hausdorff distance is that the set X need not be bounded in relation (VI-31). Thus it is often used in questions such as skeletons for non-compact sets (see Ch. XI). In Figure VI.23a the concentric zones of points $x \in X^c$, with values of $\rho = 1$ and 2, are drawn together with zones of points $x \in X$ with distance $\rho = 1, 2$ from X^c (denoted $-1, -2$). These zones are just the boundary sets $\partial\,(X \oplus nH)$ and $\partial\,(X \ominus nH)$, $n = 1, 2$.

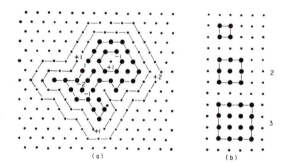

Figure VI.23.   (a) first contact point distances ρ, (b) isotropic convex sets of sizes 1, 2 and 3 in the square grid.

The modification of the above to the square grid case is worth a brief mention. A difficulty arises, since the centres of the odd sized squares do not belong to $\mathbb{Z}^2$ (see Fig. VI.23b). We are therefore forced to accept a coarser description, and work only with even sized squares. Under these conditions, there are analogues to Definition B.3a α and γ of the isotropic convex set in the square case, as well as Definition B.3a β if we allow diagonal edges, as for the 8-connected graphs. Similarly to (VI-28), a Hausdorff distance can be defined. However, relation (VI-29) now becomes

$$d\,(x,\,x') = \max\,(|i_\alpha - i'_\alpha|,\,|i_\beta - i'_\beta|) \qquad \text{(VI-32)}$$

where i_α and i_β are the two (square) co-ordinates of the point $x \in \mathbb{Z}^2$. Independently of their development from the general Hausdorff distance for compact digital sets, relations (VI-29) and (VI-32) each can be used to define a distance between points, in the hexagonal as well as the 8-connected grid (for the latter case, see A. Rosenfeld, 1978).

E.3. Stereology for the hexagonal grid

(a) *Hexagonal convex sets*
The notion of a digital convex set, in terms of intersections of half spaces (rel. (VI-15)) only requires the module structure. If we now restrict the possible orientation of the half spaces to the main directions of a given grid, we generate a sub-class of convex sets stable with respect to the rotation group of the grid. In other words, we can construct a *stereology*, starting from the sub-class. This we will do, for the hexagonal grid (the results are trivially transposable to the other types of grids); let $\mathscr{H}$ denote the set class of bounded hexagonal convex sets (Fig. VI.24).

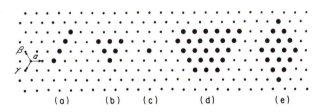

Figure VI.24. Examples of hexagonal convex sets.

There are now a few definitions to give, in terms of the sets $X \in \mathscr{H}$:

— The *perimeter* $U(X)$ of X is the number of simple boundary edges, plus twice the number of double boundary edges. For the five examples of Figure VI.24, we obtain perimeters of 6, 6, 0, 15, 12.
— *Simple and double edges*: an edge of the contour of X is made of two points x and y of X, whose neighbourhood has one of the following two forms: $x \begin{smallmatrix} 0 \\ 1 \end{smallmatrix} y$, or $x \begin{smallmatrix} 0 \\ 0 \end{smallmatrix} y$ (modulo hexagonal rotations). The first types are simple edges, while the second are double edges.
— The *projection* $I_\alpha(X)$ in direction α (resp. β, γ) is the total number of lines $\Delta\alpha$, of direction α, which hit X (resp. β, γ). For example, the fourth specimen of Figure VI.24 gives $I_\alpha = 5$, $I_\beta = 7$, $I_\gamma = 6$.
— The *sides* of X is the vector denoting the lengths of its six boundary segments. Some of these lengths could be zero. Starting from the northerly direction and turning clockwise, we find for the examples b, c, d of Figure VI.21, (2, 0, 2, 0, 2, 0), (0, 0, 0, 0, 0, 0) and (5, 0, 4, 2, 3, 1) respectively. The

vector of six sides called a side function, carries the same type of information as the *support function*, introduced in Chapter IV, Section G for the class $\mathscr{C}(\mathscr{H})$. Two sets X and X' of $\mathscr{H}$ with respective sides $(x_1, \ldots, x_6)$ and $(x_1', \ldots, x_6')$ are said to be *similar* when:

$$x_i = p y_i \quad x_i' = q y_i; \qquad y_i, q, p \in \mathbb{N}, \qquad i = 1, \ldots, 6$$

If the integers p and q are relatively prime, the convex set $(y_1, \ldots, y_6)$ is said to be of unit *size*, and p and q are called the sizes of X and X' respectively. The regular hexagons, as well as union of segments parallel to directions α, β, γ, give families of similar hexagonal convex sets. In the first case, the size is equal to one-sixth of the perimeter, while in the second, the size is the number of edges. When the set X becomes one point, its size is zero.

The main stereological results on the class $\mathscr{H}$, are presented with proofs omitted (they are relatively straightforward, and usually based on induction).

Dilations, erosions, and the side function. Let X and X' be two elements of $\mathscr{H}$, with respective side functions $(x_1, \ldots, x_6)$ and $(x_1', \ldots, x_6')$. Then the dilated set $X \oplus \check{X}'$, also belongs to $\mathscr{H}$ and has a side function $(x_1 + x_1', \ldots, x_6 + x_6')$. Moreover, if $x_i < x_i'$, $i = \{1 \ldots 6\}$, then the eroded set $X \ominus X'$ belongs to $\mathscr{H}$ and has side function $(x_1 - x_1', \ldots, x_6 - x_6')$; otherwise $X \ominus \check{X}' = \varnothing$.

Structuring elements for size distributions. We know that a sequence of morphological openings X_{B_n} indexed by a positive integer n, generates a size distribution if and only if the associated structuring elements $\{B_n\}$ are such that B_n is open with respect to all $B_p, p \leq n$ (Ch. X, Sect. C.1). The side function permits a metric interpretation of this condition when the $\{B_n\}$ belong to the class $\mathscr{H}$. Now B_n is open with respect to $B_p, p \leq n$, if and only if each side of B_p is smaller than or equal to the corresponding side of B_n, i.e. $x_{p,i} \leq x_{n,i}, p, n \in \mathbb{N}$, $p \leq n$, i.e. $1, \ldots, 6$. In particular, similar hexagonal convex sets indexed by their sizes, generate families $\{B_n\}$ suitable for size distributions (they are not the only possible ones, see the octagons and dodecagons (see Ex. VI-9)).

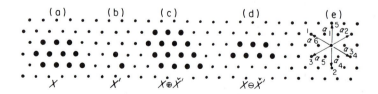

Figure VI.25. (a), (b), (c), (d) Minkowski addition and subtraction (compare the support function of X, X', $X \oplus X'$ and $X \ominus X'$). (e) barycentric characterization of the set (Fig. VI.24d).

Barycentric characterization. Modulo a translation, knowing the six bounded positive integers $(x_1, \ldots, x_6)$ defines and characterizes a hexagonal convex set if and only if the vector sum $\Sigma \, \vec{\alpha}_i \, x_i$ equals zero, where the $\vec{\alpha}_i$ are the unit vectors of the dual grid (see Fig. VI.25e).

Symmetrical elements of $\mathscr{H}$. The set $X \in \mathscr{H}$ is said to be symmetrical when its opposite sides are equal. It can be shown that $X \in \mathscr{H}$ is symmetrical if and only if it is the Minkowski sum of three segments, parallel to the principal direction (for example segments, rhombuses, but not triangles). (In the square grid, all square convex sets are symmetrical!)

Connectivity. All the elements of the class $\mathscr{H}$ are simply connected, and their connectivity number $N(X)$ is equal to 1.

Steiner's Formula. Let X be an element of $\mathscr{H}$ and iH the regular hexagon of size i. The area and perimeter of the dilated set $X \oplus iH$ are respectively:

$$\left. \begin{aligned} A(X \oplus iH) &= A(X) + i\,U(X) + 3i^2 + 3i \\ U(X \oplus iH) &= U(X) + 6i \end{aligned} \right\} \tag{VI-33}$$

in particular:

$$A(iH) = 3i^2 + 3i + 1 \quad \text{and} \quad U(iH) = 6i \tag{VI-34}$$

A dilation result. This is a generalization of Steiner's formula. Let X and B_α be two elements of $\mathscr{H}$. The orientation of X is kept fixed, while that of B_α is uniformly distributed over the six orientations of the hexagonal grid. The rotation averages of the dilated set area and perimeter are given by:

$$\left. \begin{aligned} E[A(X \oplus B_\alpha)] &= A(X) + \frac{U(X) + U(B)}{6} + A(B) - 1 \\ E[U(X \oplus B_\alpha)] &= U(X) + U(B) \end{aligned} \right\} \tag{VI-35}$$

Crofton's formula. For every $X \in \mathscr{H}$, the sum of the intercepts in the three principal directions is equal to the perimeter of X, plus 3, and also that of X^c, minus 3:

$$I_\alpha(X) + I_B(X) + I_\gamma(X) = U(X) + 3 = U(X^c) - 3 \tag{VI-36}$$

(b) *The convex ring $\mathscr{S}(\mathscr{H})$*

The class $\mathscr{H}$ is useful for dealing with digital structuring elements and for understanding their logical interactions; nevertheless it is not rich enough for the modelling of digital images that are met in practice. Therefore, by imitating the continuous case, we can extend $\mathscr{H}$ by introducing finite unions of elements $X \in \mathcal{H}$; denote the result by $\mathcal{S}(\mathcal{H})$:

$$Y \in \mathcal{S}(\mathcal{H}) \Leftrightarrow Y = \bigcup_{i=1}^{k} X_i; \qquad X_i \in \mathcal{H} \qquad \text{(VI-37)}$$

The convex ring $\mathcal{S}(\mathcal{H})$ is stable under all such operations as finite unions, intersections and cross sections by straight lines Δ_α of the grid:

$$X, Y \in \mathcal{S}(\mathcal{H}), \ \Delta_\alpha \in \mathcal{T}(\mathbb{Z}^2) \Rightarrow X \cup Y; X \cap Y; X \cap \Delta_\alpha \in \mathcal{S}(\mathcal{H}) \quad \text{(VI-38)}$$

These relations show that $\mathcal{S}(\mathcal{H})$ possesses a ring structure with respect to the two basic sets operators; hence its name. Just as in $\mathbb{R}^2$, the ring is stable under convex dilations, but more: $\mathcal{S}(\mathcal{H})$ is stable under *any* Hit or Miss transformation, which is certainly not the case for $\mathcal{S}(\mathcal{X})$ in $\mathbb{R}^2$. Hence all bounded hexagonal graphs are elements of $\mathcal{S}(\mathcal{H})$ and so satisfy the properties stated in this section. Nevertheless, if the graph (X, D) is unbounded, in any bounded mask Z the set $X \cap Z \in \mathcal{S}(\mathcal{H})$ and it can be analysed with any transformation respecting the principle of local knowledge.

The main stereological properties of $\mathcal{H}$ are extendable to $\mathcal{S}(\mathcal{H})$. We have already discussed the connectivity number $N(X)$ of any $X \in \mathcal{S}(\mathcal{H})$ in terms of two Hit or Miss transformations; see formula (VI-22). Crofton's formula (VI-36) becomes:

$$I_\alpha(X) + I_\beta(X) + I_\gamma(X) = U(X) + 3N(X) \qquad \text{(VI-39)}$$

or equivalently

$$\mathcal{N}_\alpha(0 \quad 1) + \mathcal{N}_\beta(0 \quad 1) + \mathcal{N}_\gamma(0 \quad 1) = U(X) + 3N(X) \qquad \text{(VI-40)}$$

where the perimeter is still defined as for the hexagonal convex sets. This links stereological parameters with structuring elements which are parts of the *smallest* regular hexagon H.

Notice that the left-hand side of (VI-40) does not change when we replace X by X^c in the relation, whereas the right-hand side becomes $U(X^c) + 3[N(X^c) - 1]$. Taking relation (VI-24) into account, we obtain:

$$U(X^c) - U(X) = 6N(X) \qquad \text{(VI-41)}$$

This result and relation (VI-24), show that all Minkowski functionals of X^c are immediately derivable from those of X (the area $A(X^c)$ is trivially infinite).

The digital version of Steiner's formulae (VI-31) defined on $\mathcal{H}$, has an extension in $\mathcal{S}(\mathcal{H})$ that under certain conditions, make $\mathcal{S}(\mathcal{H})$ look more like the regular model $\mathcal{R}(\mathcal{X})$ of the continuous case. Indeed, formula (VI-33) is true for $X \in \mathcal{S}(\mathcal{H})$ when X is simply connected and closed with respect to the regular hexagon $(i-1)H$ (X is morphologically closed with respect to H_{i-1} when X^c is a union of $(i-1)H$ type hexagons). Similarly, the dilation result can be extended to the class $\mathcal{S}(\mathcal{H})$; relations (V-35) are again applicable to the set

$X \in \mathscr{S}(\mathscr{H})$ if X is closed with respect to $B_{\alpha_1} \oplus B_{\alpha_2} \oplus \ldots B_{\alpha_6}$, or with respect to symmetrical $B = B_\alpha \oplus B_\beta \oplus B_\gamma$.

F. DIGITAL MORPHOLOGY IN $\mathbb{Z}^3$

The digital structures defined over $\mathbb{Z}^3$ do not fundamentally differ from those of $\mathbb{Z}^2$. The basic features are more numerous, and the visualization of the graphs less immediate than in $\mathbb{Z}^2$, but the approach followed for studying $\mathbb{Z}^3$ remains unchanged, and so do the major definitions.

The module structure has already been presented in $\mathbb{Z}^n$. Therefore, we no longer comment on the morphological operation in $\mathbb{Z}^3$ which require only the module structure (Hit or Miss, convexity). We need only develop the notions associated with the concepts of graphs and of grids. Particular attention will be paid to the *rhombohedric* grid which is the most powerful one.

F.1. Principal planes, principal directions

We will first develop the 3-D generalization of Theorem VI-4 and its corrollary. Consider the vectors $\vec{\alpha}$, $\vec{\beta}$, $\vec{\gamma}$. Let $P_1 = P(\vec{\beta}, \vec{\gamma})$, $P_2 = P(\vec{\gamma}, \vec{\alpha})$ and $P_3 = P(\vec{\alpha}, \vec{\beta})$ be the three submodules containing the origin and generated by vectors $(\vec{\beta}, \vec{\gamma})$, $(\vec{\gamma}, \vec{\alpha})$ and $(\vec{\alpha}, \vec{\beta})$ respectively. $P(\vec{\beta}, \vec{\gamma}, u)$ denotes the plane parallel to $P(\vec{\beta}, \vec{\gamma})$ going through the point u. Then

THEOREM VI-6. *The three planes $P(\vec{\beta}, \vec{\gamma}, u)$, $P(\vec{\gamma}, \vec{\alpha}, v)$, $P(\vec{\alpha}, \vec{\beta}, w)$ have a common point, regardless of u, v, w, iff vectors $\vec{\alpha}$, $\vec{\beta}$ and $\vec{\gamma}$ define a base in $\mathbb{Z}^3$ (proof based on Theorems VI-1 and VI-2).*

P_1, P_2 and P_3 are just the co-ordinate planes of the basis $(\vec{\alpha}, \vec{\beta}, \vec{\gamma})$ (Fig. VI.26). A set $\{Q^i\}$ of parallel planes is said to be a family of *principal planes* when each element of the family has a common point with two of the three planes $P(\vec{\beta}, \vec{\gamma}, u)$, $P(\vec{\gamma}, \vec{\alpha}, v)$ and $P(\vec{\alpha}, \vec{\beta}, w)$, $\forall u$, v, $w \in \mathbb{Z}^3$. By construction, planes parallel to P_1, P_2 and P_3 constitute the principal families. If there exists another family $\{Q^i\}$, then by hypothesis the intersection of the Q^i's with each plane of co-ordinates must satisfy an equation of the type (VI-17).

By choosing all the ε's of this equation to be positive, we get the three bisector planes of the unit cube, i.e. $Q_1 = (\vec{\alpha}, \vec{\omega})$, $Q_2 = (\vec{\beta}, \vec{\omega})$ or $Q_3 = (\vec{\gamma}, \vec{\omega})$ where $\vec{\omega}$ is the diagonal vector $(1, 1, 1)$ of the unit cube. Thus the planes parallel to the diagonal planes Q_1, Q_2, Q_3 and to the co-ordinate planes P_1, P_2, P_3 represent all possible families of principal planes in $\mathbb{Z}^3$ (the other solutions of equation (VI-17), with some negative ε, would have led us to replace ω by one of the three other diagonals of the unit cube). The group of the four directions $\vec{\alpha}, \vec{\beta}, \vec{\gamma}, \vec{\omega}$ is that of the principal directions of $\mathbb{Z}^3$. Note that the intersection of two principal planes is not necessarily a principal direction.

F.2. Grids and rotations

A grid in $\mathbb{Z}^3$ is a translation-invariant subset of the principal planes. With each grid we can associate a group $\mathcal{R}$ of rotations that leave it invariant. The four basic grids of $\mathbb{Z}^3$ are the cubic, prismatic, octahedric and rhombohedric grids.

(a) The *cubic* grid is made up of the three families of principal planes, parallel to the planes axis. The translation group is that of $\mathbb{Z}^3$ and is given by the rotation matrix:

$$M_{-\omega} = \begin{pmatrix} 0 & 1 & 0 \\ 0 & 0 & 1 \\ 1 & 0 & 0 \end{pmatrix} = \begin{pmatrix} 1 & 0 & 0 \\ 0 & 0 & 1 \\ 0 & -1 & 0 \end{pmatrix} \begin{pmatrix} 0 & 1 & 0 \\ -1 & 0 & 0 \\ 0 & 0 & 1 \end{pmatrix};$$

$$\text{Cycle } P_1 \rightarrow P_2 \rightarrow P_3 \rightarrow P_1$$

i.e. $M_{-\omega}$ can be thought of as either an axes permutation matrix (first eq.), or as a product of square rotations of axes ox and oz (second eq.) Note that the permutations are an anti-displacement in $\mathbb{Z}^2$ ($\Delta = -1$), but a displacement in $\mathbb{Z}^3$ ($\Delta = +1$). The unit isotropic Steiner polyhedron is the cube $\alpha \oplus \beta \oplus \gamma$.

(b) The *prismatic* grid is obtained by adding to P_1, P_2, and P_3 one of the planes containing ω, say Q_1. It leaves regular hexagonal prisms with axes $\vec{\alpha}$ invariant. The translation group is that of $\mathbb{Z}^3$; rotation matrices are:

$$M_2 = \begin{pmatrix} 1 & 0 & 0 \\ 0 & 1 & 1 \\ 0 & -1 & 0 \end{pmatrix} \quad \text{and} \quad M_3 = \begin{pmatrix} 1 & 0 & 0 \\ 0 & 0 & 1 \\ 0 & -1 & 1 \end{pmatrix}$$

depending on the sign of ω. The unit isotropic Steiner polyhedron is the hexagonal prism $\alpha \oplus \beta \oplus \mu \oplus \omega$.

(c) The *octahedric* grid is obtained by adding two of the planes containing ω, say $Q_1 = (\vec{\alpha}, \vec{\omega})$ and $Q_3 = (\vec{\gamma}, \vec{\omega})$. Rotation matrices must leave the six vectors $\vec{\alpha}, \vec{\beta}, \vec{\gamma}, \vec{\mu}, \vec{\xi}$ and $\vec{\omega}$ globally invariant, which leads to the four matrices $M_{P_1}, M_{P_3}, M_{Q_1}, M_{Q_3}$. For example, in M_{P_3} the plane P_3 is globally invariant under permutation of the three directions α, β, and $-\xi$ which generate the hexagonal group of P_3, i.e.:

$$M_{P_3} = \begin{pmatrix} -1 & -1 & 0 \\ 1 & 0 & 0 \\ 0 & 1 & 1 \end{pmatrix}; \quad \text{cycles:} \begin{cases} P_1 \rightarrow & P_3 \\ P_1 \rightarrow & Q_1 & \rightarrow -Q_3 \rightarrow P_1 \\ P_2 \rightarrow -(\xi, \mu) \rightarrow & Q_2 \rightarrow P_2 \end{cases}$$

We see that this rotation group does not map the five principal planes onto themselves, since (ξ, μ) is added to the cycles. The intersections of the latter plane with P_1, P_3, Q_1 and Q_3 are principal directions of the grid ($\vec{\xi}$ and $\vec{\mu}$) but not that with Q_2 (a similar thing happens in $\mathbb{Z}^2$ with the octagonal grid). The translates of $(\vec{\alpha}, \vec{\gamma})$; $(\vec{\xi}, \vec{\mu})$ and $(\vec{\omega}, \vec{v})$ will intersect if and only if the system $x_2 = k$, $x_1 + x_2 + x_3 = k'$; $x_1 - x_3 = k''$ has a solution in $\mathbb{Z}^3$, i.e. when k, k' and k'' are multiples of 2. Thus only even translations are compatible with octahedric rotations.

The simplest rotation invariant polyhedron is the regular octahedron, with faces parallel to P_1, P_3, Q_1 and Q_3 and edges parallel to $\vec{\alpha}, \vec{\beta}, \vec{\gamma}, \vec{\mu}, \vec{\xi}$ and $\vec{\omega}$. The unit Steiner polyhedron, obtained by dilating these six unit vectors by each other, is the tetrakaidecahedron.

(d) *The rhombohedric* grid is obtained by taking all six possible principal planes. With the two groups $(\vec{\alpha}, \vec{\beta}, \vec{\gamma}, -\vec{\omega})$ and $(\vec{\mu}, \vec{v}, \vec{\xi})$, the vectors play symmetrical roles. Thus we can use the rotation matrix $M_{-\omega}$ of the cubic grid, whose cycles are $P_1 \rightarrow P_2 \rightarrow P_3 \rightarrow P_1$ and $Q_1 \rightarrow Q_2 \rightarrow Q_3 \rightarrow Q_1$, as well as the three other rotation matrices obtained by permuting α, β, γ and $-\omega$. For example, we can write matrix M_α as:

$$M_\alpha = \begin{pmatrix} 1 & 0 & 0 \\ 0 & 0 & 1 \\ -1 & -1 & -1 \end{pmatrix} \quad \text{Cycles} \quad \begin{cases} P_2 \rightarrow -Q_1 \rightarrow \quad P_3 \rightarrow P_2 \\ P_1 \rightarrow -P_3 \rightarrow -Q_2 \rightarrow P_1 \end{cases}$$

The translation group is that of Z^3. The unit Steiner polyhedron associated with this grid is $\vec{\alpha} \oplus \vec{\beta} \oplus \vec{\gamma} \oplus \vec{\omega}$ = the rhombododecahedron. Note that in each case the Steiner polyhedra partition the space; in fact there is no other possible space partition into identical regular polyhedra. The cubic and prismatic grids are rather poor, their Steiner polyhedra being quite different from the sphere. The latter is useful when distance between serial sections is large compared to the 2-D grid step, or when the material under study exhibits a directional anisotropy.

F.3. Graphs

Let $X \in \mathscr{P}(\mathbb{Z}^3)$ be a digital set in $\mathbb{Z}^3$. Exactly as in $\mathbb{Z}^2$, each grid $\mathscr{T}$ induces on X a graph (X, D). The vertices of (X, D) are the points of X, its edges are the edges of $\mathscr{T}$ where both extremities belong to X (eventually the cubic grid could be "doped" by the addition of diagonal configurations, such as the eight-square grid in $\mathbb{R}^2$; such a refinement would result in particularly heavy algorithms for no substantial morphological advantage, since even when the vertices A D G F are 1's, in Figure VI-26, the diagonals AG and DF must not be edges). The intersection of X by any principal plane of the grid $\mathscr{T}$ induces a planar graph whose faces are the faces of the 3-D graph (X, D). The smallest polyhedron

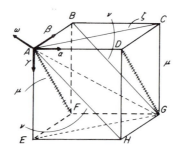

Figure VI.26. Principal planes and vectors arising from the 3-D grids.

delineated by faces of (X, D) define the elementary blocks of the graph. They are cubes in the cubic grid and rhombohedra in the rhombohedric grid; in the latter case each elementary block contains one inside edge (a diagonal of the edge).

The definitions of paths, equivalent paths, loops, connectivity, simple connectivity, boundary (by using isotropic erosion) are the same as in $\mathbb{Z}^2$. Homotopy has been introduced directly in $\mathbb{Z}^n$, in Section D.5.

F.4. Rhombohedric graphs

We now study the rhombohedric graphs in more detail, by drawing them isotropically in the *Euclidean space*. The unit Steiner polyhedron is the rhombododecahedron R represented in Figure VI.27.

R is made of:

	Vertices	Edges	Faces	Blocks
External	14	24	12	4
Internal	1	8	6	0

The internal vertex is point O of Figure VI-27, the internal edges link O to points A, L, C, G and E H N J (there is no edge from O to the vertices B, I, K, D, and F, M). The 32 edges are parallel to the four principal directions $\vec{\alpha}, \vec{\beta}, \vec{\gamma}$, $\vec{\omega}$, and the 18 faces are paralled to the six principal planes of the grid. The decomposition into four blocks is shown in Figure VI.27b, each block x turns out to be the Steiner rhombohedron obtained by dilating three of the four principal directions (the last one appearing is an internal diagonal edge). All the points of the grid are centres of rhombododecahedra equivalent to R.

There are two ways of accessing rhombohedric graphs from serial sections, depending on whether we take them normal to directions OA or OB. In the

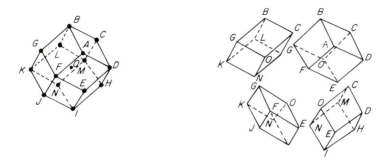

Figure VI.27. The unit rhombododecahedron R and its decomposition into elementary (rhombohedric) blocks.

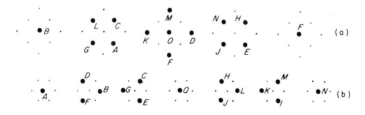

Figure VI.28. (a) Square sectioning of R. (b) Hexagonal sectioning of R. (Equidistant serial sections).

first case, R is displayed along six consecutive sections (Fig. VI.28a) equipped with hexagonal grids; in the second case, we obtain five sections only; each of them possesses a square grid interlaced (in projection) with the next one (Fig. VI.28b).

Each connected component of ∂X of X has a genus G_i defined as in $\mathbb{R}^3$ (rel. (V-9)) by the maximum number of loops which can be drawn on ∂X without disconnecting it; the sum of the quantities $(1 - G_i)$ is the digital connectivity

Configuration number	Vertices to be 1	Vertices to be 0	Vertices without condition	Number of possible orientations
1	A	$C, E, G, H, I, J, K, L, M, V$	B, D, F	8
2	$A\ B$	E, H, I, J, K, M, N	C, G, L, D, F	24
3	$A\ B\ C\ D$	$N\ I\ J\ K$	$H\ E\ F\ G\ L\ M$	12
4	$A\ B\ D\ F$	$N\ I\ K\ M$	$H\ E\ J\ G\ L\ M$	8
5	$B\ A\ C\ L\ G$	$I\ J\ N\ H\ E$	$D\ F\ K\ M$	6

Figure VI.29. Configurations of ∂R which preserve the homotopy of X.

number $N^{(3)}(X)$ of (X, D). If f, v, and e denote the numbers of faces, vertices and edges of (X, D), we have the relationship:

$$N^{(3)}(X) = f + v - e \qquad \text{(VI-43)}$$

which allows a local calculation of $N^{(3)}(X)$.

A point $x \in X$ may be removed from X without modifying its homotopy if the boundary ∂R of the neighbourhood R centred at x possesses exactly two connected components. There are more than 40 such neighbourhoods ∂R; the most interesting are the most symmetrical, i.e. modulo a rotation (Fig. VI.29).

G. EXERCISES

G.1. Imaginative entertainment

Draw examples of digital sets which are:

— convex,   non convex

and

— bounded,   unbounded

and

— connected,   unconnected

and

— isotropic,   anisotropic (for square and hexagonal lattices).

G.2. Boundary sets

Suppose (X, D) is a hexagonal graph. Prove that:

— When the interior of (X, D) is empty, X is identical to its boundary set.
— If (X, D) and $(X \ominus H, D)$ (H is the smallest regular hexagon) are simply connected, then the hexagonal graph associated with ∂X is a loop with neither double vertex nor double edge.

G.3. Orthogonality and rotations in $\mathbb{Z}^n$

Two vectors in $\mathbb{Z}^n$ are said to be orthogonal when their scalar product is zero. Show that in $\mathbb{Z}^2$ and in $\mathbb{Z}^3$ the rotations do not preserve orthogonality. [Use rotation matrices.]

G.4. Planar graphs and counting logics (J. Serra, 1967, 1969)

(a) Prove relation (VI-22), which says that for the hexagonal graph, the connectivity number $N(X)$ is given by:

$$N(X) = \mathcal{N}\begin{pmatrix} 0 & 0 \\ & 1 & \end{pmatrix} - \mathcal{N}\begin{pmatrix} & 0 & \\ 1 & & 1 \end{pmatrix}$$

[Apply the digital version of Euler's formula (VI-20), where the number of faces is $\mathcal{N}\begin{pmatrix} & 1 & \\ 1 & & 1 \end{pmatrix} + \mathcal{N}\begin{pmatrix} 1 & & 1 \\ & 1 & \end{pmatrix} + 1$, the number of edges is $\mathcal{N}(1 \quad 1)$ $+ \mathcal{N}\begin{pmatrix} & 1 \\ 1 & \end{pmatrix} + \mathcal{N}\begin{pmatrix} 1 & \\ & 1 \end{pmatrix}$ and the number of vertices is $\mathcal{N}(1)$. Simplify the resulting equality.]

(b) Demonstrate the rotation invariance of $N(X)$, namely:

$$\mathcal{N}\begin{pmatrix} 0 & 0 \\ & 1 & \end{pmatrix} - \mathcal{N}\begin{pmatrix} & 0 & \\ 1 & & 1 \end{pmatrix} = \mathcal{N}\begin{pmatrix} 0 & 1 \\ & 0 & \end{pmatrix} - \mathcal{N}\begin{pmatrix} & 1 & \\ 0 & & 1 \end{pmatrix}$$

$$= \mathcal{N}\begin{pmatrix} 1 & 0 \\ & 0 & \end{pmatrix} - \mathcal{N}\begin{pmatrix} & 1 & \\ 1 & & 0 \end{pmatrix}$$

(c) Calculate the connectivity number for the square graph. Prove that

$$N(X) = \mathcal{N}\left\{ \begin{matrix} 1 & 0 \\ 0 & 0 \end{matrix} \right\} + \mathcal{N}\left\{ \begin{matrix} 1 & 0 \\ 0 & 1 \end{matrix} \right\} - \mathcal{N}\left\{ \begin{matrix} 1 & 1 \\ 1 & 0 \end{matrix} \right\};$$

$$N(X^c) = \mathcal{N}\left\{ \begin{matrix} 1 & 1 \\ 1 & 0 \end{matrix} \right\} + \mathcal{N}\left\{ \begin{matrix} 0 & 1 \\ 1 & 0 \end{matrix} \right\} - \mathcal{N}\left\{ \begin{matrix} 1 & 0 \\ 0 & 0 \end{matrix} \right\}$$

Interpret the numbers $\mathcal{N}\{\cdot\}$ in terms of square rotations of the normal. Is the sum $N(X) + N(X^c) = 0$? Interpret why [parasite diagonal configurations]. Prove that if the set X is re-interpreted in terms of an 8-connected graph, and if $N^*(X)$ and $N^*(X^c)$ denote the 8-connectivity number of X and X^c, we have:

$$N^*(X) = \left\{ \begin{matrix} 1 & 0 \\ 0 & 0 \end{matrix} \right\} - \left\{ \begin{matrix} & 1 \\ 1 & 0 \end{matrix} \right\} \quad \text{and} \quad N^*(X^c) = -N^*(X)$$

(d) *Curve count.* Suppose that the set X is a finite union of simple arcs, each with two distinct extremities, and that we superimpose X onto a tessellation of R^2 into equal squares (Fig. VI.30a). The squares which hit X are denoted by 1, otherwise by 0. Prove that the number of curves is $N(X)$:

$$N(X) = \mathcal{N}\left[\begin{array}{|c|c|} \hline 0 & 1 \\ \hline 0 & 0 \\ \hline \end{array} \right] - \mathcal{N}\left[\begin{array}{|c|c|} \hline 0 & \\ \hline 0 & 1 \\ \hline \end{array} \right]$$

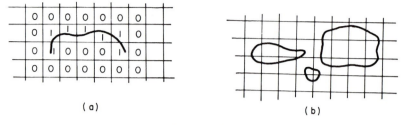

(a)   (b)

Figure VI.30.

(e) *Loop count.* Suppose that the set X is a set of loops (or equivalently a partition of $\mathbb{R}^2$ determined by Poisson lines or Voronoï segments). We superimpose a square grid (Fig. VI.34b). Let $\mathcal{N}(\cdot\frac{\xi}{2}\cdot)$ denote the number of horizontal edges which hit the set X. Prove that the number of loops is $N(X)$:

$$N(X) = \mathcal{N}\left[\overset{\cdot\cdot\cdot}{\underset{\xi\cdot}{\frown}}\right] + \mathcal{N}\left[\boxed{\square}\right] - \mathcal{N}\left[\underset{\cdot}{\lfloor\underline{\quad}}\right]$$

G.5. Digital image simplification

Starting from a digital set X, construct a new set X_1 so that X_1 and X_1^c are both open with respect to the elementary hexagon H (resp. square) and so that X_1 has digital area "close to" that of X. Answer the question again, now with the supplementary condition: $X_1 \ominus H$ and $X_1 \oplus H$ must have the same homotopy as X_1 itself (see Ex. XII-7).

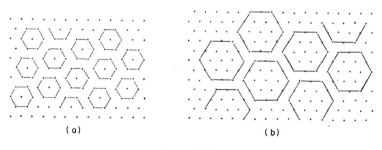

(a)   (b)

Figure VI.31.

[Partition the $\mathbb{Z}^2$ following the rules shown in Figures VI.31a and b. Assign all the points of each cell with the median value of the 7, or 19, values of X in the cell. In case (a) X_1 is open and closed; in case (b) its homotopy is also preserved after erosion and dilation by H.]

G.6. Preservation of homotopy (square grid) (Rosenfeld, 1970)

What are the Hit or Miss transformations which have structuring elements which are sub-sets of the 3×3 square, and which preserve octagonal homotopy of the picture X?

G.7. The digital version of Steiner's formulae (hexagonal grid)

(a) Steiner's formula for dilation. Prove the relation (VI-33) using the following steps ($X \in \mathcal{H}$):

(α) Minkowski additivity for the number of intercepts I_α:

$$I_\alpha(X \oplus H) = I_\alpha(X) + 2 \quad \text{(Similarly for } I_\beta \text{ and } I_\gamma\text{)}$$

(β) After iteration and rotation averaging:

$$U(X \oplus iH) = U(X) + 6i \qquad \text{(VI-33)}$$

(γ) Finally, for $A[X \oplus iH] = A(X) + iU(X) + 3i^2 + 3i$, proof is by induction.

(b) Steiner's formula for erosion. Similarly prove that:

$$\left. \begin{array}{l} U(X \ominus iH) = U(X) - 6i \\[2mm] A(X \ominus iH) = A(X) - iU(X) + 6i(i-1) \end{array} \right\} \qquad (1)$$

provided the eroded set is not empty.

G.8. Digital dodecagons

(1) Using Steiner's decomposition (Ch. IV, G.2), show that every isotropic (hexagonal grid) dodecagon D can be written as the following minkowski sum:

$$D = (H)^{\oplus i} \oplus (H^*)^{\oplus j} \qquad i, j \text{ integers}$$

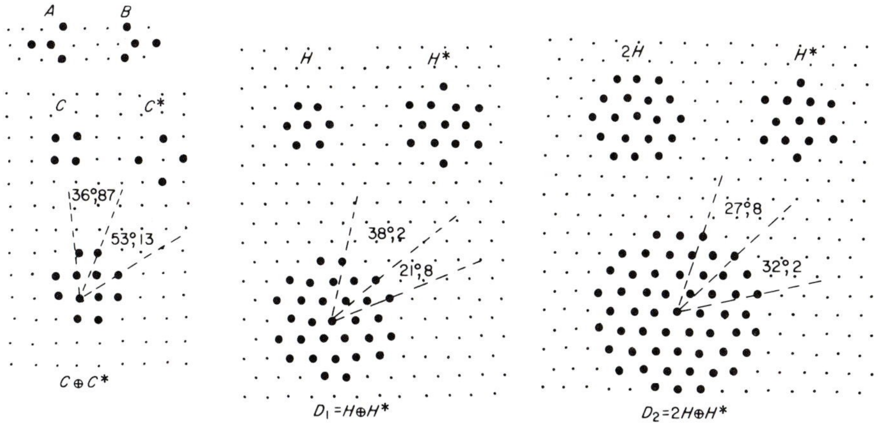

Figure VI.32

where H and H^* are the sets represented on Figure VI.32; H^* is called the conjugate elementary hexagon.

(2) A and B are the two subsets of the elementary hexagon H, shown in Figure VI.32. Check that $H = A \cup B$ and $H^* = A \oplus B$. Derive the expressions for the erosion and the opening of X by the dodecagon D, using only A and B.

G.9. Size increments for digital polygons

Show that the digital octagon (square lattice of size i) is the Minkowski sum of i in terms of $C \oplus C^*$. Similarly the type D_1 and type D_2 dodecagons of size i are Minkowski sums of i terms of respectively, $H \oplus H^*$, and $2H \oplus H^*$ (see Fig. VI.32).

Calculate the areas of the square, the hexagon, the octagon and the dodecagon of size i:

$$
\begin{bmatrix}
\text{Square} & : & i^2 + 2i + 1 \\
\text{Hexagon} & : & 3i^2 + 3i + 1 \\
\text{Octagon} & : & 7i^2 + 4i + 1 \\
\text{Dodecagon } \theta_1 & : & 24i^2 + 6i + 1 \\
\text{Dodecagon } \theta_2 & : & 45i^2 + 9i + 1
\end{bmatrix}
$$

G.10. Connectivity numbers in the cubic grid (J. Serra, 1969)

Calculate the connectivity number $N(X)$ in $\mathbb{Z}^3$ for the cubic grid and prove that:

$$
N(X) = \mathcal{N}\left[\begin{array}{c} \end{array}\right] - \mathcal{N}\left[\begin{array}{c} \end{array}\right] + \mathcal{N}\left[\begin{array}{c} \end{array}\right] - \mathcal{N}\left[\begin{array}{c} \end{array}\right]
$$

[Apply relation (VI-65) for the cubic graphs, and simplify.]

G.11. Tetrakaidecahedron

Our aim in this exercise is to perform 3-D tetrakaidecahedral erosions (or dilations) from serial sections. The elementary digital regular tetrakaidecahedron is constructed by using a sequence of four planes, making it possible to work plane by plane and finishing with the intersection of the successive planar erosions.

(1) Let T' be a tetrakaidecahedron in $\mathbb{R}^3$ with distances C and $3h$ respectively between opposing square and hexagonal faces (Fig. VI.33a). The edge length is denoted as b. Show that all the vertices of the polyhedron are contained in four planes $\Pi_1, \ldots, \Pi_4$, of successive distance h apart and parallel to the two opposing hexagonal faces (Fig. VI.33a).

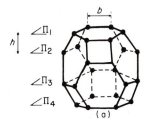

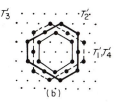

Figure VI.33.

Express h and c as a function of the edge length b.

$$[h = \sqrt{\tfrac{2}{3}}\, b \simeq 0.815b;\ c = 2\sqrt{2}b]$$

(2) Prove that the projection of $T'_i = T' \cap \Pi_i$ $(i = 1, \ldots, 4)$, onto a plane parallel to the Π_i are displayed as shown in Figure VI.33b. Deduce a digital representation T of T'. How many points define T? Is T centred?

[To digitalize T', take four hexagonal lattices $\mathscr{L}_i(a)$ in the Π_i. The spacing $a = h(\sqrt{2})^{-1}$ and the lattices are translated from each other perpendicularly to the Π_i. The three intersections T'_1 (or T'_4), T'_2 and T'_3 can be digitalized on the three different sub-lattices associated with $\mathscr{L}$, which results in T being defined using 38 points.]

(3) Show that, in a regular tetrakaidecahedron of $\mathbb{R}^3$ the 14 normals to the faces lie in a set of three orthogonal planes (for example planes (ACEG), (RDHF), (ABCD) (Fig. VI-26)). Start from three orthogonal cross-sections with normals θ_1, θ_2 and θ_3 (Fig. VI.34). Measure the 2-D intercept numbers in the six θ-directions and the eight λ-directions. The θ's (resp. λ's) are normal to the hexagonal (resp. square) faces of the associated tetrakaidecahedron. Thus the data in the λ-directions will be weighted by $3\sqrt{\tfrac{3}{2}}$, an approximation to the ratio of the two types of solid angles. This property may serve if we want to

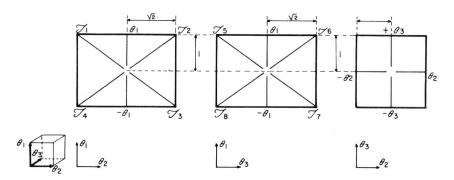

Figure VI.34. Estimation of the 14 directions in $\mathbb{R}^3$ from 3 orthogonal plane sections.

estimate the spatial distribution of the anisotropies in a 3-D material, by using the intercept numbers from three orthogonal cross-sections (and not serial sections).

G.12. Cube–octahedron (S. Sternberg, 1980)

The cube–octahedron is not Steiner and no grid can be constructed from it. However, for operations which do not require grids such as dilations, openings, etc. . . . , it turns out to be extremely useful, since its shape is closer to the sphere than those of the square and the rhombododecahedron. Show that it can be constructed from the same successions of interlaced square grids as the rhombododecahedron (Fig. VI.35). What is the spacing between planes? Show that the smallest centred octahedron is made of 13 points (instead of 15 for the rhombododecahedron and 27 for the cube), and that consequently it performs more progressive dilations. Which group of rotations leaves it invariant?

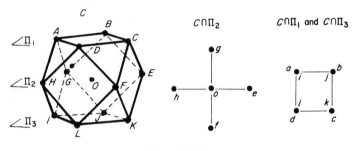

Figure VI.35.

G.13. Last

Go out and buy the Japanese game "Go" and have fun!

VII. Digitalization

This Chapter (Sect. C) treats a topic which until now has received very little attention and bridges the gap between the digital and the Euclidean space, yielding at the same time a precise concept of digitalization. Four basic theorems on increasing morphological mappings, on connectivity, on convexity and on stereological parameters, result. Within this framework, the only sensible type of question to ask is: "For *what* transformations, acting in *what* set families, is the difference between digital and Euclidean versions negligible?". Thus in terms of digitalization it is the *pair* "set transformations—set model" which has a physical meaning; separated, they are uninterpretable.

A. REPRESENTATIONS

We must accept the unfortunate fact that the two spaces $\mathbb{R}^2$ and $\mathbb{Z}^2$ are not isomorphic, since there are several different ways to interpret the same module in Euclidean space. The notion of representation will be the go-between that links the two spaces.

Definition. Let $X \in \mathscr{F}(\mathbb{R}^2)$ and k a positive integer. Given k, we associate the representation $\mathscr{T}(X, k)$ where $\mathscr{T}$ is a mapping from $\mathscr{F}$ into itself such that the range of the mapping is isomorphic to $\mathscr{P}(\mathbb{Z}^2)$; i.e. there exists a one-to-one and onto mapping $\mathscr{S}$ of the range of $\mathscr{T}$ in $\mathscr{F}$ and $\mathscr{P}(\mathbb{Z}^2)$. Therefore $\mathscr{S}^{-1}$ exists and we have:

$$\forall \hat{X} \in \mathscr{P}(\mathbb{Z}^2), \quad \exists X \in \mathscr{F}, k \in \mathbb{N}^+ : \mathscr{S}^{-1}(\hat{X}) = \mathscr{T}(X, k) \qquad \text{(VII-1)}$$

A few examples of the most common representations is the best way to proceed. To avoid confusion, throughout this section we will put a "hat" on sets and transformations that relate to $\mathscr{P}(\mathbb{Z}^2)$.

A.1. The lattice representation

Let u and u' be two independent vectors of $\mathbb{R}^2$ starting from the origin O. A lattice $\mathscr{L}$ in $\mathbb{R}^2$ is the set of all of the extremities x that satisfy:

$$\vec{O}x = \hat{x}_1 \vec{u} + \hat{x}_2 \vec{u'} \quad \text{where} \ (\hat{x}_1, \hat{x}_2) \in \mathbb{Z}^2 \tag{VII-2}$$

The two sets of Figure VII.1a and b show the basic ways of constructing lattices, namely the square lattice $(u = (o, a); \ u' = (a, o))$ and the hexagonal lattice $(u = (a, o), u' = (a/2, a\sqrt{3}/2))$, where a is a positive constant called the lattice unit. Replacing a with $a_k = a \cdot 2^{-k}$ results in a new lattice $\mathscr{L}_k$ that contains the original one, as well as all intermediary lattices of units $a_i (1 \leq i \leq k)$. Associate with lattice $\mathscr{L}_k$ a small cell $C_k(o)$ centred at the origin and congruent to $\vec{w}_k \oplus \vec{u}'_k$; i.e. a rhombus or a square depending on the type of lattice. Let $\mathscr{L}_k(y)$ denote $\mathscr{L}_k$ translated by a $y \in C_k(o)$. Over all y in $C_k(o)$, the union of the $\mathscr{L}_k(y)$'s covers the plane $\mathbb{R}^2$. Similarly, we can turn the lattice through an angle $\alpha \in [0, 2\pi]$, resulting in $\mathscr{L}_k(y, \alpha)$. As far as digitalization problems are concerned, the three parameters a, y and α are held fixed, and k varies. In what follows, no distinction is made between the hexagonal or the square mode when the property under study is valid in both cases. To construct a representation from $\mathscr{L}_k$ is now straightforward. For any $X \in \mathscr{F}$ we have:

$$\mathscr{T}(X, k) = X \cap \mathscr{L}_k \quad \text{and} \quad \mathscr{S} = \{(\hat{x}_1, \hat{x}_2) : \hat{x}_1 \vec{u}_k + \hat{x}_2 \vec{u}'_k \in \mathscr{T}(X, k)\} \tag{VII-3}$$

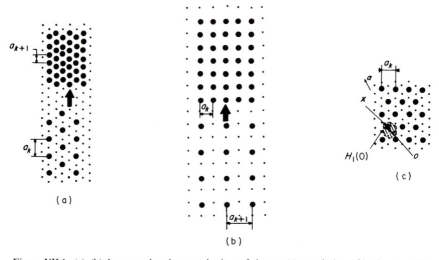

(a)

(b)

(c)

Figure VII.1. (a), (b), hexagonal and square lattices of size a_k, (c) translation of lattice over the basic cell $C_k(o)$.

A.2. The covering representation

The covering representation is commonly used in image analysis to digitally code the curves in $\mathbb{R}^2$ (H. Freeman, 1971); our main interest in it is that it is exactly right for digitalizing the increasing mappings. Start from the lattice $\mathscr{L}_k$. With each point $x \in \mathscr{L}_k$, associate the elementary cell $\{x\} \oplus C_k(o) = C_k(x)$ which is translated from the origin to the point x. Over all x in $\mathscr{L}_k$, the union of the $C(x)$'s covers $\mathbb{R}^2$. Then the covering representation $\mathscr{T}(X, k)$ of $X \in \mathscr{F}$ is defined to be the union of *the cells $C_k(x)$ which hit X*:

$$\mathscr{T}(X, k) = \{ \bigcup C_k(x); \quad x \in \mathscr{L}_k, C_k(x) \Uparrow X \}$$

$$= [(X \oplus C_k) \cap \mathscr{L}_k] \oplus C_k \tag{VII-4}$$

The covering representation satisfies the two important properties:

(i) For any $X \in \mathscr{F}$, $\lim_{k \to \infty} \mathscr{T}(X, k) = X$ in $\mathscr{F}$. (Note that $\mathscr{T}(X, k) \supset \mathscr{T}(X, k + 1)$ and that $\mathscr{T}(X, k) \downarrow X$ and apply Proposition II-21.) Figure VIII.2 illustrates this point.

(ii) With respect to the dilations in $\mathbb{R}^2$ and in $\mathbb{Z}^2$, we have:

$$\mathscr{T}(X \oplus B, k) = [(X \oplus C_k) \cap \mathscr{L}_k] \oplus [(B \oplus C_k) \cap \mathscr{L}_k] \oplus 2C_k$$
$$= \mathscr{S}^{-1} \{ \mathscr{S}[\mathscr{T}(X, k)] \hat{\oplus} \mathscr{S}[\mathscr{T}(B, k)] \} \oplus C_k$$

The covering representation of $X \oplus B$ is the inverse image of a digital dilation of the digital images of X and B, modulo a dilation by C_k.

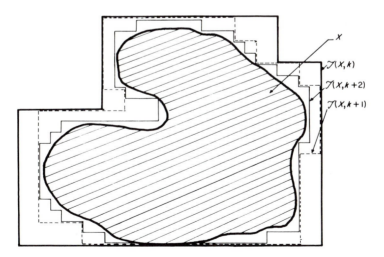

Figure VII.2. Decreasing sequence of covering representations of X.

A.3. The area-weighted representation

A third representation $\mathscr{T}(X, k)$ is made up of all cells $C(x)$, whose centre belongs to the lattice $\mathscr{L}_k$ and hits the set $X \in \mathscr{K}'(\mathbb{R}^2)$:

$$\mathscr{T}(X, k) = (X \cap \mathscr{L}_k) \oplus C_k \qquad\qquad \text{(VII-5)}$$

Its interest is that it leads to unbiased estimates of measurements of areas (Ch. VIII).

A.4. The planar graph representation

Strictly speaking, the planar graph representation is just an extension of the lattice representation. In $\mathbb{Z}^2$, we constructed grids on the modules; when going over to $\mathbb{R}^2$, the modules become lattices, but what do the grids become? The associated notion is that of a *planar graph*. Recall briefly certain ideas from graph theory.

In $\mathbb{R}^2$, any subset of the plane that is homeomorphic to the open segment $]0, 1[$ is called an edge. Let D be a collection of edges which are all disjoint from each other. The plane is assumed to have been compactified, i.e. it contains its point at infinity. Thus each edge has two extremities, called *vertices adjacent to the edge* (the same vertex may be adjacent to several edges). When its two extremities coincide, the edge is a loop. Consider the set X_1 the vertices that are adjacent to the edges D, as well as the second set X_2 of isolated points, and put $X = X_1 \cup X_2$. The planar graph is just the pair (X, D). So in a planar graph edge closures can only cross each other at vertices. Define a *face* to be any connected component of that set obtained by removing all edges D and vertices X from the plane. These three sets, namely edges, vertices, and faces yield a partition of the plane.

A planar graph is said to be locally finite if any bounded set hits a finite number of edges and vertices. Moreover, if the set of vertices is bounded, the graph is said to be finite. Denote, by e, v, f, and v respectively, the numbers of edges, vertices, faces, and connected components of a finite graph. These four quantities satisfy Euler's famous relation (proof in Ex. VII-8):

$$f + v = e + v + 1 \qquad\qquad \text{(VII-6)}$$

When the graph is only locally finite, relation (VI-48) remains true for the intersection of (X, D) with any bounded set K (of boundary ∂K), where $\frac{1}{2}$ is counted for those faces, edges and connected components of (X, D) which hit ∂K.

When we represent a module by the lattice $\mathscr{L}_k$ in the plane, we also represent the edges added to the module (that enlarges it into a grid) by straight line segments joining the corresponding points on the lattice $\mathscr{L}_k$. Indeed this was exactly the way we went when drawing the grids of Figure VI.6; they are actually planar graphs in the plane of the sheet of paper! Now consider a set $X \in \mathscr{K}'(\mathbb{R}^2)$, its lattice representation $\mathscr{T}(X, k) = X \cap \mathscr{L}_k$, and the associated digital set $\mathscr{S}(\mathscr{T}(X, k))$. We call the planar graph representation of X the planar graph $(\mathscr{T}(X, k), D)$ induced on the vertices $\mathscr{T}(X, k)$ by a given digital graph that acts on $\mathscr{S}(\mathscr{T}(X, k))$. For X and $\mathscr{L}_k$ fixed, the planar graph representation depends upon the choice of a digital graph (e.g. based on hexagonal, square or octagonal grids). Notice that we can also associate planar graphs with the covering representation; the set of vertices becomes $(X \oplus C_k) \cap \mathscr{L}_k$.

B. THE DIGITALIZATION CONCEPT

B.1. To digitalize is not as easy as it looks

The means are to hand; it just remains to bridge the gap from digital to continuous morphologies. The following three examples illustrate where the problems lie.

(a) *Is a realization of Poisson points digitalizable?*
A first approach would be to consider the intersection of a Poisson point realization X with the sequence of lattices $\mathscr{L}_k, k \in \mathbb{N}^+$. We might hope that as k increases, X becomes better and better known. Unfortunately, this is not so; the intersection $X \cap \mathscr{L}_k$ stays completely empty, no matter how large k! The lattices only "catch" those points with rational co-ordinates, and these have zero probability in the Poisson process. On the other hand, by starting from the covering lattice, we catch all the Poisson points by dilating them. How sure are we that this "parasite" dilation will not upset any subsequent morphological transformations?

(b) *Connectivity number for the Boolean model*
Consider a very simple set X, a 2-D Boolean realization with convex primary grains, together with a very simple parameter N, the connectivity number of the compact subset $X \cap Z$. Via probabilistic means we can calculate the mathematical expectation of N, as well as that of the digital connectivity numbers N' and N'' of the hexagonal and 8-connected graphs, associated with $X \cap Z \cap \mathscr{L}_k$ (Ch. XIII, Sect. B.6). It is curious that as the lattice spacing a_k becomes infinitesimally small, the two numbers $E(N')$ and $E(N'')$ tend towards different limits that are both distinct from the true value $E(N)$! Relative errors are enormous (as much as 200 per cent) (Fig. XIII.4). Thus the digitalizability

of a given class of sets is not a property that can be stated in the absolute, but depends on the transformation we have in mind.

(c) *Continuity of the representations*

All the types of representations that have been proposed involve an intersection with $\mathscr{L}_k$. But the intersection is only a semi-continuous mapping. Therefore, if we fix $\mathscr{L}_k$ and consider a sequence $\{X_i\}$ of sets tending toward a limit X, there is no guarantee that there will be a corresponding convergence of the representation. For example, take the square lattice $\mathscr{L}$ of unit spacing containing the origin, and the sequence $X(1 - 1/i)$ of compact disks centred at O with radii $1 - 1/i$ (Fig. VII.3). As $i \to \infty$ the lattice representation $X(1 - 1/i) \cap \mathscr{L} = \{0\}$, is always just the origin O, although the limit set $X(1) \cap \mathscr{L}$ consists of five points of $\mathscr{L}$. On the other hand, the sequence $X(1 + 1/i) \cap \mathscr{L}$ converges from above, to its limit $X(1) \cap \mathscr{L}$.

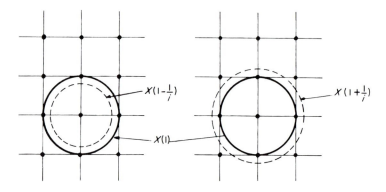

Figure VII.3. Semi-continuity of the representation.

B.2. The definition of digitalization

The first example in C.2a showed that the various representations are not equivalent in $\mathbb{R}^2$, although they are all isomorphic to $\mathscr{P}(\mathbb{Z}^2)$. The conclusion from the third example is that we cannot observe convergences in $\mathbb{R}^2$ when the unit a_k of the representation is fixed. Finally, the lesson learn from the second example is probably the deepest: the question is not to know whether a set of $\mathbb{R}^2$ is digitalizable or not—the only thing with a physical meaning is the pair "morphological criterion—set model" (again "Istina" and "Pravda"!) The subject of digitalization stems from consideration of *only* this dual relationship. The more demanding the criterion is, the less numerous are accessible sets.

In reality, the structure of the physical space is unknown to us; we can be sure that it is neither $\mathbb{R}^2$ (or $\mathbb{R}^3$) nor $\mathbb{Z}^2$. At ultra-microscopic scales, the rocks and the biological structures appear as molecules, themselves sophisticated organizations of substructures, etc. In fact, the space itself changes. But that is not our problem here; in morphology, we simply want to define chemical, physical or physiological properties of the bodies which correspond to their geometry at a given scale, or for a limited range of scales. Some of these properties are relevant to the Euclidean space (partial differential equations for mechanics, hydrodynamics, etc.; optical magnifications; mechanical rotations of a microscope stage) while others are relevant to the digital mathematical structures studied in Section B. A digitalization theory should make the two space models coherent with respect to each other, rather than deciding which fits best with the physical world. The following definition reflects the approach.

Definition: Let $\mathscr{U}(\mathscr{F})$ be a sub-class of the closed sets and Ψ a class of morphological transformations mapping $\mathscr{U}$ into $\mathscr{F}$. A pair $(X, \psi) \in \mathscr{U} \times \Psi$ is said to be digitalizable when there exists a representation $\mathscr{T}(\mathscr{U}, \mathbb{N}^+)$ and a digital algorithm $\hat{\psi}$ from $\mathscr{P}(\mathbb{Z}^2)$ onto itself, such that:

$$\forall (X, \psi) \in (\mathscr{U} \times \Psi),$$

$$k \to \infty \Rightarrow \mathscr{S}^{-1} \hat{\psi} \mathscr{S}(\mathscr{T}(X, k)) \to \psi(X) \quad \text{in } \mathscr{F} \qquad (\text{VII-7})$$

The diagram of Figure VII.4 illustrates the steps involved in the digitalization approach.

Notes: (1) Conversely, a class $\hat{\Psi}$ of digital algorithms has a Euclidean interpretation if there exists a class $(\mathscr{U}, \Psi)$ and a representation $\mathscr{T}$ such that for each pair $(X, \psi) \in (\mathscr{U}, \Psi)$, relation (VII-7) is satisfied.

(2) The preceding definition is immediately extendable to parameters which are mappings $\mathscr{U} \to \mathbb{R}$ (the real numbers) in the Euclidean case, and mappings $\mathscr{P}(Z^2) \to \mathbb{N}$ (the integers) in the digital case.

Equipped with this definition, we now try to answer the following four important questions:

(1) If we take $\mathscr{U}$ to be the largest possible class, i.e. $\mathscr{F}$ itself, what are the transformations ψ such that the pair (X, ψ), $X \in \mathscr{F}$, is digitalizable?

(2) If the transformations $\psi \in \Psi$ depend only on the homotopy of X, what is the class $\mathscr{U}(\mathscr{F})$ which is digitalizable?

(3) Is the notion of a convex hull digitalizable?

(4) Are the basic stereological parameters digitalizable?

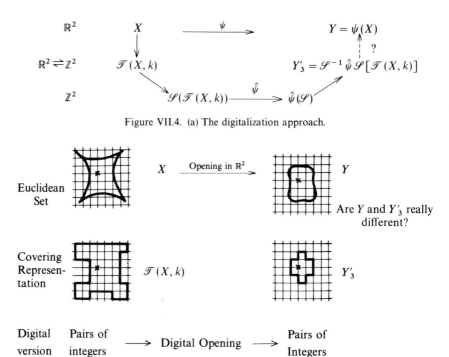

Figure VII.4. (a) The digitalization approach.

Figure VII.4. (b) Example: digitability of the opening. If $\to Y$ when the spacing of the grid tends towards zero, then the opening is digitalizable.

C. DIGITALIZABLE TRANSFORMATIONS FOR THE CLASS ($\mathscr{F}$)

While the following theorem does not generate all possible digitalizable transformations on the class $\mathscr{F}$, those resulting from it cover a broad class.

THEOREM VII-1. *Any mapping ψ from $\mathscr{F}$ into itself that is increasing, and satisfies the four principles of morphology, is digitalizable. In particular, erosion, dilation, opening, closing and the size distributions, are digitalizable.*

Proof: First of all, we must be sure that when we transform X piece by piece, using bounded information $X \cap Z_i$ ($Z_i \in \mathscr{K}'$), upon putting back together these local transforms, we obtain a result relevant to X itself. A result of G. Matheron (1975, p. 224) proves that any translation-invariant, upper semicontinuous (u.s.c.) mapping ψ_1 from $\mathscr{K} \to \mathscr{F}$ admits a smaller u.s.c. extension ψ from $\mathscr{F} \to \mathscr{F}$ if and only if ψ_1 satisfies the local knowledge principle (I, 4). And obviously, if ψ is digitalizable, then *a fortiori* ψ_1 is also. Thus we can work directly with ψ. We now proceed with two possibilities, depending upon whether ψ is a simple dilation or an arbitrary increasing morphological mapping.

If ψ is the dilation of X by B ($X \in \mathscr{F}$, $B \in \mathscr{K}'$), then relation (VII-4) implies that as $k \to \infty$ then $\mathscr{T}(X, k) \downarrow X$, $\mathscr{T}(k, k) \downarrow B$ and $\mathscr{T}(X \oplus B, k) \downarrow X \oplus B$; also $\mathscr{T}(X \oplus B, k) \to X \oplus B$ in $\mathscr{F}$. More generally, according to a theorem of Matheron (1975, Prop. 8-2.2), an increasing, translation invariant mapping ψ from $\mathscr{F}(\mathbb{R}^3)$ into $\mathscr{P}(\mathbb{R}^2)$, is u.s.c. from $\mathscr{F}$ into itself if and only if it has a representation

$$\psi(X) = \bigcap_{j \in J} (X \oplus B^j) = \bigcap_{j \in J} \psi^j(X) \qquad \text{(VII-8)}$$

for a family $\{B_j\}$ closed in $\mathscr{K}$. Each individual dilation ψ_i is a continuous increasing mapping from $\mathscr{F}(\mathbb{R}^2) \times \mathscr{K}(\mathbb{R}^2) \to \mathscr{F}(\mathbb{R}^2)$, and is digitalizable (as we just saw). In other words:

$$\psi(X) = \bigcap_j (X \oplus B^j) = \bigcap_j \bigcap_k \mathscr{T}[(X \oplus B^j), k] = \bigcap_k \bigcap_j \mathscr{T}[X \oplus B^j, k]$$

and thus $\bigcap_j \mathscr{T}(X \oplus B^j, k) \to \psi(X)$ in $\mathscr{F}$. Take the digital algorithm $\hat{\psi}$ to be:

$$\hat{\psi} = \bigcap_j \{ \mathscr{S}[\mathscr{T}(X, k)] \,\hat{\oplus}\, \mathscr{S}[\mathscr{T}(B^j, k)] \}$$

Taking relation (VII-4) into account we see that:

$$\mathscr{S}^{-1}\hat{\psi} = \left\{ \bigcap_j [(X \oplus C_k) \cap \mathscr{L}_k] \oplus [(B^j \oplus C_k) \cap \mathscr{L}_k] \right\} \oplus C_k$$

$$\subset \bigcap_j \left\{ [(X \oplus C_k) \cap \mathscr{L}_k] \oplus [B^j \oplus (k) \cap \mathscr{L}_k] \oplus 2C_k \right\}$$

$$= \bigcap_j \mathscr{T}(X \oplus B^j, k)$$

Now since dilation is continuous, as $k \to \infty$, $\mathscr{S}^{-1}\hat{\psi} \to \lim \bigcap_j \mathscr{T}(X \oplus B^j, k)$ in $\mathscr{F}$, so that finally $\mathscr{S}^{-1}\hat{\psi} \to \psi(X)$.   Q.E.D.

Remarks: (1) Physically speaking, Theorem VII-1 means that the increasing mappings do not further "complicate" the initial set X, and even often free it from small details. As a result, all of the sets, even the most tortuous, such as fractal sets, can be investigated by using dilations, erosions and size distributions.

(2) Although the family $\{B_j\}$ is potentially infinite, the $\mathscr{T}(B^j, k)$ involved in the digital algorithm $\hat{\psi}$ overlap, so that for each k, one has a finite digital family.

(3) We might wonder whether a similar theorem is possible if we replaced the increasing condition and the semi-continuity condition with one of continuity. Unfortunately not, as is well known by experimenters who have tried to digitalize the rotations.

(4) There are independent repercussions of Theorem VII-1 on random closed sets. By Choquet's Theorem (XIII, G.2) a random closed set is characterized from dilations, which are digitalizable operations. Here statistical inference for random sets must avoid non increasing transformations such as connectivity number.

D. HOMOTOPY AND DIGITALIZATION

THEOREM VII-2. *Let* $X \in \mathcal{F}(\mathbb{R}^2)$ *be a closed set,* X^c *be its complement, and* $(\mathcal{F}(X), D)$ *and* $(\mathcal{F}(X^c), D)$ *their respective planar graph representations induced by a hexagonal lattice with origin* y, *orientation* α *and spacing* α_k. *These representations preserve the homotopy of* X *if and only if* X *belongs to the regular class* $\mathcal{R}(\mathcal{K})$, *of radius* $r = a_k$ (Ch. V-C).

Proof: First, sufficiency: we show that for each pair $x_1, x_2 \in X$, linked by a path $[x, x_c] \subset X$, there is an associated path on the graph $(\mathcal{F}(X), D)$. Indeed, suppose that there exists a $B_z(a_k)$ which contains x_1 and x_2 and, which is embedded in X. Now $B_z(a_k)$ also contains at least one point x of the lattice. The path $[x_1 \; x][x \; x_2]$ is homotopic with $[x_1 \; x_2]$. By convention the path becomes the point x itself, and hence the smallest connected components of X appear as isolated points on the lattice. Suppose that one cannot find a $B_z(a_k)$ which contains x_1 and x_2. Then there exists z_1 and z_2 such that $x_1 \in B_{z_1} \subset X$ and $x_2 \in B_{z_2} \subset X$ (definition of $\mathcal{R}(\mathcal{K})$), where $B_{z_1} \cap B_{z_2} = \varnothing$, and where $[z_1 \; z_2] \subset X \ominus \check{B}(a_k)$ is a path such that $[x_1, x_2]$ and $[x_1 \; z_1] \; [z_1 \; z_2][z_2 \; x_2]$ are homotopic (property no. 3 of $\mathcal{R}(\mathcal{K})$, Ch. V-C). Denote T_i to be the smallest topologically closed triangles of side a_k, constructed on the vertices of the hexagonal lattice $\mathcal{L}_k$. Consider the smallest covering T of $[z_1 \; z_2]$ by triangles T_i. One can order the $T_i \subset T$ by running along $[z_1 \; z_2]$ from z_1 to z_2. The index set of these T_i's are finite, and T_i and T_{i+1} are adjacent to each other either by the edge or the vertex that $[z_1 \; z_2]$ hits when crossing from T_i to T_{i+1}. The distance from any point of T to the path $[z_1 \; z_2]$ is $\leq a_k$; hence the covering T is contained in $[z_1 \; z_2] \oplus \check{B}$, and thus in X. Therefore, we can construct a path homotopic to $[x_1 \; x_2]$ and included in X by linking:

— x_1 with one of the vertices, ξ_1, of the first of the T_i's,
— x_2 with one of the vertices, ξ_2, of the last of the T_i's,
— and ξ_1 with ξ_2 by a path constructed on $(\mathcal{F}(X), D)$.

As before, by convention the paths $[x_1, \xi_1]$ and $[x_2, \xi_2]$ become the points ξ_1 and ξ_2 respectively. Thus with each path on X we have associated a homotopic path on $(\mathscr{T}(X), D)$ (the proof is the same for X^c and the dual graph $(\mathscr{T}(X^c), D)$.

Now take two vertices ξ_1 and ξ_2 of $\mathscr{T}(X)$ which are extremities of a path in $(\mathscr{T}(X), D)$. They are also extremities of an equivalent path in X, since $X \oplus B(a_k)$ contains the vertices and the edges of $(\mathscr{T}(X), D)$, and has the same homotopy as X. Hence X and $(\mathscr{T}(X), D)$ are homotopic.

Conversely, we can prove that the hypotheses of $\mathscr{R}(\mathscr{X})$ are necessary by means of counter examples. The set X represented in Figure VII.5a, consisting of two tangent disks, is opened by $B(a_k)$. That of Figure VII.5b is closed by $B(a_k)$, and has the same homotopy as X eroded by $B(a_k)$ and X dilated by $B(a_k)$.

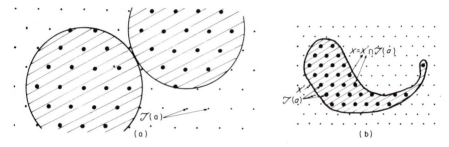

Figure VII.5. (a) X is connected and opened by $B(a_k)$, but the graph $(\mathscr{T}(X), D)$ is disconnected. (b) X^c is opened by $B(a_k)$; X, $X \oplus B(a_k)$ and $X \ominus B(a_k)$ exhibit the same connectivity. Nevertheless, $(\mathscr{T}(X), D)$ may be disconnected.

Remarks: (1) Theorem VII-2 generalizes to 2-D sets the well known result due to Shannon (in the U.S.A.) and to Kotelnikov (in the Soviet Union) that the highest harmonic detectable in a signal has a frequency equal to one half the sampling frequency.

(2) Neither the origin y nor the orientation α of the lattice $\mathscr{L}_k$ appeared in the proof, and so Theorem VII-2 remains valid for any displacement of $\mathscr{L}_k$. If α is fixed, then the disk $B(a_k)$ can be replaced by the hexagon of side a_k. If the lattice is square, the theorem can be transposed by taking $a_k \sqrt{2}$ for the radius of B. In other words, for a given sampling frequency a_k, the hexagonal lattice tolerates a weaker hypothesis on the structure of the object than does the square lattice (or equivalently, it tolerates a larger spacing, given equal hypotheses).

(3) Quite apart from it giving us a class of sets that are digitalizable with respect to homotopy, Theorem VII-2 can also be used for digital images (to condense data for example, see Ex. VI-5).

COROLLARY: *The class of compact sets satisfying conditions* (i) *and* (ii) *of Theorem VII-2 is digitalizable for the morphological transformations depending only on the homotopy of X.*

Proof: The compactness of X, plus condition (i), implies that the homotopy numbers of the connected components of X and of X^c are finite. Therefore the limit in relation (VII-7) must be attained for a finite value of k. Theorem VII-2 shows that it is effectively the case if we take a grid spacing small enough.

E. CONVEXITY AND DIGITALIZATION

Central to the whole notion of convexity is the convex hull, a global operation. Hence for this section we assume all the sets under study, whether digital or not, to be bounded. The mapping $\psi: X \to C(X)$ which takes $X \in \mathcal{K}$ ($\mathbb{R}^2$) into its convex hull is increasing and continuous on $\mathcal{K}$ (Prop. IV-5) and so is digitalizable with respect to the covering representation. On the other hand, the notion of a *digital* convex hull has been defined in Section VI, B.5. It remains therefore to give a Euclidean meaning to this digital operation.

THEOREM VII-3. *Let X be a bounded part of the lattice $\mathcal{L}$, and $\mathcal{S}(X)$ its digital interpretation in $\mathbb{Z}^2$. The convex hulls $C(X)$ of X in $\mathbb{R}^2$ and $\hat{C}$ of $\mathcal{S}(X)$ in $\mathbb{Z}^2$ are related by:*

$$\mathcal{S}^{-1}(\hat{C}) = \mathcal{L} \cap C(X) \qquad \text{(VII-9)}$$

Proof: From the results of Chapter VI, B.2, a digital half plane $E(k)$ of $\mathbb{Z}^2$ is formed by the points x such that $\{x: \sum \gamma_i x_i \leq k\}$, γ_i relatively prime, k integer. Its lattice representation on $\mathcal{L}$ is the restriction to $\mathcal{L}$ of the closed half space H_j of $\mathbb{R}^2$ defined by: $\{y: \sum \gamma_i y_i \leq k\}$. By definition:

$$\mathcal{S}^{-1}[\hat{C}] = \mathcal{S}^{-1}[\bigcap E_i, E_i \supset \mathcal{S}(X)] = \mathcal{L} \cap [\bigcap \{H_i, H_i \supset X\}]$$

But the convex set $\bigcap \{H_i; H_i \supset X\}$ contains $C(X)$ since it contains X. Hence, we have $\mathcal{S}^{-1}(\hat{C}) \supset \mathcal{L} \cap C(X)$. Conversely, let $z \in \mathcal{L}$, $z \notin C(X)$. To prove equality, we have to show that $z \notin \mathcal{S}^{-1}(\hat{C})$. For simplicity, let us put the origin at z; thus $0 \notin C(X)$. Then there exists a straight line in $\mathbb{R}^2$, namely $\{\sum a_i y_i = 0\}$, containing 0 and such that $\sum a_i x_i > 0$ for all $x \in X$. Choose sequences $\{r_i(n)\}_{n \geq 1}$ of rational numbers such that $\lim_n r_i(n) = a_i$. Since X is bounded, for n large enough $\sum r_i(n) x_i$ is strictly positive for all $x \in X$. We can write $r_i(n) = \gamma_i^{(n)}/N$ where the $\gamma_i^{(n)}$ are relatively prime, and where N is an integer. So $\sum \gamma_i^{(n)} x_i > 0$ for all $x \in X$ and, in addition we must have $\sum \gamma_i^{(n)} x_i \geq 1$ ($\gamma_i^{(n)}$ and x_i integers). Thus the lattice representation of the half space of $\mathbb{Z}^2$ defined by $\sum \gamma_i^{(n)} x_i \geq 1$ contains X but not 0, so $0 \notin \mathcal{S}^{-1}(\hat{C})$.

COROLLARY: *Let X be a bounded part of the lattice $\mathscr{L}$. Then:*

(i) $\mathscr{S}(X)$ *is a convex set in $\mathbb{Z}^2$ if and only if $X = \mathscr{L} \cap C(X)$*

(ii) $\mathscr{S}(X)$ *is a convex set in $\mathbb{Z}^2$ if and only if X is the restriction to $\mathscr{L}$ of a convex set of $\mathbb{R}^2$.*

Proof: Using Theorem VII-3, (ii) is trivial. For (i), the sufficient condition is also trivial. Conversely, suppose $X = \mathscr{L} \cap K$, where $K = C(K)$; thus $C(X) = C(\mathscr{L} \cap K) \subset K$. But $X \subset (\mathscr{L} \cap C(X)) \subset (\mathscr{L} \cap K) = X$. So $X = \mathscr{L} \cap C(X)$.

Remarks: (1) Notice that it is essential to assume X to be bounded. If not, the proof fails. As a counter example, suppose $X \subset \mathscr{L}$ is defined as follows: $X = \mathscr{L} \cap \{y_1 \leq \alpha y_2\}$ α irrational. $\mathscr{S}(X)$ is not included in any half space of $\mathbb{Z}^2$ and its convex hull is $\mathbb{Z}^2$ itself. However, $C(X) = \{y_1 \leq \alpha y_2\}$ and $X = \mathscr{L} \cap C(X)$.

(2) The theorem guarantees us that by taking *all* possible intersections of digital half planes E_i, we get the digital version of the convex hull in $\mathbb{R}^2$. In practice, one is usually satisfied with an approximation based on six, or twelve directions (hexagonal lattice). The larger the curvatures of X (if they exist!) the better the approximation. For example, when digitalizing a disk of radius r, a bit of contour such as M_1 M_3 (Fig. VII.6a) can occur only if r is small enough with respect to the spacing a; if not, $M_1 M_3$ will be replaced by the segments $M_1 M_2$ and $M_2 M_3$ (of "dodecagonal" slopes). More precisely, we need

$$r < \frac{a}{2}\sin\theta = \frac{a}{2}\sin\left[\operatorname{tg}^{-1}\frac{5}{3} - \operatorname{tg}^{-1}3\right] = 2.65a$$

However in practice, even when this theoretical upper bound is exceeded, non-dodecagonal slopes rarely appear: in Figure VII.6b, $r > 5a$, although the

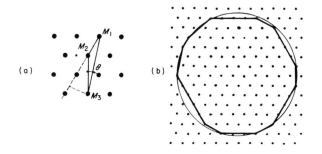

Figure VII.6. (a) a configuration where dodecagonal convex hull is insufficient. (b) digitalization of a circular contour.

exact hull is almost dodecagonal. For square lattices, the octagonal hull has roughly the same quality as the hexagonal hull; i.e. is quite poor (it would not be suitable for the disk of Figure VII.6b for example). Some image analysers allow up to 16 or more directions for the structuring element (for example, the conic erosion of grey tone images performed by S. Sternberg's Cytocomputer).

F. MINKOWSKI FUNCTIONALS AND DIGITALIZATION

F.1. The area and connectivity number

Since the Lebesgue measure is increasing and upper semi-continuous on $\mathcal{F}(\mathbb{R}^2)$, the area of $X \in \mathcal{F}$ is digitalizable with respect to the covering representation. Nevertheless, it is preferable to use the area weighted representation since it leads to unbiased estimates. If $\hat{A}$ denotes the number of points of $X \cap \mathcal{L}_k$, the area weighted estimates of $A(X)$ are

$$\frac{\sqrt{3}}{2} a_k^2 \hat{A} \text{ (hexagonal lattice)}, \qquad a_k^2 \hat{A} \text{ (square lattice)} \qquad \text{(VII-10)}$$

In the next chapter, we will see that when the origin y of the lattice $\mathcal{L}_k$ is uniformly distributed over the central cell $C_k(0)$, A becomes a random number and the expected value of the two estimates (VII-10) is just the area $A(X)$ (assuming X to be compact).

The set model associated with the connectivity number digitalization is the regular model and the corresponding digital algorithms are relation (VI-22), for the hexagonal graph and those of Exercise VI-4 for the 8-connected and square graphs. The same framework is also used for the convexity numbers (Ex. X-14). Note that the conditions of Theorem VII-2 exclude the convex ring as a digitalizable set model for connectivity and convexity numbers.

F.2. The perimeter

The perimeter is a notion that we defined only on the regular class $\mathcal{R}(\mathcal{K})$ and on the extended convex ring $\overline{\mathcal{S}}(\mathcal{K})$. In both cases we can interpret it, via Crofton's formula, as the rotation average of the total projection in direction α; this turns out to be just the right link needed to go from the perimeter to its digitalization.

THEOREM VII-4. *Let $X \in \overline{\mathcal{S}}(\mathcal{K})$ be a set of the extended convex ring, $D(X, \alpha)$ its total projection in direction α, and $\mathcal{L}_k$ a lattice having α as a principal direction; X is represented by the covering representation, and $\mathcal{N}_\alpha(0,1)$ is the intercept* *number of its digital version in the direction* α; *Then* $D(X,\alpha)$ *is digitalizable and, as* $k \to \infty$,

$$a_k \frac{\sqrt{3}}{2} \mathcal{N}_\alpha(0,1) \to D(X,\alpha) \quad \text{(hexagonal lattice)}$$

$$a_k \mathcal{N}_\alpha(0,1) \to D(X,\alpha) \quad \text{(square lattice)} \qquad \text{(VII-11)}$$

Proof: Let $\Delta(y,\alpha)$ be a direction α test line in $\mathbb{R}^2$, with ordinate y on an axis $\perp\alpha$. By definition (rel. (V-11))

$$D(X,\alpha) = \int_{-\infty}^{+\infty} N^{(1)}[X \cap \Delta(y,\alpha)]\,dy \qquad \text{(VII-12)}$$

$X \cap \Delta(y,\alpha)$ is a finite collection of disjoint compact segments. If we take the covering step a_k smaller than the smallest interval separating these segments, then the covering representation and the subsequent digital image contain exactly $N^{(1)}$ disjoint digital segments; hence $N^{(1)}[X \cap \Delta(y,\alpha)]$ is the number of consecutive pairs (0 1) on the digital version of $X \cap \Delta(y,\alpha)$. Thus the quantity $a_k(\sqrt{3}/2)\,\mathcal{N}(0,1)$ (resp. $a_k\,\mathcal{N}(0\ 1)$) turns out to be an approximation to the Riemann integral (VII-12), where dy varies in steps of $a_k\sqrt{3}/2$ (resp. a_k). Hence the theorem.

Remarks: (1) The theorem remains true for the regular class $\mathcal{R}(\mathcal{K})$.

(2) The digital quantities involved in relations (VII-11) are only asymptotically unbiased estimators of $D(X,\alpha)$. For k fixed, a narrow isthmus, a small particle and a "scalping" intercept all result in an underestimation of $\Delta(X,\alpha)$ (see Fig. VII.7a).

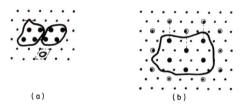

(a)   (b)

Figure VII.7. (a) the three types of digitalization bias when estimating $D(X,\alpha)$ (dotted segments), (b) use of one of the conjugate lattices to estimate $D(X,\alpha)$.

By Crofton's formula, the perimeter $U(X)$ is the integration of $D(X,\alpha)$ from $\alpha = 0$ to $\alpha = \pi$ which can be approximated by averaging over intersections with lattices with different directions α_p. For example, if the α_p are uniformly distributed on the unit circle:

$$U(X) = \lim_{\substack{p \to \infty \\ k \to \infty}} \frac{\pi}{P} \sum_{p=1}^{P} a \frac{\sqrt{3}}{2}[\mathcal{N}_{\alpha_p}(0,1)] \qquad \text{(VII-13)}$$

In this sense, the perimeter is strictly speaking not digitalizable (the successive lattices are not nested in one another); is it possible to obtain estimates of $U(X)$ from fixed orientation lattices? We consider here only the hexagonal case (the square case is left to Ex. no. VII-7); let α; $\beta = \alpha\,(2\pi/3)$; $\gamma = \beta + 2\pi/3$ denote the three main directions of $\mathscr{L}_k$. The first approximation of (VII-13) is made by restricting the directions α_p to α, β and γ, resulting in the estimate

$$U^*(X) = a_k \frac{\sqrt{3}}{2} \cdot \frac{\pi}{3} [\mathscr{N}_\alpha(0\;\;1) + \mathscr{N}_\beta(0\;\;1) + \mathscr{N}_\gamma(0\;\;1)] \qquad \text{(VII-14)}$$

Notice that this estimate can be written as a function of the two digital perimeters $\hat{U}$ and $\hat{U}^c$ of the digital version of X and its component in $\mathbb{Z}^2$. From (VI-40) and (VI-41) we obtain

$$U^*(X) = \frac{\pi}{4\sqrt{3}} a_k [\hat{U} + \hat{U}^c]$$

A possible refinement of the estimate $U^*(X)$ is to double the number of test directions by using the perpendiculars to α, β, γ (Fig. VII.7b), and applying relation (VII-13). The new estimate is now $\frac{1}{2}[U^*(X) + U^{**}(X)]$, where:

$$U^{**}(X) = \frac{a_k}{6}\left[\mathscr{N}\left\{ \begin{matrix} \cdot & 0 \\ & \cdot \\ 1 & \end{matrix} \right\} + \mathscr{N}\left\{ 0 \begin{matrix} & 1 \\ \cdot & \end{matrix} \right\} + \mathscr{N}\left\{ \begin{matrix} 0 & \\ \cdot & \cdot \\ & 1 \end{matrix} \right\} \right] \qquad \text{(VII-15)}$$

We cannot always be certain that the quality of estimation can be improved by adding $U^{**}(X)$; the bias carried by U^{**} is worse than that of U^* (due to a larger elementary step), which suggests that we go no further with digital directions (except if the set X is really oversampled by an extraordinarily fine lattice).

We will close this section with a few comments.

(i) Freeman (1973) or Kulpa (1977) methods of chaining, Grant and Reid $\sqrt{n^2 + 1}$ edge corrections (1977), Quantimet and Bausch and Lomb stacking algorithms, give a practical perimeter estimate which can be as good as the practical values obtained by Crofton formula. However, their Euclidean meaning is more difficult to establish than that of estimate (VII-14).

(ii) Due to Crofton's formula we were able to estimate the perimeter $U(X)$ without having to localize, or digitalize, the boundary ∂X, which is a happy situation to be in. For even if we could build a digital contour tending toward ∂X, it would not necessarily mean that its length tends toward that of ∂X (semi-continuity of length, the perimeter, Chapter V, A). The primary purpose of the contour following techniques is to derive information regarding discrete individuals. Such contour followings are not necessary at all for estimating a perimeter within a field. The algorithms proposed here (VII-14) and (VII-15) are applicable equally well to individuals.

(iii) Algorithms (VII-14) and (VII-15) are set up so that the perimeter measurement is subsequent to the digitalization of X. By performing digital treatment before the measurement, we introduce hexagonal, or square distortions in the object, which causes a directional bias. Formulae (VII-14) and (VII-15) can fail. For instance, this happens if one skeletonizes long particles before measuring their length (see Exs. VII-5 and XI-10).

(iv) The more complicated is the set $X \in \mathscr{F}(\mathscr{K})$ (or $\in \mathscr{R}(\mathscr{K})$), the better the approximations (VII-14) and (VII-15) are likely to become, since the boundary portions will exhibit each orientation more often. Frequently, the most tortuous structures are isotropic, or possibly possess an elliptic isotropy, and come within the framework of Exercise VII-4. Moreover, for the regular set model, the rose of directions is itself digitalizable.

(v) The relationships between area and perimeter of X and the number A of points in $X \cap \mathscr{L}$ (digital area) have been carefully studied when $X \in \mathscr{C}(\mathscr{K})$. In J. Hammer (1977), one can find an extensive review of this question, to which Hammer himself contributed a great deal. We will just quote a few important results. Minkowski proved that if the symmetric compact convex set X in $\mathbb{R}^n$, is centred at the origin of a unit step lattice $\mathscr{L}$, and contains no supplementary point of $\mathscr{L}$, then its n-volume is smaller than 2^n. This result has been the starting point of numerous developments; important contributions are E. Erkhart, G. Fejes Toth, H. Hadwiger, and J. Hammer. In $\mathbb{R}^2$, M. Nozarzewski (1948) proved that

$$A(X) - \tfrac{1}{2}U(X) \le A \le A(X) + \tfrac{1}{2}U(X) + 1$$

for every unit square lattice, and for every $X \in \mathscr{C}(\mathscr{K})$. Again in $\mathbb{R}^2$, D. G. Kendall (1948) estimated the number of points of a randomly moved lattice hitting an $X \in \mathscr{C}(\mathscr{K})$.

G. CONCLUSIONS

This chapter could have been written in the more general setting of $\mathbb{R}^n$. The notion of a planar graph could also have been generalized by adding faces and hyperfaces to the edges; the other representations also have obvious extensions to $\mathbb{R}^n$. Similarly the four Theorems VII-1 to VII-4 are really n-dimensional results (Ths. VII-1 and VII-4 remain valid for LCS spaces). In order to obtain results convenient representations were adopted in each theorem. These might seem rather arbitrary, but this is not the case; we cannot go from $\mathbb{Z}^2$ to another, richer, space, without adding supplementary information. In fact, the representations are particularly chosen to be commensurate with experimental techniques. For example when a metallograph realizes grain boundaries, its technology actually produces a covering representation with an enlargement of the edges (though this is often badly done, since the boundaries may appear as dotted lines). In general, we have avoided defining set classes by the use of conditions on the curvatures of ∂X. For example, we have introduced the regular model $\mathscr{R}(\mathscr{K})$ by demanding invariance under both small spherical (or circular) openings and closings. The idea of curvatures is very poorly suited to digital images; robust digital tests on the regularity of ∂X are well nigh impossible. In contrast to this, set invariance under small openings, although not completely verifiable, results in some good workable tests:

(α) When possible, change the magnification factor before digitalization. If perimeter estimates such as (VII-14) (or the slope of the covariance for grey tone images) increase slightly with the magnification, then the regular model, or convex ring are acceptable models. If not, then one must restrict oneself to using increasing transformations and to area measurements.

(β) For a given magnification, move and turn the object by analogue means, then digitize and compute the connectivity number. If it stays constant for any bounded portion of the object that is completely contained in the digitized field, then one can assume that the conditions of Theorem VII-2 are fulfilled, and hence perform homotopy analyses. If not, then forget about homotopy, or try to see if the connectivity number stays invariant under displacements followed by a small digital opening, or by the simplification technique proposed in Exercise VI-5.

(γ) If the image is closed and open with respect to a small digital convex set, then it is oversampled and one would do better by reducing the magnification.

The problems of digitalization have not been exhaustively treated in this chapter. The variety of results given here does show however that all the criteria studied in Chapter IX and Chapter X are digitalizable, as well as all the functional moments of the random set models in Part Four. However, there is no general guarantee. Some sequential algorithms are digitalizable for convenient set classes, others not, for any class of sets as in the general notion of a skeleton for example. It is more often the case, that good, widely applicable set classes are not as yet known.

H. EXERCISES

H.1. Bias in the number of intercepts of a circle

Let $\mathscr{L}$ be a square lattice with a spacing 2a, and X' be a circle with a radius r. Consider the intercepts of X' with the horizontal lines of the lattice. What is the

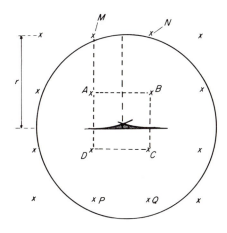

Figure VII.8.

probability of a missing intercept, such as MN on Figure VII.8?

[Fix the lattice, and uniformly implant the centre of the circle. In terms of the top intercept MN, the critical position of the centre is in the zone (dashed lines) delineated by the parallel to MN at distance r, and by the two circles of radius r centred in M and N. The area of this zone is $s = r^2[2(u - \sqrt{1-u^2})$ $-\arcsin u]$ with $a = ru$. To take into account the other end PQ, double this area assuming no overlap of the critical zones. The proportion of critical circles is $p = s/2u^2$. For $u = 2^{-i}(i = 1, \dots, 4)$, we find $p = 0.089, 0.042, 0.041, 0.016$. So, when the diameter $2r$ is eight times the spacing $2a$, the bias is again 4 per cent].

H.2. Lattice translations

Let $\mathcal{L}(x)$ be a square or hexagonal lattice, where the origin x can be anywhere in the associated elementary cell C_0, and let $X \in \mathcal{P}(\mathbb{R}^2)$ be a set of $\mathbb{R}^2$. Prove that for every set $B \subset \mathcal{L}(o)$, it is equivalent either to erode (or dilate) X by B in $\mathbb{R}^2$, or to erode (or dilate) $X \cap \mathcal{T}(x)$ by B in $\mathbb{Z}^2$ and then take the union of the eroded (dilated) sets over all x in C_0; i.e.

$$X \ominus B = \left[\bigcup_{x \in C_0} X \cap \mathcal{L}(x) \right] \ominus B = \bigcup_{x \in C_0} \{[X \cap \mathcal{L}(x)] \hat{\ominus} B\} \qquad (1)$$

$$X \oplus B = \left[\bigcup_{x \in C_0} X \cap \mathcal{L}(x) \right] \oplus B = \bigcup_{x \in C_0} \{[X \cap \mathcal{L}(x)] \hat{\oplus} B\} \qquad (2)$$

[In (1) for example, the right hand side is trivially included in the central set. Conversely if $z \in X \ominus B$, it belongs to one and only one $X \cap \mathcal{L}(x)$.]

H.3. Sub-lattices and rotations

Start from the square lattice $\mathscr{L}(o)$ with spacing a and direction given by the horizontal. How many sub-lattices can we extract from $\mathscr{L}(a)$, with a direction α = arc cot i (i is an arbitrary integer)? What is the value of the corresponding spacing a_i. As an example, Figure VII.9 shows two possible sub-lattices for $i = 1$ and 2. Notice the improvement in angular resolution is at the expense of size resolution.

(a)   (b)

Figure VII.9.

H.4. A fast characterization for anisotropy

Let X' be a bounded set in $\mathbb{R}^2$, $X = \mathscr{L} \cap X'$ its digital picture on the hexagonal lattice of spacing a, and I_α, I_β, I_γ the number of intercepts of X' in the three principal directions of $\mathscr{L}$. As an approximation, we assume that, as α varies in $\mathbb{R}^2$, I_α describes an ellipse E with axes of length $2d_1$, $2d_2$, and main directions θ, $\theta + \pi/2$. The eccentricity $e(e = 1 - \sqrt{d_2^2/d_1^2})$ is a measure of the anisotropy of X' (see Fig. VII.10).

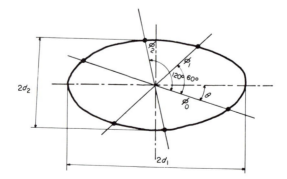

Figure VII.10.

Define:

$$\Delta_1 = (I_\alpha)^{-2} + (I_\beta)^{-2} + (I_\gamma)^{-2}; \quad \Delta_2 = (I_\alpha)^{-4} + (I_\beta)^{-4} + (I_\gamma)^{-4}$$

$$\Delta_3 = \tfrac{1}{2}(\Delta_1^2 - \Delta_2) \tag{1}$$

(1) Show that the curve defined by $(I_\alpha, I_\beta, I_\gamma)$ results in a conic satisfying the following equation:

$$ay^2 + bxy + cx^2 + f = 0 \tag{2}$$

with half axes:

$$d = \frac{1}{-2f}\left[a + c \pm \sqrt{b^2 + (a-c)^2} \right]^{1/2} \tag{3}$$

(2) Using definition 1, deduce from equations (2) and (3) that:

(i) when $I_\alpha = I_\beta = I_\gamma$, X' is isotropic (and E is a circle).
(ii) when $\Delta_2 - 2\Delta_3 \leq 0$, the elliptic approximation is unrealistic (the conic is a parabola or a hyperbola).
(iii) when $\Delta_2 - 2\Delta_3 > 0$, the main directions θ and $\theta + \pi/2$ are the solutions to the equation:

$$\tan 2\theta = \sqrt{3}\,\frac{(I_\gamma)^{-2} - (I_\beta)^{-2}}{(I_\beta)^{-2} + (I_\gamma)^{-2} - 2(I_\alpha)^{-2}}$$

(iv) the main axes lengths (eq. 3) are:

$$2\sqrt{3}[\Delta_1 \pm 2\sqrt{\Delta_2 - \Delta_3}]^{-1/2} \times a\frac{\sqrt{3}}{2}$$

(v) the eccentricity is given by $e = (4\sqrt{\Delta_2 - \Delta_3})(\Delta_1 + 2\sqrt{\Delta_2 - \Delta_3})^{-1}$

H.5. Interference between orientations of the lattice and of the figure (E. Frossard, 1978)

In this exercise it is shown that Crofton's formula no longer holds when the borders of X' are consistently in the directions of the lattice. Let X' be an equilateral triangle of side l. For any compact convex set of $\mathbb{R}^2$, define the coefficient of circularity $\rho' = U^2(X')/A(X')$, the ratio of the squared perimeter to the area. It is known that ρ' is minimum when X' is a disk.

(a) Compute the coefficient of circularity for a disk and for the triangle X' [$\rho'(\text{disk}) = 12.6$; $\rho'(X') = 20.8$].
(b) Let $X = X' \cap \mathcal{L}_a(\alpha)$ be the digital version of X' on a hexagonal lattice of spacing a, and orientation α. Compute the digital circularity coefficient $\rho = U^{*2}(X, \alpha)/A(X)$ (where U^* is given by algorithm (VII-14) and $A(X)$ by algorithm (VII-10)) for the digital versions of X' in Figure VII.11.

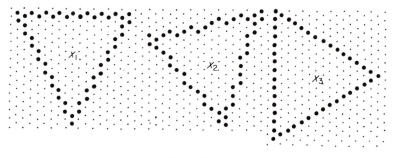

Figure VII.11.

Interpret the result. $[\rho(X_1) = 21.6; \rho(X_2) = 21.1; \rho(X_3) = 16.1$—in the last case the bias is serious.$]$

(c) In the case of X_3 above, instead take the digital boundary set multiplied by a, for the perimeter. What is the new coefficient ρ? Interpret. $[\rho = 19.6$, which is now acceptable.$]$

(d) Answer (b), but now for the hexagon H_5 parallel to the lattice directions. $[\rho(H_5) = 9.4$, which is smaller than the theoretical minimal value 12.6 for the disk!.$]$

H.6. A criticism of the perimeter algorithm

Let X' be a compact set in $\mathbb{R}^2$ and $X = $ the hexagonal lattice representation associated with X'. To estimate the perimeter $U'(X')$ of X', consider the following three algorithms:

(i) $U'(X') \sim a[A(X \oplus H) - A(X)]$ — Outside perimeter

(ii) $U'(X') \sim a[A(X) - A(X \ominus H)]$ — Inside perimeter

(iii) $U'(X') \sim \dfrac{\pi a}{2\sqrt{3}} [I_\alpha(X) + I_\beta(X) + I_\gamma(X)]$ — Crofton's formula

where a is the spacing of the hexagonal lattice and I_α and A are respectively the number of intercepts in direction α, and the digital area.

Compute these three algorithms in the case of Figure VI.15. Are the relative differences negligible? Which are biased? Under what conditions could they give the same numerical result?

H.7. Perimeter estimation using a square lattice

Prove that the perimeter $U'(X')$ of a compact set $X' \in \mathscr{S}(\mathscr{K})$ (or $\mathscr{R}(\mathscr{K})$) in $\mathbb{R}^2$ can be estimated on a square lattice with a spacing a, by the formula:

$$U'(X') \sim a\left[\mathscr{N}(0\ \ 1) + \mathscr{N}\binom{1}{0}\right] + \frac{a}{\sqrt{2}}\left[\mathscr{N}\begin{pmatrix}1\\0\end{pmatrix} + \mathscr{N}\begin{pmatrix}0\\1\end{pmatrix}\right]$$

[The derivation is similar to that used for relation (VII-15).]
   Extend the result to the orientations of the type shown in Figure VII.9b.

H.8. Euler's formula in $\mathbb{R}^2$

The concept of a planar graph was introduced by Cauchy, simply to prove
relation (VII-6) called Euler's formula, and valid in $\mathbb{R}^2$. Prove this relation:

$$f + v = e + v + 1 \qquad\qquad\text{(VII-6)}$$

where f, v, e and v are respectively the number of faces, vertices, edges and
connected components of a finite planar graph Y.
   [Proof by induction: let $E_1, \ldots, E_i, \ldots, E_e$ be the edges of Y. The sub-
graph $Y_i \subset Y$ is the graph with edges $\{E_1, \ldots, E_i\}$ and with vertices given by
the adjacent vertices. Show that relation (VII-6) is satisfied for Y_1, and also for
Y_i when satisfied for Y_{i-1}. Treat the isolated vertices separately. Compare the
Euclidean relation (VII-6) with the digital one (VI-20).]

H.9. Euler's formula in $\mathbb{R}^3$

(1) Prove Euler's formula in three dimensions:

$$f + v = e + p \qquad\qquad\text{(1)}$$

(f, v and e as above, p is the number of polyhedra) for every finite spatially
connected graph Y. Proceed in five steps.

(a) Each polyhedron Π of the graph is homeomorphic to a sphere or a
   compactified plane. Thus prove that

$$v(\Pi) + f(\Pi) = e(\Pi) + 2 \qquad\qquad\text{(2)}$$

and
$$e(S) + p(S) = f(S) + 2 \quad \text{for every vertex } S \in Y, \qquad\qquad\text{(3)}$$

   where $v(\cdot)$ denotes the number of vertices of $(\cdot)$, etc. . . .
   [Apply (VII-6), and duality.]
(b) Prove that for every edge E and every face F of the graph:

$$\left.\begin{array}{ll} v(E) = 2 & f(E) = p(E) \\ p(F) = 2 & e(F) = v(F) \end{array}\right\} \qquad\qquad\text{(4)}$$

(c) Put $\sum_v p(V) = v \cdot p_v$, where p_v is the mean number of polyhedra associated
   with a vertex (and so on). Derive from relations (2) and (3) the following:

$$\left.\begin{array}{l} p \cdot v_p + p \cdot f_p = p \cdot e_p + 2p \\ v \cdot e_v + v p_v = v f_v + 2v \end{array}\right\} \qquad\qquad\text{(5)}$$

and from relation (4):

$$ev_e = 2e \quad ef_e = ep_e \\ fp_f = 2f \quad fe_f = fv_f \Big\} \tag{6}$$

(d) Show that

$$sp_s = ns_v; \, vf_v = fv_f = ef_e = fe_f = ep_e = pe_p \tag{7}$$

(e) Finally, derive Euler's relation (1), from (5), (6) and (7).

(2) Prove Euler's formula in $\mathbb{R}^3$; i.e. for every finite spatial graph Y

$$f + v = e + v + p \tag{8}$$

where v is the number of connected components of Y and p is now the number of *bounded* polyhedra of Y.

[The set of vertices is assumed to be bounded. Thus, one and only one polyhedron is unbounded. By applying relation (1) to each connected sub-graph $Y_i \subset Y$ $(i = 1, \ldots, v)$ one finds $f + v = e + p + 1$, where the 1 is for the unbounded face. Then sum over i.]

H.10. A tale of three houses and three factories

Consider three houses a, b, c and three factories producing water, gas, and electricity. Can we link each house with each factory, so that two different pipes never cross?

[No! If it were possible, the resulting planar graph would have $f = 2 - 6 + 9 = 5$ faces (rel. (VII-6)). Moreover, each face would have at least four edges (two houses or two factories cannot be adjacent), i.e. $2e \geq 4f$; but the edge number is $e = 9$. Hence the contradiction.]

VIII. Random Sampling

A. INTRODUCTION

A.1. There is no such thing as a random phenomenon

After the process of digitalization, we enter the second phase of partial knowledge. The question is now that of piecemeal sampling of a spatio-temporal phenomenon. When an anatomist wants to know the surface area of lung alveolae, he will sample the lung by a few histological thin sections of several square centimetres (Fig. I. 10) and then perform his analysis. Not only is there a problem due to his not knowing the object in its totality, but he must also estimate a 3-D parameter (volumic percentage of the stringers) from 2-D cross sections.

Sampling always means a loss of information. However, at the same time there is always some sort of redundancy in the structures under study; the spatial distribution of stringers in steel will have a uniform appearance throughout the metal plate. Out of this conflict, a probabilistic approach emerges which utilizes the redundancies in structural data to compensate for the sampling loss. Each structural criterion, such as "$B_x \subset X$", is now called an event and is assigned a probability. Statistical estimation then refers to these probabilities, and is not dependent on whether such and such an event actually occurred during a particular realization. For example, if you are told that you have a 20 per cent chance of seeing your eightieth birthday, this information pertains to a particular population of which you are one (e.g. a man of 38, currently in good health), but cannot be applied to you directly since you might fall down the stairs tomorrow.

We open with a parenthetic remark. From what has gone before, it is clear that it is the *method* which is probabilistic and not the *natural object*. There are no random (or deterministic) physical phenomena, but simply probabilistic models that present differing degrees of suitability. This basic distinction can be illustrated via an example. Since the time of Kepler and Newton, we have been able accurately to predict the eclipses of the sun with a deterministic model. However it is not out of the question to interprèt them in the context of a Markov type process, and to estimate the probability of experiencing a solar eclipse here and now, by taking into account the happenings of previous years. It is quite likely that the ancient Chinese in fact analysed them in this way for centuries. A change in method obviously does not imply that the eclipses themselves have changed their nature. Therefore there should be no real difficulty in handling the coexistence of the very geometric morphology of Part I, and a probabilistic version of the same criteria.

In morphology, probability is used at two different levels. The deeper level concerns the genesis of sets and functions, and their easy formalization; these structures are studied in Chapter XIII where the elements of random set theory are introduced. At the second level, namely estimation, probabilities are not as imperative since we could interpret all the results in terms of an approximate calculus of integrals. However, we study them because in practice they turn out to be useful.

A.2. Local-global; individual-textural

The general randomization procedure is governed by the double distinction between global and local, and individual and textural analyses.

(a) *Local-global*
The local-global opposition has been treated at length in Part I. Remember that a set transformation $X \to \Psi(X)$ (or a set parameter) is local when, for every $x \in \mathbb{R}^n$, it suffices to know X in a bounded neighbourhood of x to calculate $\Psi(X)$ at point x. Otherwise, the transformation is global. This constraint is very strict for set transformations, and we effectively need to know X completely to be able to find its convex hull, for example. But, as far as *additive* parameters are concerned, we can satisfy ourselves with a sampling of X; then, we shall say that X is globally approached when the convex hull of the sampling mask Z covers X (two examples of such masks are given in Figure VIII. 7). This weakened version of the globality will be sufficient for the purposes of the current chapter.

(b) *Individual-textural*
An analysis is textural when it considers that all the points $x \in X$ belong to the same plan, namely the set X, where as it is individual when sub-components X_i' of X are looked at globally. The X_i''s may be connected components of X, but not necessarily. If one wants to count the number of mitoses (cells in division) in a cell culture, each individual is a group of 46 chromosomes, more or less

clustered. If one wants to analyse the neighbouring relations in a mosaïc of metallic grains (Ch. Lantuéjoul (1978)), the ith individual i turns out to be the group with the ith grain and its adjacent neighbours, etc. . . .

The reader might be astonished by such and apparent inroad of subjectivity into the morphological treatment: when, and according to whom, can we decide that a given cluster of particles must be considered as *one* individual? There is in fact no subjectivity at all in the choice, since it is based on the distinction between local and global algorithms (and not on sets). To carry out an *individual* analysis on an image always implies the following two steps:

— individual extraction by use of local algorithms (such as particle marking algorithm (XI, 34) for example)
— individual morphology by use of global algorithms working on the extracted individuals.

A.3. Edge effects

Before any randomization we must be able to select the zones of the space in which measurements can effectively occur. The two oppositions local-global and individual-textural yield three different possibilities

(i) if X is assumed to be known everywhere, the mask Z is the whole space and the question of edge corrections is vaccuous;
(ii) if we apply a local textural transformation Ψ to X inside the mask Z, then, assuming that we can decompose Ψ into unions and sequences of Hit or Miss transformations, the mask in which $\Psi(X)$ is known is determined by relations (II-17) to (II-19) (Ex. in Fig. II. 8);
(iii) if we apply a transformation Ψ to individuals defined on X, then the only individuals accessible in $X \cap Z$ are those which miss ∂Z: there is no guarantee that the other individuals present in $X \cap Z$ do not extend outside of Z (Ex. in Fig. XI. 23).

These distinctions are rather ticklish, and ignoring them often leads to errors. As an example, consider the very simple question of counting the number v connected components of a 2-D set X. We take for X the digital set represented in Figure VIII.1a. If we decide to consider X globally (hypothesis (i) above) then v equals the connectivity number $N^{(2)}(X)$ (the particles have no holes) and

$$v = N^{(2)}(X) = \mathcal{N}\begin{pmatrix} & 1 & \\ 0 & & 0 \end{pmatrix} - \mathcal{N}\begin{pmatrix} 1 & & 1 \\ & 0 & \end{pmatrix} = 15 - 6 = 9.$$

Suppose now we modify our approach and interpret X as a part of a larger set Y (hypothesis (ii)). Then we can only estimate the *proportion* of connected components of Y per unit area:

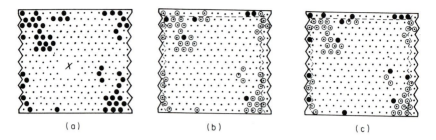

Figure VIII.1. (a) set X; (b), (c) Hit or Miss transformations of X by the structuring elements B and B' of Fig. II.2, leading to the connectivity number of X.

— by computing $N^{(2)}$ in the portion of Y which lies inside the eroded mask $Z \ominus (B \cup B')$ (edge correction formula II-18), i.e. inside the dotted lines in Figure VIII.1;

— and by dividing this number by the area $A[Z \ominus (B \cup B')]$ of the eroded mask, i.e.

$$v = \frac{\left[\mathcal{N}\begin{pmatrix} & 1 & \\ 0 & & 0 \end{pmatrix} - \mathcal{N}\begin{pmatrix} 1 & & 1 \\ & 0 & \end{pmatrix} \right] \text{ in the eroded mask}}{A[Z \ominus (B \cup B')]} = \frac{6-4}{256}.$$

Finally if we consider locally the connected components of the larger set Y as individuals, then after having removed the particles which hit the edges of the mask, only one grain remains and $v = 1$. In this latter case, we still have to infer from the local v to the total number of particles of Y (rel. VIII-16).

In what follows we always assume that the samples under study have geometrically been corrected of eventual edge effects, according to the individual or textural approach adopted.

A.4. Five key situations

We have seen that the random aspect was introduced in the analysis, not by the object but by *the mode of operation*. In our case, putting a sample here rather than there without any preference is translated as choosing the centre of the sample according to a *uniform* distribution over the region under study. Depending upon whether we emphasize either the local or global, and the individual or textural aspects of the morphological investigation, the general randomization procedure divides into four very different situations. Moreover the local individual analyses set two distinct problems, that we prefer to separate for the sake of clarity; finally the chapter will comprise the five sections of Figure VIII.2. Each situation can be refined to include rotation averaging and the possibility of many realizations of the object under study as

	Local	Global
Textural	*Sit I (sect. B)*	*Sit III (sect. D)*
individual	Sit II (sect. C) (Bias in extraction of individuals)	Sit IV (sect. E)
	Sit V (sect. F) (stereology)	

Figure VIII.2. Table of the key situations in estimation problems.

extra sources of random variation. In order to average over the rotations we always (uniformly) choose *one* direction (or one orthogonal basis), and *then* make sections and eventually local measurements in the obtained direction. Theoretically, this procedure is not the only possible one, but it does model the step-by-step techniques of the experimenter. For pedagogical reasons, we will not base the analyses on a potential collection of objects but rather speak of "the" set under study. Indeed, as seen in Chapter I, Section E.2, "the" lung actually investigated can always be said to have been taken from some population of lungs, necessarily finite. Basically though, there is no substantial change in estimation methods.

Spatio-temporal phenomena can also be integrated into our approach. Each object is now defined in an abstract space, which is the product of the Euclidean space by time. In general, an object's time evolution cannot be experimentally followed, since we block temporal development by taking a sample at time t. However the images turn out to be sections of a population of sets which are defined in the space × time continuum. In this sense, they can be handled by our approach.

We conclude this presentation with a remark about the set models used in the five situations. They are always deterministic, and belong to the class $\mathcal{K}(R^3)$, i.e. the class of compact sets, when the estimation procedure involves increasing mappings or Lebesgue measures. When other Minkowski functionals are involved, we then assume that they belong to the classes $\mathcal{R}(\mathcal{K})$ or to $\mathcal{F}(\mathcal{K})$ of Chapter V.

A.5. A bibliographical survey

Strictly speaking, it would be best to start with situation O, where there is no sampling at all. It is exactly in this framework that L. A. Santalo (1953) and W. Von Blaschke (1936) defined, via space integrals and "kinematic densities", the important parameters seen in Chapter V. Now if we sample according to these densities, we find we are working with random variables having uniform laws. This approach is that of classical geometrical probability, which goes back to Buffon (1777) and Laplace (1812), the father of Monte Carlo methods. In fact, most of the results presented in Situation IV below generalize the theorems initially proved by M. W. Crofton (1860) concerning the intersections of a 2-D compact convex set with random lines (see Ex. VIII-3). Later, D. Deltheil (1926), M. G. Kendall and P. A. Moran (1962), H. Giger (1970) L. A. Santalo (1976) added their investigations.

At one end, these methods are limited by the global point of view, which we need for dealing with *integral* parameters over sets. We saw in Chapter V how we could relax this constraint in the deterministic morphology by introducing Minkowski measures. But how can the local point of view be introduced when randomizing?

The approach which a priori seems to be the most direct one, is to base the quantitative description on the notion of a stationary random set. All the authors quoted immediately above, as well as many others, used this approach, sometimes using relatively restrictive hypotheses based on Boolean sets or on isolated compact convex particles, and sometimes a more general framework (G. Matheron (1967, 1975), K. Krickeberg (1974), R. V. Ambartzumian (1974), D. Stoyan (1979), J. P. Rasson (1977), R. Miles (1975), J. Mecke (1979), etc.). Their contribution to random set theory is considerable, and we shall have the opportunity to present it in more detail in Chapter XIII. However the common basis of their approach poses a real problem in estimation. The notions of stationarity and ergodicity are properties of *random models*, and not characteristics of the physical phenomena. Must we test them, or consider them as indiscernable by experimentation, just as we did for the topological closure? If we have to test them, how do we do it?

This is a very ticklish question; here the notion of mathematical infinity intervenes just a bit too much for our liking. In order to overcome it, R. Miles and P. Davy (1976, 1977) proposed an intermediary path between local and global. The specimen under study, bounded and fixed, is modelled by a deterministic compact set X which contains the set Y (i.e. the phase which it is of interest to analyse). Planes, lines, or measuring masks are randomly implanted and generate random sections in which one measures ratios of parameters relative to Y and X respectively. General formulae are established whose limits are studied when X and Y increase indefinitely. We agree completely with their starting point of a compact deterministic set, uniformly sectioned, but we do not follow them beyond this point. In particular:

(i) they do not incorporate into their methodology the notion of different scales for the object and the sample. Situation I shows how crucial such a notion is;

(ii) they decided to study ratios of random variables. We prefer to stay in the line of integral geometry and analyse absolute parameters in the global case, transposing them in terms of specific (or relative) parameters when we are in the framework of situation I.

B. SITUATION I

B.1. Definition

The first situation occurs when:

— the object under study is *very large* with respect to the portions actually analysed;
— morphological investigations are not performed on particles or individuals which exist (at a suitably small scale) in the object, but are only concerned with its *texture* (that is, the image transformations are defined, and performed, at the level of each point). Furthermore, for the sake of simplicity, we temporarily assume that these morphological operations do not involve rotations.

This situation is very common, and often appears in the examples of this book. The wood anatomy in Chapter IX, the petrographic examples in Chapter V and X, the metallography of sinters (Ch. XIII), the structure of the lung (Ch. I), the elimination of artefacts on cervical smears (Ch. XI) etc. . . . all belong to this first type. Our first approach is to make more precise the notion of a "very large" object, which leads us to consider four levels of fields, ordered by successive inclusions. For the sake of clarity we illustrate these four levels by using the example of the Caucasian dolerites, presented in Chapter I.

(i) we start with the definition field R of the object X. Just as for any physical phenomenon, R is necessarily bounded, even if it looks infinite at the working scale. In the example, R is taken as a particular volcanic region of the Causasus, with some defined geographical or geological boundary.
(ii) the *working domain D* is a limited portion of R, which by construction we give ourselves; the study is assumed to be exclusively restricted to a dolerite zone with depth of about 100 m. and covering a few hectares; it is the zone where one wants to build a dam. Now D is not exhaustively investigated, but is sampled, here by boring, to give neighbourhoods W_x.
(iii) the neighbourhoods W_x are the zones of the space where the estimations hold. Since the algorithms which we use are translation invariant, the W_x are the translate of a canonical $W = W_0$ centred at the origin.
(iv) finally we come to relatively small zone Z_x in which the morphological analysis is actually performed. In geneneral Z_x is the union of disjoint

elementary measuring masks Z_x^i. In our example, cylinders 10 cm thick (the W's) have been sectioned in the borings, they are used for physical measurements such as the speed of sound, crushing strength, etc. In parallel, morphological analyses of the microstructures are performed in $Z_x = \bigcup Z_x^i$ (taken on the two adjacent faces) with a view to correlating them with the physical properties. Therefore the morphometric parameters (opening fractions, area percentages . . .) measured on Z_x, are considered as estimates of the mean values of the same parameters *over* W_x, and not over all the borings, or over D (see also Fig. VIII.3).

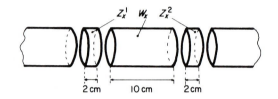

Figure VIII.3. A drill section from a geological rock structure. Physical measurements are made on the central zone W_x(neighbourhood). The microstructure is examined inside the masks Z_x^i taken on the two adjacent faces.

B.2. Randomization

Let $k_{X \cap D}$ be the indicator function of the set $X \cap D$. Random points x, random sets X and random functions f are respectively denoted by $\dot{x}$, $\dot{X}$ and $\dot{f}$. The point x is randomized by attributing the uniform probability law $[\text{Mes}(D \oplus \check{W})]^{-1} k_{D \oplus \check{w}}(x)\,dx$ to the dilate $D \oplus \check{W}$ of the working domain D. We then define the random closed set $\dot{X}(h)$ in W by the relationship:

$$\dot{k}_{\dot{x}}(h) = k_{X \cap D}(\dot{x} + h) \qquad h \in W \qquad \text{(VIII-1)}$$

The various realizations of the random closed set $\dot{X}$ are the portions of $X \cap D$ which we see through the window $W_{\dot{x}}$, when $\dot{x}$ is distributed in $D \oplus \check{W}$ according to the uniform law.

By Choquet's Theorem (Ch. XIII, Sect. G.2), a random closed set is completely determined by the datum of the probability $\Pr\{\dot{X} \pitchfork B\}$, that $\dot{X}$ is hit by B, where B describes the class $\mathcal{K}$ of compact sets; this probability turns out to be a Choquet capacity. Moreover the random closed set $\dot{X}$ is said to be stationary when $\Pr\{\dot{X} \pitchfork B\}$ is

$$P(B) \equiv P_r\{\dot{X} \pitchfork B\} = \Pr\{\dot{X} \pitchfork B_h\} \qquad \forall h, \forall B \in \mathcal{K} \qquad \text{(VIII-2)}$$

However, from one standpoint, we use a more restrictive type of stationarity, proposed by G. Matheron for mining estimation (1979) and defined as follows. $\dot{X}$ is said to be *stationary in W* when, for any compact set B and any h

such that $B \cup B_h \subset W$, relation (VIII-2) is satisfied. In other words, the structuring elements B used here for stationarity in W, are not only bounded (which would be sufficient to fulfil the local knowledge principle (I-4)) but they are also small enough to satisfy the relationship

$$(W \ominus \check{B}) \cap (W \ominus \check{B})_{-h} \neq \varnothing \Leftrightarrow B \cup B_h \subset W \qquad \text{(VIII-3)}$$

THEOREM VIII-1. *The random closed set $\dot{X}(h)$ is stationary in W.*

Proof: From the definition of the set $\dot{X}$ and from relation (VIII-2), we must prove that the following two sets have the same probability:

$$\{x : \dot{X} \Uparrow B_x; \ x \in D \oplus \check{W}\}; \quad \{x : \dot{X} \Uparrow B_{x+h}; \ x \in D \oplus \check{W}\}$$

They are respectively equal to:

$$[(X \cap D) \oplus \check{B}] \cap (D \oplus \check{W}); \quad \{(X \cap D) \oplus \check{B}]_{-h} \cap (D \oplus \check{W})$$

From $B \cup B_h \subset W$ and $X \cap D \subset D$, we deduce

$$[(X \cap D) \oplus \check{B}] \cup [(X \cap D) \oplus \check{B}]_{-h} \subset D \oplus \check{W}$$

and

$$\left[[(X \cap D) \oplus \check{B}] \cap (D \oplus \check{W}) \right]_{-h} = [(X \cap D) \oplus \check{B}]_{-h} \cap (D \oplus \check{W})$$

Since the Lebesgue measure is translation invariant, the two associated probabilities are equal. Q.E.D.

As we see, stationarity in W is due solely to properties of the criteria (translation invariance, local knowledge, semi-continuity) and to the uniform mode of sampling. We can write symbolically:

$$\begin{matrix} \text{Superimposition} \\ \text{of scales} \end{matrix} + \begin{matrix} \text{Morphological} \\ \text{principles I, III, IV} \end{matrix} + \begin{matrix} \text{Randomization} \\ \text{by uniform law} \end{matrix} \Rightarrow \begin{matrix} \text{Local} \\ \text{stationarity} \end{matrix}$$

Therefore local stationarity will no longer appear as a property of the *objects* but only of our *modes of operation*. So we need not test stationarity, but rather we need only test whether the implantation of the samples occurs according to a uniform distribution. We see once more the approach typical in mathematical morphology, of transferring to the criteria a property well understood for the set models (remember the semi-continuity Ch. I, B-4, or the concept of digitalization). Up to now no assumptions have been made about X. Is it implied that such a "stationarity" technique will always work? Not exactly, since it has been necessary to surround the region of interest. $X \cap D$, with the black stripe $D \oplus \check{W}/D$ (as shown in Fig. VII.4) and to make the randomization

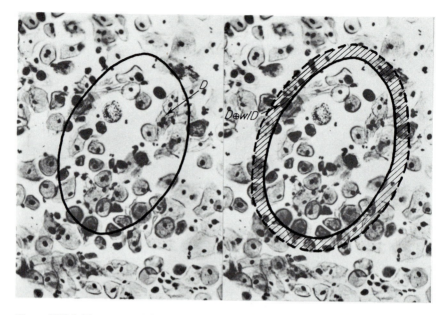

Figure VIII.4. The zone used for randomization combines a part of the object $(X \cap D)$ with a black stripe around it.

hold on both structures. The influences of the working field D and of the neighbourhood W also appear in the value of the probability $P(B)$; indeed we have:

$$P(B) = \int \frac{\text{Mes}\,[(X \cap D \oplus \check{B}) \cap W_x]}{\text{Mes}\,W \cdot \text{Mes}\,(D \oplus \check{W})} \times k_{D \oplus \check{W}}(x)dx$$

$$= \int \frac{(k_{(X \cap D) \oplus \check{B}} * k_W)(x) \times k_{D \oplus \check{W}}(x)dx}{\text{Mes}\,W \cdot \text{Mes}\,(D \oplus \check{W})}$$

$$P(B) = \frac{\text{Mes}\,[(X \cap D) \oplus \check{B}]}{\text{Mes}\,(D \oplus \check{W})} \qquad \text{(VII-4)}$$

As for W, relation (VIII-4) and Figure VIII.4 show that its effect on $P(B)$ is as an edge effect which in the limit completely vanishes. For example, if D and W are compact convex sets, and if D is large with respect to W (a realistic hypothesis), Steiner's formula allows us to write the limited expansion:

$$P(B) = \frac{\text{Mes}\,[(X \cap D) \oplus \check{B}]}{\text{Mes}\,D}[1 + \varepsilon], \qquad \text{(VII-5)}$$

$$\varepsilon \to 0 \quad \text{when} \quad \frac{\text{Mes}\,W}{\text{Mes}\,B} \to 0$$

More interesting is the possibility of starting from relation (VIII-4) to define a sort of physical stationarity for the set X. Apply the relation to larger and larger working domains D_i, all taken inside the definition field R of X. If the $P_{D_i}(B)$'s are approximately the same, we can say that $P(B)$ represents something *intrinsic* to the phenomenon under study, independent of the working domains D_i. This last interpretation is implicit when, without indicating any working domain, we speak in local terms such as volumes or area proportions (V_V, A_A), specific numbers N_A, N_L, N_V, local covariance $C(h)$ or linear erosion $P(l)$ etc. . . . However, independently of the final change of scale, if we restrict ourselves to one working domain D and to structuring elements small with respect to it, Theorem VIII-1 gives us a perfectly good framework for using the theory and vocabulary of stationary random sets. (For example V_V is the $P(B)$ corresponding to B reduced to a single point.)

B.3. Generalization

(a) *A unique or reproducible object?*
It may happen that the object under study is taken from a population of similar phenomena. Consider for example the horizontal section of an oak, and suppose that you have to characterize how the microscopic anatomy of this particular tree differs from that of the others oaks. The species "oak" is somewhat vague, since it covers an unlimited number of trees which have not even yet begun to grow. The populations must be made finite and some of specific: "consider oaks which are adult this year in a given forest, and for each tree, take horizontal sections at one metre", for example. We give each tree the same probability of being chosen, and then sample a certain number for the experiment. The rest of the analysis is carried out as before, independently for each tree.

Notice that in the first step we affect the same weight to each specimen (law in number), although we proceed with analyses that may eventually use different working domains. In itself, Situation I is indifferent as to whether the phenomenon is reproducible or not.

(b) *Rotations*
The rotations appear either as a component of the structuring elements in certain iterative transformations (the skeleton for example), or in the averaging of a parameter that depends on the orientation of the space. The first case concerns 2-D analyses, and changes nothing of Situation I: at each point x, the union (or a more complex combination) of p set transforms according to B_i; $i = 1, \ldots, p$, generate a random set that is stationary in W as before. The B which now intervenes in relation (VIII-3) is $B = B_1 \oplus \ldots \oplus B_p$.

We now return to one B only, and assume that it depends on directional variables denoted by θ (an angle α in 2-D, a solid angle ω in 3-D, more generally an orthogonal basis in $\mathbb{R}^3$). The space of variation Θ of θ is compact; thus it has one and only one probability invariant under rotations, which is denoted by $d\theta/\mathrm{Mes}\,\Theta$. To each $\dot\theta$ with this distribution we associate a $\dot W_\theta$ given by a canonical W centred at the origin with rotation $\dot\theta$. The probability $P(B)$ given by relation (VIII-4) becomes $P(B|\theta)$ conditional on θ. The a priori probability $P(B)$ is then just the rotation average of the $P(B|\theta)$. Practically speaking, this means that we can treat each direction separately, using correctly oriented θ-neighbourhoods and θ-measuring masks and average over θ afterwards. Good estimators of the a priori probability $P(B)$ will have to use discrete patterns of directions regularly displayed in the plane or the space.

We see then that the two generalizations of rotations and of reproducible objects, do not essentially modify Situation I. Also the number of dimensions of the space does not really intervene. All these advantages come from the basic decision to look at *local textures* and not at individual sub-components of X. Thus, the possibility of studying a 3-D X from 2-D sections is solely a function of the dimension of B (which must be 1 or 2), the possibility of rotation averaging depends only on the non-isotropy of B, etc. . . . On the other hand, as soon as we wish to study individual sub-components of X, the "responsibilities" of B in the analysis will decrease rapidly, as we see below in Situation II.

B.4. Estimators

The neighbourhood W and its associated mask Z are in a fixed position with respect to each other and are moved within $D \oplus W$ by the same translate $\dot x$. When they occupy the respective locations $W_{\dot x}$ and $Z_{\dot x}$, the sampling problem consists of estimating a morphological transform $\Psi(X \cap W_{\dot x})$ from investigations performed on $X \cap Z_{\dot x}$. More precisely, we want to construct an unbiased estimator of $\mathrm{Pr}\{y \in \Psi(X \cap W_{\dot x})\}$. The general form of Ψ is a sequence of hit or miss transformations (see the various examples in Part Three; Chs. XI and XII in particular). The elementary steps are dilations and set differences, so Choquet's theorem shows that $\Psi(X \cap W_{\dot x})$ is stationary in W.

With $Z_{\dot x}$ we can associate a mask $Z'_{\dot x}$ satisfying the local knowledge principle (I-4), i.e. in which $\Psi(X)$ is known without bias. Due to the stationary of $X \cap (X_{\dot x} \cup Z_{\dot x})$ and to the translation invariance of Ψ, all the points y play an identical role as estimators. Thus the ratio

$$\frac{\mathrm{Mes}\left[\Psi(X \cap Z_{\dot x}) \cap Z'\right]}{\mathrm{Mes}\,Z'} = \mathrm{Pr}^*\{y \in \Psi(X \cap W_{\dot x})\} \qquad \text{(VIII-6)}$$

is an unbiased estimator of $\mathrm{Pr}\{y \in \Psi(X \cap W_{\dot x})\}$.

Remarks:

(i) The mask Z' depends only on the transformation Ψ and is provided by the edge correction formulae (rels. (II-11) and (II-17) to (II-19)).

(ii) We could also weight each point of $\Psi(X) \cap Z'_x$ differently, by kriging for example (Journel (1978)), which would reduce the estimation variance. However such a refinement is generally unnecessary in morphological studies.

(iii) Relation (VIII-6) show that various Z', regardless of their sizes, their shapes or their dimensions, provide unbiased estimators of the same probability, as soon as they satisfy relation (I-4). This is exactly Delesse's stereological "principle"; no longer is it a principle, but rather a consequence of an adequate formalization of Situation I, given in B-1 above.

(iv) The problem of definition and estimation of specific parameters such as S_V or M_V is slightly more sophisticated, but leads to similar results. To simplify notation, we denote by X the set actually investigated, even if it has previously been transformed by Ψ. Before randomization we defined over $X \cap D$ and its projections the two Minkowski measures $s(dy)$ and $a^{(3)}(dy, \omega)$, whose integrals are $S(X \cap D)$, and $A^{(3)}(X \cap D; \omega)$ respectively (notation illustrated by Fig. V.6). These two measures are linked via the relationship:

$$\frac{\pi}{4} s(dy) = \frac{1}{4\pi} \int_\Omega a^{(3)}(dy, \omega) d\omega \qquad \text{(VIII-7)}$$

(proof is by identification of their integrals over an arbitrary compact set).

When randomizing, $s(dy)$ and $a^{(3)}(dy, \omega)$ become stationary random measures in W, and their mathematical expectation in any compact set $Z \subset W$ is given by:

$$E[s(Z)] = \int \Pr\{y \in X \cap Z\} s(dy)$$

(Matheron 1975, Theorem 2-5.1); a similar relation is obtained by replacing s with $a^{(3)}$. Due to stationarity $E[s(Z)]$ and $E[a^{(3)}(Z)]$ do not depend on the size or shape of Z, nor on its location in W. They are two parameters intrinsic to the structure, namely the specific surface S_V and the specific total projection $A_V^{(3)}(\omega)$ in direction ω.

$$E[s(Z)] = S_V, \qquad E[a^{(3)}(Z, \omega)] = A_V^{(3)}(\omega) \qquad \forall Z \subset W$$

The geometrical interpretation (VIII-4), applied to S_V and $A_V^{(3)}(\omega)$, implies:

$$S_V = \frac{S[X \cap D]}{\text{Mes}(D \oplus \check{W})} \qquad A_V^{(3)}(\omega) = \frac{A^{(3)}(X \cap D, \omega)}{\text{Mes}(D \oplus \check{W})}$$

and by applying Cauchy's formula (V-15), we have

$$\frac{\pi}{4} S_V = \frac{1}{4\pi} \int_\Omega A_V^{(2)} (\omega) \, d\omega \qquad \text{(VIII-8)}$$

This integral allows us to estimate S_V via $A_V^{(3)} (\omega)$. To estimate the latter, take for Z, segments with total length $L(Z)$ and total number of inputs in $X \cap D$, $N^{(1)} (X \cap Z, \omega)$. Then the ratio

$$\frac{N^{(1)} (X \cap Z, \omega)}{L(Z)} = A_V^{*(3)} (\omega) \qquad \text{(VIII-9)}$$

is an unbiased estimate of the mathematical expectation $A_V^{(3)} (\omega)$. By repeating these estimations in different directions ω, averaging in ω, and applying (VIII-7), we obtain unbiased estimates of S_V.

Now all the stereological relations holding on the Minkowski functionals on deterministic compact sets can be given in terms of stationary random sets, just as we have done with relation (VIII-8). As above, the parameters $N_L(\omega)$, $U_A(\omega)$, $N_A(\omega)$ and M_V are interpreted as mathematical expectations of random measures. The first one is just $A_V^{(3)}(\omega)$ and thus is estimated by relation (VIII-9); U_A and N_A are estimated by:

$$\left. \begin{aligned} \frac{U(X \cap Z; \omega)}{A(Z)} &= U_A^*(\omega) \\[2mm] \frac{N^{(2)}(X \cap Z, \omega)}{A(Z)} &= N_A^*(\omega) \end{aligned} \right\} \qquad \text{(VIII-10)}$$

where Z is a plane compact set with normal ω, and belonging to W. (Note that when measuring the perimeter U or the 2-D connectivity number $N^{(2)}$ one *must* apply the edge corrections defined in Chapter II-C.)

By measuring in different directions, the two estimates (VIII-10) allow access to S_V and M_V, due to the relationships:

$$\left. \begin{aligned} \frac{1}{4} S_V &= \frac{1}{4\pi} \int_\Omega N_L(\omega) \, d\omega = \frac{1}{\pi} \cdot \frac{1}{4\pi} \int_\Omega U_A(\omega) \, d\omega \\[2mm] \frac{1}{2\pi} M_V &= \frac{1}{4\pi} \int_\Omega N_A(\omega) \, d\omega \end{aligned} \right\} \qquad \text{(VIII-11)}$$

C. SITUATION II

C.1. The estimation problem in Situation II

Situation II occurs under the following three conditions:
— a large object X (exactly in the sense of Situation I);

— morphological investigations concern individual sub-components of X;
— lack of stereological constraints.

As a corollary to the dealing with individuals, we are led to ratios of estimates. Therefore we could like to start this section by quoting a well known theorem of probability theory, which will constantly be used in Situations II, IV and V.

THEOREM VIII-2. *Let X_n and Y_n be two sequences of random variables independently converging in probability towards the random variables X and Y respectively. Then, the ratio X_n/Y_n converges in probability towards the limit X/Y:*

$$X_n \overset{Pr}{\to} X ; Y_n \overset{Pr}{\to} Y \Rightarrow \frac{X_n}{Y_n} \overset{Pr}{\to} \frac{X}{Y}$$

In other words, if X_n and Y_n represent the arithmetic mean of unbiased estimators calculated in n different measuring zones Z_n, by applying the law of large number we see that X_n/Y_n is an asymptotically unbiased estimator of the deterministic ratio X/Y (in brief: a.u. estimator).

As in Situation I, we delimit a working domain D inside the definition field R of the phenomenon. By construction, the domain D contains a finite integer number p of individuals X_i $(i = 1, \ldots p)$ whose union generates X. The X_i's may overlap. Unlike Situation I, the notion of a neighbourhood no longer exists, since the question now consists of estimating average characteristics of the population of size p in D, from the individuals visible in a few measuring masks Z_x. Consequently, we take the randomization to hold on the centre x of the moving mask Z and as before, we say that the random variable $\dot{x}$ is uniformly distributed in $D \oplus \check{Z}$.

Assume that a previous step of the analysis provides us with all the individuals contained in the mask $Z_{\dot{x}}$. The symbol $f(X_i)$ denotes an arbitrary global parameter associated with the individual X_i. In Situation II, the first goal consists of constructing an estimate of the mean value $\bar{f}$ of the $f(X_i)$ for the p individuals X_i of D:

$$\bar{f} = \frac{1}{p} \sum_{i=1}^{p} f(X_i) \tag{VIII-12}$$

In order to do this, we first calculate the mathematical expectation of the sum $\sum f(X_i)$ for the X_i's included in $Z_{\dot{x}}$. The locus of the points $\dot{x}$ such that $Z_{\dot{x}}$ contains X_i is equal to $Z \ominus \check{X}_i$, up to a translation. Thus, we may write:

$$E\left[\sum_{X_i \subset Z_{\dot{x}}} f(X_i) \right] = \sum_{i=1}^{p} f(X_i) \cdot \frac{\text{Mes}\,(Z \ominus \check{X}_i)}{\text{Mes}\,(D \oplus \check{Z})} \tag{VIII-13}$$

Since relation (VIII-13) is satisfied for any functional f, we may apply it to the function $g(X_i)$:

$$g(X_i) = \begin{cases} f(X_i) \cdot \dfrac{\text{Mes}(D \oplus \check{Z})}{\text{Mes}(Z \ominus \check{X}_i)} & \text{if Mes}(Z \ominus \check{X}_i) > 0 \\[2mm] 0 & \text{if not,} \end{cases}$$

yielding the fundamental Miles–Lantuéjoul formula:

$$\text{Mes}(D \oplus \check{Z}) \cdot E\left[\sum_{X_i \subset Z_{\dot{x}}} \frac{f(X_i)}{\text{Mes}(Z \ominus \check{X}_i)} \right] = \sum_{\text{Mes}(Z \ominus \check{X}_i) > 0} f(X_i) \qquad \text{(VIII-14)}$$

In this relation, the quantity between brackets is experimentally accessible from measuring masks $Z_{\dot{x}}$. However, it does not directly correspond to the mean characteristics of the individuals, for three reasons:

(i) the condition $\text{Mes}(Z \ominus \check{X}_i) > 0$ selects individuals small enough to be visible in the measuring masks, and extracts a subset of p' individuals from the total population p of X_i's present in the working domain D

(ii) the number p' (as well as p) is unknown

(iii) the multiplicative factor $\text{Mes}(D \oplus \check{Z})$ is a useless parasite

The first restriction, which cannot be overcome, is intrinsic to the approach. The other two are partially answered by applying relation (VIII-14) to the indicator functional of the individuals, i.e. the parameter

$$k(X_i) = \begin{cases} 1 \text{ if } X_i \subset D, \text{ and } \text{Mes}(Z \ominus \check{X}_i) > 0 \\ 0 \text{ otherwise,} \end{cases} \qquad \text{(VIII-15)}$$

which gives:

$$\text{Mes}(D \oplus \check{Z}) \cdot E\left[\sum_{X_i \subset Z_{\dot{x}}} \frac{k(X_i)}{\text{Mes}(Z \ominus \check{X}_i)} \right] = p' \qquad \text{(VIII-16)}$$

If, by some indirect means, we know the quantity $\text{Mes}(D \oplus \check{Z})$, then relation (VIII-16) provides an unbiased estimate of the number p' of the accessible individuals of the domain D. If not, dividing (VIII-14) by (VIII-16), we find

$$\frac{E\left[\displaystyle\sum_{X_i \subset Z_{\dot{x}}} \frac{f(X_i)}{\text{Mes}(Z \ominus X_i)} \right]}{E\left[\displaystyle\sum_{X_i \subset Z_{\dot{x}}} \frac{k(X_i)}{\text{Mes}(Z \ominus X_i)} \right]} = \frac{1}{p'} \sum_{i=1}^{p'} f(X_i) \qquad \text{(VIII-17)}$$

Each of the mathematical expectations of the left hand side can be separately estimated from the various measuring masks used in an experiment; the ratio of such estimates turns out to be a biased estimate of the mean characteristic f, over the *accessible* individuals of D, and not for all the X_i's of D as we initially wished. Note that the bias tends to zero as the number of individuals analysed is increased.

We focused the description of Situation II at the level of the working domain D. As for the meaning of the results in D relative to the definition domain R, we return to the discussion of Situation I, which remains valid in the present case.

One final remark: in what precedes, the working domain D and the definition field may be equally interpreted as bounded sets

— in $\mathbb{R}^1$, $\mathbb{R}^2$ or $\mathbb{R}^3$;
— in a product of Euclidean space by a space of rotations;
— in $\mathbb{R}^n \times \mathbb{R}^+$, where $\mathbb{R}^+$ represents a time evolution (grains in a sintering process, for example);
— in $\mathbb{R}^n \times \mathbb{N}$, where $\mathbb{N}$ denotes a finite collection of possible experiments; . . . or as any combination of these interpretations.

C.2. An example of Situation II

In the following example, due to Ch. Lantuejoul (1977), we study the number of edges of metallic grains seen in section. The definition domain R is a block of sintered alumina prepared under given metallurgic conditions. The working domain D is made of a few polished sections on R. The total population of individuals is *the set of all possible cross sections* of the 3-D metallic grains. Measurements are performed in rectangular masks Z of about $300 \times 400\,\mu^2$. The photograph (VIII-5a) gives a view of one polished section intersected by such a mask Z_i. The first step is to eliminate the particles hitting the mask edges, according to the procedure described in Chapter XI, G.2. Consider a fixed measuring mask, say Z_x. With each individual $X_i \subset Z_x$, we associate the indicator $\delta(i, n)$ which equals one when X_i has n sides, and zero when not. Then we calculate the following two sums:

$$\Sigma_1 = \sum_{X_i \subset Z_i} \frac{\delta(i, n)}{A(Z \ominus X_i)} \quad \text{and} \quad \Sigma_2 = \sum_{X_i \subset Z_{\dot{x}}} \frac{k(X_i)}{A(Z \ominus X_i)}$$

($k(X_i)$ is defined in (VIII-15)). The ratio f^* of the arithmetic averages of Σ_1 and Σ_2 over all the measuring masks is an a.u. estimate of frequency of the n-sided grains (rel. (VIII-17)).

Figure VIII.5 shows the corresponding histogram calculated for more than 1,000 grains and compares it with the one obtained *without performing the Miles–Lantuéjoul correction*, i.e. by calculating the ratio f^{**} of the arithmetic average of $\sum_{X_i \subset Z_{\dot{x}}} \delta(i, n)$ and $\sum_{X_i \subset Z_{\dot{x}}} k(X_i)$.

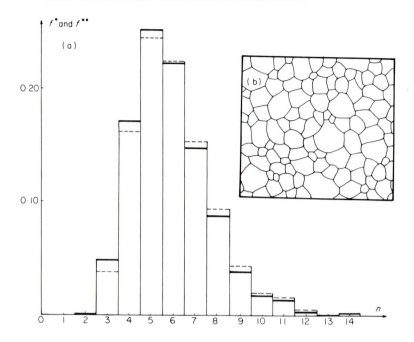

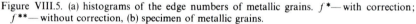

Figure VIII.5. (a) histograms of the edge numbers of metallic grains. f^*—with correction; f^{**}—without correction, (b) specimen of metallic grains.

In particular the mean values of the edge numbers are equal to 5.895 (without correction), and 5.998 (with correction). Remember that for thermodynamic reasons, the theoretical value of this mean is 6.

Remarks:

(i) the reader may find the above correction negligible. In fact, the larger the range of individual sizes, the more important are the conditions, since they counterbalance the fact that the small individuals have more chance to be included in measuring masks.

(ii) the experimental technique for computing the number of sides of a grain Z by using morphological transformations is given in Ch. XI(H.1). As for the area $A(Z \ominus X_i)$, which might present some difficulty since the object itself plays the role of a structuring element, it is in fact an easy computation. We treat the case where the elementary mask Z is a rectangle. Then we have:

$$Z \ominus X_i = Z \ominus T(X_i) \qquad \text{(VIII-18)}$$

where $T(X_i)$ is the smallest rectangle enclosing X_i, with edges parallel to those of Z. To prove relation (VIII-18), consider Z to be the intersection of two strips, say D_1 and D_2 (Fig. VIII.6).

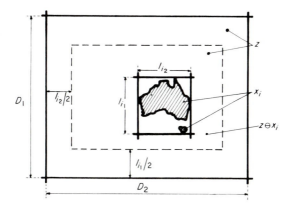

Figure VIII.6. Erosion of a rectangular mask by the individual X_i.

We can write:

$$Z \ominus X_i = (D_1 \cap D_2) \ominus X_i = (D_1 \ominus X_i) \cap (D_2 \ominus X_i);$$

but $D_j \ominus X_i = D_j \ominus T(X_i)$ $j = 1, 2$

and formula (VIII-18) results. Consequently, if L_1 and L_2 are the two lengths of the sides of Z, and $l_{i,1}$ and $l_{i,2}$ those of $T(X_i)$, then the area $A(Z \ominus X_i)$ is simply equal to $(L_1 - l_{i,1})(L_2 - l_{i,2})$.

D. SITUATION III

D.1. Definition

Situation III concerns estimation for only *one* individual; it is characterized by the following two conditions:

— the bounded object X is partially known by I samples taken in the zones $Z_i (i = 1, \ldots I)$ of the space, where the convex hull of the union of the Z's covers X;
— we are only interested in estimating the integral $\mu(X)$ of a measure $\mu(dx)$ defined over X.

Moreover, for the sake of simplicity we shall assume the sampling to be periodic (for more sophisticated sampling patterns, see the stratified random networks in M. David (1975) or in J. Serra (1967)). One might wonder whether the masks Z_i themselves have to be bounded, or not. In fact the distinction between bounded and unbounded samples is not a basic one, and can be overcome in the following way.

Consider, in the (Ox_1, Ox_2) plane, two possible samplings, one by (unbounded) stripes (Fig. VIII.7a), and one by rectangular masks (b). In the first case, associate with the measure $\mu\,(dx_1, dx_2)$ its integral $\mu_1\,(dx_1)$ along the x_2-axis; that is,

$$\mu_1\,(dx_1) = \int_{-\infty}^{+\infty} \mu\,(dx_2; dx_1)$$

The support of the 1-D measure $\mu_1\,(dx_1)$ is the projection of X on the axis dx_1. Sampling $\mu\,(dx_1, dx_2)$ by stripes Z_i with a thickness d, or $\mu_1\,(dx_1)$ by segments of length d are two identical operations. These segments are just 1-D versions of the rectangles of Figure VIII.7b, or more generally parallelepipeds in $\mathbb{R}^n$. Therefore, without loss of generality, we perform the analysis only for the case of bounded masks Z_i, but leave free the number of dimensions of the space.

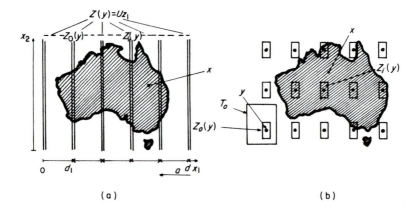

Figure VIII 7. Two possible modes of sampling. (a) unbounded cross-sections, (b) bounded measuring masks.

D.2. Randomization

We first examine the regularizing effect of sampling. Z is modelled to be a bounded open set. The moving average $(1/\mathrm{Mes}\,Z)\,\mu\,(Z_x \cap X)$, centred at point x, generates a *function* (and no longer a measure) $\mu_Z\,(x)$. One calls it the regularized version of μ by Z. It can be expressed as a convolution product of μ by the indicator function $k_{\check{Z}}$ of the transposed $\check{Z}$ of the mask centred at the origin. We have:

$$\mu_Z(x) = \frac{1}{\text{Mes } Z} \mu(X \cap Z_x) = \frac{1}{\text{Mes } Z} \int_{\mathbb{R}^n} \mu(dy) k_Z(y-z)$$

$$= \frac{1}{\text{Mes } Z} \int_{\mathbb{R}^n} \mu(dy) k_{\check{z}}(x-y) = \frac{\mu * k_{\check{z}}}{\text{Mes } Z}$$

(VIII-19)

Consequently, our problem reduces to the sampling of a (bounded with a bounded support) function using a regular grid of points i, which denotes the centres of cells T that partition the space. Let us denote by T_0, the cell of type T centred at the origin of the axes, by x the location within T_0 of the origin point of the grid, and by i the vector joining x, the origin of the grid to the grid point no. i (see Fig. VIII.7b).

The assertion "The pattern of samples Z_i is uniformly implanted over the object" will be translated into probabilistic terms by interpreting the origin x of the grid as a random point $\dot{x}$ uniformly distributed on the cell T_0. Then $\mu_Z(\dot{x})$ becomes a random variable, as well as the sum $\sum_i \mu_Z(\dot{x}+i)$ of the values of μ_Z for all the grid points. The mathematical expectation of this sum is easily derived from (VIII-19):

$$E\left[\sum_i \mu_Z(\dot{x}+i)\right] = \frac{1}{\text{Mes } T_0} \sum_i \int_{T_0} \mu_Z(x+i)dx$$

$$= \frac{1}{\text{Mes } T_0} \int_{\mathbb{R}^n} \mu_Z(x)dx = \frac{\mu(X)}{\text{Mes } T_0}$$

In other words, if we assume a uniform implantation of the grid origin, then the quantity

$$\frac{\text{Mes } T_0}{\text{Mes } Z} \cdot \sum_i \mu(X \cap Z_i)$$

(VIII-20)

is an unbiased estimate of the integral $\mu(X)$.

Remarks

(i) we used the measure formalism above so that we could include into Situation III measurements such as perimeters, surface areas, convexity numbers, etc. To obtain these parameters requires us to model the sampling zones Z_i as open sets rather than thinking of them as points. However the approach remains *a fortiori* valid when the measure $\mu(dx)$ is a function of $\mu(x)dx$. Then the Z_i can be reduced to points, and the quantity $\text{Mes } T \cdot \sum_i \mu(i)$ replaces (VIII-20) as an unbiased estimator of $\int \mu(x)dx$;

(ii) we randomized in Situation III by keeping the orientation of the samples fixed. The extension of the results to a uniform orientation of the sampling pattern is immediate. When the measure under study, let us say $\mu(dx, \alpha)$, is anisotropic (in $\mathbb{R}^2$ the convexity number of X in direction α for example), one associates with each $\mu(X, \alpha)$ its average over rotations, $\mu(X)$:

$$\mu(X) = E[\mu(X, \alpha)] = \int_{\alpha} \mu(X, \alpha) \cdot \Pr(d\alpha) \qquad \text{(VIII-21)}$$

where $\mu(X, \alpha) = \int_{\mathbb{R}^n} \mu(dx, \alpha)$

By reversing the order of integration, we see that

$$\mu(X) = \int_{\mathbb{R}^n} \int_{\alpha} \mu(dx, \alpha) \Pr(d\alpha) \qquad \text{(VIII-22)}$$

where the integral over α is the isotropized version of $\mu(dx, d\alpha)$. From a practical point of view, relation (VIII-22) means that we can fix the orientation of the sampling pattern, and average in each mask Z_i of the pattern. The integral over $\mathbb{R}^2$ of the resulting measure is then estimated as previously. Of course, in each mask the average over α forms the subject of a further estimation, since digitalization demands that we replace $\Pr(d\alpha)$ by probabilities concentrated on the main directions of the digital lattice of picture points.

(iii) Relations (VIII-21) and (VIII-22) are equally valid in $\mathbb{R}^2$ or in $\mathbb{R}^3$. However, orientation averaging in $\mathbb{R}^3$ is hardly possible on a single 3-D specimen: to section a specimen in the first chosen direction automatically destroys it! Thus a realistic extension of the situation consists of enlarging the analysis to N objects of the same type (e.g. N ovaries) and considering each of them as a different realization of the same random set. Then $\mu(X)$ as defined in (VIII-22) becomes a random variable. Each of the N objects is sectioned in a different direction, according to the law $\Pr(d\alpha)$, and $\mu(X, \alpha)$ is calculated. The quantity $E[\mu(X)]$ (mathematical expectation over the N realizations) is then estimated by the arithmetic mean of the $\mu(X, \alpha)$'s.

E. SITUATION IV

E.1. Definition

Consider images such as the alumina grains (Fig. VIII.5a) or the section of the clusters of cells in an embryonic ovary (Ch. IX, E). In all these studies, we are concerned with 3-D individuals, namely the population alumina crystals, or the neighbouring cells around each ovogonia (where the *individual* is the disconnected pack of the surrounding cells). These situations have the following two common features:

(i) each sampled individual is recognized by one section only;
(ii) the observations refer to a large population of individuals.

To what extent can we counterbalance only one sectioning per object with the fact that we only wish to estimate average characteristics of the population (such as the mean volume of the individual, or its mean 3-D connectivity number, for example)?

E.2. Randomization

Denote by X'_i ($i = 1, \ldots, I$) the I individuals under study and their union by X. The X''_i's are bounded (thus topologically modelled to be compact) and disjoint from each other.

$$X = \bigcup_i X'_i \qquad X'_i \cap X'_j = \varnothing, \qquad i \neq j \qquad \text{(VIII-23)}$$

Basically, the mode of randomization that we shall adopt does not differ from that of Situation III. Firstly, we fix the direction ω of the section plane $\Pi(x, \omega)$, whose abscissa x is measured along an axis $O\omega$ parallel to direction ω (Fig. VIII.8).

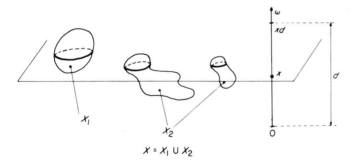

$$X = X_1 \cup X_2$$

Figure VIII.8. 2-D section of individuals.

Let d be a positive number greater than the largest diameter $D_\omega(X)$ of X in direction ω. Choose the origin O in such a way that the projection of X on the axis $O\omega$ is completely embedded in the segment $[O, x_d]$ (x_d is explained by Fig. VIII.8). Take x to be a random point $\dot{x}$ uniformly distributed over $[O, x_d]$, and apply the randomization described in Situation III, using parallel sections with spacing d. By construction the only plane which hits X is $\Pi(\dot{x}, \omega)$; therefore the surface area of its intersection with X, multiplied by d, provides an unbiased estimate $V^*(X)$ of the volume of X:

$$V^*(X) = d \cdot A[X \cap \Pi(\dot{x}, \omega)] \qquad \text{(VIII-24)}$$

If, instead of X, we consider any of the X_i''s, we obtain a similar relation by replacing X by X_i' in relation (VIII-24). Define now the individual average volume $E[V(X')]$ of the X_i''s by putting:

$$E[V(X')] = \frac{1}{I} \sum_{i=1}^{I} V(X_i') = \frac{V(X)}{I} \qquad \text{(VIII-25)}$$

Relation (VIII-25) enables us to construct the following unbiased estimate of $E[V(X')]$:

$$E^*[V(X')] = \frac{d}{I} A[X \cap \Pi(\dot{x}, \omega)] \qquad \text{(VIII-26)}$$

This fundamental equation clearly indicates the limits of the one section analysis. The total number I of individuals is unknown, and the spacing d has some arbitrariness about it. We can only play on ratios of similarly structured estimators, in order to eliminate d/I by division. We shall now work towards this end by distinguishing whether or not the individuals can be recognized on the sections.

E.3. General results

By "general", we mean that results are obtained without knowing whether or not two points of the section $X \cap \Pi$ belong to the same individual.

(a) *Anti-extensive transformations*
Each set X_i' is transformed by an anti-extensive mapping Ψ, locally calculable from 2-D sections (erosion by a disk for example). Define the reduced volume v of the transform as the volume of $\Psi(X_i')$ divided by the volume of the original set:

$$v_\Psi(X_i') = \frac{V[\Psi(X_i')]}{V(X_i')}$$

We are interested in the average value $v_\Psi(X')$ over the X_i''s, where each term is volume weighted:

$$v_\Psi(X') = \frac{\sum_i V(X_i') \cdot v_\Psi(X_i')}{\sum_i V(X_i')} = \frac{V[\Psi(X)]}{V(X)}$$

Applying (VIII-24) to $V[\Psi(X)]$ and to $V(X)$ we obtain:

$$v_\Psi^*(X') = \frac{V^*[\Psi(X)]}{V^*(X)} = \frac{A[\Psi(X \cap \Pi(\dot{x}, \omega))]}{A[X \cap \Pi(\dot{x}, \omega)]} \qquad \text{(VIII-27)}$$

which is an a.u. estimator of $v_\Psi(X')$ (Th. VIII-2). The reader will find illustrations of this situation in the material on the size distributions (Ch. X, Ex. 11).

(b) *Cells with nuclei*
Each X_i' contains an individualized part Y_i', such as a biological cell with its nucleus for example. The mean volume $E[V(Y')]$ of the nuclei is defined by transposing to the Y_i''s the relation (VIII-25) which defines $E[V(X')]$. Then we derive from (VIII-26):

$$\frac{E^*[V(Y')]}{E^*[V(X')]} = \frac{A[Y \cap \Pi(\dot{x}, \omega)]}{A[X \cap \Pi(\dot{x}, \omega)]} \qquad (VIII-28)$$

where $Y = \bigcup_i Y_i'$.

(c) *Reduced Minkowski functionals*
When dealing with individuals, one often wants to express their surface, or their norm, per unit volume of the individual; i.e. to consider the ratios

$$\frac{S(X_i')}{V(X_i')} \quad \text{and} \quad \frac{M(X_i')}{V(X_i')}$$

instead of the absolute parameters $S(X_i')$ and $M(X_i')$. To what degree can we estimate such ratios?

We will present the construction of an estimator in the case of the norm M. Integral geometry provides us with relationships that associate the norm with integrals of 2-D connectivity numbers (rels. (V-13) and (IV-18)) as follows:

$$2\pi M(X_i') = \frac{1}{4\pi} \int_\Omega D^{(3)}(X_i', \omega)\,d\omega$$
$$= \frac{1}{4\pi} \int_\Omega d\omega \int_{-\infty}^{+\infty} N^{(2)}[X_i' \cap \Pi(x, \omega)]\,d\omega \qquad (VIII-29)$$

The simplest randomization approach here is to interpret the pair (x, ω) as a random point uniformly distributed in the product space $[0, x_d] \times \Omega$ of the segment of length d by the unit sphere Ω. Then relation (VIII-29) shows that the quantity

$$\frac{d}{2\pi I} N^{(2)}[X \cap \Pi(\dot{x}, \dot{\omega})]$$

is an unbiased estimator of the average norm $E[M(X')] = (1/I)\sum M(X_i')$. Dividing by the average volume estimator we find:

$$\frac{E^*[M(X')]}{E^*[V(X')]} = \frac{1}{2\pi} \cdot \frac{N^{(2)}[X \cap \Pi(\dot{x}, \dot{\omega})]}{A[X \cap \Pi(\dot{x}, \dot{\omega})]} \qquad (VIII-30)$$

Similarly, starting from the integral geometry relations (IV-17) and (IV-19) for the surface $S(X_i')$, and averaging over i, we obtain the estimator:

$$\frac{E^*[S(X')]}{E^*[V(X')]} = \frac{\pi}{4} \cdot \frac{U[X \cap \Pi(\dot{x}, \dot{\omega})]}{A[X \cap \Pi(\dot{x}, \dot{\omega})]} \qquad \text{(VIII-31)}$$

($U(X) = $ perimeter of X).

Relations (VIII-30) and (VIII-31) are trivially extended to the case (b) of cells with nuclei, by replacing X with Y in the numerators.

To estimate the integral over Ω in (VIII-29) is a less trivial question. If we could simultaneously take sections in different directions without destroying the material, as the rotating scanners do in radiology, then the integral could be estimated by independent uniform random directions. However this seems rather utopian, and a more realistic way would be to section several samples, each one in a different direction (see Ex. VIII-5). But even this multi-sampling technique is over-sophisticated for 90 per cent of our practical purposes. Experience has shown us that either the individuals are oriented in every possible direction and their disorder makes up for the multi-sectioning problems (it suffices therefore to apply relation (VIII-30)), or they all present anisotropies in the same directions, and then the sections are systematically made normal or parallel to these preferential directions. In this anisotropic case, the right-hand side of formula (VIII-30) must be read as a.u. estimating

$$\frac{1}{2\pi} \frac{E[D^{(3)}(X', \omega)]}{E[V(X')]}$$

Knowledge of the diameters $D^{(3)}$ in the key directions is here more instructive than knowledge of the norm, which is an isotropic parameter.

(d) *Comments*

The three sub-situations (a) to (c), described immediately above, share some common features.

(i) It is not necessary that the random plane Π hits all the individuals X_i' that make up X. Empty intersections are implicitly taken into account in the formalism.

(ii) the ratio of the mathematical expectations of two random variables x and y is in general not the mathematical expectation of the ratio x/y, although it is when the random variable y is a constant. If this condition holds, all the formulae (VIII-27) to (VIII-31) can be re-interpreted. Take for example relation (VIII-28): when all the cells X_i' all have the same volume, the right hand side of the relation is an unbiased estimator of $E[V(Y')/V(X')]$. This observation bridges the gap between our approach and that of R. E. Miles and P. Davy (1976). They consider a single set Y embedded in a single set X and assume that after sectioning by one random plane $\Pi(\dot{x}, \dot{\omega})$ the specimen can be glued back together for a new sectioning, according to other values of the pair $(\dot{x}, \dot{\omega})$. Moreover, they want to use the random ratio $A[Y \cap \Pi(\dot{x}, \dot{\omega})]/A[X \cap \Pi(\dot{x}, \dot{\omega})]$ to estimate the volume ratio, which leads them to replace the uniform probability on $\Omega \times [0, x_d]$ with another, weighted by $A[X \cap \Pi(x, \omega)]$. This finally amounts the use of the random ratio. In their problem, it would have been logically equivalent to separately estimate the numerator and the denominator of the random ratio (as we did above in (b) and (c)) by using uniform probabilities. Indeed, by taking all the identical Y_i'''s and all the identical X_i''s in our approach, we again end up with their results. Moreover, their system of probabilities is really difficult to implement in practice (The experimenter must firstly pick a point at random inside the object, then choose a randomly oriented section that goes through this point).

(iii) similar results are easily obtained by 1-D sections of 2-D or 3-D structures (see Ex. VIII-5). In all these cases, a fundamental limitation comes from the fact that the individuals are not in general recognizable on section. We now show that new results are possible for which this weakness no longer exists.

E.4. Individuals identifiable on sections

As before, X is the union of disjoint individuals X_i', but we now assume that upon taking two arbitrary points in $X \cap \Pi$ we are always able to say whether or not they belong to the same X_i'. If the individuals are convex particles, it obviously suffices to look at the segment joining the two points and for more complex individuals other criteria may serve (the distance between the two points; the fact that they are connected after a morphological closing, etc.). We will not discuss this pattern recognition question any further, but will simply make the above assumption.

(a) *V-weighted volumes*
In the relationship (VIII-25), all the volumes $V(X_i')$ are given the same weight $1/I$. Alternatively, we could also weight each of them by its own value; the procedure defines the *V*-weighted average volume $\mathscr{M}(X')$ as follows:

$$\mathscr{M}(X') = \frac{\sum_i V^2(X_i')}{\sum_i V(X_i')} = \frac{\sum_i V^2(X_i')}{V(X)} \qquad (\text{VIII-32})$$

This average is automatically performed by certain sizing techniques such as sieving for example (Ch. X, Sects. B and D). Modulo a constant factor, the numerator of expression (VIII-32) can be derived from the set covariance of B.

Put:

$$K(h) = V(X \cap X_h) \qquad K_i(h) = V(X'_i \cap X'_{i,h})$$

where X_h and $X'_{i,h}$ are the translates of X and X'_i respectively, and h is a vector with modulus $|h|$ and direction α. For all directions α lying in the plane $\Pi(x, \omega)$ (i.e. $\alpha \perp \omega$), $K_i(h)$ is estimated by the quantity

$$K_i^*(h) = dA[X'_i \cap X'_{i,h} \cap \Pi(\dot{x}, \dot{\omega})] \qquad \text{(VIII-33)}$$

and

$$\sum_i K_i^*(h) = d \sum_i A[X'_i \cap X'_{i,h} \cap \Pi(\dot{x}, \dot{\omega})] \qquad \text{(VIII-34)}$$

The summation on the right-hand side is extended to all the X'_i's, even those whose section by $\Pi(x, \omega)$ is empty. This means that one must construct the intersections $X'_i \cap X'_{i,h} \cap \Pi(x, \omega)$ only for those individuals visible on $\Pi(x, \omega)$, which is fortunately possible when we make the above identification assumption. The corresponding areas are then summed up (note that the result is *not* the area $A[X \cap Y_h \cap \Pi(\dot{x}, \dot{\omega})]$!).

From relation (IX-3) we know that

$$\int_{R^3} K_i(h)\,dh = 2 \int_\Omega d\alpha \int_0^\infty |h|^2 K_i(|h|, \alpha)\,d|h| = V^2(X'_i) \quad \text{(VIII-35)}$$

For $|h|$ and α varying in the plane $\Pi(\dot{x}, \dot{\omega})$, the integral (VIII-35) is estimated from (VIII-33). For the variation of α we meet again the situation described in Section D, 2c, and in Exercise VIII-5. For the sake of simplicity, we shall be satisfied with only one section plane, i.e. with the following unbiased estimator:

$$\sum_i [V^2(X'_i)]^* = 2d \sum_i \sum_{l=1}^{2\pi/\Delta\alpha} l\,\Delta\alpha \sum_{s=1}^{\infty} s^2(\Delta h)^2 K_i^*(s\,\Delta h, l\,\Delta\alpha) \quad \text{(VIII-36)}$$

where Δh and $\Delta\alpha$ are the increments in the discrete version of $K_i(|h|, \alpha)$, and α varies perpendicularly to ω in the section plane $\Pi(\dot{x}, \dot{\omega})$. Dividing (VIII-36) by (VIII-24), we eliminate the parameter d and, according to (VIII-32), we get an a.u. estimate of the V-weighted moment $\mathscr{M}(X')$.

Using a proof similar to that described in Chapter X, D.3 we can also estimate the non-centred V-weighted moment of order two $\mathscr{M}(X'^2)$ from plane section measurements, but not the moments of higher order. Therefore, we are able to test whether $\mathscr{M}(X'^2) \sim (\mathscr{M}(X'))^2$. If so, the X'_i's practically all have the same volume and the ratios of estimates involved in formulae (VIII-27) to (VIII-32) become a.u. estimates of the limiting expected ratios.

(b) *Mean sections and projections*
Being able to identify the individuals on sections enables us to define and to measure their average section areas as follows. Denote by $L[X'_i | \Delta(\omega)]$ the

length of the projection of X'_i on an axis Δ of direction ω (this is *not* the total diameter $D^{(3)}(x', \omega)$ that we see in the norm). The volume $V(X'_i)$ is given by the integral of the area of the plane section of X'_i by $\Pi(x, \omega)$ as x sweeps over this projection. Therefore the average section $E_x[A(X'_i \cap \Pi(x, \omega))]$ of X'_i by $\Pi(x, \omega)$ is just the expression:

$$E_x[A(X'_i) \cap \Pi(x, \omega)] = \frac{1}{L[X'_i|\Delta(\omega)]} \int A[X'_i \cap \Pi(x, \omega)]\,dx$$

$$= \frac{V(X'_i)}{L[X'_i|\Delta(\omega)]}$$

where E_x is the mathematical expectation over x, for i and ω fixed. To remove i, we must weight the above by the probability an individual 2-D section of direction ω belongs to X'_i, i.e. by

$$L(X'_i|\Delta(\omega))/\sum_i L(X'_i|\Delta(\omega))$$

We have

$$\underset{i}{E}\,\underset{x}{E}[A(X'_i \cap \Pi(x, \omega))] = \frac{\sum_i V(X'_i)}{\sum_i L[X'_i|\Delta(\omega)]}$$

$$= \frac{E_i[V(X'_i)]}{E_i[L(X'_i|\Delta(\omega))]}$$

(where $E_i[L(X'_i|\Delta(\omega))] = (1/I)\sum_i L(X'_i|\Delta(\omega))$). That is, the individual mean section area equals the average volume divided by the average 1-D projection in direction ω.

Formula (VIII-37) lends itself to an isotropic version by unconditioning over ω. Denoting by $\bar{*}$ the *a priori* mathematical expectation of $*$, e.g.

$$\bar{L}(X'|\Delta) = \underset{i}{E}\,\underset{\omega}{E}[L(X'_i|\Delta(\omega))] = \frac{1}{4\pi I}\sum_i \int_\Omega L(X'_i|\Delta(\omega))\,d\omega$$

and

$$\bar{A}[X' \cap \Pi] = \underset{i}{E}\,\underset{\omega}{E}\,\underset{x}{E}[A(X'_i \cap \Pi(x, \omega))]$$

we have:

$$\bar{A}(X' \cap \Pi) \times \bar{L}(X'|\Delta) = E[V(X')] \tag{VIII-38}$$

That is, the individual mean volume is the product of the individual mean 2-D section by the individual mean 1-D projection.

The same approach gives a series of similar formulae of which the two most important are the following:

$$E[V(X')] = \bar{A}[X'|\Pi] \times \bar{L}[X' \cap \Delta] \tag{VIII-39}$$

$$\bar{A}[X'|\Pi] = \bar{L}[X' \cap \Delta] \cdot \bar{L}[X'|\Delta] \tag{VIII-40}$$

The relationships (VIII-38) to (VIII-40) are immediately transposable in terms of a.u. estimators. Take, for example: relation (VIII-37). Firstly, notice that the number $I[X \cap \Pi(\dot{x}, \omega)]$ of individuals present on the section $\Pi(\dot{x}, \omega)$ allows us to construct an unbiased estimator of the individual mean 1-D projection in direction ω, as follows:

$$I \cdot E_i^* [L(X_i') | \Delta(\omega)] = d \cdot I[X \cap \Pi(\dot{x}, \omega)] \qquad \text{(VIII-41)}$$

Then combine with (VIII-26), to give:

$$E^*_{i,x} [A(X_i' \cap \Pi(\dot{x}, \omega))] = \frac{A[X \cap \Pi(\dot{x}, \omega)]}{I[X \cap \Pi(\dot{x}, \omega)]} \qquad \text{(VIII-42)}$$

Relation (VIII-41) is interesting in itself, since it links the 3-D number of individuals with the apparent 2-D number. It will be given a more thorough formulation in the stationary case (rel. (VIII-43)).

Remarks

1. All the comments already made about multi-directional sectioning (Situation III, 2c) are again valid here.

2. No stronger result comes from the supplementary assumption that the X_i's are convex grains. Such a hypothesis is only of interest in that via Cauchy's formula, a geometric interpretation of the mean projections in terms of surface areas and norms is possible (see Ex. VIII-3).

F. SITUATION V

This last situation deals with individual analysis in the local stationary case. The scales being used are the same as in Situation I, and the sizes of the samples (the neighbourhoods) are small with respect to the whole structure under study. Moreover, each sample is investigated within 2-D measuring masks, such as rectangular T.V. frames for example, and as in Situation II, we assume the individuals to be inscribable in the measuring masks.

The randomization in Situation V is governed by the local stationarity Theorem VIII-1. Indeed, the task is more to transpose the results already established for Situation IV than to find new ones. As before, we shall distinguish between whether or not individuals are identifiable in section and when they are, we shall give special attention to the specific number of induced grains.

F.1. Individuals identifiable on sections: specific numbers

In the local approach, formula (VIII-41) is replaced by the simpler formula:

$$I_V(X) \cdot E[L(X_i') | \Delta(\omega)] = I_A[X \cap \Pi(\omega)] \qquad \text{(VIII-43)}$$

That is, the number $I_V(X)$ of individuals per unit volume, multiplied by the mean projection length of an individual on an axis $\Delta(\omega)$ of direction ω, equals the number of induced individuals per unit area, in direction ω. (We do not use the symbols N_A and N_V here, since we are not dealing with Euler–Poincaré constants.)

Proof: Let X be a stationary random set in $\mathbb{R}^3$ which is sectioned by the plane $\Pi(\omega)$. Denote by (x_1, x_2, x_3) the co-ordinates of a point x in the space, and take the x_2 and x_3 axes in $\Pi(\omega)$; the direction of the x_1 axis then coincides with ω. Suppose X is made up of disjoint individuals X_i'. With each of them, we associate a point $x_i = (x_{i1}, x_{i2}, x_{i3})$, called the germ of X_i' (e.g. its centre of gravity, or its highest point in direction ω, etc.). Consider a small cylinder with a base $dx_2\, dx_3$ and a direction ω. Suppose that a grain placed at (x_1, x_2, x_3) induces by section, another grain in $\Pi(\omega)$ with a probability $\Pr(x_1)$. Summing up such events along a small cylinder of direction ω and base $dx_2\, dx_3$, we obtain:

$$I_A[X \cap \Pi(\omega)]\, dx_2\, dx_3 = I_V(X) \cdot dx_2\, dx_3 \int_{-\infty}^{+\infty} \Pr(x_1) dx_1$$

Noting that the integral on the right hand side is just the mathematical expectation of the projected length of an individual in direction ω, we finally obtain relation (VIII-43). Q.E.D.

The above result concerns a stationary random set. However because of Theorem VIII-1, we can easily apply it to our actual situation by specifying a few restrictions. Let $Z(x)$ denote the 2-D measuring mask oriented in direction ω, and let ρ be the largest diameter of the individuals. We take $W(x)$ to be the neighbourhood of an arbitrary set which contains the dilate $Z(x) \oplus \rho B$ of Z by the ball of radius ρ. Thus the individuals which are not completely inside $W(x)$ cannot hit $Z(x)$. We now randomize x according to Situation I: $W(\dot{x}) \cap X$ and $Z(\dot{x}) \cap X$ become local stationary random sets, to which relation (VIII-43) is applicable. The intersection $X \cap \Pi$ can be replaced by $X \cap Z$, provided that we use Miles–Lantuéjoul edge correction of Situation II.

As for the passage $\mathbb{R}^1 \to \mathbb{R}^3$, we have a similar result and the surface number I_L of individuals induced on a line satisfies the relationship:

$$I_V(X) \cdot E[A(X_i')|\Pi(\omega)] = I_L[X \cap \Delta(\omega)] \qquad \text{(VIII-44)}$$

F.2. Other formulae

We have just shown, with a detailed examination in one example, how the transposition from global $\to$ local proceeds for an individual analysis. More generally:

(a) All the estimates established from relations (VIII-24) to (VIII-31) (non-identifiable individuals) remain valid by replacing $X \cap \Pi(\dot{x}, \dot{\omega})$ with $X \cap Z(\dot{x}, \omega)$ in the right-hand side of the formulae. Here, the edge corrections when computing $N^{(2)}$ and U, are those of Chapter II, Section C.2e.

(b) In the case of identifiable individuals, we can also substitute $Z(\dot{x}, \dot{\omega})$ for $\Pi(\dot{x}, \dot{\omega})$, but we must be careful to systematically apply the Miles–Lantuéjoul edge correction (in formulae (VIII-33) and (VIII-37) for example).

G. CONCLUSION

G.1. The limits of Situations IV and V

Looking beyond the technicalities of a quantitative description, Situations IV and V look like variations on a failure. We obtained several non-trivial results, but there is always one missing piece of information, namely either the mean projection of an individual on a straight line, or on a plane, or the 3-D number I. If this information is not available from some external source, relations (VIII-25) to (VIII-42) or (VIII-44) do not allow estimation of $E[V(X')]$ or of $I_V(X)$.

Sometimes the information is available from techniques other than plane sections (e.g. in powder technology, the particles are spread out on a piece of glass looked at in transmitted light). However, if our experimental access to the structure is made solely via plane sections, the only possible resort is the morphological model. In particular if the X_i''s:

— have the same volume (Ch. VIII, E.4b, Ch. VIII, C.1)
— are balls (Ch. X, F)
— are Poisson polyhedra (Ch. XIII, D)

then $E[V(X')]$ is accessible from plane measurement. Note that each of these three hypotheses can be tested from sections. If one of these fails, the V-weighted mean volume $\mathcal{M}[V(X')]$ and the corresponding variance are always available, and serve as very robust descriptors of the volumes.

G.2. A critique of the uniform distribution

As an initial comment, we would like to emphasize that using a uniform distribution does not imply that the experimental material is necessarily sampled by *independent* sections. It could equally be accessed by parallel serial sections, with an origin which is uniformly distributed (VIII, D.3).

Throughout the chapter we formalized the sampling modes of operation by uniform distributions in suitable domains. The resulting probabilities transcribe into mathematical terms the sentence "the sample was chosen randomly from the structure" and they provide unbiased estimates of the Minkowski measures defined either on the initial image, or after a morphological transformation. But there is nothing sacred about their use. If the experimenter knows for example that the structure contains a central axis and he wants to have it in the section plane, then let him have it there. Also consider the biopsy samplings in the human bone. They are often used in the diagnosis of osteoporoses and osteomalaciae. The biopsies must obviously be taken from a place where damage to the bone is kept to a minimum, and consequently the middle of the iliac bone is chosen. The statistical suggestion that one should work with a random bone or even a random position on the iliac bone, is unacceptable. Quantitative diagnosis is the horse and sampling theory is the cast. The final goal is then to perform reproducible analyses of the structures, and to make comparisons. Unbiased estimates of the volume or proportions are only one of many ways, and are never an end in themselves.

H. EXERCISES

H.1. Bertrand's paradox

What is the probability Pr that a "random chord" of a disk of unit radius has a length greater than $\sqrt{3}$, the side of an inscribed equilateral triangle?

(a) Any chord intersects the circle at two points; assume that these two points are uniformly distributed over the circle, and prove that $\mathrm{Pr}^{(1)} = \frac{1}{3}$.
(b) Now assuming that the orientation α of the chord and its distance ρ to the centre are both uniform, prove that $\mathrm{Pr}^{(2)} = \frac{1}{2}$.
(c) Finally assuming that the mid-point of the chord is uniformly located in the disk, prove that $\mathrm{Pr}^{(3)} = \frac{1}{4}$.
(d) Which is the "best" probability? Discuss the three results above by showing that the expression "random chord" has been interpreted in three different ways. Express the three probability densities as functions of the distance ρ and of the orientation α.

$$[d\,\mathrm{Pr}^{(1)} = d\rho \cdot d\alpha [2\pi \sqrt{1-\rho^2}]^{-1}; \qquad d\,\mathrm{Pr}^{(2)} = (2\pi)^{-1} d\rho \cdot d\alpha$$
$$d\,\mathrm{Pr}^{(3)} = (\pi)^{-1} \rho\,d\rho\,d\alpha]$$

H.2. Infinite averages

The purpose of this very simple exercise is to show why we go to great lengths to avoid unbounded set models, and consequently general stationarity for random sets. Construct in $\mathbb{R}^1$ a set of points by putting down successive segments whose lengths are independent random variables with probability density

$$f(l) = \begin{cases} 0 & l < 1 \\ \dfrac{1}{l^2} & l \geq 1 \end{cases}$$

(a) Prove that the random process is stationary.
(b) Calculate the average distance separating two successive points, and interpret the result.

H.3 Convex grains

(a) Carry over the results of sections E and F, by assuming that the individuals are compact convex particles of $\mathbb{R}^3$, and show that:

Relation   (VIII-38)   becomes $\overline{A}(X' \cap \Pi) \cdot E[\Pi(X')] = 2\pi E[V(X')]$
Relation   (VIII-39)   becomes $\overline{L}(X' \cap \Delta) \cdot E[S(X')] = 4E[V(X')]$
Relation   (VIII-40)   becomes $\overline{L}(X' \cap \Delta) \cdot E[M(X')] = \pi/2 E[S(X')]$
Relation   (VIII-43)   becomes $N_V(X) \cdot E[M(X')] = 2\pi N_A[X \cap \Pi]$
Relation   (VIII-44)   becomes $N_V(X) \cdot E[S(X')] = 4N_L[X \cap \Delta]$

where $E[\cdot]$ and I_V mean "averages over all the particles", and $\overline{L}, \overline{A}, N_A$ and N_L mean "averages over all the sections of particles in all the directions".
(b) Similarly show that from $\mathbb{R}^1 \to \mathbb{R}^2$ we have:

$$\overline{L}[X' \cap \Delta] \cdot E[U(X')] = \pi A[X']$$

(This result goes back to Crofton.)

H.4. 1-D Section of 2-D Individuals

Carry over the results of sections E and F to the passage from $\mathbb{R}^1 \to \mathbb{R}^2$, proving that

relation (VIII-24)   becomes $E[A^*(X')] = dL[X \cap \Pi(\dot{x}, \alpha)]$

and that $\pi d N^{(1)}[X \cap \Pi(\dot{x}, \omega)]$ estimates the average perimeter $E[U(X')]$.

H.5. Multisampling procedures

The total population X of individuals is made up of J different samples $X_j(j = 1, \ldots, J)$ each in turn made up of I_j individuals $X'_{i,j}$. Sample j is sectioned by the plane $\Pi(\dot{x}_j, \dot{\omega}_j)$ whose random direction $\dot{\omega}_j$ is uniformly distributed on the unit sphere. Construct an estimator which generalizes relation (VIII-30) to the present case.

[Denote by X' the mean grain averaged over the $\sum_{j=1}^{J} I_j$ individuals. Thus by definiton:

$$E[M(X')] = \frac{\sum_{i,j} M(X'_{ij})}{\sum I_j} = \frac{\sum_j M(X_j)}{\sum I_j}$$

$$E[V(X')] = \frac{\sum_{i,j} V(X'_{ij})}{\sum I_j} = \frac{\sum_j V(X_j)}{\sum I_j}$$

Exactly as in Chapter VIII, E.3c, the quantity $(d/2\pi)N^{(2)}[X_j \cap \Pi(\dot{x}_j, \dot{\omega}_j)]$ is an unbiased estimator of $M(X_j)$, and the ratio

$$\frac{E^*[M(X')]}{E^*[V(X')]} = \frac{1}{2\pi} \frac{\sum_j N^{(2)}[X_j \cap \Pi(\dot{x}, \dot{\omega})]}{\sum_j A[X_j \cap \Pi(x, \omega)]} \qquad (\text{VIII-45})$$

is an a.u. estimator of $E[M(X')]/E[V(X')]$.]

H.6. An imaginative exercise

Try to find natural 3-D objects which you systematically misinterpret when you know them from 2-D sections only [For example, by sectioning a bunch of grapes you find isolated circles. If you had only random sections of the grapes, could you really go back to the 3-D structure?]

H.7. Perimeter and mean curvature in $\mathbb{R}^2$

The purpose of this very simple exercise is to point out the ambiguity of the adjective "mean" when applied to a morphological parameter.

(a) Let $y = f(x)$ be a 2-D curve, defined in the interval $[x_1, x_2]$. Find its mean curvature. [If we take a uniform law in x, we obtain:

$$E_1\left[\frac{1}{R}\right] = \frac{1}{x_2 - x_1} \int_{x_1}^{x_2} \frac{dx}{R} = \frac{\sin\theta_2 - \sin\theta_1}{x_2 - x_1}$$

which depends on the choice of the axes. If we take a uniform law proportional to the curve length, then:

$$E_2\left[\frac{1}{R}\right] = \frac{1}{u_2 - u_1} \int_{u_1}^{u_2} \frac{du}{R} = \frac{\theta_2 - \theta_1}{u_2 - u_1} \qquad (u = \text{curvilinear coordinates}).]$$

(b) Calculate the mean radius of curvature.
[For an easier computation, assign the uniform probability to the rotation angle of the normal. Then:

$$E_3[R] = \frac{1}{\theta_2 - \theta_1} \int_{\theta_1}^{\theta_2} R\,d\theta = \frac{u_2 - u_1}{\theta_2 - \theta_1}; \qquad \text{note that } E_2\left[\frac{1}{R}\right]\cdot E_3(R) = 1.]$$

PART III.   THE CRITERIA

Introduction to Part III

In the literature, the expression "Texture Analysis" is often used in a very broad sense, by authors who want to check the morphology of the object under study, generally to compare it with non-geometrical data (physical properties of a material) or to segment it into meaningful subsets (aerial photographs). With this in mind, the task of Mathematical Morphology is to provide a small number of tools with as little redundancy as possible. The four chapters of this third part are an answer to this task. The three types of information extracted from the object via order two probabilities (Ch. IX), size distribution (Ch. X), and the connectivity criteria (Ch. XI), are radically different from one another (Ch. XII extends these criteria from sets to functions).

As we saw in Chapter II, D, these various criteria differ by the shapes of the structuring elements they use, but fundamentally they are also governed by four underlying general properties which are the key to any morphological analysis. Indeed, the power, the "character" of a given transformation depends on its ability to be *increasing, extensive* (or *anti-extensive*) *idempotent* and to *preserve homotopy*. These four factors are the corner stone for a typology of the criteria, and must always be kept in mind by the reader, if he wants to master the third part of this book. As an illustration of that idea, may we suggest that the reader do the following exercise, once before and once after reading Part III?:

Find set transformations which are translation invariant and which satisfy the following criteria:

Criteria	Properties			
	Increasing	Extensive (or anti-extensive)	Idempotent	Preserve homotopy
1	No	No	No	No
2	Yes	No	No	No
3	No	Yes	No	No
4	Yes	Yes	No	No
5	No	No	Yes	No
6	Yes	No	Yes	No
7	No	Yes	Yes	No
8	Yes	Yes	Yes	No
9	Yes	No	No	Yes
10	No	Yes	Yes	Yes

[1. Complementation, Hit or Miss; 2. Erosion and dilation when $0 \notin B$, median filtering; 3. Thickening and thinning; 4. Erosion and dilation when $0 \in B$; 5. Boundary; 6. Projection; 7. Sequential thinning, skiz, skeleton, cond. bisector, watershed, ultimate erosion; 8. Opening, closing, convex hull, size distribution, thresholding, umbra; 9. Displacement, similarity, affinity, symmetry; 10. Homotopic and sequential thinning and thickening.]

IX. The Covariance

A. A SIMPLE, BUT EFFECTIVE, TOOL

The simplest structuring element beyond the trivial case of one point is the pair of points B. The measure of the corresponding eroded set is *the covariance*. This function is essentially a morphological tool for disentangling different structures imbricated within each other. We can quote for example, its uses in:

— the analysis of *clusters* of points or particles (Fig. IX.14);
— the description of structures where several populations of objects coexist with *rather distinct size distributions* (see for example the mixture of fibres, radii, and vessels in the wood histological section, Fig. IX.17);
— *the superposition of scales.* Consider a microscopic structure. Beyond the elementary field seen by a T.V. camera, there may exist interfield correlations, due to macrostructures much larger than the camera field;
— *periodicities*, or pseudo-periodicities. They can occur at the scale of the elementary mask or in the macrostructures. They can be affected, and even hidden by other regionalizations;
— *noise* due to the technology (mechanical, optical or electronic), but also due to the existence of substructures which are microscopic at the scale of investigation. Using a suggestive terminology suggested from the mining world, this noise is called the a "nugget effect";
— *multiphased structures.* A particular case of imbricated structures when the set X may be partitioned into subsets, such as the metallographic phases in Figure V.10. Then the covariance measures the degree of dependence between each pair of phases;
— *grey tone functions.* Considered as sets in $\mathbb{R}^n \times \mathbb{R}^+$, they appear as the limit of multiphased structures, and are susceptible to the same approach as above;
— *anisotropies*, in the sense of anisotropies of the whole texture at a given

scale, and not of the elongation of individual particles (for the covariance, the union of segments, uniformly randomly oriented).

— This is one of the most important applications of the covariance, since it opens the way to the *estimation variances* of samplings. The theory which studies this aspect of the covariance is the geostatistic (G. Matheron 1965–1980) (A. Journel, 1978).

The best way to give an idea of the morphological potential available is to show a few very representative examples of actual studies. This we will do in the last section of the chapter (Section E), but we will precede it with a brief introduction to the formal properties of the covariance. We have encountered several of them already in the previous chapter; for completeness we will put them all together in Section B. We follow this up in Section C, with some algorithms involved in covariance computations, and in Section D with a few probabilistic models.

B. MORPHOLOGICAL PROPERTIES

B.1. Global representation

(a) *Deterministic compact sets*

In this section, X is a deterministic compact set in $\mathbb{R}^n$, and $B = \{0, h\}$, the union of the two extremities of the vector $\overrightarrow{Oh}$ (modulus $|h|$, direction α). By definition the covariance $K(h)$ is the measure of the set $X \ominus \check{B}$, which is X eroded by B. Consider the translate B_x of B by x. The point x belongs to the eroded set $X \ominus \check{B}$ if and only if x and $x + h \in X$. In other words:

$$X \ominus \check{B} = X \cap X_{-h} \qquad \text{(IX-1)}$$

If $k(x)$ is the indicator function associated with the set X, then the covariance $K(h)$ is given by

$$K(h) = \mathrm{Mes}(X \ominus \check{B}) = \int_{\mathbb{R}^n} k(x) \cdot k(x + h) dx \qquad \text{(IX-2)}$$

From this relation, we immediately derive the following properties of $K(h)$:

$$\left. \begin{array}{ll} K(o) = \mathrm{Mes}(X); \; K(\infty) = 0 \\ K(h) = \mathrm{Mes}(X \ominus \check{B}) = \mathrm{Mes}(X \ominus B) = K(-h) \\ K(h) \le K(o) & (X \cap X_{-h} \subset X) \\ \displaystyle\int_{\mathbb{R}^n} K(h) dh = [\mathrm{Mes}\, X]^2 \end{array} \right\} \qquad \text{(IX-3)}$$

When X is composed of several grains, the covariance at small distances affects the individual behaviour of grains more than at larger distances, where

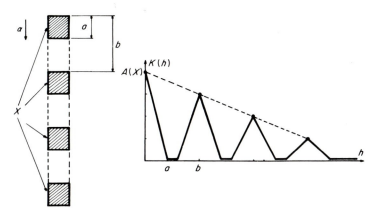

Figure IX.1. The set X and its covariance $K(h)$ in the direction α.

it is telling us about their spatial distribution. Look for example at Figure IX.1; the set X is the union of five separate rectangles. The covariance $K(h)$ increases when some particles of the shifted X_h overlap other particles of X. The envelope of the local maxima of $K(h)$, in dotted lines, corresponds to the convex hull of X, i.e. the macrostructure generated by X. The smallest a such that $K(h) \equiv 0$ for $|h| > a$ is called the *range* of the covariance in direction α.

(b) *Generalization*
The preceding results are easily generalized to functions. Let $f(x)$ be a summable function in $\mathbb{R}^n$ (i.e. $\int_{\mathbb{R}^n} f(dx) < \infty$), and non-zero in a bounded domain. Then the relation

$$g(h) = \int_{\mathbb{R}^n} f(x) \cdot f(x+h) dx \qquad (\text{IX-4})$$

generalizes the set covariance $K(h)$ of relation (IX-2). The covariogram $g(h)$ satisfies the following properties of relation (IX-3):

$$\left. \begin{aligned} g(o) &= \int_{\mathbb{R}^n} f(x)dx, \ g(\infty) = 0 \\ g(h) &= g(-h) \le g(o) \\ \int_{\mathbb{R}^n} g(h)dh &= \left[\int_{\mathbb{R}^n} f(x)dx \right]^2 \end{aligned} \right\} \qquad (\text{IX-5})$$

A given function $\varphi(h)$ can play the role of a covariance $K(h)$, or $g(h)$, if and only if $\varphi(h)$ is a positive definite function (i.e. the Fourier transform of a summable positive measure).

We now return to sets. Instead of demanding that the two points of B_x belong to X, we could ask for those x such that $x \in X$ and $x + h \in X^c$. The corresponding Hit or Miss transform is M:

$$M = \{x : x \in X, x + h \in X^c\} = X \cap (X_{-h})^c = X/(X \ominus \check{B}) \qquad \text{(IX-6)}$$

and the associated measure is the global variogram $K(o) - K(h)$, whose datum is obviously equivalent to that of $K(h)$. Figure (IX.2) shows the geometrical relationship between the two notions.

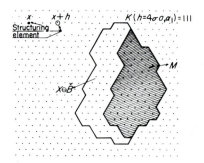

Figure IX.2. Correspondence between covariance and variogram: $K(h) = A(X \ominus \check{B})$, variogram $= A(M)$, where A is the area, i.e. the number of pixels.

The notions of global covariance and variogram for sets and for functions, have probabilistic versions. The set X is now taken as (almost surely) the compact random set. Let us write, for example, the covariance $K(h)$:

$$K(h) = E \cdot \int_{\mathbb{R}^2} k(x) \cdot k(x+h) dx = \int E[k(x) \cdot k(x+h)] dx$$

Hence,

$$K(h) = \int \Pr\{x, x+h \in X\} dx \qquad \text{(IX-7)}$$

For deterministic or random sets, the limit $-[K(o) - K(h)]/|h|$, when $|h| \to 0$ and α is fixed, may be infinite (see Ex. IX-4), but for the regular model and the convex ring, this limit exists. In $\mathbb{R}^3$ and $\mathbb{R}^2$, it is respectively equal to the total projection $A^{(2)}(X, \alpha)$ and $D^{(2)}(X, \alpha)$ (Figs. V.7 and V.8), or to mathematical expectations of these quantities:

$$
\begin{aligned}
-K'_\alpha(o) &= E[A^{(2)}(X, \alpha)] \quad \text{in } \mathbb{R}^3 \\
-K'_\alpha(o) &= E[D^{(2)}(X, \alpha)] \quad \text{in } \mathbb{R}^2
\end{aligned}
\qquad \text{(IX-8)}
$$

By averaging in α, we have:

$$
\left.
\begin{aligned}
-\frac{1}{4\pi} \int_{4\pi} K'_\alpha(o)d\alpha &= \frac{E[S(X)]}{4}, \quad \text{in } \mathbb{R}^3 \\
-\frac{1}{2\pi} \int_0^{2\pi} K'_\alpha(o)d\alpha &= \frac{E[U(X)]}{\pi}, \quad \text{in } \mathbb{R}^2
\end{aligned}
\right\}
\tag{IX-9}
$$

Moreover, (IX-8) may be used for calculating the rose of directions (see Sect. C.2, below).

B.2. Local representations

(a) *Covariance for sets and for functions*
Since the formalism for studying local representations is a probabilistic one, we now assume that X is a realization of a stationary random set. In the theory of probabilities, the covariance Cov(h) of a stationary random function $f(x)$ is given by the classical definition:

$$
\begin{aligned}
\text{Cov}(h) &= E\{[f(x)-p][f(x+h)-p]\} \\
&= E[f(x)f(x+h)] - p^2
\end{aligned}
\tag{IX-10}
$$

where

$$
P = E[f(x)]
$$

Let $f(x)$ be the indicator function of a random set X. The local equivalent of relation (IX-7) is now:

$$
\begin{aligned}
C(h) &= \text{Cov}(h) + p^2 = P\{x \in A, x+h \in A\} \\
p &= P\{x \in A\}
\end{aligned}
\tag{IX-11}
$$

We will denote Cov(h) $+ p^2$ by $C(h)$.

In terms of erosion, relation (IX-11) means that $C(h)$ is the proportion, or the first moment, of the stationary random set $X \ominus B$, where $B = \{0, h\}$.

The relations (IX-5) are easily transposable, and give:

$$
\left.
\begin{aligned}
C(o) &= p \qquad C(\infty) = p^2 \\
C(h) &= C(-h) \\
C(o) &\geq C(h)
\end{aligned}
\right\}
\tag{IX-12}
$$

In the same way, one easily transposes the relationships (IX-8), (IX-27) concerning the boundary of X and the rose of directions by changing areas and perimeters into specific areas and specific perimeters. Note the difference between $C(\infty) = p^2$ and $K(\infty) = 0$, which expresses the passage local $\rightarrow$ global.

(b) *The range*

The notion of a range, as defined for global representations, is not easily adapted to the present local case. By assuming that the set X is not bounded, we allow the possibility of covariances $C(h)$ which are never equal to zero (for example, $C(h) = \exp[-\lambda h]$). It seems more appropriate to transfer the integral $\int K(h) \, dh$ of the global case. Thus we take for the range a_n, the quantity:

$$a_n = \frac{1}{p - p^2} \int_{R^n} [C(h) - p^2] \, dh \qquad \text{(IX-13)}$$

The integral (IX-13) may be defined in the whole space $\mathbb{R}^n$, or in its restriction to a hyperplane of dimension i. For example in $\mathbb{R}^3$, the range of the covariance in the plane section with normal ω is:

$$a_2(\omega) = \frac{1}{p - p^2} \int_0^{2\pi} d\alpha \int_0^\infty [C(r, \alpha) - p^2] r \, dr, \qquad |h| = r$$

(c) *Rectangular covariance*

Let us consider the following example. A thin section of silicate is analysed. It is made up of the direct sum of n mineralogic phases (amphiboles, pyroxenes . . .). Each one has a given percentage a_i of magnesium. What is the covariance of the magnesium on the thin section? We denote by $M(x)$ the percentage of Mg at the point x, and by $k_i(x)$ the indicator function of the mineralogic phase number i ($i = 1, \ldots, N$), so that

$$M(x) = \sum_{i=1}^{N} a_i k_i(x) \qquad \text{(IX-14)}$$

The non-centred covariance $C(h)$ of Mg is given by:

$$C(h) = E[M(x)M(x+h)] = \sum_{i,j} a_i a_j E[k_i(x)k_j(x+h)] \qquad \text{(IX-15)}$$

The quantities $C_{ij}(h)$

$$C_{ij}(h) = E[k_i(x)k_j(x+h)] \qquad \text{(IX-16)}$$

have a morphological interest; C_{ij} describes the spatial relations between the phases i and j (A. Haas *et al.*, 1967). At the origin

$$C_{ij}(o) = p_i \delta_{ij} \qquad (\delta_{ij} = \text{krönecker's delta})$$

and when $|h| \to \infty$, $C_{ij}(h) \to p_i p_j$ where $p_i = E[k_i(x)]$.

The matrix $C_{ij}(h)$ is antisymmetric: $C_{ij}(h) = C_{ji}(-h)$, and in general $C_{ij}(h)$ differs from $C_{ji}(h)$. This difference reflects an asymmetry in the relative dispositions of phases i and j (for example a cell with an eccentric nucleus). At the origin, the slope of the tangent $|C'_{ij}(o)|_\alpha$ is finite (for regular sets considered in practice) but may be equal to zero if the phases i and j have no contact in the direction α.

We can define the specific contact area σ_{ij} between the phases i and j; these are related to the derivatives of C_{ij} by the relations:

$$\text{in } \mathbb{R}^3: \sigma_{ij} = \frac{1}{4\pi} \int_{S_3} |C'_{ij}(o)|_\alpha d\alpha$$

$$\text{in } \mathbb{R}^2: \sigma_{ij} = \frac{1}{2\pi} \int_0^{2\pi} |C'_{ij}(o)|_\alpha d\alpha$$

Conversely, the matrix of covariances $C_{ij}(h)$ immediately gives the covariance of any phase l, by the relationship (IX-17):

$$C_l(h) = p_l - \sum_{\substack{i=1 \\ i \neq l}}^{N} C_{li}(h) \tag{IX-17}$$

Let us now return to the algorithm (IX-15). Its purpose was to reach indirect parameters (Mg for example), by appropriate weighting of the visible ones (mineralogic phases). However it may be used in another way. In various applications, the object investigated does not directly correspond to a particular set, i.e. to a black or white function. This situation often occurs in biology. The optical densities of nuclear acids in a nucleolus for example, vary continuously. In such cases, we can partition the density of light $f(x)$ in classes by thresholds of successive levels:

$$f(x) \simeq \sum_{i=1}^{N} a_i k_i(x) \qquad (k_i(x) \cdot k_j(x) \equiv \delta_{ij}) \tag{IX-18}$$

According to a well known result, every continuous function $f(x)$ is the limit of step functions, such as the $\sum a_i k_i(x)$. Even when $f(x)$ is not continuous, another representation of f this time by integral thresholds (see below, algorithm (IX-44)), facilitates a similar approach. In fact the matrix $C_{ij}(h)$, which appears as a covariance matrix from the set point of view, gives "second order" probabilities of the random function

$$C_{ij}(h) = \Pr\{a_{i-1} < f(x) < a_{i+1}; a_{j-1} < f(x+h) < a_{j+1}\}$$

Instead of summing up these probabilities to give the covariance (IX-15), we could also consider them individually as basic pieces of information. We will not pursue this further, but instead refer the reader to the works of R. Haralick (1972, 1978), and those of O. Faugeras (1977) who very thoroughly investigated this method of textural description.

(d) *The covariance and the power spectrum*

The famous Wiener–Khintchine Theorem links the covariance with the power spectrum of a stationary random function by the relationship

$$C(h) - p^2 = \int \exp(2i\pi vh) g(dv) \tag{IX-19}$$

where $g(dv)$ is the energy associated with the set of frequencies dv. For $h = 0$, (IX-19) becomes:

$$C(o) - p^2 = p(1 - p) = \int_{\mathbb{R}^n} g(dv) \tag{IX-20}$$

that is, the variance of $f(x)$ equals the integral of the power spectrum. In the same way, we can write:

$$g(o) = \int_{\mathbb{R}^n} [C(h) - p^2] dh = C(o) \cdot a_n$$

i.e. the range a_n is proportional to the value of the power spectrum at the origin. We see here a well-known phenomenon of Fourier transformations, namely an exchange of the behaviour at the origin with that at infinity. Consider for example the exponential covariance in $\mathbb{R}^1$. Noticing that $C(h)$ and $g(dv)$ are even functions, we may write:

$$C(h) - p^2 = \int_0^\infty g(dv) \cos(2\pi vh)$$

Hence, if $C(h) - p^2 = 1/2\theta \exp(-|h|/\theta)$, then

$$g(dv) = \frac{dv}{1 + 4\pi v^2 \theta^2}$$

From this relation the first moment of the spectrum $g(dv)$ does not exist. Unfortunately, this is generally the case for the all the $0 - 1$ functions. The derivative of the covariance at the origin is discontinuous (although the left and right derivatives are negatives of each other), which implies an infinite average value of the Fourier transform, and consequently, the impossibility of estimating specific surfaces from the power spectrum.

We now consider the particular but important, case of isotropic stationary functions in $\mathbb{R}^3$. Their covariance $C(h)$ depends only on the scalar $r = h$. Likewise, the power spectrum $G(d\rho)$ depends only on the frequency ρ of spherical harmonics. Hankel's formula shows that:

$$C(r) - p^2 = 2r^{-1} \int_0^\infty \rho \sin(2\pi \rho r) G(d\rho)$$

If $G(d\rho) = g(\rho)d\rho$ has a density, the formula can be inverted, and we find:

$$g(\rho) = 2p^{-1} \int_0^\infty r\sin(2\pi\rho r)\left[C(r) - p^2\right]dr \qquad \text{(IX-21)}$$

Relation (IX-21) is well suited for a numerical computation of $g(\rho)$ from the covariance. Notice that the harmonics are *spherical*, though the covariance is estimated from *linear* traverses of $f(x)$.

We can fit a model to the covariance; suppose we choose the exponential model:

$$C(r) - p^2 = \frac{1}{\theta^3}\exp\left(-\frac{r}{\theta}\right)$$

Then (IX-21) gives

$$g(\rho) = \left[\frac{2\pi}{1 + (2\pi\rho\theta)^2}\right]^2$$

Although they carry equivalent information about the stationary random function, the covariance and the power spectrum are usually determined by relying on profoundly different technologies. The latter is associated with analogic optical designs, the former with digital analysers. Optical methods have a considerable advantage in that they are instantaneous and bi-dimensional, whereas the digital covariance is computed direction by direction. On the other hand, we saw that for the power spectrum the behaviour near the origin is associated with the less ergodic parameters of the texture (the ranges), whereas the simple porosity p requires an integration of the power spectrum, and the number of intercepts, $-C'_\alpha(o)$, is not accessible. Also, digital methods allow the computation of covariance matrices of the n-phase sets, and second order probabilities. Nothing equivalent can be derived from the power spectrum.

(e) *Regularization, variogram*
The variogram, introduced by G. Matheron in 1955 is a notion which generalizes the covariance. In micrographic structures, the porosity $p(y)$, calculated from a scanned field centred at point y, is not a constant but varies slowly when the sampling frame is moved over the specimen. The variance of this local porosity will typically undergo large increases with the size of the specimen. In such a case the covariance and the variogram are not equivalent notions, the latter being less demanding on the underlying structure of the specimen. Consequently, if we suspect that the spatial correlations in the specimen are more expansive than the zone defined by the elementary scan field (or than the elementary specimen, when serially sampled), then the variogram is a more suitable tool for a correlation study.

The point variogram $\gamma(h)$ is the probability that $x \in X$ and $x + h \in X^c$:

$$\gamma(h) = \Pr\{x \in X; \quad x + h \in X^c\} = \Pr\{x \in X^c; \quad x + h \in X\} \qquad \text{(IX-22)}$$

which is obviously equal to the difference, $C(o) - C(h)$, when $C(o) < \infty$, but which remains defined when $C(o)$ is unbounded in contrast to the covariance. The point variogram is the probability that the phase at point x (black or white) is not the phase at point $x + h$. Just as for the covariance, the variogram depends on the orientation α, of the vector h, and may be extended to multiphased sets. In terms of indicator functions, we immediately see that relation (IX-22) is equivalent to:

$$\gamma(h) = \tfrac{1}{2}E[k(x) - k(x + h)]^2 \qquad \text{(IX-23)}$$

The regularization algorithm allows us to go from field to interfield correlations. Define the regularized variogram $\gamma_Z(h) \, y$:

$$\gamma_Z(h) = \tfrac{1}{2}E[p(y) - p(y + h)]^2 \qquad \text{(IX-24)}$$

where $p(y) = [1/A(Z)] \int_{Z_y} k(x)dx$ is the proportion of 1's in the mask centred at point y. Denote by $K_Z(h)$ the (global) covariance of mask Z (alg. IX-2). Then

$$\gamma_Z(h) = (\gamma * K)_h - (\gamma * K)_o \qquad \text{(IX-25)}$$

i.e. the regularized variogram is the convolution product of the point variogram by the covariance of the mask Z. This convolution product takes a simpler form when the structure of X exhibits two distinct scales and $\gamma(h)$ is the sum of two terms $\gamma^1(h)$ and $\gamma^2(h)$. If we choose Z such that γ^1 reaches its asymptote for values of $|h|$ smaller than the diameter of Z in direction α, then $\gamma^2(h)$ grows very slowly, and is virtually linear in the range of the dimensions of Z. Under this assumption, relation (IX-25) becomes:

$$\gamma_Z(h) = m + \gamma^2(h) \qquad |h| > \text{diam}(Z); \; m \text{ positive constant}$$

which relates the point variogram γ^2 to the regularized version γ_Z, and allows a continuous transition over the successive scales of structural heterogeneity.

C. A FEW ALGORITHMS

C.1 Parameter estimation for the point covariance $C(h)$

The interpretation techniques presented below may also be applied to the point variogram, since $\gamma(h) = C(o) - C(h)$, and more generally to the monotonic curves met in mathematical morphology, such as the size distribution $P(l)$, the openings, etc.

(a) *Calibration of the data*

Let Z be the digital mask in which we measure the covariance of the digital set $X \cap Z$. We use as estimate $C^*(k)$ of $C(k)$, the following ratio

$$C^*(k) = \frac{\mathcal{N}\{\overbrace{1, \ldots, 1}^{k}\} \text{ in } X \cap Z}{\mathcal{N}\{\underbrace{1, \ldots, 1}_{k}\} \text{ in } Z} = \frac{A[(X \cap Z) \ominus B(k)]}{A[Z \ominus B(k)]}$$

where the integer k is the number of *intervals* that separate the two points of the digital structuring element $B = \{1, \ldots, 1\}$. This relation, which is just the mask erosion formula (II-11), simply means that we count the number of configurations occurring in the mask Z, and divide by the number of possible configurations; A denotes the digital area.

If we can perform the measurements in several masks $Z_\lambda (1 \leq \lambda \leq g)$, the averaged covariance is obtained by weighting the $C_\lambda(h)$ in proportion to the areas $A(Z_\lambda)$, giving:

$$C^*(k) = \frac{\sum\limits_{\lambda=1}^{g} A(Z_\lambda) \cdot C_\lambda^*(k)}{\sum\limits_{\lambda=1}^{g} A(Z_\lambda)}$$

(b) *Slope of the tangent at the origin*

Figure IX.3 shows an example of a point covariance $C^*(k)$. The first problem that arises is to estimate the slope of the tangent at the origin. It will be solved by fitting a straight line via least squares, using only the covariance points of abscissae 1, 2, 3 and 4.

The general formulation for fitting a straight line using least squares is as follows: let (x_k, y_k) be the co-ordinates of the experimental points $(k = 1, \ldots, p)$ and let

$$y = \alpha x + \beta$$

be the estimated line. The two parameters α and β satisfy the two relationships:

$$\left. \begin{array}{l} \alpha = \dfrac{p\left(\sum_k x_k y_k\right) - \left(\sum_k x_k\right)\left(\sum_k y_k\right)}{p\sum_k x_k^2 - \left(\sum_k x_k\right)^2} \\[6mm] \beta = \dfrac{\left(\sum_k y_k\right)\left(\sum_k x_k^2\right) - \left(\sum_k x_k\right)\left(\sum_k x_k y_k\right)}{p\sum_k x^2 - \left(\sum_k x_k\right)^2} \end{array} \right\}$$

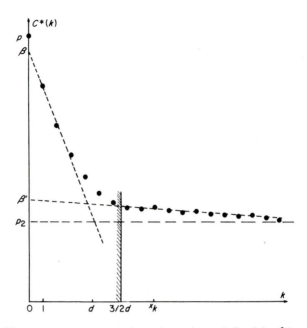

Figure IX.3. Measurements on an experimental covariance. Left of $k = \frac{3}{2}d$ are points for estimating the slope at the origin; right of $k = \frac{3}{2}d$ are points for measuring the slope of the asymptote.

In the case of the slope of the tangent at the origin, we have $p = 3$.

(c) *Estimation of the noise*

When we extend the estimated straight line to cut the y-axis, its β-value at the origin differs from the beginning p of the experimental covariance. The difference $p - \beta$ must be positive, and represents the variance due to noise (electronic noise as well as from particles reduced to only one picture point).

(d) *Slope of the asymptote*

The covariance, minus its noise, starts from $p - \beta$ and approaches asymptotically the value $p^2 - \beta^2$ (where β^2 is often negligible in practice).

To estimate the slope by which the covariance reaches its asymptote, we must first define criteria that will limit the data to be considered. These criteria are as follows:

— compute the abscissa d of the intersection of the estimated line at the origin with the horizontal $p^2 - \beta^2$ (see Fig. IX.3).
— use only the data whose abscissae are larger than $\frac{3}{2}d$.

The covariance is then fitted to a straight line through these data by least squares, using formula (IX-26). Let us denote this estimator by:

$$y = \alpha' x + \beta'$$

When no structure is larger than the maximal step measured, α' has to be zero.

(e) *The range of the covariance*
The algorithm for the range is given by relation (IX-12), in which we must substitute $p - \beta - p^2$ for the initial $p - p^2$, since $p - \beta$ is the porosity of the image corrected for its noise. Numerically, it is impossible to take the argument k up to infinity, but only to a given k_{max}. We will consider the one, two and three dimensional cases separately; e.g. for an isotropic covariance, measured from a grid with spacing a, formula (IX-13) may be written:

$$\left.\begin{array}{ll}
\text{in } \mathbb{R}^1 & a_1^* = \dfrac{2a}{p - \beta - p^2} \displaystyle\sum_{k=1}^{k_{max}} [C(k) - p^2] \\[2em]
\text{in } \mathbb{R}^2 & a_2^* = \dfrac{2\pi a}{p - \beta - p^2} \displaystyle\sum_{k=1}^{k_{max}} k[C(k) - p^2] \\[2em]
\text{in } \mathbb{R}^3 & a_3^* = \dfrac{4\pi a}{p - \beta - p^2} \displaystyle\sum_{k=1}^{k_{max}} k^2[C(k) - p^2]
\end{array}\right\} \qquad \text{(IX-26)}$$

It would not be wise to fit a mathematical model to the covariance in order that integrations beyond k_{max} could be performed. We have *absolutely no idea* of the actual behaviour of a covariance at scales larger than those we can observe.

C.2. The rose of directions

(a) *Theory in $\mathbb{R}^2$*
The rose of direction is the angular distribution of the perimetric measure $u(dx)$ in $\mathbb{R}^2$. An example is given by Figure IX-5. Indeed, the domain of existence of the perimetric measure does cover the set classes for which $K_\alpha'(0)$ exists, i.e. in particular $\mathcal{R}(\mathcal{K})$, $\mathcal{S}(\mathcal{K})$, $\mathcal{S}(\mathcal{F})$ (but neither $\mathcal{K}$ ($\mathbb{R}^2$) nor $\mathcal{F}$ ($\mathbb{R}^2$)). For making the notation simpler, we treat the question in the deterministic and global case, and put $D(\alpha)$ for the 2-D total projection $D^{(2)}(X; \alpha)$.

Let $u(d\theta)$ be the unknown perimetric measure of the 2-D set X, and suppose we know the total projection $D(\alpha)$ of X in every direction α; we easily see that:

$$D(\alpha) = \int_\theta^{\theta + \pi} \sin(\theta - \alpha) u(d\theta)$$

By differentiating twice in α, we find:

$$[D(\alpha) + D''(\alpha)]\, d\alpha = u[d(\alpha)] + u[d(\alpha + \pi)] \qquad \text{(IX-27)}$$

where the derivatives are taken here as derivatives of measures, since $[D''(\alpha)]$ is not always a function. Notice that relation (IX-27) treats the roles of directions α and $\alpha + u$ symmetrically. When measuring a set of faults say, this does not matter, but if we want to describe all the contacts of two phases in a three-phased medium, then in general $u(d\alpha) \neq u(d\alpha + \pi)$, and hence we can only reconstruct the rose of directions up to a symmetry. On the other hand, the numerical techniques (least squares) smooth the perimetric measure into a function. Consequently, relation (IX-27) can be replaced in the experiments by $2u(\alpha) = D(\alpha) + D''(\alpha)$.

(b) Digital implementation

According to relation (IX, 9), the quantity $-D(\alpha)$ is nothing but the slope of the tangent of the covariance at the origin. Therefore, it can be estimated by the technique developed in Section C.1b. More simply, we can also use relation (VI-53). Then, up to the multiplicative constant a (square grid) or $a\sqrt{3}/2$ (hexagonal grid), the rose of directions $u(\alpha)$ is given by the formula

$$2u(\alpha) = I(\alpha) + I''(\alpha) \qquad (I(\alpha), \text{ intercept number in direction } \alpha)$$

We will assume that we can turn the grid (superimposed on the object) by analogous means, producing the successive directions α_i at spacings of $\Delta\alpha$. To estimate the second derivative $I''(\alpha)$ at point α_0, we fit a parabole to the $2k + 1$ values $I(\alpha_i) (-k \leq i \leq k)$ by least squares. Put:

$$\sigma_2 = \sum_{i=-k}^{+k} I(\alpha_i) \times i^2, \qquad \sigma = \sum_{i=-k}^{+k} I(\alpha_i)$$

The parabola centred at point α_0, is denoted by:

$$(\alpha - \alpha_0)^2 \, a(\alpha_0) + (\alpha - \alpha_0)b(\alpha_0) + c(\alpha_0)$$

Then, the optimal estimation of $a(\alpha_0)$ and $c(\alpha_0)$, in the least squares sense, is:

$$\left.\begin{array}{l} a(\alpha_0) = \dfrac{15}{(\Delta\alpha)^2 (2k+3)(2k+1)(2k-1)} \left[\dfrac{3\sigma_2}{k(k+1)} - \sigma \right] \\[4mm] c(\alpha_0) = \dfrac{3}{(2k+3)(2k+1)(2k-1)} \left[(3k^2 + 3k - 1)\sigma - 5\sigma_2 \right] \end{array}\right\} \qquad \text{(IX-28)}$$

Note that $c(\alpha_0)$ and $I(\alpha_0)$ should have about the same value. The difference $I(\alpha_0) - c(\alpha_0)$ indicates whether the neighbourhood $2k + 1$ is large enough (or too large). The rose of directions itself, is estimated by u^* (we see that $b(\alpha_0)$ plays no role):

$$2u^*\left(\alpha_0 + \frac{\pi}{2}\right) = 2a(\alpha_0) + c(\alpha_0)$$

The computation is performed by sliding neighbourhoods. The choice of the increment $\Delta\alpha$ and of the neighbourhood size $2k + 1$ obviously depends on the type of image under study. However, our experience has been that excellent results are practically always obtained by taking a rotation step $\Delta\alpha = 2°$ and a size $k = 4$, which gives for relation (IX-28):

$$\left.\begin{aligned} a(\alpha_0) &= -17.78\sigma + 2.67\sigma_2 \\ c(\alpha_0) &= 0.255\sigma - 0.021\sigma_2 \end{aligned}\right\} \qquad \text{(IX-29)}$$

where $\Delta\alpha$ is expressed in *radians*, the unit involved in the derivative $I''(\alpha)$. As an example, Figure IX.4 shows the simulation of a set of faults (F. Conrad, 1972) used for studying the permeability of petroleum rocks. For pedagogical reasons, it is more enlightening to study a simulation than a natural object, since we already know the answer. We see on the plotted rose (Fig. IX.5) that even for such a complex medium, the estimated values of the rose obtained by applying relation (IX-29) are not too bad.

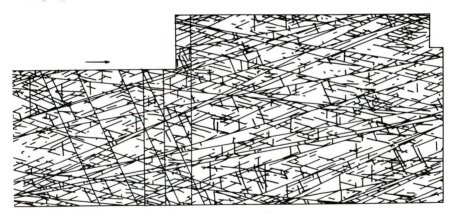

Figure IX.4. Simulation of geological faults.

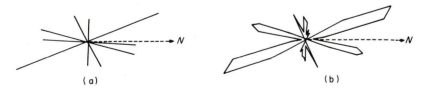

(a)   (b)

Figure IX.5. Rose of directions of the simulation on Fig. IX-4(a), and its estimation (b).

Finally, we will present a counter-example, to show the finesse of the method. One might guess that the function $D(\alpha)$, which potentially contains the same amount of information as $u(d\alpha)$, is itself a good descriptor of anisotropies. This is right, but the type of reasoning is dangerous. Consider two groups of segments with the same total length, where each member in one group is orthogonal to every member of the other group. From Figure IX.7 we see the function $D_1(\alpha)$ associated with the horizontal group, and by addition, the $D_2(\alpha)$ of the set union of two groups. Paradoxically, the maxima of $D_2(\alpha)$ are located at $\pi/4$ and $3\pi/4$, when the directions of elongation of the structure are either 0 or $\pi/2$!

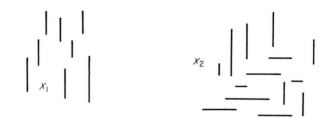

Figure IX.6. Sets X_1 and X_2.

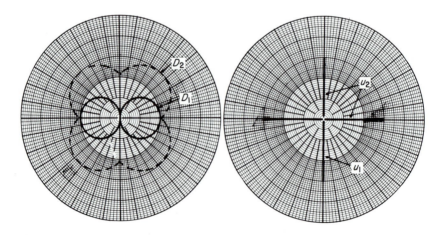

Figure IX.7. Total projections D_1 and D_2 of X_1 and X_2 (a), and their roses of direction u_1 and u_2.

In practice, $D(\alpha)$ is used as a descriptor of anisotropy only when we know, a priori, that the set X under study possesses *one* main direction. Then a very fast algorithm (Ex. VI-18) will compute it (on a hexagonal grid) from only three $D(\alpha)$'s. In particle analysis, such fast determinations of anisotropy are sometimes useful.

D. A FEW PROBABILISTIC MODELS

The models that we develop here allow us to look critically at how useful the covariance is. We shall see how cautious we must be in interpreting experimental covariances.

D.1. Cyclic binary processes

(a) *The basic variogram equation*
In this section, we stay in $\mathbb{R}^+$. Consider a stochastic process where each point $x > 0$ can belong to one of two states, e_0 or e_1. If x is interpreted as time, then the process describes the random evolution of a physical system. We could also imagine the succession of e_0 and e_1's as defining a random set in $\mathbb{R}^1$. At the beginning of each e_0 state, the process is assumed to start afresh (i.e. the future is independent of the past). In other words, if Y^i is the length of the cycle of two successive segments e_0^i and e_1^i, then the Y^i are independent of each other, but not e_0^i of e_1^i. We assume that the probability law of all the cycles is the same, and that its mathematical expectation $E\{Y\}$ is finite. G. Matheron (1968) proved that under such conditions the process is ergodic (the origin can be moved to minus infinity) and that one can find an initial law for which it is stationary. He then calculated the relation (IX-31) that will be presented below (see also J. Kingman (1972) and D. R. Cox (1962)).

We wish to calculate the variogram $\gamma(h)$ of the process. The random segments e_0, e_1, and Y have densities which are respectively, f_0, f_1 and f, with means m_0, m_1 and m, and have Laplace transforms denoted by Φ_0, Φ_1 and ψ, and have distribution functions F_0, F_1 and F. According to relation (IX-22), the variogram $\gamma(h)$ equals the probability of the event "$0 \in e_1$ and $h \in e_0$". This event occurs in one of two ways:

— either the segment e, containing 0, ends in $y < h$, and the next segment e_0 contains h; this occurs with probability

$$p_1 \int_0^h \frac{1 - F_1(y)}{m_1} [1 - F_0(h-y)] \, dy = p_1 \frac{1 - F_1}{m_1} * (1 - F_0)$$

(p_1 = probability of belonging to e_1)

— or one or more complete cycles (e_0, e_1) are inserted between 0 and h. Let l be the length of these cycles $(l < h)$, and a and $a + l$ be its two extremities. The position of its starting point a, plays no role in the probability calculation: we can equivalently remove an intermediary portion of length l from the segment $(0, h)$ and then join together the two ends a and $a + l$ of the intermediary cycle. Thus the conditional probability of the event, "$0 \in e_1$ and $h \in e_0$", given l, is $\gamma(h-l)$, giving the unconditional probability $\int_0^h f(l)\gamma(h-l)\,dl$.

The following equation for $\gamma(h)$ results:

$$\gamma(h) = p_1 \int_0^h \frac{1 - F_1(y)}{m_1} [1 - F_0(h - y)] \, dy$$

$$+ \int_0^h f(l)\gamma(h - l) \, dl \tag{IX-30}$$

Recall that $\gamma'(o)$ is the number of cycles (number of intercepts) and is equal to $(m_0 + m_1)^{-1}$. Upon evaluating the derivative with respect to h at $h = 0$ in (IX-30), we find that:

$$\gamma'(o) = \frac{p_1}{m_1}$$

The proportion of e_1 per unit length is therefore:

$$p_1 = \frac{m_1}{m_0 + m_1}, \quad \text{and by duality,} \quad p_0 = \frac{m_0}{m_0 + m_1}$$

For further uses of the convolution equation (IX-30), we will take Laplace transforms, allowing us to solve for (the Laplace transforms of) γ:

$$\chi(\lambda) = \frac{1}{m_1 + m_0} \frac{(1 - \Phi_0)(1 - \Phi_1)}{\lambda^2 (1 - \psi)} \tag{IX-31}$$

where $\chi(\lambda)$ is the Laplace transform of the variogram $\gamma(h)$.

(b) *A mirage of periodic structures*
An exactly periodic process has all intermediate cycles periodical, the densities f_0 and f_1 are two Dirac measures, and the variances of the "grains" and of the "pores" are both equal to zero. Now imagine that the lengths of the successive segments e_0 and e_1 are *independently* drawn, respectively from two size distributions f_1 and f_0, taken to be:

$$f_0(x) = f_1(x) = a^2 \times \exp(-ax); \qquad p_1 = p_0 = \tfrac{1}{2}$$

Then relation (IX-31) implies:

$$\gamma(h) = \tfrac{1}{4}(1 - \exp(-ax)\cos ax) \quad \text{or} \quad C_{11}(h) - p_1^2 = \tfrac{1}{4}\exp(-ax)\cos ax$$

The centred covariance turns out to be the damped cosine function!

We can easily design similar pseudo-periodicities. As another example, imagine a sedimentary process. At each point of a constant rate Poisson process, we "break" the line, and insert a particle (state e_1) of constant length a. The distance separating two consecutive particles is thus a random variable having a distribution law $F(x) = 1 - \exp(-bx)$. We see from relation (IX-31)

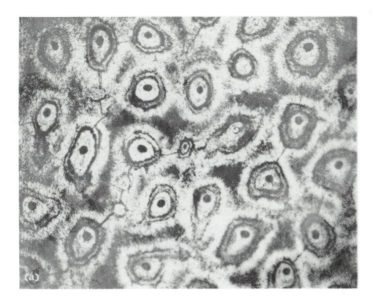

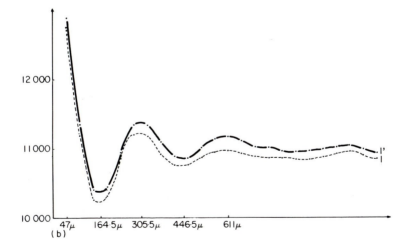

Figure IX.8. (a) Micro-fossil of a polypodium from the upper cretaceous. (b) and its covariance in two orthogonal directions. (Note that the fossil is *both* almost periodic and isotropic!)

that the variogram exhibits fluctuations of period a. The oscillations are entirely due to the fixed length of the grains a, and in no way result from any genetic periodicity!

Although a periodic covariance results from a periodic process, an "almost" periodic covariance may come from a completely aperiodic structure. This should emphasize that a covariance analysis must firstly take a *look* at the image under study. If it obviously does possess periodic structures, then quantitative information about them will be brought out by the covariance. But, such an interpretation of covariance periodicities is valid *only* when the visual confirmation exists; see Figure IX.8 for example.

(c) *Calcite migration or the hole effect*

The best way to introduce this second model is probably to put it in a geological context. In some layers of the Lorraine iron orebody, there appear limestone concretions (in white on the photograph, Figure IX.9) set on a (dark) background of iron ore. These concretions often seem underlined by shales. An interesting feature is that when concretions are thick, the inter-concretion distances are large and vice-versa, which to the geologists suggests a probable genesis: the concretions are the result of migrations of initially disseminate

Figure IX.9. Limestone concretions in an iron mine in Lorraine.

limestone particles, and the vehicle of this migration is the classical equilibrium which relates the free CO_2 to $CaCO_3$:

$$Ca(HCO_3)_2 \rightleftharpoons CaCO_3 + CO_2 + H_2O$$

Depending on the point of equilibrium, the limestone precipitate $CaCO_3$ can alternate with $Ca(HCO_3)_2$, which is present in solution. During their chemical evolution (diagnosis phase), the sediments are impregnated with water, and parts of them, such as the shales and the organic products in decomposition, then emit CO_2 which shifts the equilibrium to the right and generates a local deposit of $CaCO_3$ (L. Bubenicek, 1963).

This genetic description can be transposed into the framework of random set models (J. Serra, 1968). We will start with the one-dimensional space $\mathbb{R}^1$; suppose e_1 is the migrating element, i.e. consider points uniformly distributed on $\mathbb{R}^1$ and covering a constant proportion. We now implant migration centres M_i according to a Poisson process with a parametew θ, and each germ M_i gathers to its right all the migrating matter e_1 lying between M_i and M_{i+1}; if L is the intergerm distance, we therefore observe successively the segments of length $p_1 L$ and $p_0 L$ of the two phases e_1 and e_0 ($p_1 + p_0 = 1$), as shown on Figure IX.10a. Since the length L of the cycle follows the exponential law, $1 - F(x) = \exp(-\theta\lambda)$, the length distributions of both phases e_1 and e_0 are

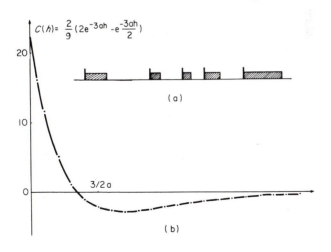

Figure IX.10. Migration process in $\mathbb{R}^1$ and its covariance.

also negative exponentials with modified parameters:

$$1 - F_0(x) = \exp\left(-\frac{\theta x}{p_0}\right); \qquad 1 - F_1(x) = \exp\left(-\Delta\frac{\theta x}{p_1}\right)$$

Thus:

$$\psi = \frac{\theta}{\theta + \lambda}; \qquad \Phi_0 = \frac{\theta}{\theta + p_0 \lambda}; \qquad \Phi_1 = \frac{\theta}{\theta + p_1 \lambda}$$

By applying (IX-31) and inverting the Laplace transform, we eventually derive the centred covariance $C_{11}(h) - p_1^2$, to be

$$C_{11}(h) = p_1^2 = \frac{p_0 p_1}{p_1 - p_0}\left[p_1 \exp - \theta h/p_0 - p_0 \exp\left(-\theta h/p_1\right)\right] \quad \text{(IX-32)}$$

The behaviour of this covariance is worth a closer look. Choosing $\frac{1}{3}$ for the grade p_1, which is roughly the area proportions of the concretions on Figure IX.11a, we see from the covariance plot of Figure IX.11b, that the curve drops well below its asymptote. In fact, if we calculate the range a_1 of this covariance, we obtain:

$$a_1 = \frac{2}{p_0 p_1}\int_0^\infty \left[C_{11}(h) - p_1^2\right]dh = 0$$

The negative spatial correlations exactly counterbalance the positive ones!

The stochastic model presented here obviously simplifies the reality of the situation. In a slightly more sophisticated version, we have also calculated the case of migrations from the left *and* from the right, in $\mathbb{R}^1$ (Serra, 1968). The resulting covariance shows a similar shape. In two dimensions, the calculations quickly become intractible, but we do have recourse to simulations (Fig. IX.11a). For example, start from a realization of a Poisson point process in $\mathbb{R}^2$. Place the mid points of horizontal segments X_i (the shales) with lengths given by a density law $f(l)$, at the Poisson points. Let X be the union of the horizontal segments. Construct the skiz $S_z(X)$ in the sense of Lantuéjoul (see Ch. XI), which delimits the zones of attraction Y_i of the shales X_i and associate with each Y_i the similar set Y_i' centred at point i, but reduced in scale by the factor $p_1 < 1$. The simulation results in random set realizations such as in Figure IX.11a, where parameters have been determined from the experimental data of thirty earlier photographs similar to Figure IX.9. As one can see, the mean experimental covariance does not coincide with the one taken from the simulation (Fig. IX.11b). Indeed, the actual geological process is three dimensional, and the various assumptions of isotropic and stationary migrations are partially wrong. This kind of conclusion is often reached when modelling structures with random sets or random functions. The real strength

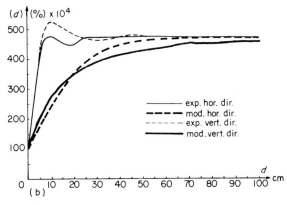

Figure IX.11. (a) simulation of a 2-D migration process, which models Figure IX.9, (b) covariance of the simulation and of the actual limestone migration.

of these models is heuristic: they carry morphological ideas, and suggest criteria (here a covariance). However, to a statistician they will seem weak, with a formulation which is not amenable to statistical tests.

(d) Range and macrostructures

Fortunately, statistical tests are not all that difficult to construct in a morphological context, as is shown in this section. Suppose we intersect a 2-D or 3-D medium X with a test line Δ of direction α, with the purpose of detecting whether macrostructures exist in direction α. On the line Δ, the sets X and X^c

delimit alternate segments $X_1 \subset X$ and $X_0 \subset X^c$. Assume that this 1-D structure is a realization of a cyclic binary process, where m_0 and m_1, σ_0^2 and σ_1^2, and σ_{01} are the mathematical expectation, the variance and the covariance of the lengths of the two segments of the cycle. From definition (IX-13) of the range a_1 (applied here in $\mathbb{R}^1$) and from the Laplace transform (IX-31) of the variogram, we derive:

$$a_1 = \frac{2}{p_1 p_0} \int_0^\infty [p_0 p_1 - \gamma(h)] \, dh = \frac{2}{p_0 p_1} \lim_{\lambda \to 0} \left[\frac{p_0 p_1}{\lambda} - \chi(\lambda) \right] \qquad \text{(IX-33)}$$

This limit is easily calculable if we use in $\chi(\lambda)$ limited expansions for the quantities $\Phi_0(\lambda)$, $\Phi_1(\lambda)$ and $\psi(\lambda)$, such as:

$$\Phi_0(\lambda) = 1 - \lambda m_0 + \tfrac{1}{2}\lambda^2 (m_0^2 + \sigma_0) + \dots$$

After simplification, relation (IX-33) gives:

$$a_1 = \frac{m_0 m_1}{m_0 + m_1} \left[\frac{\sigma_0^2}{m_0^2} + \frac{\sigma_1^2}{m_1^2} - \frac{2\sigma_{01}}{m_0 m_1} \right] \qquad \text{(IX-34)}$$

In practice, the most interesting case is the one of total lack of macrostructure, not only between cycles, but also within each cycle (i.e. between e_0 and e_1). Then $\sigma_{01} = 0$ and (IX-34) becomes

$$a_1 = p_1 \frac{\sigma_0^2}{m_0} + p_0 \frac{\sigma_1^2}{m_1} \qquad \text{(IX-35)}$$

In relation (IX-35), the left-hand side depends only on the covariance $C_{11}(h)$ while the right-hand side depends only on the two moments $P_1(l)$ and $P_0(l) \cdot P_1(l)$ is the probability that a segment of length l is included in the phase e_1; see next chapter. Thus the two sides of the relation can be experimentally determined by independent pieces of information. Relations (X-16) and (X-17) of the next chapter allow us to rewrite (IX-35) directly in terms of these pieces of information, giving:

$$\frac{2}{p_1 p_0} \int_0^\infty [C_{11}(h) - h^2] \, dh = 2 \int_0^\infty [P_0(l) + P_1(l)] \, dl + \frac{p_0^2 + p_1^2}{P_0^1(0)} \qquad \text{(IX-36)}$$

The statistical use of algorithm (IX-36) requires some care. The two integrals are estimated from erosion moments which are experimentally accessible only in bounded zones. We must replace the two infinite upper integration limits of relation (IX-36), with the maximum distance r_{max} involved in the computations of the covariance and the moments $P(l)$. Thus relation (IX-36) no longer remains a strict equality, but only an approximate relationship. Nevertheless, the comparison of the two sides of this new relation is sufficient for practical detection of macrostructures smaller than r_{max}, in the direction α.

D.2. What is a covariance blind to?

We will now present two quite simple models, and subject them to critical examination. Given a random set X, or a random function $f(x)$, will the probabilities of order two (i.e. $\Pr\{f(x_1) < a_1, f(x_2) < a_2\}$) be sufficient *in practice*, if not theoretically, to describe the function? Of course, the datum of the spatial law is incomparably more comprehensive than the order two characteristics, but if these latter carry 90 per cent of the information about the function, then in practice they are an exhaustive summary of the whole process. Julesz et al. (1973) experimented with this idea in various simulations, and finally came to the conclusion that texture fields having the same second order 2 statistics but differing in their higher orders could not be discriminated by a human viewer. These conclusions leave us rather sceptical. Second order probabilities are important but not exhaustive as we will now see.

Rather than embarking on a huge number of simulations, we will pose the problem slightly differently: can we construct random sets, or random functions, which have the same order two probabilities, but which in spatial law, are significantly different?

This proves to be possible. A first very simple approach is the following: take an arbitrary random set X, say a Boolean process, and add to X a locally finite number of isolated points and segments, denoted by Y. The spatial laws of the indicator functions of X and $X \cup Y$ are absolutely identical. The only difference occurs with the moments associated with non denumerable sets of points B (the probability that a disk is included in the pores, for example). The example could also be transposed in terms of random functions, by adding a finite constant to the function at a locally finite number of points. The spatial law is unchanged, while the supremum of the function over compact sets B such that $\mathrm{Mes}(B) \neq 0$, is deeply perturbed.

Here is a second example, start from a Poisson line process realization in $\mathbb{R}^1$ with a density θ_1 (Fig. IX.12a). Assign the same value to all the points of each Poisson polygon, independently from each other. The numerical values are taken at random from the density $f(x)$. The resulting random function $G_1(x)$ looks like a proscenium arch. We now construct a second random function $G_2(x)$, as follows. We start, once again, from a realization of a Poisson line process with density θ_2. Consider each polygon Π_i of the tessellation individually, and intersect with it a realization of a Poisson line process with density θ_2'. From one polygon to the other, the realizations are independent (Fig. IX.12b). We finish the construction as before, by assigning to each sub-polygon a random value taken from the density $f(x)$. From Figure IX.12b, it is obvious that there are two scales of construction, which differ sharply from the function $G_1(x)$.

To the human eye, which sees in two dimensions, there is clearly a difference.

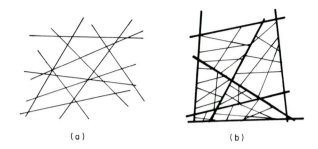

(a)   (b)

Figure IX.12. The two sets (a) and (b) induce the same process on any straight line.

However on any arbitrary linear intercept the two functions induce the *same* process, as soon as the density θ_1 equals $\theta_2 + \theta_2'$! Consequently all the probabilities pertaining to linear structuring elements are identical for the two functions $G_1(x)$ and $G_2(x)$. In particular the laws of order two, the run lengths, the linear size distributions, and the *sup* or *inf* of the function on segments are all identical. Note the paradox: the covariance, which is a tool especially adapted for detecting nested structures, is particularly blind to this one! The reader can easily extend the example to $\mathbb{R}^3$. For instance, start from a partition of the space into Poisson polyhedra; in each of them take a realization of a Poisson line process (in $\mathbb{R}^3$) restricted to the said polyhedron, and repeat the procedure independently for all the polyhedra of the partition. Consider the cross section of all these Poisson segments by an arbitrary plane. You get exactly a Poisson point process! The total lack of structure in the planes does not mean at all that macrostructures do not exist in the space.

E. A FEW EXAMPLES

We would not like to leave the reader with a negative impression about the covariance. The following four examples show the utility of a covariance analysis, and not only serve as numerical illustrations, but in addition each makes use of a new algorithm. Many further references could be given. In particular, R. Haralick (1972, 1973) has made very clever use of the order two probabilities for remote sensing from aerial photographs. Also, in O. Faugeras (1977), we see a critique of the performance of the covariance in characterizing textures, and in D. Jeulin (1978) there is a comprehensive analysis of multiphased petrographic specimens based on the rectangle covariances.

E.1. Clusters and stationarity

This example is taken from a study in embryogenesis, (J. Serra, H. Digabel, J. C. Marianna and P. Mauleon, 1973).

We will study the spatial distribution of oogoniae in the embryonic ovary of a rat. On the surface, the embryonic ovary of mammals does not exhibit any structure, although the proved existence of a physiological scheme of cell divisions will result in some organization of the germinal tissue. Can we detect the beginning of structuring via covariance analysis?

To simplify the problem, we will settle on an age of 15.25 days of foetal life; at this time, the only germinal cells present in the gonade are the ovogoniae. This presence is only a few per cent, and unfortunately they cannot be marked by a specific staining. Therefore the biologist must identify them on micrograph enlargements of size 40 × 30 cm, which cover whole cross sections of the ovary. One of the resulting documents is shown on Figure IX.13a.

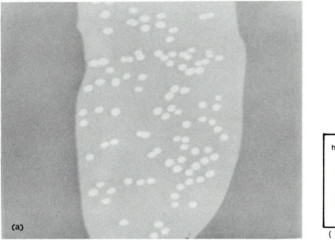
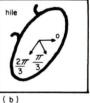

(a)

(b)

Figure IX.13. (a) partial view of an embryo ovary, (b) schematic display of the ovary and the Hile channel.

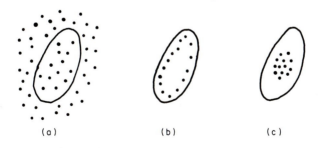

(a)   (b)   (c)

Figure IX.14. (a) arbitrary contour on a random set; (b), (c) examples of edge dependence.

The experimental data are 8 serial sections of 5μ thickness each, at a distance from each other of 15μ. The plane sections are parallel to the Hile channel (Fig. IX.13) which connects the ovary to the rest of the embryo, and they sample the ovary from end to end. The three directions o, $\pi/3$ and $2\pi/3$ of Figure IX.13b are investigated on each section.

We will ask two questions about the spatial distribution of the oogoniae:

— are they clustered?
— are they uniformly distributed within the ovary? (Fig. IX.14).

(a) *Global or local?*

The second question can be formulated in another way: must we calculate the global or the local covariance? The structure under study is as it stands, ambiguous; we need more information to decide between the two possible approaches.

Generally speaking, the notion of independence of two point regionalized variables f_i and f_j defined inside a bounded field Z and taking value zero outside (global approach) may be expressed in terms of integrals and devoid of probabilistic language (Matheron, 1965). If $g_i(h)$, $g_j(h)$, and $K_Z(h)$ are the global covariances of f_i, f_j, and of the indicator k_Z, and if $g_{ij}(h)$ is the rectangle covariance of f_i and f_j, the independence of the point variables f_i and f_j is by definition, given by:

$$\int dy \int f_i(x) f_j(x)\, dx = \int f_j(y)\, dy \times \int f_i(x)\, dx$$

or equivalently:

$$g_{ij}(o) = \frac{g_i(o)\cdot g_j(o)}{K_Z(o)} = m_i m_j K_Z(o) \qquad (IX-37)$$

(where the mean $m_i = [\int f_i(x)\, dx] \times [\int k_Z(x)\, dx]^{-1}$).

Algorithm (IX-37) expresses *point* regionalized variable independence. It does not imply that moving averages of f_i and f_j are independent of each other. If we want to extend the independence to all the regularized versions of f_i and f_j, it is easy to show that relation (IX-37) must be replaced by

$$g_{ij}(h) = m_i m_j K_Z(h) \qquad (IX-38)$$

which is very much stronger than the previous relation.

We can apply this to our particular problem by taking for $f_i(x)$ the set indicator k_1 of the oogoniae, and for f_j the indicator of the field Z itself, i.e. the ovary ($f_j = k_Z$). Relation (IX-38) simplifies, giving:

$$K_{1,Z}(h) = m_1 K_Z(h) \qquad (IX-39)$$

We can make this relationship more geometrical by introducing the mean value $m_1(h)$, which represents the oogoniae area proportion in the field $Z \cap Z_h$:

$$m_1(h)K_Z(h) = \int k_1(x)k_Z(x - h)\,dx = k_1 * \check{k}_Z \qquad (IX\text{-}40)$$

By noting that $K_{1,Z}(h) = \frac{1}{2}(k_1 * \check{k}_Z + \check{k}_1 * k_Z)$ and substituting in relation (IX-40), we finally have:

$$m_1 = \frac{1}{2}[m_1(h) + m_1(-h)] \qquad (IX\text{-}41)$$

Thus there is internal independence between the variable k_1 and its field Z if and only if the average of k_1 in the intersected fields $Z \cap Z_h$ and $Z \cap Z_{-h}$ remains constant and equal to the mean of k_1 in Z.

This approach puts us in the happy position of being able to test for the ovary/oogoniae independence in the three scanned directions. Rewriting algorithm (IX-41) in terms of covariances, we obtain:

$$\frac{K_{1,Z}(o)}{K_Z(o)} = \frac{1}{2}\left[\frac{K_{1,Z}(h) + K_{1,Z}(-h)}{K_Z(h)}\right] \qquad (IX\text{-}42)$$

In each direction, we can now plot the ratio of the right side over the left side of (IX-2) as a function of h, averaging over the eight sections. The result (Fig. IX.15) leaves us in no doubt: the central zone of the ovary is denser in oogoniae than the periphery. The effect is virtually identical in the two directions $\pi/3$ and $2\pi/3$, and weaker for $\alpha = 0$ (the standard deviations of the fifteen ratios for $h = 10$ and $\alpha = \pi/3$ is $\sigma = 1.1 \times 10^{-3}$ and for $h = 60$, $\alpha = \pi/3$, $\sigma = 0.9 \times 10^{-3}$). The structure must therefore be studied globally.

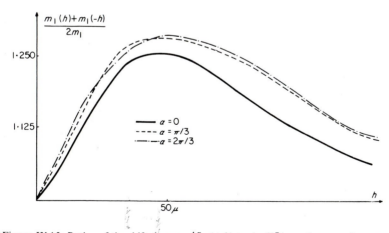

Figure IX.15. Ratios of the shifted means $\frac{1}{2}[m(+h) + m(-h)]$ over the general mean m.

(b) *Clusters of oogoniae*

The mean point covariance $K_1(h)$ of oogoniae, averaged over the sections and the directions, is plotted on Figure IX.16. The decrease at the start of the curve reflects the intersections of each oogonia with its own translate. There is no macroscale, but some hole effect, or pseudo-periodic effect near 15μ. The direction by direction analysis provides more accurate information with regard to this point (Fig. IX-16). We averaged separately the odd and the even sections (a) and (b) respectively, to test whether the oscillations were artefacts. This does not seem to be the case. For $\alpha = 0$ and $\alpha = \pi/3$, the first maximum appears at $h = 19\mu$, which means that the distance between the two centres of neighbouring oogoniae is 19μ (their mean diameter is approximately 10μ). For $\alpha = 2\pi/3$, the maximum is reached when $h = 35\mu$. Going back to the actual sections we notice that due to the division of the initial mother

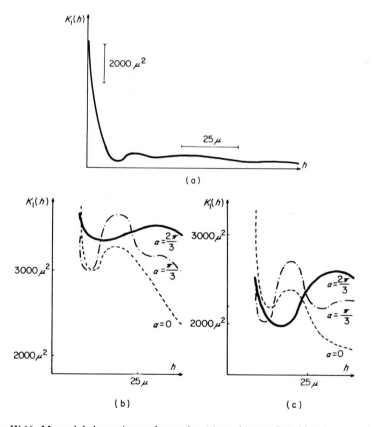

Figure IX.16. Mean global covariance of oogoniae: (a) at a large scale (odd and even sections); (b) even sections; (c) odd sections.

oogoniae, there are many groups of two cells. This division is probably an isotropic phenomenon. However the volume of the ovary doubles in one day of foetal life, and increases more quickly parallel to the long axis of the organ, whose direction is exactly $\alpha = 2\pi/3$.

Thus by use of the covariance, we are able to derive from the morphometry considered at a particular moment of the foetal life, the time gradient of the morphogenetic activity that governs the structuring of the ovary.

E.2. Scales and anisotropies

(a) *Wood anatomy*
The following study asks of Mathematical Morphology a very particular question. Consider the wood transverse section of Figure IX.17a: it was cut off perpendicular to the trunk axis and presents a complex structure. This wood, called "Aucoumea klaineana", is one of twenty tropical varieties on which we wish to relate morphology to an important physical property: the anisotropy of shrinkage during drying. We obtained this property by calculating the shrinkage ratio during drying in two directions, radial and tangential (Fig. IX.17a). Let us briefly describe the sequence of the wood structures as we change our scale of observation. The wood anatomy is made up of a set of cells that form a skeleton of vertical tubes made of rigid walls surrounding cavities that are different in dimensions and shape. At a ten microns scale, we see the fibres (the smallest cells) lined up in strings. Their direction is parallel to the rays (the long slim cells); this direction is the one which radiates from the trunk centre (Fig. IX.17b). Both these series of cells induce a strong anisotropy in the

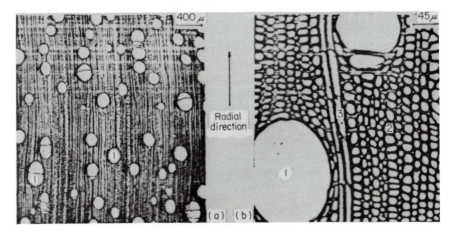

Figure IX.17. (a) transverse section of Aucoumea (× 25) with vessels (1) structure; (b) Same section (× 216) after thresholding. Fibres (2) and rays (3).

wood structure. At a hundred micron scale, the vessels appear, round and isolated, big tubes conducting the sap. Then they are more or less regrouped in small clusters and, at a millimetric scale, we can see their density slowly change (Fig. IX.17a). At the same scale, we can also observe slow variations of darkness in the fibres; these variations, called "tree-rings", are due to a regionalized flattening of the fibres and a thickening of their walls. The difficulty of the study under consideration is that previous analyses have shown that neither the number, nor the shape, nor the dimension of the vessels, play a role in the variations of the shrinkage anisotropy. Note that it is only the vessels which appear as structures of isolated "particles"; the others, the rings and fibre strings, look much like a continuous medium. Thus we need an investigation method

  (i) independent from the notion of particles;
 (ii) involving a directional character which is required by the structure anisotropy;
(iii) able to scan the continuous succession of scales from the fibre strings to the rings, to ensure that no possible correlation with the physical parameter is overlooked.

In terms of Mathematical Morphology, the simplest structuring element able to satisfy these three requirements is the pair of points, the associated measurement of which is the variogram.

(b) Measurements on the point variogram

By thresholding, the cavities and walls of Figure IX.17b are transformed into a medium of two phases X and X^c. We have computed the point variogram in two directions, for $\alpha = 0$ (tangential) and $\alpha = \pi/2$ (radial). For $h = 0$, $\gamma(h)$ is obviously equal to zero. As h increases, so the scales of structure that are investigated become larger. Now, the possibility that the variogram may grow indefinitely (starting from $\gamma(o) = 0$) makes it more able to describe the superpositions of structures than the closely related notion of covariance. For each of the twenty woods, the two variograms were calculated by taking their average value on 36 fields of the type given by Figure IX.17b, the fields being set on the points of a square grid with grid spacing of $1,280\mu$. The maximal step h of the point variogram is 80μ.

Several structures are detected by the variogram of Figure IX.18.

(a) The slopes at the origin, $\tan \theta_r$ and $\tan \theta_x$, are equal to the specific number of intercepts of the whole texture, for $\alpha = 0$ and for $\alpha = \pi/2$. Their mean value multiplied by π provides an estimate of the specific perimeter.

(b) After 30μ the variograms do not reach a sill, but continue to grow slowly. This second scale is where the elliptic vessels appear. For this structure, the

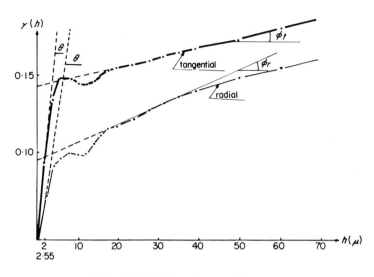

Figure IX.18. Point variogram of Aucoumea.

equivalents of the slopes at the origin are now $\tan \varphi_r$ and $\tan \varphi_t$, which exhibit a different anisotropy from that of $\tan \theta_r$ and $\tan \theta_t$. Therefore we can indirectly measure the specific perimeter of the vessels without having to isolate them in the picture.

(c) When comparing the two variograms after 25μ, notice that they essentially have the same behaviour, apart from a vertical shift. This shift, called zonal anisotropy, is proportional to the surface percentage occupied by the wood radii. Essentially, the smaller ordinate in the radial direction means that when one of the two extremities of a radial $\overrightarrow{h}$ falls in a white stripe (wood radii) the other is also likely to fall in the stripe, even for high values of h.

(d) The initial growth from 0 to 5μ is followed by an oscillation called the "hole effect" which corresponds to a pseudo-periodicity of the fibres. The abscissa of the minimum α_r or α_t gives the average distance between fibres in the described direction. In the case of the radial direction, the cavities are aligned in strings and hence the abscissa is also equal to the average diameter of these fibres. The interpretation of the maximum is more delicate: one needs to simulate the given structure from a probabilistic model, and proceed from there. The difference of ordinates between minimum and maximum explains the degree of regularity in the period-icity of distances between the fibres (Δ_r and Δ_t).

The point variograms are limited by the size of the elementary field; it is too small to attain the scale of the yearly rings. To continue the investigation we

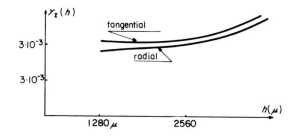

Figure IX.19. Regularized variogram of Aucoumea.

therefore have to use the regularized variogram; see relation (IX-24), where $p(y)$ is the proportion of white in the field centred at point y. The 36 fields of each specimen form a square grid (6×6) having a spacing of 1280μ. The new variograms (Fig. IX.19) again exhibit the vertical shift, which is due to zonal anisotropy. The rings themselves are described by the covariance increasing afresh after $3{,}000\mu$. We notice a parabolic behaviour at the beginning of the curve; this is due to the smoothing caused by the regularization, and probably also due to the progressive differentiation of the rings in the wood (i.e. zones with more vessels).

After computing a regression of the shrinkage anisotropies versus the variogram data, it appears that 55 per cent of the variance regression is explained by the ratios Δ_r/α_r and α_t/α_r, i.e. by the regularity of the display of fibres in the wood matrix. Finally, these fibres play exactly the same role as columns in a Greek temple: the more regularly they are implanted, the more resistant to shrinkage is the wood.

E.3. A sum of independent structures

We continue on with the examples, this time changing the scale of the object and its nature, but not completely its structure. The object is not an image (we cannot see it) nor a set, but a positive function. Consider the red layer of the Lorraine iron orebody, which has already been encountered when we looked at limestone concretions. The mineralized part is a horizontal flat layer three metres thick (Fig. IX.9 is a vertical section of it). In the two neighbouring mines of Hayange and Bois d'Avril, the red layers spread out over nonoverlapping areas of 7 and $5\,\mathrm{km}^2$ respectively. The mines were sampled using 160 and 120 samples (small drills taken from the mine levels) from an almost hectometre square grid. For mining estimation purposes, we have calculated the variogram of the iron grades in the two main directions of the grid (R. Carpentier, J. Serra, 1963); the result is the curves of Figure IX.20. Unlike the wood variogram, there is no longer a zonal anisotropy nor a larger scale comparable

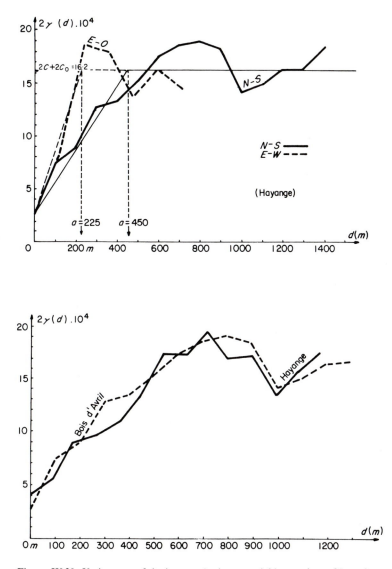

Figure IX.20. Variograms of the iron grades in two neighbour mines of Lorraine.

to that of the vessels. On the contrary, there is a smaller scale due to the limestone concretions. Its order of magnitude is in metres; thus in practice it appears on the curves as a nugget effect.

Nevertheless, the main features of the curves are identical for both mines: linear in the beginning, followed by one oscillation (one maximum and one minimum) before the sill. In fact, the actual underlying regionalizations are similar: the iron ore layers of Lorraine are marine sedimentary deposits caused by fresh water flowing from a river mouth into a shallow sea (L. Bubenicek, 1967). One can observe nowadays the same kind of phenomena in the large estuaries of the Rhine and the Meuse flowing into the North Sea. The conflict between marine and river flows creates sand banks separated by channels of running water (Fig. IX.21a): the elliptic elongation of the mineralized lens is seen on the variogram (Fig. IX.20). Modulo an affinity for the co-ordinates of space to be in the direction NE–SW, the variogram in all directions becomes isotropic. The thus "isotropized" stratum then possesses exactly the same structure as the fibre sections of the wood in photograph IX-17b. There is one difference however; generally we cannot map it, or visualize it. During the successive periods of time of the deposit, about twenty such strata with an elementary thickness of 15 cm, are *independently* piled onto each other, so generating the mining layer. So when we sum up the quantity of iron oxide along a given vertical drill, then from one end of the deposit to the other the contours of the elementary stratum are completely jumbled (Fig. IX.21b).

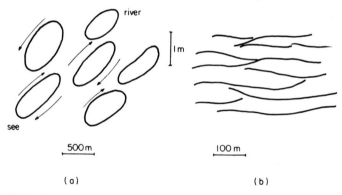

Figure IX.21. Sedimentation of sand banks. (a) horizontal view of elementary stratum, (b) vertical section of the mineralized layer.

Now what about the variograms? We assume that each stratum (no. i) of the layer has a constant thickness l_i; let $f_i(x)$ be its grade on the vertical line at point $x \in \mathbb{R}^2$, and $\gamma_i(h)$ the associated variogram. The grade $f(x)$ of the whole layer at point x is

$$f(x) = \frac{\sum_i l_i f_i(x)}{\sum_i l_i} \qquad (1 \le i \le N)$$

which implies that if the strata are independent:

$$\gamma(h) = \frac{1}{\left(\sum_i l_i\right)^2} \sum_{i=1}^{N} l_i^2 \gamma_i(h)$$

$(\gamma(h) =$ variogram of $f(x))$. In particular, if the N strata have the same thickness, and the same structure, i.e. $\gamma_i(h) = \gamma_1(h)$, the resulting $\gamma(h)$ becomes:

$$\gamma(h) = \frac{1}{N}\gamma_1(h)$$

Thus up to a multiplicative constant, the variogram of the sum of super-imposed independent realizations of the same random function is the same as the variogram of each of the elementary phenomena. This is as expected: exactly as for the power spectrum, the variogram ignores phases and concentrates on the amplitude of the Fourier expansion of the structure.

E.4. Clusters for random functions

In this last example, we stay with structures represented by bounded positive functions, but return to the microscopic scale where $f(x)$ denotes the grey tone at the point x. The phenomenon under study is of bounded size, so we are able to use the covariance (instead of the variogram). It is large enough however to involve estimation from several fields, which suggests a local approach.

The photograph (Fig. IX.22) shows a thin section cut normal to the elongation direction of the fibres of a striated muscle of a rabbit. A particular staining technique allows us to gauge the speed of contraction of each fibre by detecting the activity of the myofibrillar adenosine triphosphatase (the muscle myosine destroys this enzyme, so the darker the fibre appears, the slower it contracts).

From the histological point of view, two levels of structure are present here: the elementary fibre and the bundle of fibres (a pack of about fifteen fibres surrounded by intersticial tissue). Both can be seen clearly on Figure IX.22. But physiologically, are these the only two structures? Is the behaviour of each fibre independent of the others? Do clusters of fibres exist, and are there clusters of the clusters?

Before presenting and discussing the results, we will say a few words about the most suitable algorithms for computing such a covariance. In practice, $f(x)$ is the step function (IX-18), however the $k_i(x)$ are not the best medium for carrying the information. Finally there is some redundancy present, since for every x we have $\sum k_i(x) = 1$. Secondly, their use implies that the image is split into neatly joined stripes of greys, without voids or overlappings. In the less sophisticated image analysers, the image is not stored, and the grey stripes are accessed (depending on the logical requirements of the operation) by successive scannings. So noise will put uncertainties into the idea of a neatly joined image. For these two reasons, we will instead work with integral grey

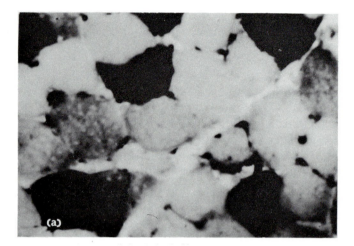

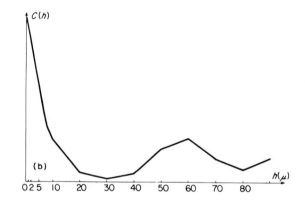

Figure IX.22. (a) fibres of an elongated rabbit muscle, (b) the corresponding covariance.

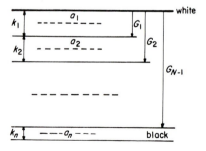

Figure IX.23. Integral grey levels G_i.

levels $G_i(x)$ (see Fig. IX.23), defined as follows:

$$G_i(x) = \sum_{j=1}^{i} k_j(x) \qquad i = 1, \ldots, N-1$$

$$G_0(x) = 0 \quad \text{and} \quad G_N(x) = 1$$

Of course, the G_i are directly experimentally accessible without having to sum the k_i. Associate with the G_i, the covariance matrix $D_{ij}(h)$:

$$D_{ij}(h) = \frac{1}{A[Z \cap Z_h]} \int_{\text{mask } Z} G_i(x)G_j(x+h)\,dx \qquad \text{(IX-43)}$$

From these data we easily derive (see Ex. IX-9) an estimate p^* of the mean value $p = E[f(x)]$:

$$p^*(Z) = a_x + \sum_{i=1}^{N-1} \frac{(a_i - a_{i-1})}{A(Z)} \int_{\text{mask } Z} G_i(x)\,dx$$

and also an estimate $\text{Cov}(h)^*$ of the centred covariance $\text{Cov}(h)$ of relation (IX-10):

$$\text{Cov}(h)_* = \sum_{i=1}^{N-1} \sum_{j=1}^{N-1} (a_i - a_{i+1})(a_j - a_{j+1})D_{ij}(h)$$

$$- \left[\sum_{i=1}^{N-1} (a_i - a_{i+1})D_{ii}(o) \right]^2 \qquad \text{(IX-44)}$$

We can see from Figure IX-22b what this looks like for the fibre covariance of the rabbit muscle. Firstly, the reader should notice an oblique tangent at the origin; this corresponds to the mean discontinuity measures of the specimen. The grey value is almost constant inside a given fibre, and jumps to another value when crossing to a neighbouring fibre. The slope of the tangent at the origin is equal to the length of the boundary per unit area, where each perimeter element ds is weighted by the square gradient of grey tones across the boundary (modulo the coefficient π). It turns out to be a contrast factor. A second interesting thing to notice is that the value of the range a is equal to 30μ, and is bigger than the mean size of the elementary fibre; the elementary biological structure (in terms of the speed of contraction) is not the elementary fibre itself, but a small group of three or four joined fibres. Notice also a pseudo-periodic oscillation in the covariance. The fast fibres, which are responsible for the strength of the muscle, seem to be consistently surrounded by slower ones, as if they were draining the activity of the muscle. A last point is worth mentioning: the bundle structure hardly shows itself on the covariance (the fact that the covariance does not actually reach the x-axis is presumably due to the bundles). This point confirms the conclusions of Section IX, D.2

about the models Figure IX-12. If the intersticial tissue were in the limit extremely thin, then the covariance would be completely blind to the bundle structure.

Remark: The optical devices often overlighten the central regions of the measuring field. That is creating a bias on the covariance estimate; it can be corrected by replacing the term $-[\cdot]^2$ in relation (IX-44) by the quantity $p^*(Z \cap Z_h) \cdot p^*(Z \cap Z_{-h})$ (theory of Universal Kriging, in Journel, 1977).

F. EXERCISES

F.1. Global covariance for sets

Draw the geometric covariogram of Figure IX.24 in the two directions α and β. Interpret the value and the slope at the origin, as well as the maxima and minima. What can you say about the ratios between the successive maxima when the number of squares tends to infinity?

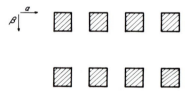

Figure IX.24.

F.2. Local set covariance (P. Hersant *et al.*, 1976)

In the three following sets, the same pieces of puzzle are differently displayed. Interpret the covariances of the sets, via the ordinate at the origin, the slope of the tangent at the origin, the superimposition of scales and the range reached by upper and lower values (Fig. IX.25).

F.3. Set covariance of a ball

Calculate the set covariance of a ball X with a diameter D, in $\mathbb{R}^3$.

$$[K(h) = \frac{\pi D^3}{6} \left(1 - \frac{3}{2} \frac{|h|}{D} + \frac{1}{2} \frac{|h|^3}{D^3} \right) \quad |h| < D$$

$$= 0 \qquad\qquad\qquad |h| > D$$

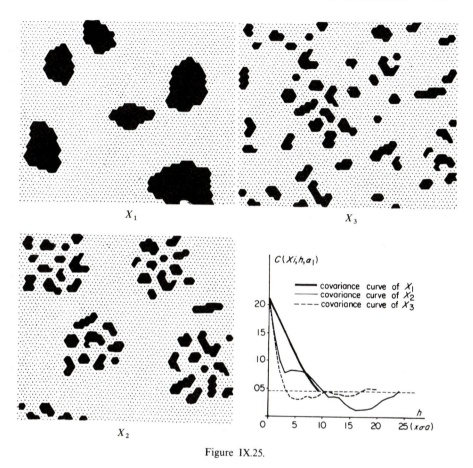

Figure IX.25.

Start from the apparent contour of the intersection $X \cap X_h$, namely $K'(h)$ $= (\pi/4)(D^2 - h^2)$; then integrate, remembering that $K(o) = \pi D^3/6$]

F.4. Fractal sets in $\mathbb{R}^1$

Let us consider the set X in $\mathbb{R}^1$, defined by the indicator function $k(x)$.

$$k(x) = 0 \quad 2^{-(2i+2)} < x < 2^{-(2i+1)} \quad \text{(pore no. } i+1)$$
$$k(x) = 1 \quad 2^{-(2i+1)} \leq x \leq 2^{-2i} \quad \text{(grain no. } i)$$

(the support of the set X is the segment $[0, 1]$ and i varies from 0 to ∞) (Fig. IX.26).

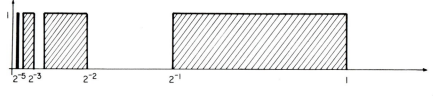

Figure IX.26.

Prove that the covariance $K(h)$ of $X \oplus \delta r$ satisfies

$$- K'_{X \oplus \delta r}(o) = n, \quad \text{when } 2^{-2n-1} < \delta r < 2^{-2n}$$

Conclude that $- K'_X(o)$ equals infinity.

F.5. Three-phase structures and rectangle covariances

Consider the biological cell of Figure IX.27, where the nucleus is represented by white circles and the cytoplasma by dark ones. Compute and draw the three rectangle covariances in directions α and β. Interpret the behaviour at the origin, and at infinity.

F.6. White noise on a covariance

In $\mathbb{R}^1$, the interval $[0, b]$ is divided into n equal intervals of length $a = b/n$. Consider n independent random variables x_i with the same expectation m and the same covariance σ^2. Put $f(x) = X_i$ when $(i-1)a < x \le ia, i = 1, 2, \ldots, n$, and $f(x) = 0$ outside $[0, b]$. Find the covariance of $f(x)$ (Fig. IX.28).

$$[\sigma^2(b - |h|) + m^2(b - |h|) \text{ when } |h| \le a, \, m^2(b - |h|)$$

$$\text{when } a < |h| \le b, \, 0 \text{ otherwise.}$$

When a is small, or n large, this scheme corresponds to a nugget effect].

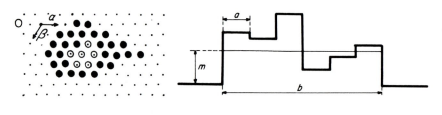

Figure IX.27.   Figure IX.28.

F.7. Isotropic covariances

(a) Let $p(r)$ be a 3-D positive summable function with spherical symmetry. What is its covariance $g(h)$, as defined by relation (IX-4)?

[Compute the volume of the intersection of two spherical crowns, with radii r_1 and r_2, and thicknesses dr_1 and dr_2. Then derive the expression:

$$g(h) = \frac{\pi}{h}\left\{(2\breve\varphi - \varphi)\Phi - 2\left[\Phi^2(\infty) - \int_0^\infty \varphi^2(r)\,dr\right]\right\}$$

where $\varphi(r) = rp(r)$ and $\Phi(r)$ is the primitive of φ.]

(b) Apply the preceding result to calculate the covariance of the ball.

[Take $p(r) = 1, 0 \le r \le a$ and $p(r) = 0, r > a$; the previous result implies:

$$K(h) = g(h) = \frac{4}{3}\pi a^3 - \pi h a^2 + \frac{\pi}{12}h^3 \qquad h \le a$$

$$K(h) = g(h) = 0 \qquad\qquad\qquad h > a.]$$

F.8. Hole effect

(a) Let X and Y be two compact sets such that $X \subset Y$. What is the covariance $K(h)$ of the set difference X/Y? (When X and Y are convex, X/Y is the grain X with the hole Y inside).

$$[K(h) = \text{Mes } A(Y \cap Y_h) - \text{Mes}(Y_h \cap X) - \text{Mes}(X_h \cap Y) + \text{Mes}(X \cap X_h)]$$

(b) An application: take for X and Y two concentric balls with radii r_1 and r_2.

[Denote by K_X and K_Y the geometric covariance of the balls X and Y calculated in Exercise IX-7.

$$K(h) = K_X(h) + K_Y(h) - 2K_X(o) \quad \text{for } 0 \le h \le r_2 - r_1$$

$$= K_X(h) + K_Y(h) - 2K_X(o) + K_X\left[\frac{r_2^2 - r_1^2 - h^2}{h}\right] - K_Y\left[\frac{r_2^2 - r_1^2 + h^2}{h}\right]$$

$$\text{for } r_2 - r_1 \le h \le \sqrt{r_2^2 - r_1^2}$$

$$K(h) = K_X(h) + K_Y(h) - K_X\left[-\frac{r_2^2 + r_1^2 + h^2}{h}\right] - K_Y\left[\frac{r_2^2 - r_1^2 + h^2}{h}\right]$$

$$\text{for } \sqrt{r_2^2 - r_1} \le h \le r_2 + r_1$$

$$K(h) = K_X(h) \quad \text{for } r_2 + r_1 \le h.]$$

F.9. The covariance for grey tone functions

(a) Prove relation (IX-44) given Cov $(h)^*$ as a linear form of the $D_{ij}(h)$.

[Start from $\text{Cov}(h)^* = \sum_{i=1}^{N}\sum_{j=1}^{N} a_i a_j E\{[G_i(x) - G_{i-1}(x)][G_j(x+h) - G_{j-2}(x+h)]\}$, and simplify.]

(b) The global case: let the covariance $g(h)$ be defined by relation (IX-4), and the $G_i(x)$ as previously. By putting:

$$D_{ij}(h) = \int_{\mathbb{R}^2} G_i(x) \cdot G_j(x+h)\,dh$$

prove that

$$g(h) = \sum_{i=1}^{n-1} \sum_{j=1}^{n-1} (a_i - a_{i+1})(a_j - a_{j+1})\Delta_{ij}(h)$$

Note the difference between this and the local case result (rel. (IX-44)).

F.10. Rectangle covariance computation

(a) Let X be a random stationary partition of the space into N phases $1, \ldots, i, \ldots, N$, and let its cross-covariance be the C_{ij} of relation (IX-16). Using the D_{ij} defined in algorithm (IX-43), prove that C_{ij} may be estimated by:

$$C_{ij}^*(h) = D_{ij}(h) - D_{i,j-1}(h) - D_{i-1,j}(h) + D_{i-1,j-1}(h)$$

and that

$$i = 1 \Rightarrow C_{i-1,j}(h) \equiv 0, \qquad \forall j$$

$$i = N \Rightarrow C_{i,j} = E[G_j] = \sum_{k=1}^{j} p_k$$

(b) Show that for $N = 2$ (grains and pores), the datum of one of the covariances $C_{11}, C_{00}, C_{10}, C_{01}$ gives all the others.
  [We have $p = C_{11}(o) = C_{11}(h) - C_{10}(h)$, and a similar relation for $C_{00}(h)$. Thus $C_{11}(h) = p - q + C_{00}(h)$, and $C_{10}(h) = C_{01}(h)$. The covariance for the grains is the same as for the pores, but with $p - q$ added. The two variograms $C_{10}(h)$ and $C_{01}(h)$ are identical.]
(c) Show that for $N = 3$, knowing the $C_{ij}(j \neq i)$ is equivalent to knowing the C_{ii}. Give the relationship between the two.
  [$C_{11}(h) = C_{11}(o) - C_{12}(h) - C_{13}(h)$, plus the two other relations derived by cyclic permutation.]
(d) Show that for $N > 3$ the $C_{ii}(h)$ are known when the $C_{ij}(h), i \neq j$, are given, but that the converse is false.
(e) For $N = 3$ and $N = 4$ express the $C_{ij}(h)$ as functions of the $D_{ij}(h)$.

F.11. Probabilistic models for the triangular and exponential covariance

(a) An origin x_0 being chosen at random on $(0, a)$, the straight line is divided into segments of length a with subdivision points $x_0 + ka$ (ka positive or negative integer). Let $f(x)$ be a random function taking for each of these segments of length a, a random constant value. The values are drawn by chance independently from one segment to another according to the same law of probability, with mean value m and variance σ^2. Calculate the probability P that x and $x + h$ belong to the same segment a. Derive from it the covariance $C(h)$ of $f(x)$.

[$P = 1 - |h|/a$ when $|h| \leq a$, 0 otherwise; $C(h) = \sigma^2$.]

(b) Same question as in (a), but the lengths of the segments are now randomly distributed, independently from each other, and follow the same exponential law $\exp(-\lambda h)$ (in other words, the points of discontinuity constitute a Poisson process.

[$\Pr\{x$ and $x + h \in$ same segment$\} = \exp(-\lambda h)$. Then $\text{Cov}(h) = \sigma^2 \exp(-\lambda h)$.]

F.12. Nugget effect in the pure state

We are given nugget sets at points distributed in space according to a Poisson scheme (definition: the number $N(X)$ of nuggets contained in X is a Poisson random variable of expectation $\lambda V(X)$; if X and X' are disjoint, $N(X)$ and $N(X')$ are independent).

(a) work with volumes v, and put $f(x) = N(v_x)$, v_x denoting the translate of v located at point x. Show that the covariance of $f(x)$ and $f(x + h)$ is the variance of $N(X_x \cap X_{x+h})$ or $\lambda K(h)$ where $K(h)$ denotes the geometric covariogram of the volume $V(X)$.

(b) The nuggets are now assumed to have random independent weights (mean p_0, variance σ_p^2), and $f(x)$ is the sum $P_1 + \ldots + P_N$ of the weights of the $N = N(X_x)$ nuggets contained in X_x. Calculate the mean and the variance of $f(x)$. [$E(f) = \lambda V p_0$, $\text{Var}(f) = \lambda_v(p_0^2 + \sigma_p^2)$.] What is the covariance of $f(x)$ and $f(x + h)$? [Replace V by $K(h)$.]

F.13. Probabilistic dilution model

We are given Poisson points as in Exercise 12. Let $p(x)$ be a function and $f(x) = \sum_i p(x - x_i)$, x_i denoting the location of the different nuggets. Thus $f(x)$ is a dilution of these nucleii. Find the covariance $\text{Cov}(h)$ of $f(x)$.

[$\lambda g(h)$, with $g = p * \check{p}$. Only the case where p has a compact support can be examined: consider two points x_0 and $x_0 + h$ and a bounded domain X containing the translates by x_0 and $x_0 + h$ of the support of p. First reason conditionally when the number n of Poisson points falling within X is fixed, then unconditionally with respect to n. Notice also how interpretation using the covariance measure $\lambda \delta$ of the Poisson points makes this result intuitive.]

F.14. Physical interpretation of the range

The stationary random function $f(x)$ is the Poisson dilution of random compact convex grains X, having a set covariance $K(h)$, and a volume $V(X)$. Interpret the meaning of the range a_n and the ranges a_i, associated with the restriction of $f(x)$ to sub-spaces E_i of $\mathbb{R}^n$ ($1 \leq i \leq n$). What about the variance formulae for a large specimen?

$$\left[a_n = \frac{1}{C} \int_{R^n} \mathrm{Cov}(h)\, dh = \frac{1}{\lambda E(V)} \int \lambda K(h)\, dh = \frac{E[V^2]}{E[V]} = \mathcal{M}[V] \right.$$

where $\mathcal{M}[V]$ is a moment weighted in measure.
Similarly

$$a_i = \frac{1}{C} \int_{E_i} C(h)\, dh = E[\mathrm{Mes}(X \cap E_i)^2]/E[V]$$

The variance $\sigma^2(Y)$ of a large specimen Y with a volume $V(Y)$ is

$$\sigma^2(Y) \sim \lambda \frac{E[V(X)^2]}{V(Y)} = \lambda \frac{E[V(X)]}{V(Y)} \cdot \mathcal{M}[V(X)]$$

If Y is contained in a sub-space E_i;

$$\sigma^2(Y) = \lambda \cdot \frac{E[\mathrm{Mes}(X \cap E_i)]}{V(Y)} \cdot \mathcal{M}[\mathrm{Mes}(X \cap E_i)]. \left. \right]$$

F.15. Covariance or variogram? (G. Matheron, 1968)

For a stationary random function of order two possessing a mean m and a centred covariance $\mathrm{Cov}(h)$, the datum of $\mathrm{Cov}(h)$ is obviously equivalent to that of the variogram $\gamma(h)$, since $\gamma(h) = \mathrm{Cov}(o) - \mathrm{Cov}(h)$. But what happens when only a variogram exists, and nevertheless we compute experimental covariances?

Let us consider a realization $f(x)$ of Brownian motion in $\mathbb{R}^1$, known in the segment $[O, L]$. We have:

$$E[f(x+h) - f(x)] = 0 \quad \text{and} \quad \gamma(h) = \tfrac{1}{2} E[f(x+h) - f(x)]^2 = |h|$$

(a) Estimate the hypothetical covariance $\mathrm{Cov}(h)$, which in fact does not exist here, from the experimental quantities:

$$\bar{f} = \frac{1}{L} \int_0^L f(x)\, dx \qquad \mathrm{Cov}^*(x, y) = (f(x) - \bar{f})(f(y) - \bar{f})$$

by putting $\text{Cov}^*(h) = \dfrac{1}{L-h} \displaystyle\int_0^{L-h} \text{Cov}^*(x+h, x)\, dx$

Show that $E[\text{Cov}^*(x, y)] = \dfrac{2}{3}L + \dfrac{x^2 + y^2}{L} + 2\sup(x, y)$

derive $E[\text{Cov}^*(h)] = \dfrac{1}{3}L - \dfrac{4}{3}h + \dfrac{2h^2}{3L} \quad (0 \le h \le L)$

Comment on the result. [An apparent variance $E[\text{Cov}^*(o)] = \tfrac{1}{3}L$ is found, depending on the length L of the segment considered. It is a pure artefact, since the true variance is infinite. Although there is no covariance (the variogram is linear), the biases introduced by this procedure of estimation result in an apparent confirmation of the existence of a covariance (with a range!). It will be noted that the structure of the phenomenon is extremely distorted: not only is the straight line replaced by a parabola, but even *the slope at the origin is changed* ($\tfrac{4}{3}$ instead of 1). Thus $\text{Cov}^*(h)$ represents almost nothing of the true structure].

(b) Show that the experimental variogram

$$\gamma^*(h) = \frac{1}{2(L-h)} \int_0^{L-h} [f(x+h) - f(x)]^2\, dx$$

has expectation $E[\gamma^*(h)] = \gamma(h) = |h|$, and so does not run into the same bias problems as $\text{Cov}^*(h)$.

Comment. [When the experimental variance of a phenomenon increases with the size of the zone investigated, without tending towards a horizontal asymptote, it is not wise to fit the structure under study with a stochastic model possessing a covariance. However in such a case, the variogram still exists and its estimation is significant. Consequently it provides a *safer* method than the covariance.]

X. Size Criteria

A. WHAT DOES "SIZE" MEAN?

A.1. Four classical sieving techniques

The word "size" is a simple one-syllable word which is used all the time in everyday speech. The question is, can we clarify and quantify what we mean by size in order to build a morphological criterion on it? We will start by reviewing the most usual size distribution procedures used in physical sciences.

The most frequent sizing technique is probably *sieving*: small solid particles are classified according to a series of sieves with decreasing mesh openings. Firstly one performs the classification by letting the population of grains work its way through the sieves (in effect, this is a transformation). Then one weighs the contents of each sieve (set measurement). Note that in this procedure, each grain is measured in proportion to its own weight. The corresponding techniques in biology are typified by the Coulter countings (named after a family of measuring devices). Here the cells, carried in a fluid, run one by one through a capillary tube where their apparent contour (or a more sophisticated parameter) is electrically measured. The final histogram expresses the *number* of cells (not their weight) versus the size parameter. Hence these two techniques differ due to their system of weighting, but are both based upon the notion of an individual particle.

There are other procedures that are not based on this individualized particle idea. For example, in the Purcell method used in the petroleum industry, mercury is injected into specimens of porous rocks (sandstones . . .). For a given pressure, the law of capillarity associates with the mercury a minimum meniscus. Hence if all the pores are accessible from the outside, the injection is precisely a spherical "opening" of the porous medium. When the pressure increases the minimum meniscus decreases, and finer and finer zones of the porous medium are reached. The associated measure of size is the volume occupied by the mercury as a function of the pressure. A second example of a method not based on connected particles is provided by the linear intercept distributions in image analysis. In performing this very classical technique (it goes back to Crofton!) one considers all the possible chords drawn on the set X, and then constructs a histogram; all chords are treated identically regardless of which connected component they belong to.

This brief review shows obvious variations in size analyses, pertaining mainly to questions of connectivity and weighting. In fact these questions will arise again and again in the course of the chapter. However the real interest for us here, is to find the *common* features of these four techniques, upon which the notion of sizing can be generalized.

A.2. Matheron's axioms

Giving a simple example is the best approach to take here. Consider a population of five persons, consisting of three men and two women. They make up a set X of five elements (Fig. X.1). We give to each of them a number, namely his (her) height, which ranges from 1.20 m to 1.80 m. To classify them according to their size means to define a set transformation family $\psi_\lambda(X)$ depending on a positive parameter $\lambda \geq 0$. For example "$\psi_{1.4}(X) =$ the subset of those people taller than 1.4 m". This set transformation $\psi_\lambda(X)$ must be:

| 1·2 | 1·4 | 1·5 | 1·7 | 1·8 |

Figure X.1. Set X and the corresponding sizes.

(a) *Anti-extensive*

$$\psi_\lambda(X) \subset X \qquad \forall \lambda \geq 0 \qquad\qquad (\text{X-1})$$

(people taller than λ are a subset of the original set of people).

(b) *Increasing*:

$$Y \subset X \Rightarrow \psi_\lambda(Y) \subset \psi_\lambda(X) \qquad \forall \lambda \geq 0 \qquad (\text{X-2})$$

(a subset, e.g. the women, taller than λ, are a part of all the people taller than λ).

These two conditions are illustrated on Figure X.2 for our elementary example.

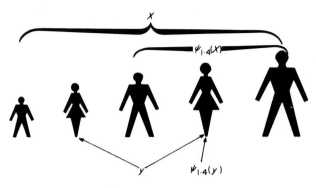

Figure X.2. Every sizing procedure is anti-extensive and increasing.

(c) *The stronger sieve is the most important*

$$\psi_\lambda[\psi_\mu(X)] = \psi_\mu[\psi_\lambda(X)] = \psi_{\mathrm{Sup}(\lambda,\mu)}(X) \qquad \forall \lambda, \mu \geq 0 \qquad \text{(X-3)}$$

This condition is a little more subtle than the preceding ones. We see on Figure X.3 what we mean by $\psi_{1.6}[\psi_{1.4}(X)]$. We first transform the initial population into $\psi_{1.4}(X)$, which results in the set made up of the three people taller than 1.4 meters. On *this resulting set* we apply the selection "$\psi_{1.6}$ = taller than 1.60 meters". Therefore axiom (X-3) quantifies the *stability* aspect of sizing procedures. We could search, among the three people taller than 1.40 m, those who are also taller than 1.60 m (i.e. $\psi_{1.6}[\psi_{1.4}(X)]$), however we could

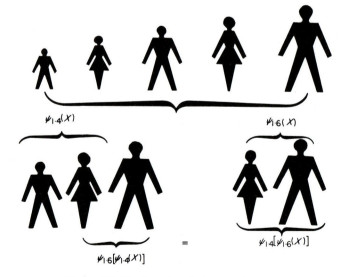

Figure X.3. The finer sieve is the most important.

equally search directly from the initial population, those taller then 1.60 m., or even invert the order of the operations. As a particular case, when $\lambda = \mu$ in (X-3), we find:

$$\psi_\lambda[\psi_\lambda(X)] = \psi_\lambda[X] \qquad\qquad \text{(X-4)}$$

which says that a sizing procedure is *idempotent*.

We bring to the reader's attention the parsimony of the system of the logical relationships (X-1) to (X-3). The first orders a transform with respect to the initial set, and the second and the third are respectively order relations holding either on the sets (when λ fixed) or on the λ's (when X fixed). Following G. Matheron (1975) we state that by definition a *sizing criterion* is a criterion which satisfies the three axioms (X-1) to (X-3). The set operation involved in a size criterion $\psi_\lambda(X)$ may then be followed by a measurement $F(\psi_\lambda)$, where F is an increasing positive function of λ. In practice, it will be the Lebesgue measure or the counting measure for discrete populations X.

A.3. Why take an axiomatic approach?

We can answer this question in several different ways.

(a) There is a mountain of literature on size distribution (see for example the reviews of D. V. Little (1974) and P. A. P. Moran (1972)) written by researchers who do not know if they speak about the same concepts or not, and when we compare this barrage of methods (and tricks, sometimes) to the simplicity of Matheron's three axioms, the advantage of an axiomatic approach becomes clear.

(b) The size axioms, exactly as for the classical distance (or metric) axioms, start from an intuitive notion of the everyday world, which is then translated into a logical form. Such a conceptualization is not often possible in science (try to do it with the concept of "shape", for example). Here however, one is able to describe all the common properties of the sizing procedures and consequently, to set out their different implications in a precise way.

(c) The study of the properties derived from the three size axioms, plus the four principles of mathematical morphology, yields an initial fundamental result. *Every morphological size distribution is a union of (morphological) openings* (see a more precise statement in section G below). In other words, we have been directed towards a particular mode of analysis from a fairly general axiomatic approach (this theorem reminds us of the link between increasing mappings and dilations, already met in Chapter II). More specifically, the size distribution mapping $\psi_\lambda(X)$ can be written as a single opening $X_{B(\lambda)}$ if and only if the $B(\lambda)$'s are similar convex sets.

(d) Among other common properties, morphological size distributions are all digitalizable (increasing semi-continuous mappings).

(e) A random version of size analysis is possible. A size criterion applied to a random set generates a measurable random function (see Sect. F. 2e).

(f) The final comment is more heuristic than formal. Matheron's three axioms, and the versions of them when stated for morphological openings scrutinise the objects and images from an aspect which is radically new compared to the second order analysis met in the last chapter.

A.4. Plan of the chapter

The only mathematical section of this chapter is the last one (Sect. F) which concentrates on a brief review of the general properties of morphological size distributions. In Sections B to E, we specify four sizing techniques frequently used in image analysis, with each technique having its own separate section. These sizing methods, which are relatively simple, do not cover the entire field of possibilities (we could also define size distribution depending on *several* parameters for example, or study multiphased sets, etc. . . .).

The four techniques really depend on just four algorithms, namely linear erosion, 2-D opening, connected component labelling, and triangular erosion. Rather than allow the algorithms to govern the order of treatment in the chapter (which would be the computer's point of view), we have decided that the order will be determined by the form of the sets under study (which is from the viewpoint of the image). In Section B, we assume that we can only look at the object along linear probes. Applications of the approach may be found in the study of (geological) bore holes, or in the sampling taken from forestry paths for inventory purposes (Marbeau, 1973). Due to its 1-D nature, this case reduces to a single morphological analysis (linear erosion). Section C is concerned with the operation of opening, theoretically in $\mathbb{R}^n$, but in practice in $\mathbb{R}^2$. The emphasis is put on the fact that the openings combine shape and size descriptions of erosion without being based on the existence of individuals. This appears in Section D; in Section E the assumptions become even more specific, namely that the particles are balls. Thus the distinction between textural and individual analyses governs the development of this chapter. The introductory examples also bring attention onto another distinction, namely the question of weights. Each section will have its systematic treatment of this question. We believe that by progressively making the set model more specific, the experimenter has the opportunity, at each step, to use those algorithms already defined for the previous steps. The approach can be set out in the following table:

Set Models

		Sets in $\mathbb{R}^n$ $n \geq 1$	Sets in $\mathbb{R}^n$ $n \geq 2$	Individuals	Convex particles	Balls
Algorithms	Linear erosion	X	X	X	X	X
	2-D opening		X	X	X	X
	Individual labelling	X (Consequence of linear erosion)		X	X	X
				X	X	X
	Triangle erosion				X	X (useless)

B. 1-D SIZE DISTRIBUTIONS

In the space $\mathbb{R}^1$, the four size distribution criteria of Sections C to F (i.e. openings, connectivity, convexity, and balls) are all based on the single notion of a *linear erosion*.

B.1. Linear erosion

In the exposition that follows, we emphasize the local approach, preferring to present the global version with particle size distributions (Section D-2). The set X may belong to $\mathbb{R}^1$ or more generally to $\mathbb{R}^n$. We begin by fixing a direction α going through the point ξ of a plane Π_α normal to α, to intersect with X: each particular intersection $X \cap \Delta(\alpha, \xi_0)$ will be used independently of other $X \cap \Delta(\alpha, \xi_1)$ intersections. So in this sense, we can say that we are concerned only with a 1-D analysis. Of course this does not stop us from exploring the stereological implications of the transformations and measurements as α varies, but only as a *second* step. Let B be a segment of length l and direction α, and define B_x to be that B whose right extremity is the point x. We will erode the set X under study. If X is just a single grain, for example as in Figure X.4, the eroded X is represented by the obliquely shaded area of the figure.

More generally, X may be an infinite set of features, or a connected medium (pores in a porous medium, metallic matrix in dendrite . . .). In this case we interpret the phenomenon as a realization of a random stationary set, and take

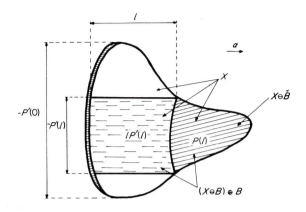

Figure X.4. Linear erosion and opening (global approach).

a local point of view. The measure Mes $(X \ominus B)$ is replaced by the proportions $V_V(= A_A = L_L)$ associated with the eroded set $X \ominus B(l, \alpha)$; this is denoted by

$$P_\alpha(l) = \text{volume (or area, or line) fraction of the set } X \ominus B(l, \alpha)$$
$$= \Pr\{0 \in X \ominus B(l, \alpha)\} \qquad \text{(X-5)}$$

It is interesting to compare $P_\alpha(l)$ with the set covariance $C_\alpha(l)$. Both depend only on a length l and a direction α (and are obviously the same quantity when X is a compact convex set). Starting from the covariance structuring element (i.e. the couple of points) and taking its *convex* hull, we get the segment $B(l, \alpha)$ used for linear erosion. In fact certain properties of $P_\alpha(l)$, which we now present, are strongly associated with the convexity of B.

Firstly, note that the behaviour of $C_\alpha(l)$ and $P_\alpha(l)$ near the origin are identical. (For the Hit or Miss topology, where a set X is recognized only by its topological closure $\overline{X}$, erosion by a segment or by its two extremities become equivalent when $l \to 0$):

$$P_\alpha(o) = C_\alpha(o) = L_L(X) = p(X)$$
$$P'_\alpha(o) = C'_\alpha(o) = N_L(X, \alpha) \qquad \Big\} \qquad \text{(X-6)}$$

Remarks: A plot of $P(l, \alpha)$ against l in any direction, will always start at the same ordinate, namely the proportion p of X, or equivalently the expected linear fraction L_L of the set X intersected by a uniform (in ξ) random line $\Delta(\alpha, \xi)$. The derivative at the origin $P'_\alpha(o)$ depends upon the direction α, and is equal to the total projection of the set per unit of volume (resp. area, or line) in direction α (see Ch. V, rel. (V-21)). When l increases, we have:

$$
\left.
\begin{aligned}
C_\alpha(l) &\geq P_\alpha(l) \geq 0 & &\forall \alpha \\
P_\alpha(l_1) &\geq P_\alpha(l_2) & &l_1 \geq l_2, \forall \alpha \\
P_\alpha(\infty) &= 0 & &\forall \alpha
\end{aligned}
\right\} \qquad \text{(X-7)}
$$

Moreover, the two basic morphological parameters L_L and N_L associated with the 1-D sets, satisfy the relationships:

$$
\left.
\begin{aligned}
P_\alpha(l) &= L_L[(X \cap \Delta_\alpha) \ominus B(l, \alpha)] = p(X \ominus B) \\
P'_\alpha(l) &= N_L[(X \cap \Delta_\alpha) \ominus B(l, \alpha)]
\end{aligned}
\right\} \qquad \text{(X-8)}
$$

A geometrical illustration of $P'_\alpha(l)$ is given in Figure X.4 (although there the approach is global), for a planar set X. The last two relations in (X-7), which are not true for the covariance, show the difference between $P_\alpha(l)$ and $C_\alpha(l)$. Another difference appears through the fact that linear erosion of the grains X give no information on the linear erosion of the pores X^c; large grains can equally be surrounded by large or small pores. In other words, if we denote by $Q_\alpha(l)$ the equivalent of $P_\alpha(l)$ for the pores X^c, no functional relation exists between $Q_\alpha(l)$ and $P_\alpha(l)$ as for the covariance (see Ex. X-10b). Notice that the quantity $1 - Q_\alpha(l)$ is just the proportion of the dilated set $X \oplus B(l, \alpha)$:

$$
Q_\alpha(l) = \Pr\{0 \in X^c \ominus B(l, \alpha)\} = 1 - \Pr\{0 \in X \oplus B(l, \alpha)\}
$$

The notion which corresponds to the range of the covariance is the *star* St_n, which is, by definition:

$$
St_n = \frac{1}{p} \int_{\mathbb{R}^n} P_\alpha(l) l^{n-1} \, dl \, d\alpha \qquad \text{(X-9)}
$$

The concept of a star will play a fundamental role in size distributions of convex particles (Sect. E); in general in $\mathbb{R}^n$, the star St_n has stereological implications for measure weighted moments.

The moment $P_\alpha(l)$ not only gives the probability that a point belongs to the linearly eroded set $X \ominus B$, but it also allows us to calculate $\Pr\{0 \in B_{(l, \alpha)}\}$, the probability that a point belongs to the *linear opening* of X by the segment $B(l, \alpha)$. By the definition of opening, $\Pr\{0 \in X_{B(l, \alpha)}\}$ is the probability that the origin 0 belongs to a segment having a length $\geq l$ which is contained in X. This event occurs when:

— either the interval $(0, l)$ is contained in the grain phase; this occurs with probability $P_\alpha(l)$

— or the interval $(-h, -h + l)$ is in the grain phase, but $-h + l + dh$ is in the pore phase; this occurs with probability

$$
-\int_0^l P'_\alpha(l) \, dh = -l P'_\alpha(l)
$$

Adding then these two mutually exclusive possibilities, we get

$$\Pr\{0 \in X_{B(l,\,\alpha)}\} = P_\alpha(l) - l\,P'_\alpha(l) \qquad (X\text{-}10)$$

The drawing in Figure X.4 illustrates this relationship for the global approach; for simplicity we have used the same notations $P_\alpha(l)$ and $-P'_\alpha(l)$, which here denote the area and projection of the eroded set of Figure X.4. The obliquely shaded area is equal to $P_\alpha(l)$; then after dilation of $X \ominus \check{B}$ by B, the horizontal shaded zone of area $-l\,P'_\alpha(l)$ is added.

As an example of linear erosion measurement in the local case, we can compute the function $P_\alpha(l)$, for each of the true sets represented in Exercise IX-3 (Fig. IX.25). This results in the curves shown in Figure X.5. Note that the sets X_2 and X_3, which only differ from each other by the spatial organization of the same particles, yield the same $P_\alpha(l)$ (in contrast, the function $Q_\alpha(l)$ associated with the linear intercepts of the backgrounds X_2^c and X_3^c would be different).

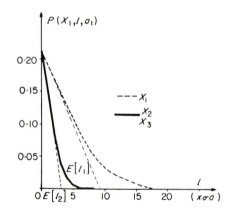

Figure X.5. Function $P_\alpha(l)$ associated with the sets X_1, X_2, X_3 of Figure IX.25. α is chosen as the horizontal direction.

B.2. Linear size distributions in number and in measure

(a) *Size distribution criteria*
For ease of notation, assume that the test line $\Delta(\alpha, \xi)$ remains fixed, so that the subscript α can be dropped. In one dimension, the following three set operations are equivalent:

(i) Suppressing the particles (i.e. the segments) of X having a length $< l$.
(ii) Performing an opening with the compact segment $B(l)$.
(iii) Removing those points of X from which one "sees" a zone of the 1-D space whose measure is smaller than l.

The mechanism of elimination of the particles by opening appears in Figure X.6. The three operations are obviously equivalent in $\mathbb{R}^1$ because here the compact connected components and convex sets are segments. However they become very different as soon as the space has dimension bigger than one. Thus we will for the moment base the size distributions on only one of those three transformations, for example the opening, (ii).

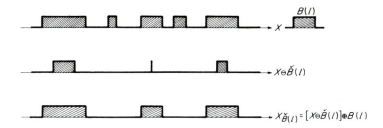

Figure X.6. Linear opening in R^1 and elimination of particles.

In $\mathbb{R}^1$, as in $\mathbb{R}^n$, openings by convex sets satisfy Matheron's axioms (see Sect. C below). Finally, these set transformations must be acted on to give a positive morphological parameter which increases with l. This prescription is filled by the two Minkowski functionals of $\mathbb{R}^1$ (local version) which lead to the two distribution laws:

$$1 - F\,(l) = \frac{N_L(X_{B(l)})}{N_L(X)} \tag{X-11}$$

$$1 - G\,(l) = \frac{L_L(X_{B(l)})}{L_L(X)} \tag{X-12}$$

We devote (b) and (c) below to their derivation and study.

(b) *Size distribution law in number*
Start from the frequency distribution $f(l)$ of the traverse length of the set X; the function $f(l)\,dl$ represents the probability that a *given* traverse, which begins in the interval $(x - dh_1, x)$ ends in the interval $(x + l, x + l + dh_2)$ (this is a conditional probability). The *a priori* probability that a traverse begins in the interval $(x - dh_1, x)$ is $-P'(o)\,dh_1$, and the *a priori* probability that the traverse begins in $(x - dh_1, x)$ and ends in $(x + l, X + h + dh_2)$ is $P''(l)dh_1\,dh_2$. To see this, note that $P(l)$ is the probability that the segment $(x, x + l)$ is contained in the grain phase. This event is the sum of the two mutually exclusive events:

(1) The segment $(x, x + l + dh_2)$ is contained in the grain phase; which happens with probability $P(l + dh_2)$.

(2) The segment $(x, x + l)$ is in the grain phase, but there is at least one end of the grain in $(x + l, x + l + dh_2)$; this happens with probability A. Thus,

$$P(l) = P(l + dh_2) + A$$
$$A = P'(l) dh_2$$

Similarly, by imposing an infinitesimal change dh_1 at the other extremity of the segment $x, x + l$, we finally obtain the above interpretation of $P''(l) dh_1 \, dh_2$. Thus, applying the definition of conditional probability, we have:

$$f(l) = -\frac{P''(l)}{P'(o)} \qquad (X\text{-}13)$$

i.e. the size distribution (in number) of the traverses; possesses a density proportional to the second derivative of the function $P(l)$. The distribution function $F(l)$, i.e. the cumulative distribution function of $f(l)$ (from zero to l) is deduced from (X-13) by integration, giving:

$$1 - F(l) = \frac{P'(l)}{P'(o)} \qquad (X\text{-}14)$$

But this is precisely relation (X-11), since the eroded set $X \ominus B(l)$ and the open set $X_{B(l)}$ have the same specific number of grains, which, according to relation (X-6), is equal to $-P'(l)$ (rel. (X-14) could be proved directly, without use of the possibly non-existent density $f(x)$).

From (X-14) and (X-6) we are able to calculate the average value of the intercept $E[l]$

$$E[l] = \int_0^\infty [1 - F(l)] \, dl = -\frac{P(o)}{P'(o)} = \frac{L_L(X)}{N_L(X)} \qquad (X\text{-}15)$$

So the average traverse length, multiplied by the specific number of intercepts, is equal to the length proportion $L_L(X)$. Note that relation (X-15) suggests a graphic construction for $E[l]$, via the behaviour at the origin of $P(l)$ (see Fig. X-5). More generally, all the moments of the distribution are expressed as functions of $P(l)$ by the relation:

$$E[l^k] = \frac{k(k-1)}{-P'(o)} \int_0^\infty l^{k-2} P(l) \, dl \qquad (X\text{-}16)$$

Instead of traversing the grains, we can alternatively consider traversing the pores; denote this traverse length by l_0. A morphological analysis identical to the above can be performed. In particular the mean pore traverse length $E[l_0]$ is

$$E[l_0] = \frac{Q(o)}{-Q'(o)} = \frac{1 - P(o)}{-P'(o)} \qquad (X\text{-}17)$$

From (X-15) and (X-17) we deduce:

$$-P'(o) = -C'(o) = N_L(X) = N_L(X^c) = \frac{1}{E[l_0 + l_1]}$$

$$P(o) = L_L(X) = \frac{E[l_1]}{E[l_0 + l_1]}$$

$$Q(o) = L_L(X^c) = \frac{E[l_0]}{E[l_0 + l_1]}$$

where l_1 is the grain traverse length.

(c) *Size distribution in length*

Let us return to the open set $X_{B(l)}$. We now measure the proportion of X that has been eliminated by the opening process, and denote it by $G(l)$. Thus by definition, $G(l)$ satisfies relation (X-12). $G(l)$ can be interpreted as a distribution function whose density $g(l)$ is related to $f(l)$. Formally speaking, we see that

$$g(l) = -G'(l) = \frac{1}{P(o)}[P(l) - P'(l)]' = \frac{l}{E[l]}f(l) \qquad (X–18)$$

This has the following morphological interpretation: associate with each point $x \in X$ the maximal size of a segment B contained in X and containing the point x. Thus all the points x of a segment $L \subset X$ of length l are classified in the same slice of the histogram $g(l)$. Consequently the *segment* L (and no longer the *points* which constitute it) appears in $g(l)$ with a weight proportional to its length l. The normalizing constant $E[l]$ simply ensures that $\int_0^\infty g(l)dl = 1$; relation (X-18) results.

Note once more that the basic piece of information remains the datum of $P(l)$, even for the determination of the size distribution in length, $g(l)$. To avoid confusion, we use $\mathcal{M}$ to denote the moments associated with $g(l)$:

$$\mathcal{M}[l^k] = \int_0^\infty l^k G(dl)$$

Taking (X-16) and (X-18) into account we easily have the moments of $g(l)$ from integrals of $P(l)$; thus

$$\mathcal{M}[l^k] = \frac{k(k+1)}{P(o)} \int_0^\infty l^{k-1} P(l)dl$$

which shows a formal link with the star St_n, defined in relation (X-9). In fact it is more than formal, as we will now demonstrate.

(d) *Stereology for linear size distributions*

Any closed set X of $\mathbb{R}^n$ can be analysed by its traverse length distributions. When the position ξ of the test line $\Delta(\alpha, \xi)$ is uniformly chosen, the linear intercepts of $X \cap \Delta(\alpha, \xi)$ give an unbiased sample of the linear intercepts of the set X itself. In other words, an image such as Figure X.7a is split into thin strips parallel to α, which can be studied in arbitrary sequence. What morphological characteristics of X itself can be derived from its linear size distributions? There are several aspects to be considered in answer to this broad question, for example the relationship between the traverse length distributions of X and its connected components X_i. This point will be discussed in detail in section D. Here we concentrate firstly on the questions of the angularity of X and then on the stereological meaning of the star.

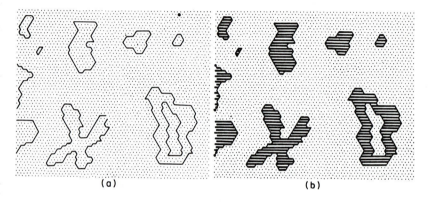

(a)   (b)

Figure X.7. (a) set X, (b) image of set X used in determining linear size distributions.

(i) Angularity. Consider a two-dimensional set X belonging to the regular model $\mathscr{R}(\mathscr{K})$ (local version); i.e. X possesses a finite curvature at any point of its boundary ∂X. Contributions from the small values of l in $f(l)$ come from the neighbourhoods of those points $x_\alpha \in \partial X$ that have a tangent of direction α. Let R be the radius of curvature at such a point. The corresponding neighbourhood gives a contribution proportional to $l^2/8R$, to the total number of traverses smaller than l, which itself comes from the first term of $R - \sqrt{R^2 - (l/2)^2}$. The number of points x_α per unit area is by definition the specific convexity number $N_A^+(\alpha)$ (see Ch. V.B5), and the number of traverses smaller than l, per unit area, is $P'(l) - P'(o) = -P'(o)F(l)$. Thus we can write:

$$N_A^+(\alpha) \cdot \frac{l^2}{8} E\left[\frac{1}{R}\right] = -P'(o) \cdot F_\alpha(l)$$

where $E[1/R]$ denotes the average value of the radii of curvature taken over the tangent points x_α. Therefore the density law $f_\alpha(l)$ is zero when $l = 0$, and we have:

$$
\begin{cases}
f_\alpha(o) = 0 \\
f'_\alpha(o) = -\dfrac{N_A^+(\alpha)}{4N_L(\alpha)} \cdot E\left[\dfrac{1}{R}\right]
\end{cases}
$$

The lack of angular boundary regions and narrow isthmuses we assumed for the regular model $\mathscr{R}(\mathscr{K})$ is reflected by a *linear behaviour* of the histogram $f(l)$ near the origin, the slope being proportional to the average curvature of the boundary. (Note that the above relation, derived for a given direction α, remains true after averaging over α.) If the set X belongs to the regular model in $\mathbb{R}^3$, similar considerations lead to the same conclusions, where the slope $f'_\alpha(o)$ is proportional to the average mean curvature.

An example of this linear behaviour of $f_\alpha(l)$ near $l = 0$ is provided by Figure X.8; here the two functions $f(l)$ and $g(l)$ of the set X of Figure X.7 are presented. (For a discussion of the digitalization problem see Ex. X-5).

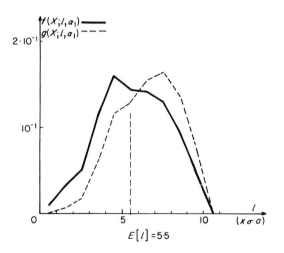

Figure X.8. The two traverse densities $f.(l)$ and $g(l)$. Note the difference in behaviour at the origin and the shift in the mean value.

Suppose now that the set X possesses v_A angles per unit area, whose bisecting lines have a direction uniformly distributed between 0 and 2π. We denote the magnitude of an angle (in radians) by u, and the mean angle is $E[u]$. Consider that part of the traverse length distribution $F(l)$, averaged over α, which is due to the neighbourhoods of the angular zones (for small l); call it

$F^*(l)$; we see from Exercise X-4 that $F^*(l)$ has a representation

$$F^*(l) = \frac{v_A E[\pi - u]}{-2\pi P'(o)} [1 + E(\pi - u) \operatorname{Cot}(u)] \times l + \ldots$$

$$= K \cdot l + \ldots$$

Therefore the first term of the limited expansion of $F^*(l)$ for small l is proportional to l, so that $f^*(l)$ tends towards a constant $K \neq 0$ when $l \to 0$. Now, when u varies from 0 to π, the quantity $(\pi - u) \cot u$ decreases from $+\infty$ to -1, going through zero for $u = \pi/2$. Thus the expression $1 + E[(\pi - u) \cot u]$ equals infinity when all the angles u are null (i.e. extremely sharp), equals one when all the angles $u = \pi/2$, and equals zero when they are all π (i.e. flat). The expression $[1 + E(\pi - u) \cot u]$ gives the proportionality constant K an angularity interpretation: when angles u increase from 0 to π, K decreases from $+\infty$ to zero, and has value $v_A/4 P'(o)$ for a population of angles all equal to $\pi/2$.

Similar, but formally more complicated considerations, can be developed in $\mathbb{R}^3$. In practical applications, the jump at the origin of the intercept histogram is particularly clear when studying populations of metallic grains, since they are in essence angular polyhedra.

(ii) **The stereological meaning of the star.** Consider in $\mathbb{R}^n$, the zone of X which is seen directly from the point $x \in X$. For example, in $\mathbb{R}^2$, when X is the background drawn in Figure X.9, the zone is the shaded part of the background surrounding the point x (imagine that X is transparent, X^c opaque, and you place a candle at the point x). What is its average value, when the point x sweeps the set X?

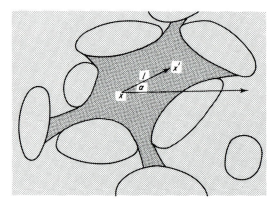

Figure X.9. Star Y_x of point x in two dimensions.

With each $x \in X$, we associate the set Y_x of the points $y \in X$ seen directly from x. If $1_x(y)$ is the indicator function of the set Y_x, we can write

$$V(Y_x) = \int_{\mathbb{R}^n} 1_x(y)\,dy \qquad \text{(X-19)}$$

which has expectation

$$E[V(Y_x)] = \int_{\mathbb{R}^n} E[1_x(y)]\,dy = \int_{\mathbb{R}^n} \Pr\{y \in X \,|\, x \in X\}\,dy$$

$$E[V(Y_x)] = \frac{1}{P(o)} \int_{\mathbb{R}^n} P_\alpha(|y - x|)\,dy \qquad (\alpha = \text{direction of } y - x)$$

which is exactly the star $\mathrm{St}_n(X)$ defined in relation (X-9).

In the isotropic case, $P_\alpha(|y - x|)$ depends only on $|y - x|$ and we obtain:

$$\left.\begin{aligned}
\mathrm{St}_n &= E[V(Y_x)] = \frac{n\omega_n}{P(o)} \int_0^\infty l^{n-1} P(l)\,dl \\[2mm]
&\omega_n = \text{volume of the unit ball in } \mathbb{R}^n \\[4mm]
\text{i.e. in } \mathbb{R}^3 \qquad \mathrm{St}_3 &= \frac{4\pi}{V_V} \int_0^\infty l^2 P(l)\,dl = \frac{\pi}{3}\mathscr{M}[l^3] \\[3mm]
\text{and in } \mathbb{R}^2 \qquad \mathrm{St}_2 &= \frac{2\pi}{A_A} \int_0^\infty l P(l)\,dl = \frac{\pi}{3}\mathscr{M}[l^2]
\end{aligned}\right\} \qquad \text{(X-20)}$$

Formulae (X-20) have important stereological implications, since they provide us with the means of estimating a three (resp. two)-dimensional parameter, from a knowledge of only the linear intercept laws of X in the various directions of the space. Note that the star and the moments in *measure* $\mathscr{M}[l^k]$ occur naturally together, which should not be surprising since the elementary parameter $V(Y_x)$ is associated with each *point* x (and not with each particle).

The notion of the star will again be discussed when analysing the size distribution of compact convex particles (see Sect. E).

C. OPENING, CLOSING, AND SIZE DISTRIBUTION IN $\mathbb{R}^n$

C.1. Opening with respect to similar convex sets

This section will be very brief, since we have in fact already covered the ground in various parts of the text; our job is now to collect together the scattered treatments.

The morphological opening X_B of a set X by the compact set B was introduced in Chapter II, Section E. We saw that this transformation was

— anti-extensive   i.e. $X_B \subset X$ $\forall B \in \mathcal{K}(\mathbb{R}^n)$
— increasing   i.e. $X^1 \subset X^2 \Rightarrow X_B^1 \subset X_B^2$ $\forall B \in \mathcal{K}(\mathbb{R}^n)$

Furthermore, for a family $B(\lambda)$ of compact sets parametrized by the positive parameter λ, we saw that if B_{λ_1} is open with respect to B_{λ_2}, then:

$$\left(X_{B_{\lambda_1}} \right)_{B_{\lambda_2}} = \left(X_{B_{\lambda_2}} \right)_{B_{\lambda_1}} = X_{B_{\lambda_1}}$$

At this point, we checked that a family generated by the openings of a set by such B_λ's satisfied Matheron's size distribution axioms. But how can the B_λ's be chosen in order to also satisfy the four principles of mathematical morphology? The opening algorithm is, by construction, invariant under translation, and for a compact B, upper semi-continuous (X is modelled to be a topologically closed set) and satisfies the *local approach* (rel. I-4) (it is a result of finite iterations of bounded erosions).

It remains then to satisfy compatibility under changes of scale (principle 2). We showed (Ch. II, Sect. C.2) that finite iterations of erosions and dilations of X by a family.B_λ was compatible under changes of scale if and only if the B_λ's were homothetics of each other.

And the final piece of the puzzle (provided by Ch. IV, Prop. IV-3) is: for a given compact set B, the homothetic set $\lambda B (\lambda \geq 1)$ is open with respect to B if and only if B is convex.

Therefore, to measure a size of X by openings, we must choose a compact convex shape B, and generate the family $\{X_{\lambda B}\}$. The (geometrical) property of convexity of the structuring element explains why the physical size distributions already presented show certain similarities to openings: they are all generated by the physical equivalent of a convex structuring element (ball for the Purcell method, elliptical cylinder for the sieves, etc. . . .).

After the set transformation by openings, what parameter should we take to measure size? In general, the only increasing morphological measurement is Lebesgue measure. (Contrary to the 1-D case, the other Minkowski functionals are no longer, in general, increasing functions of λ.)

From the upper semi-continuity of the opening and the continuity under the change of scale, we see that the mapping $(\lambda, X) \to X_{\lambda B}$ from $\mathbb{R}^+ \times \mathscr{F}(\mathbb{R}^n)$ onto $\mathscr{F}(\mathbb{R}^n)$ is upper semi-continuous. Furthermore, since Lebesgue measure is upper semi-continuous on the set of closed sets, the functions $\lambda \to \mathrm{Mes}(X_{\lambda B})$ and in the random approach, $\lambda \to \mathrm{Pr}\{x \in X_{\lambda B}\}$, are decreasing and semi-continuous from the left. Thus, by putting:

$$P_B(-\lambda) = \mathrm{Prob}\{x \in X_{\lambda B}\} = \begin{cases} A_A(X_{\lambda B}) & \text{in } \mathbb{R}^2 \\ V_V(X_{\lambda B}) & \text{in } \mathbb{R}^3 \end{cases} \qquad \text{(X-21)}$$

we define the size distribution law (cumulative distribution function) $G_1(\lambda)$ as the proportion of points $x \in X$ which have been eliminated by the opening λB, i.e.:

$$G_1(\lambda) = 1 - \text{Prob}\{x \in X_{\lambda B} | x \in X\} = 1 - \frac{P_B(-\lambda)}{P_B(o)}; \lambda \geq 0 \qquad (X-22)$$

If we focus now on the background, the closing $X^{\lambda B}$ replaces the opening $X_{\lambda B}$, and by putting

$$P_B(\lambda) = \text{Prob}\{Rx \in X^{\lambda B}\} \qquad (X-23)$$

we obtain the size distribution law of the pores:

$$G_0(\lambda) = \frac{P_B(\lambda) - P_B(o)}{1 - P_B(o)} \qquad (X-24)$$

Instead of considering the proportion of points *of X* (resp. X^c) which have been eliminated by opening (resp. closing), we could also take the proportion of points *of the space* that remain after openings and closings of X. Then by making λ vary (by convention from $-\infty$ to $+\infty$) we work directly with the function $P_B(\lambda)$ itself; the part corresponding to negative (positive) values of λ concerns openings (closings) of X. $P_B(\lambda)$ increases from zero to one as λ goes from $-\infty$ to $+\infty$, and $P_B(o)$ is equal to the volume fraction of X. Upon taking a global approach, relation (X-22) and (X-24) remain valid by replacing $P_B(-\lambda)$ by $\text{Mes}[X_{\lambda B}]$ and $P_B(+\lambda)$ by $\text{Mes}[X^{\lambda B}]$. The digital algorithms involved in all these measurements are made explicit in Exercise X-8.

The upper semi-continuity of the mapping $(\lambda, X) \to X_{\lambda B}$, and the geometrical interpretation of the opening (locus of the points x belonging to a λB included in X) imply that when sizing by opening, each *point* $x \in X$ is given a size λ, defined by the relationship:

$$\lambda_X(x) = \text{Sup}\{\lambda: \exists y \text{ such that } x \in (\lambda B)_y \subset X\} \qquad (X-25)$$

The size at the point x is the multiplicative parameter λ of *the maximum λB included in X containing x* (see examples of this size distribution in Fig. (X.12)). Practically speaking this means that if we take for X the union of disjoint particles, each of them is counted in the sizing by openings in proportion to the number of points, i.e. H is *weighted in measure*. The size $\lambda_X(x)$ is connected with $P_B(-\lambda)$ and with $\text{Mes}[X_{\lambda B}]$ by the relationship:

$$P_B(-\lambda) = \text{Pr}\{\lambda_X(x) \geq \lambda\} \qquad \text{(local version)}$$
$$\text{Mes}[X_{\lambda B}] = \text{Mes}\{x: \lambda_X(x) \geq \lambda\} \qquad \text{(global version)}$$

Among possible shapes for B, we see that the segment leads to linear size distributions. What morphological potential does a 2-D or 3-D opening possess? The 3-D ones are not frequently used since they require knowledge of the object from serial sections close together. The 2-D openings by hexagons or dodecagons for example, possess less of a stereological meaning than the linear size distribution, and their moments $\int l^n (G(l) dl$ have no 3-D interpretation similar to the star. Their major interest is to synthesize size and shape description in a single approach, as we will see now.

C.2. Size, shape and clustering

Let us start with rather simply shaped individuals, such as sand grains, or cell nuclei, etc. . . . The two notions of roundness and circularity (one could also say sphericity) are well known to the petrographers, the cytologists and others involved with the study of particles. They refer to shape features represented in Figure X.12a to d. Since the work of Wadell (1932), a number of algorithms and various standard pattern charts have been proposed as quantifications of these two notions (a review is given in Pettijohn (1970)). In practice, the most commonly used of them is the circularity factor defined by the ratio $\rho = (\text{perimeter})^2/4\pi$ area, which is equal to one for the disk. Indeed the ratio ρ does not discriminate between the two very different shapes of Figures (X.10a and b), and principally confuses roundness and circularity. When the number of cogs indefinitely increases, in Figure X.10c, $\rho \to \infty$, even if the cog-wheel tends toward a disk! (It is really a poor idea to introduce semi-continuous functionals (e.g. perimeter) in a shape factor.)

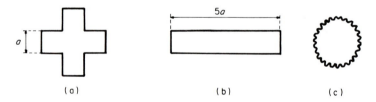

(a)   (b)   (c)

Figure X.10. Counter examples of the circularity factor $(\text{perimeter})^2/(4\pi . \text{area})$.

In individual analysis in R^2, the morphological approach consists of associating, with each particle, the *complete curve* of its 2-D openings by increasing disks, hexagons or dodecagons. The results may be presented in terms of cumulative function (Mes $X_{\lambda B}$), or of density distribution $f(d\lambda)$ (area occupied by points x of size $\lambda_X(x) \pm \frac{1}{2}d\lambda$). Note that the latter may exhibit discontinuities such as Dirac measures. If the set X is made of the union of disjoint X_i which are all similar to B, then all the points of a given particle belong to the same size class λ. For example, take for X the union of disjoint disks as in Figure X.11. The erosion $X \ominus \lambda B$ by the disk λB removes only those

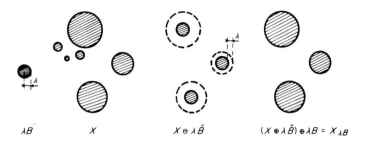

$$\lambda B \qquad\qquad X \qquad\qquad X \ominus \lambda \breve{B} \qquad\qquad (X \oplus \lambda \breve{B}) \oplus \lambda B = X_{\lambda B}$$

Figure X.11. Opening of disks by a disk: no change in the shapes.

disks smaller than λ; the others are completely reformed by the subsequent dilation: the convex structuring element cannot modify shapes similar to itself (we again meet Prop. IV-3). Therefore the density distribution $f(d\lambda)$ of a disk is reduced to a Dirac measure (Fig. X.12a). When the degree of sophistication of the shape of the particle X increases, other classes appear in the density distribution f. In particular we see in Figure X.12 that the first classes and the last one reflect the roundness and the circularity respectively. The practicioners in morphology use the negative moment

$$\alpha = \frac{1}{A(X)} \sum \frac{f(\lambda_i)}{\lambda_i}$$

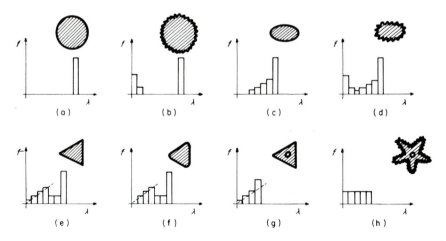

Figure X.12. Shape description and openings. (a) round, circular; (b) no round, circular; (c) round, no circular; (d) no round, no circular; (e) a few angles; (f) opening of e; (g) hole in e; (h) starfish.

for the roundness factor and the ratio

$$\beta = \frac{f(\lambda_{i\,max})}{A(X)} = \frac{\text{area of the max. inscribable disk}}{\text{area of } X}$$

for the circularity factor (Here $A(X) = \sum f(\lambda_i)$.) However the datum of all the $f(\lambda_i)$ carries more information than α and β, and can directly be incorporated to a statistical treatment (e.g. the Caucasian dolerites below).

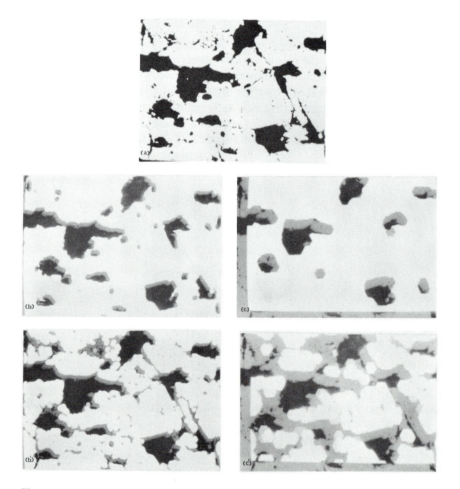

Figure X.13. Smoothing by successive hexagonal openings, and closings. (a) initial set X; (b) X_{3H}; (c) X_{8H}; (b') X^{3H}; (c') X^{8H}.

We now enlarge the point of view and are interested in the shape description for both X and its complement X^c. Start for example from the porous medium represented in black on Figure X.13a, and follow the erasing process of the opening. The initial connected components are successively smoothed and tend towards the "circular" shape (i.e. here the hexagonal shape) before suddenly disappearing (Fig. X.b–c). If they are initially very different from being convex, the opening will divide them at their isthmus regions. The larger λB is, so more and more details (such as small grains, capes and isthmuses) are filtered from the image. Suppose we now look at the same three photographs from the point of view of the solid white phase; by duality it has also been transformed. So eliminating a porous isthmus is equivalent to *clustering* two grains; erasing small porous islands and capes is the same as filling up holes and gulfs in the grains. The deformation on the grains due to opening the pores, is precisely the *closing* of the grains; it is a tool for describing packing and clustering of the particles.

We can also open the white phase, i.e. close the pores (Fig. X.13b′, c′); Particles are progressively individualized by this process. As before, the image is simplified by the removal of finer details. One sometimes says that the image is *cleaned*. Notice that when we handled images by the variogram technique (in Ch. IX, E.2), the treatments of figure and background were not distinguished (just as in a Max Escher's drawings): the variogram of the pores *is* exactly the variogram of the grains. Now with the opening–closing operations, grains and pores play symmetrical but distinct roles.

C.3. The Caucasian dolerites

Each image of the plate (Fig. X.14) is a micro-photograph of a polished section of dolerite. It belongs to a set of about fifty specimens of basaltic or dolerite rocks, sampled in several regions of the Russian Caucasus. The geologists want a quantitative description of the sizes, the shapes (angularity, roundness), the arrangements (clusters . . .) of the porous media (in black on the photographs), in order to correlate these data with resistance to crushing and with the speed of sound in such rocks (Kolomenski–Serra, 1976). The ideal morphological tool for performing such a textural "check-up" of the rocks, is obviously the 2-D opening-closing, parametrized by $P(\lambda)$, for $-\infty < \lambda < +\infty$.

For each specimen, we investigated one hundred fields of the size of the photographs; in Figure X.15, we present the corresponding mean values for eight specimens, expressed per unit area, and corrected for edge effects. The value i of the side of the hexagon is given on the x-axis; it is negative for openings (of the pores), and positive for closings. On the y-axis we indicate the

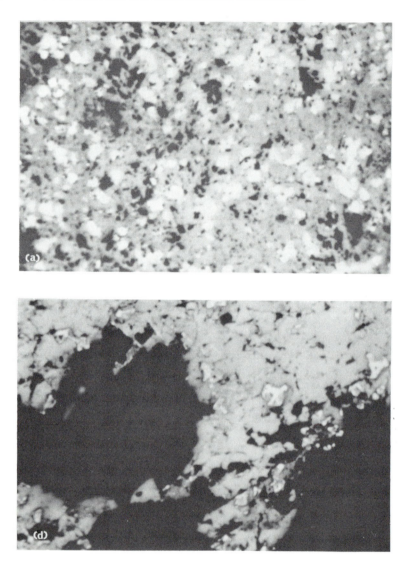

area percentage of pores after each step of image transformations, which varies from zero (huge opening) to 100 per cent (huge closure) and passes through the actual porosity of the specimen (hexagon size $i = 0$). Conventionally, on the hexagonal grid, i is the distance separating two points along one of the three main directions of the grid. Using the Texture Analyser, the total computing time lasted about 5 minutes (measured between placing the specimen under the

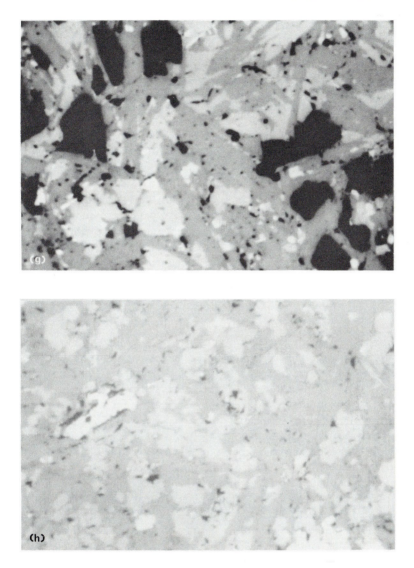

Figure X.14. Polish sections of various types of Caucasian dolerites.

microscope, and taking the output curves of Figure X.15). The instrument computed 16 steps of openings and closings for each elementary field of 420,000 picture points, calibrated the results according to the scaling table of

Exercise X-8 (local opening-closings), and took their mean value for a hundred fields per specimen.

The sides λ of the hexagons involved in the 16 steps are the following:

$$\lambda \text{ (microns)} = 0; 4; 8; 12; 16; 32; 48; 64; \infty$$

Let us denote by $p(-j; k)$ the opening area increment for specimen k, between steps $j-1$ and j. For example, $p(-2, k)$ is the proportion of space vanishing between openings of size $4\,\mu$ and $8\,\mu$. The notation $p(j; k)$ is similarly associated with closings (e.g. $p(8, k)$ is the proportion of space *added* between a closing of $64\,\mu$ and an infinite closing, which gives the whole space). The multidimensional data $p(j, k)$ define a table of 16 rows in j, and 50 columns, in k, which we can treat by factor analysis. Among the different possible methods of multivariate analysis, we took the correspondence analysis theory of J. P. Benzécri (1974). It has the advantage of being able to provide results that do not depend on the selected opening classes: if two points representing two opening increments are close together in the main factor subspaces, the two corresponding classes can be regrouped into a single class without disturbing the remaining constellation of points. The method considers the 50 specimens, characterized by 16 morphological parameters, as points of a 16-dimensional space, and the parameters as points in a 50-dimensional space. These points are provided with weights equal to $\sum_j p(j, k)$ or $\sum_k p(j, k)$ according to the space, and the systems of main axes of inertia are calculated in both spaces (the factor axes). The first two main axes define the first factor plane. Conventionally, one superimposes the projections of the two clouds of points on their first factor plane. On this simultaneous projection, each parameter point is the centre of gravity of the weighted specimen points, and conversely. Thus, if the specimen point s is near the opening point O, the opening class O is particularly developed in specimen s. Similarly, one can show that the more strongly specimen points (resp. parameter points) are clustered, the more correlated are the corresponding specimens (resp. parameters).

Figure X.15α shows the first factor plane. An easy way to see the physical meaning of these factor axes is to consider a few representative specimens. Let us take for example, the two groups of rocks (a) to (d) and (e) to (h) respectively aligned on lines parallel to the first factor axis, but with a positive (a) to (d) or negative (e) to (h) co-ordinate on the second factor axis. We see that each group forms a bundle radiating from a point, which we shall call the "pivot". The phenomenon is general for all the alignments of specimens parallel to the first factor axis. The second factor axis reflects a second type of evolution and makes the bundles around a same pivot correspond to each other by an affinity along the porosity axis (y-axis of the size distribution). If we neglect the slight shift to the left of the pivots (which are not on the y-axis but are more or less aligned parallel to it at a $5\,\mu$ distance), we see that the first factor axis simply

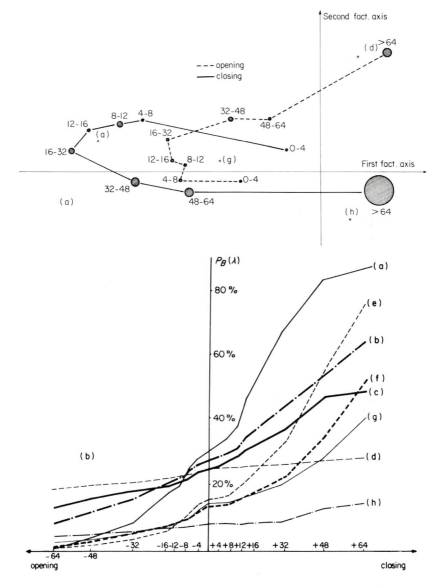

Figure X.15. (a) first factorial plane (multivariate analysis); (b) typical size distributions.

reflects a scale factor in the textures; a magnification of the specimen appears as an affinity parallel to the *x*-axis of the plotted size distributions.

The shift of the pivots comes from the finest porosity, such as microfissures or asperities on the pore boundaries. Finally, the first two factor axes, which respectively explain 50 and 23 per cent of the total variance of the textures, lead to three parameters (porosity, affinity ratio and ordinate of the pivot) which summarize 73 per cent of the morphological information extracted from the dolerite population by openings and closings.

For four of the eight specimens a to h, we have taken and photographed one elementary measuring field, qualitatively representative of the polished section, and placed them near the curves. The meaning of the first two axes is confirmed, but the existence of a pivot, and its constant value for all the rocks is less visually obvious (according to geologists consulted about this point, this small structure might be caused by a primary volcanism, acting prior to the one that produced the rest of the dolerite textures).

We will conclude the study by adding to the cloud of parameters, two physical properties measured on each specimen, namely the resistance to crushing ρ and the speed of sound v. Although they are given no weight, their proximity to morphological parameters on the factor axes reflect correlations. After examining the first five axes, we decided to calculate the linear regression ρ^* of the resistance to crushing ρ of the three size distribution classes $p(-6)$, $p(-1) + p(+1)$, $p(+7)$; the following regression results:

$$\rho^* = 16.1 - 0.38p(-6) - 1.15[p(-1) + p(+1)] + 0.12p(7)$$

The residual relative variance associated with ρ^* being equal to 0.31. The resistance to crushing therefore mainly depends on both fine and very large pores and on a particular interpore distance (68μ).

D. SIZE DISTRIBUTION AND INDIVIDUALS

D.1. Matheron's axioms

As they stand, the individual size distributions are rather trivial. The individuals are assumed to be compact sets. Any reasonable positive parameter λ_i given to each individual X_i allows a classification which satisfies Matheron's axioms (the subset ψ_λ is the union of the X_i's whose λ_i's are $\geq \lambda$). In two dimensions, λ_i might be the radius of the circumscribed circle, or the number of holes of the individual, or its Feret diameter (which may or may not be in a given direction), its perimeter, its area, etc. . . . But λ_i cannot be the distance from X_i to its nearest neighbour or the third axiom is not satisfied, since an inter-individual notion intervenes.

Because a size is no longer given to each point (as for the opening) but to each individual of the set $X = \bigcup_i X_i$, the suitable mode of convergence is that given for the skiz in Chapter XI. C5. If we only assume compactness of the X_i's, Lebesgue measure, which is semi-continuous for this mode of convergence, is a convenient size distribution parameter. The semi-continuity of other possible parameters will depend upon more precise specifications of the set model.

D.2. Size distributions weighted in measure

We now fix the dimension of the space and only consider the individual size parameter defined by the *Lebesgue measure* of each grain. For example, in $\mathbb{R}^2$ each $X_1 \in X$ has an area $A_i = A(X_i)$, from which the frequency histogram $f(A)$ of the A_i is constructed (similarly in $\mathbb{R}^1$ for lengths and in $\mathbb{R}^3$ for volumes). To give the same importance (i.e. the same weight) to each individual is one point of view. In fact this approach is usually adopted in cytology (counting techniques using flows, as well as image analysis), where physiologically speaking, all the cells of a given type have the same importance. But there are other techniques, especially the sieving of mineral grains, which proceed differently. After having sieved the grains according to their minimum cross sections, each sieve fraction is weighed and the curve of weights versus the sieve diameters is plotted. The weight W_0 of sieve fraction D_0 is global information: each particle stopped by the sieve D_0 is counted in proportion to its own weight. It is very important that this point be crystal clear to the reader, since the rest of the section, and the analogous one on spherical individuals, is based on it. An extremely simple illustration is the following: consider a population consisting of two spherical grains, with respective radii one and two. First, consider them as biological cells and plot the histogram of their sizes; Figure X.16a results, where the two classes have the same proportion. Now send the two balls through a sequence of sieves, and plot the weight of the oversize of the sieves. We now obtain the curve in Figure X.16b, since a ball with twice the radius weighs eight times more (see also Ex. X-12 for other illustrations of the same phenomenon).

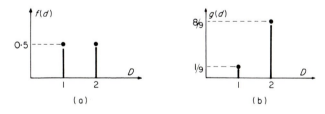

Figure X.16. Difference between size distributions in number and in volume for a population of two particles.

How can we cross from one type of size distribution to the other? We use μ to denote the Lebesgue measure of each individual (i.e. its length, area, or volume according to the space), $f(\mu)$ to denote the density of the distribution in number and $g(\mu)$ to denote the density of the distribution in measure. To use a weight proportional to the measure implies

$$g(\mu) = \frac{\mu \cdot f(\mu)}{\int \mu f(\mu) d\mu} = \frac{\mu f(\mu)}{E(\mu)} \qquad \text{(X-26)}$$

and, conversely:

$$f(\mu) = \frac{g(\mu)}{\mu} \times \frac{1}{\int [g(\mu)/\mu] d\mu} = \frac{g(\mu)}{\mu} \times \frac{1}{\mathcal{M}(\mu^{-1})} \qquad \text{(X-27)}$$

where by convention, $E(\cdot)$ and $\mathcal{M}(\cdot)$ correspond respectively to averaging using distributions in number and averaging using distributions in measure. In one dimension, relation (X-26) coincides with relation (X-18).

Finally, we can interpret the weights in measure by giving the weight $\mu(X_i)$ to each point x belonging to the individual X_i. In this way, we can again discuss the axioms of size, now at the level of the point rather than the openings. With this in mind, we are able to solve the stereological question. Consider the position where 3-D individual X_i are analysed in 2-D or 1-D space; let $g_i(A)$ be the area-weighted law of the areas of the cross sections of X_i with a uniformly distributed random plane, and let $g(A)$ be the same function defined on $X = \bigcup_i X_i$. Then

$$g(A) = \frac{\sum_i V(X_i) \cdot g_i(A)}{\sum_i V(X_i)} \qquad \text{(X-28)}$$

The passage from one individual to the population is made by weighting according to the *volumes* of the X_i, i.e. according to the Lebesgue measure of the volume in their actual space. The result is general to all size distributions weighted in measure e.g. traverse distributions of 3-D grains, or 2-D grains, openings, etc. Relation (X-35) may be estimated by the techniques developed in Chapter VIII, Section E, and in particular by relation (VIII-27).

D.3. Size distribution and convexity

In the last section we did not assume the individuals to be recognizable on cross-sections. We shall do it now. Moreover, we assume that they are convex. This latter hypothesis is not really necessary, but will allow us to show the use of the star (the generalization to the individuals recognizable on sections may be made by replacing the probability $P(l, \alpha)$ by the geometric covariogram $K(|h|, \alpha)$).

(a) *The star*

(i) The order one measure weighted moments $\mathcal{M}(V)$. Let us return to the morphological meaning of the star, as described in Section B.2c (ii) (namely, a transparent set X embedded in an opaque background X^c in $\mathbb{R}^n$; the set Y_x is the zone illuminated by a light placed at $x \in X$ (Fig. X.9)). We saw that the average value of the volume $V(Y_x)$, as x runs over X, is just the integral of the moment $P(l, \alpha)$ over the space $\mathbb{R}^n$, divided by $P(o)$; see relation (X-20). The above analysis was quite general, and independent of any assumption on the structure of X. If we now take X to be the union of disjoint compact convex sets X_i, the interpretation of the star can be pushed further. All the points X of a same X_i "see" the same zone, namely X_i, and hence are associated with the same value $V(Y_x) = V(X_i)$. In other words, in the average $E[V(Y_x)]$, the contribution of X_i is proportional to its own volume $V(X_i)$, and $E[V(Y_x)]$ turns out to be the measure weighted average $\mathscr{M}(V)$ of $V(X_i)$. Hence, relations (X-20) become

$$\mathscr{M}(V) = \frac{1}{n(n+1)} \int_\Omega \frac{E[l^{n+1}|\omega]}{E[l|\omega]} d\omega \qquad \text{(X-29)}$$

(where $E[\cdot|\omega]$ denotes the conditional expectation when direction ω is fixed). On the other hand, relation (X-26) implies that for (not necessarily convex) particles:

$$\mathscr{M}[V^k] = \frac{E[V^{k+1}]}{E[V]} \qquad \text{(X-30)}$$

Therefore, from these two relations we can derive the measure-weighted average of the volume $V(X_i)$, just from linear measurements (see relation (X-29)), and connect it with the ratio $E[V^2]/E[V]$. Suppose we particularize these relations to the plane and the space in the isotropic case; the expressions then take a simpler form. Then (X-29) and (X-30) imply:

$$\left. \begin{array}{ll} \mathbb{R}^3 & \mathscr{M}[V] = \dfrac{E[V^2]}{E[V]} = \dfrac{\pi}{3} \dfrac{E[l^4]}{E[l]} = \dfrac{\pi}{3} \mathscr{M}[l^3] \\[3mm] \mathbb{R}^2 & \mathscr{M}[A] = \dfrac{E[A^2]}{E[A]} = \dfrac{\pi}{3} \dfrac{E[l^3]}{E[l]} = \dfrac{\pi}{3} \mathscr{M}[l^2] \end{array} \right\} \qquad \text{(X-31)}$$

(Note that we get ratios, but not the two terms $E(V^2)$ and $E(V)$.)

(ii) The order two measure-weighted moment $\mathscr{M}(V^2)$. Does the interpretation of $V(X_i)$ in terms of the zone seen directly from $x \in X_i$, allow us to calculate not only the average $\mathscr{M}(V)$, but more generally all the moments $\mathscr{M}(V^k)$ of the random variable $V(X_i)$? Unfortunately, the answer is no for $k > 2$, but for $k = 2$, the interpretation is still powerful enough to provide us with a new stereological result. Start from relation (X-19) and square:

$$[V(Y_x)]^2 = \int_{\mathbb{R}^n} \int_{\mathbb{R}^n} 1_x(y) \cdot 1_x(y') \, dy \, dy'$$

This has mathematical expectation:

$$E[V(Y_x)]^2 = \int_{\mathbb{R}^n} \int_{\mathbb{R}^n} P(y-x; y'-x)\, dy\, dy' = \mathcal{M}[V^2] = \frac{E[V^3]}{E[V]} \quad (\text{X-32})$$

where $P(y-x; y'-x)$ is the probability that points y and y' are simultaneously seen from the point $x \in X$. In terms of structuring elements, this probability is just the volumetric percentage of the erosion of X by the triangle with vertices y, y' and x (since the X_i's are convex). To perform such an erosion requires only planar cross sections of X, which is feasible from an experimental point of view. But if we want to calculate, by the same method, the moment of order three (or more), we would have to consider the three points y, y', y'' able to be seen from point x and to erode by the tetrahedron y, y', y'', x.

Formula (X-32) solves the theoretical question of the calculation of $\mathcal{M}[V^2]$, but we are not yet out of the woods as we still have the digitalization problem of building all possible triangles $0\ y\ y'$, as y and y' sweep the plane.

(iii) Estimation of $\mathcal{M}[V^2]$ on the hexagonal grid. To avoid long and complicated formulae, we suppose the structure X to be isotropic (if not, we must replace the $P(l, l', \alpha)$ by their average values taken over the rotations. The structuring element associated with probability $P(y-x, y'-x)$ depends on the two lengths $l = |y-x|$ and $l' = |y'-x|$ and on the angle $\alpha = (\overrightarrow{y-x}, \overrightarrow{y'-x})$, defined modulo a rotation (isotropy). The simplest angles α that are available on a hexagonal grid are multiples of $\pi/3$. If we restrict ourselves to this rather sparse angular discretization, the digital version of relation (X-32) for $n = 2$ leads to the estimate:

$$\mathcal{M}^*[V^2] = \frac{2\pi^2}{3}\left\{\frac{6}{7}\sum_0^\infty i^3 P(i) + 4\sum_0^\infty i \sum_0^i j\left[P\left(i,j,\frac{\pi}{3}\right) + P\left(i,j,\frac{2\pi}{3}\right)\right]\right\}$$

$$(\text{X-33})$$

(Notation: $P(3, 2, \pi/3)$, $P(2)$ and $P(1, 2, \pi/3)$ for example, are the area fractions of the erosion of X by the three structuring elements of Figure X.17.)

Figure X.17. Types of structuring elements involved in the measurement of $\mathcal{M}[V^2]$.

Consider now the image on a cross section of a 3-D structure. We must make the digital α's on the plane correspond with angular variations on the unit sphere. With $\alpha = 0$ and $\alpha = \pi$, we associate the solid angle Ω_1 of the cap intersected on the sphere by a core of revolution of angle $\pi/6$. Hence, Ω_1 is equal to:

$$\Omega_1 = 2\pi \left[1 - \frac{\sqrt{3}}{2} \right]$$

With $\alpha = \pi/3$ and $\alpha = 2\pi/3$ we associate the solid angle Ω_2, where $\Omega_1 + \Omega_2 = 2\pi$, i.e. the total covers a hemisphere. According to these weights, the digital version of formula (X-32) yields the estimate:

$$\mathscr{M}^*[V^2] = 8\pi^2 \left\{ \left(1 - \frac{\sqrt{3}}{2} \right) \frac{11}{10} \sum_0^\infty i^5 P(i) \right.$$
$$\left. + \frac{\sqrt{3}}{2} \sum_0^\infty i^2 \sum_0^i j^2 \left[P(i, j, \pi/3) + P(i, j, 2\pi/3) \right] \right\} \qquad \text{(X-34)}$$

Discussion: When we substitute steps of 60° for the continuous differential $d\alpha$, are we being a bit too reckless? In fact we have absolutely no idea of the real angular accuracy demanded by an approximation of the integral (X-32). To test the validity of such an angular discretization, we suggest a random set model be chosen for which both exact and approximate values can be formally calculated; let us take the Poisson polygons, in $\mathbb{R}^2$. They are convex, with a variety of shapes (triangles, squares, pentagons, . . .) and a large range of sizes. (Indeed, they are more irregular in size and shape than many of the convex particle populations met in practice).

A Poisson polygon is an isotropic random set depending on a single parameter λ, which is a scaling factor (see Ch. XIII, Sect. D-7). Its moment $\mathscr{M}[A^2]$ is given by:

$$\mathscr{M}[A^2] = \frac{4\pi^2}{7 \lambda^4} = 0.5714 \frac{\pi^2}{\lambda^4} \qquad \text{(X-35)}$$

For the discrete estimate, assume that the lengths l and l' in $P(l, l', \alpha)$ vary continuously (which is realistic when compared to the discretization of α), but that α jumps by steps of $2\pi/3$ radians. For Poisson polygons, the probability $P(y - x, y' - x)$ is given by the law (XIII-47) which here becomes:

$$P(l, l', \alpha) = P(y - x, y' - x) = \exp\{-\lambda[|y - x| + |x - y'| + |y' - y|]\}$$

By substituting this expression in relation (X-32), we can integrate it over l and l', and sum over α. After intermediary calculation we find the estimate:

$$\mathscr{M}^*(A^2) = 0.5977 \frac{\pi^2}{\lambda^4} \qquad \text{(X-36)}$$

Thus, the relative error from the exact value (X-35) is smaller than 4 per cent. What about the *centred* second moment $\mathscr{M}[A^2] - [\mathscr{M}(A)]^2$? Its theoretical expression, for Poisson polygons, is given by

$$\sigma^2(A) = \mathcal{M}[A^2] - (\mathcal{M}[A])^2 = \frac{4}{7}\frac{\pi^2}{\lambda^4} - \frac{1}{4}\frac{\pi^2}{\lambda^4} = 0.3214\frac{\pi^2}{\lambda^4}$$

If we assume that the mean value $\mathcal{M}[A]$ has been estimated without bias, the variance $\sigma^2(A)^*$ of the discrete estimate is then:

$$\sigma^2(A)^* = \mathcal{M}^*[A^2] - \frac{1}{4}\frac{\pi^2}{\lambda^4} = 0.3477\frac{\pi^2}{\lambda^4}$$

The relative error of the variance is larger than the previous one, but the bias does not exceed 8 %, which is still acceptable. This confirmation via a Poisson model is not a proof, but is strong evidence that the approximation is valid for typical convex objects.

Up to now. we have focused our attention on convex sets, however formula (X-32) and its discrete versions, could also be used to calculate the variance of Y_x, the volume seen from x, when X is an arbitrary set (e.g. the variance of the area of the octopus in Figure X.9 as the centre x varies). However, note that the values $V(Y_x)$ given to the x generate a family of subsets (whose λth member is the set of all x such that $V(Y_x) \geq \lambda$) which satisfy Matheron's axioms only in the convex case and not for more general sets.

E. SPHERICAL PARTICLES

E.1. Size algorithms

A considerable number of papers deal with the problem of finding the size distribution of balls from their 2-D or 1-D cross-sections. In fact, the key algorithms used in answering this question (ref. X-37 and X-38) were due to the Norwegian mathematician N. Abel in the 1820's, although they were presented in a different context. More recently, an important study by S. D. Wicksell (1925) solved the problem, giving also practical algorithms. Other derivations of the same laws can be found in P. A. P. Moran and M. G. Kendall (1963) and in A. Haas et al. (1967). The influence of a non zero thickness of the sections has been studied by G. Matheron (1972) and by R. Coleman (1979). Redundancies in the literature allow us to be brief, and we shall be even briefer by only stating the main formulae. For a detailed discussion involving the accompanying use and know-how of the formulae, we refer the reader to E. E. Underwood's book (1970), which is extremely comprehensive on the subject. Our approach in what is to follow, is to rather focus attention on the *domain of validity* of the spherical hypothesis. When is one able to say that the object under study is or is not made up of balls? Working through an example on dust particles is the way we have chosen to pursue this section.

The basic relationships are given in $\mathbb{R}^3$, within the framework of the local approach. Consider a population of disjoint balls whose centres appear (in $\mathbb{R}^3$) according to a stationary random process. Let $N_V^{(3)}$ be the number of balls per unit volume, and $F_3(D)$ be the distribution function of their diameters (with corresponding density $f_3(D)$). In other words, $N_V^{(3)}[1 - F_3(D)]$ is the number of balls with diameters $\geq D$ per unit volume. These balls induce respectively disks and chords on test planes and test lines embedded in $\mathbb{R}^3$. Also $N_V^{(2)}$ (and $N_V^{(1)}$) are the number of circles (and chords) per unit area (per unit length) and $F_2(D)[F_1(D)]$ is the associated distribution function of these intersections. Note that $F_1(D)$ can either be considered as coming from the 2-D population of disks or from the balls themselves. F_1 is the law in number that we have already met in relation (X-14).

These numbers and distributions are linked by the following relationships:

— $\mathbb{R}^1 \to \mathbb{R}^3$:

$$
\left.
\begin{aligned}
N_V^{(3)} &= \frac{2}{\pi} \cdot N_V^{(1)} \cdot F_1''(o) \\[2ex]
1 - F_3(D) &= \frac{1}{D} \cdot \frac{F_1'}{F_1'(o)}
\end{aligned}
\right\}
\tag{X-37}
$$

— $\mathbb{R}^1 \to \mathbb{R}^2$:

$$
\left.
\begin{aligned}
N_V^{(2)} &= \frac{N_V^{(1)}}{\pi} \int_0^\infty \frac{F_1'(h)\,dh}{h} \\[2ex]
N_V^{(2)}[1 - F_2(D)] &= \frac{N_V^{(1)}}{\pi} \int_D^\infty \frac{1}{\sqrt{h^2 - D^2}} F_1'(h)\,dh
\end{aligned}
\right\}
\tag{X-38}
$$

Formulae for going from $\mathbb{R}^2$ to $\mathbb{R}^3$ are obtained by replacing 1 and 2 by 2 and 3 respectively in the labels of relation (X-38). Simple probabilistic proofs for relations (X-37) and (X-38) are found in Exercise X-13. Various digital versions of the integrals and derivations of relations (X-37) and (X-38) have been proposed. S. A. Saltykov (1958) bases an approximation by segmenting the histograms into classes, and estimating each class by the value of its last point. L. M. Cruz-Orive (1977) prefers to take the centres of the classes, and compares the quality of his approximation with several other ones.

In relations (X-37) and (X-38) the size distribution is expressed in number. This approach is particularly convenient here, since all the shapes are identical, and the size of a ball is just the ratio of similarity between it and the unit ball. Openings by balls, and size distribution by the convexity approach, leads to distribution in volume of the balls, with a density $g_3(\mu)$, where μ is the ball-volume. Relation (X-27) shows that $g_3(\mu)$ carries the same information as the number-weighted function $f_3(\mu)$ (where $\mu = \frac{4}{3}\pi D^3$). Hence the different types of size distribution come together here to form a single notion, all the necessary information being contained in the pair $(N_V^{(3)}, F_3)$ (or equivalently $(N_V^{(2)}, F_2)$, or $(N_V^{(1)}, F_1)$).

In practice, it is sometimes enough just to know the first few moments of a distribution. Let us denote by E_i and $\mathcal{M}_i$ the number and measure weighted moments associated with F_i ($i = 1, 2, 3$). From relation (X-37) we derive

$$E_3(D^\alpha) = \alpha \frac{E_1(D^{\alpha-2})}{f'_1(o)} \qquad \text{(X-39)}$$

$$\mathcal{M}_3(D^\alpha) = \frac{\alpha+3}{3}\mathcal{M}_1(D^\alpha) \qquad \text{(X-40)}$$

(Remember that for convex particles, the 3-D star is directly related to the moments of same order of the traverse length (rel. X-31). For the balls (X-40) generalizes this property to all the moments.)

Similarly the passage (X-38) from $\mathbb{R}^2 \to \mathbb{R}^3$, yields the inter-moment relations:

$$E_3(D^\alpha) = \frac{\pi \cdot \Gamma(1+\alpha/2)}{\Gamma(1+\alpha/2)} \cdot \frac{E_2(D^{\alpha-1})}{E_2(D^{-1})} \qquad \text{(X-41)}$$

where Γ is the Euler (or gamma) function. For $\alpha = 1, 2, 3$, we find:

$$E_3(D) = \frac{\pi}{2 \cdot E_2(D^{-1})}; \qquad E_3(D^2) = 2\frac{E_2(D)}{E_2(D^{-1})};$$

$$E_3(D^3) = \frac{3\pi}{4}\frac{E_2(D^2)}{E_2(D^{-1})} \qquad \text{(X-42)}$$

and for the measure-weighted moments:

$$\mathcal{M}_3(D) = \frac{32}{9\pi}\mathcal{M}_2(D); \qquad \mathcal{M}_3(D^2) = \tfrac{5}{4}\mathcal{M}_2(D^2) \qquad \text{(X-43)}$$

As for the passage from $\mathbb{R}^1 \to \mathbb{R}^2$, a similar analysis gives:

$$\mathcal{M}_2(D) = \frac{3\pi}{8}\mathcal{M}_1(D); \qquad \mathcal{M}_2(D^2) = \tfrac{4}{3}\mathcal{M}_1(D^2) \qquad \text{(X-44)}$$

The measure-weighted moments $\mathcal{M}_i$ are more robust than the number-weighted E_i, since they are less sensitive to fluctuations in the small classes of the size distribution (see Ex. X-2), although it is essential for their use that the largest ball of the population has a diameter smaller than the dimension of the measuring mask.

E.2. Extension of the model

Formally speaking, the problem of reconstituting spherical particles from planes and lines is well defined, and completely solved by algorithms (X-37) and (X-38). However these formal derivations give us no insight into extension of the validity of the model. Can the particles photographed in Figures X.18 and X.19 be considered as balls, at least in terms of our aim of "size reconstitution"? As a matter of fact, a strict answer to a question of this kind is impossible to give, and the applicability of model and criterion

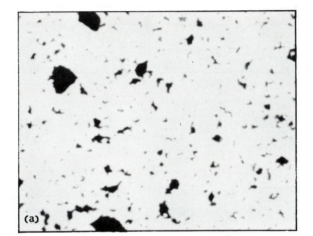

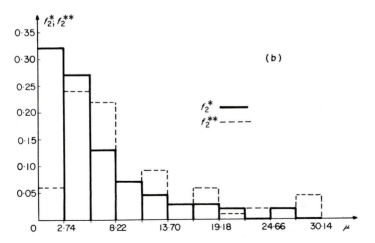

Figure X.18. (a) partial photograph of dust particle, (b) size analysis of specimen (a) using intercepts and 2-D openings. f_2^*: linear intercepts; f_2^{**}: projected areas.

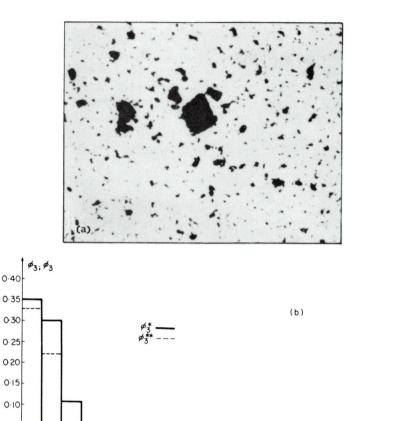

Figure X.19. (a) photograph of the second dust specimen; (b) comparative size distribution: 3-D openings performed on the projected grains Φ_3^* and a polished section Φ_3^{**}.

has to be empirically determined. The following example is a contribution of this type.

J. C. Klein (1973) studied dust particles of silica using two technologies of preparation of the specimen. In the first, he puts the particles into teepol (a type of liquid soap) and looks at the suspension through transmitted light. The second method is looking at a polished section of the particles embedded in polymerized araldite. From these data, we wish to test the second relation of (X-38), using two different methods of estimation.

In technology 1, we take for f_1^* the digital histogram of the intercepts directly measured on the suspension, and for f_2^* a digital estimate of (X-27) constructed as follows:

$$f_2^* = \left[A^*(X_{B(i)}) - A^*(X_{B(i+1)})\right] \cdot \left[(3i^2 + 3i + 1) \cdot \frac{a^2 \sqrt{3}}{2}\right]^{-1} \cdot K \quad \text{(X-45)}$$

$(A^*(X_{B(i)}))$ is the digital area of the hexagonal opening of size i; the factor $[\ ]^{-1}$ is the area, in $\mathbb{R}^2$, of the hexagon of size i; and K is the normalizing factor that makes $\sum f_2^*(i) = 1)$.

Take for a digital version of the second relation of (X-38) the following expression:

$$1 - F_2^{**}(i) = \left[\sum_i^\infty \frac{f_1^*(i)}{i}\right]^{-1} \times \sum_{i+1}^\infty f_1^*(j)(j^2 - i^2)^{-1/2} \quad \text{(X-46)}$$

which gives an alternative estimate of f_2:

$$f_2^{**}(i) = F_2^{**}(i+1) - F_2^{**}(i) \quad \text{(X-47)}$$

The two expressions (X-45) and (X-47) give two different estimators of the size distribution of the "disks" seen on Figure X.18a; these are plotted on Figure X.18b. These "disks" being projections of the 3-D particles, we can consider f_2^* and f_2^{**} as estimates of the 3-D diameter histograms. The shapes of the two curves differ considerably. The histogram $f_2^{**}(i)$ overestimates the large particles (or f_2^* underestimates them). Is the 2-D opening, or algorithm (X-46) not able to handle grains whose shapes are not exactly spherical?

We now take another specimen of dust particles of silica, this time having a range of sizes and shapes which seem even less circular than in the previous case (photograph Figure X-19). As before, we look at it through transmitted light (technology 1), perform hexagonal openings, and apply algorithm (X-45), resulting in a numerical function $\varphi_3^*(i)$. Considering that we see the apparent contour of the particles, we may interpret $\varphi_3^*(i)$ as an estimate of the density function $f_3(D)$ of the diameters of the "spherical" particles. In a second step, we make a polished section with another part of the same powder and once again perform hexagonal openings and apply relation (X-45). We denote the result by $\varphi_2^*(i)$ and consider it as an estimate of the law $f_3(D)$ of the diameters of the intercepted disks. Now, formulae (X-46) and (X-47) allow us to make the passage going from disks to balls: f_1^* is replaced by φ_2^*, and the resulting estimate is denoted as $\varphi_3^{**}(i)$. The two quantities $\varphi_3^*(i)$ and $\varphi_3^{**}(i)$ are two experimentally independent estimates of the same ball diameter density. A check on Figure X.19b shows the two histograms to indeed be comparable. As before, the deconvolution algorithm (X-46) gives too much weight to the large particles, but in a more modest fashion (we have not calculated a χ^2 test; this would hide from us whether differences are positive or negative).

What conclusions can be drawn from these experiments? In *the first case* (see Fig. X-18), we used two different shapes of structuring elements that take the angularity into account in different ways (see Ex. X-4). The opening is less sensitive to small irregularities, which explains why $\varphi_3^* - \varphi_3^{**}$ is smaller than $f_2^* - f_2^{**}$ in the small classes. In the second experiment, where the measurements are more homogeneous, the result is satisfactory.

It remains paradoxical however, how the 2-D openings, which easily detect differences in shapes between disks and the actual grains under study, give good working input (such as φ_2^*) into the deconvolution algorithms (such as (X-46)). It is a pity that J. C. Klein did not extend his experiments to size distributions of particles via particle area measurement and maximum inscribable circle. Presumably, the particle area approach would be worse than 2-D openings (the former is blind to the differences in shape), and the maximum circles would be the best of all the procedures.

In any case it is clear that the domain of use of the deconvolution algorithm (X-46) is made up of other set populations other than balls, and depends on the choice of size descriptor. To exhibit that such a domain is non-trivial makes some indirect comment about the type of study which tries to connect shape and size of say, ellipsoids, prisms, etc. with those of their cross sections (see the proceedings of the various symposia on stereology, for example). Suppose we state that the particles in photographs X-18 and X-19 are ellipsoids, or prisms, or tetrahedra, and derive from certain 2-D measurements their 3-D elongation, mean height, etc. . . . However, we see that the same particles can *also* be considered as balls, and that a 3-D size distribution can be obtained on this basis.

An algorithm given without its field of application is limited in its use; unfortunately a general theoretical approach is not the answer. Therefore, we suggest that the experimenter who wants to use spherical tools but is anxious about the sphericity of his particles, should proceed as J. C. Klein did, namely to get a feel for the algorithm by comparing say, results for section with results for projection. Figures X.18 and X.19 give us some idea about the practical class to which the spherical results can be extended.

F. THEORETICAL COMPLEMENTS

F.1. The general equation of size mappings

In this section, we summarize briefly the theory of sizing as presented by Matheron in his book (1975). For a given λ, the three axioms (X-1), (X-2) and (X-3) define an operation ψ_λ which in algebra is termed an opening (of which the morphological opening is a particular case). It is known that an algebraic opening is characterized by the family $\mathscr{B}_\lambda$ of its invariant sets; i.e.:

$$\psi_\lambda(B) = B \Leftrightarrow B \in \mathscr{B}_\lambda$$

It turns out that for any X, the transform $\psi_\lambda(X)$ is the union of the invariant sets B included in X:

$$\psi_\lambda(X) = \bigcup\{B \in \mathscr{B}_\lambda, B \subset X\} \qquad \text{(X-48)}$$

It can easily be shown that if we treat λ as varying, axiom (X-3) is satisfied iff the families $\mathscr{B}_\lambda$ are decreasing with λ, i.e. iff

$$\mu \geq \lambda \Rightarrow \mathscr{B}_\mu \subset \mathscr{B}_\lambda \qquad \text{(X-49)}$$

Thus formula (X-48), associated with condition (X-49) gives the general algebraic form of the size mappings.

F.2. Morphological size distribution

(a) *Compatibility with translation and change of scale*
We now in turn add the four principles of mathematical morphology as supplementary conditions that relation (X-48) must fulfil. We will treat separately, the more algebraic notions of compatibility with translations and changes of scale; i.e. distinct from the more topological conditions of local knowledge and semi-continuity.

The condition

$$\psi_\lambda(X_h) = [\psi_\lambda(X)]_h$$

is satisfied iff the family $\mathscr{B}_\lambda$ is closed under translation. Similarly, the relation

$$\psi_\lambda(\lambda X) = \lambda\psi_1(X)$$

is equivalent to

$$\mathscr{B}_\lambda = \lambda\mathscr{B}_1; \qquad \text{i.e.} \qquad \lambda B \in \mathscr{B}_\lambda \Leftrightarrow B \in \mathscr{B}_1$$

where $\mathscr{B}_1$ must be a family closed under scalar multiplication by factors ≥ 1. In other words, the size criterion ψ_λ is compatible with translations and changes of scale iff there exists a family $\mathscr{B}$, closed under translations and scalar multiplication by factors ≥ 1, such that

$$\psi_\lambda(X) = \bigcup\{\lambda B, B \in \mathscr{B}, \lambda B \subset X\} \qquad \text{(X-50)}$$

Denote by $\mathscr{B}_0$ a subfamily of $\mathscr{B}$ such that the topological closure of $\mathscr{B}_0$ by union, translation and scalar multiplication by factors ≥ 1, is exactly $\mathscr{B}$. Then, relation (X-50) takes the following simpler form:

$$\psi_\lambda(X) = \bigcup_{B \in \mathscr{B}_0} \bigcup_{\mu \geq \lambda} X_{\mu B} \qquad \text{(X-51)}$$

Conversely, formula (X-51) gives the general form for size criteria that satisfy the first two principles of morphology.

(b) *Compactness*

While it is morphological openings we deal with, it will be convenient to combine the two principles of local knowledge and semi-continuity into the definition of a *compact opening*: an algebraic opening ψ on $\mathscr{P}(\mathbb{R}^n)$ is compact if the following conditions are satisfied:

(i) $\mathscr{F}$, $\mathscr{G}$ and $\mathscr{K}$ are topologically closed under ψ.
(ii) The opening ψ is upper semi-continuous on $\mathscr{F}$ and on $\mathscr{K}$, and lower semi-continuous on $\mathscr{G}$.
(iii) The opening ψ is the smallest extension of its restriction to $\mathscr{K}$, in the following sense:

$$\psi(X) = \bigcup\{\psi(K), K \subset X, K \in \mathscr{K}\} \qquad \forall X \in \mathscr{P}(\mathbb{R}^n)$$

An opening that is compatible with translations is compact if and only if there exists a class $\mathscr{B}_0$, closed in $\mathscr{F}$, contained in $\mathscr{F}_{\{0\}}$, and such that:

(i) $B \in \mathscr{B}_0$ and $x \in B$ imply $B \oplus \{-x\} \in \mathscr{B}_0$
(ii) The class $\mathscr{M}_0$ of the minimal elements in $\mathscr{B}_0$ is contained in $\mathscr{K}$.
(iii) $\psi(X) = \bigcup\{X_M, M \in \mathscr{M}_0\}$ for any $X \in \mathscr{P}$.

For example, let K be a compact set, and $\mathscr{B}_0$ the class of its translates containing 0. Then $\mathscr{M} = \mathscr{B}_0$, and the opening $X \to X_K$ is compact.

We shall say that the size criterion ψ_λ is compact if for any $\lambda > 0$ the opening ψ_λ is compact. Moreover, if ψ_λ is compatible with changes of scale it is enough that the above criterion of compactness be satisfied for only one ψ_λ (Ex. $\lambda = 1$).

(c) *Upper semi-continuity*

A size criteron ψ_λ on $\mathscr{P}(\mathbb{R}^n)$, satisfying the first two principles, is said to be upper semi-continuous on $\mathscr{F}$ if $(\lambda, X) \to \psi_\lambda(X)$ is an upper semi-continuous mapping from $\mathbb{R}_+ \times \mathscr{F}$ into $\mathscr{F}$. This definition is equivalent to other possibilities.

— For any $\lambda > 0$, ψ_λ is upper semi-continuous on $\mathscr{F}$.
— The family $\lambda\mathscr{B}$ of the ψ_λ-invariant sets is closed under topological closure, and $\mathscr{B} \cap \mathscr{F}$ is closed in $\mathscr{F}$.

It is a consequence that the ψ_λ-invariant set has no isolated points.

(d) *Size distribution and morphological openings*

In practice, the most interesting case occurs when we choose a single compact set B for the family $\mathscr{B}_0$ appearing in formula (X-51). The corresponding size criterion is given by:

$$\psi_\lambda(X) = \bigcup_{\mu \geq \lambda} X_{\mu B} \tag{X-52}$$

If the compact set B is also convex, we get a very simple result. In this case, $X_{\lambda B}$ is a decreasing function of λ:

$$\lambda \geq \mu \Rightarrow X_{\lambda B} \subset X_{\mu B}$$

and we simply get:

$$\psi_\lambda(X) = X_{\lambda B}$$

Conversely, it can be shown that the mapping $X \to X_{\lambda B}$ is a size criterion if and only if the compact set B is convex. Thus, the remarkable simplicity of relation (X-52) is closely connected with the convexity of the structuring element B.

(e) *A random version of the size criteria*

Let ψ_λ be an upper semi-continuous size criterion (but not necessarily compact, nor compatible with scalar multiplication and translation), and let X be a random closed set. Then $\psi_\lambda(X)$ is a random closed set measurable on $\mathbb{R}^+ \times \mathscr{F}$. Moreover,

$$\Lambda(x) = \mathrm{Sup}\{\lambda : x \in \psi_\lambda(x)\}, \qquad x \in \mathbb{R}^n$$

is a measurable random function, and $\{\Lambda(x) \geq \lambda\} = \{x \in \psi_\lambda(X)\}$, for every $x \in \mathbb{R}^n$. At each $x \in \mathbb{R}^n$, define the size distribution of X with respect to the size criterion ψ_λ, as the function $\lambda \to 1 - G_x(\lambda)$, where:

$$1 - G_x(\lambda) = \Pr\{\Lambda(x) \geq \lambda\} = \Pr\{x \in \psi_\lambda(X)\}, \qquad \lambda > 0$$

The realizations of the random function $x \to \Lambda(x)$ are upper semi-continuous on $\mathbb{R}^n$.

We will now examine what happens when we apply the same size criterion ψ_λ to the background X^c, restricting ourselves only to size criteria satisfying the four principles of morphology.

Let ψ_λ, $\lambda > 0$, be a morphological size criterion (compatible with translations and changes of scale, compact and upper semi-continuous). Let X be a random closed set, and Pr its probability. Then, for any $\lambda > 0$, $\psi_\lambda(X^c)$ is a measurable random open set, and for any measure $\mu \geq 0$ on $\mathbb{R}^n$, we have:

$$E[\mu(\psi_\lambda(X^c)] = \int \mu(dx)\, \Pr[x \in \psi_\lambda(X^c)]$$

Moreover, $x \to \Lambda'(x) = \mathrm{Sup}\{\lambda : x \in \psi_\lambda(X^c)\}$ is a measurable random function and $\{\Lambda'(x) \leq \lambda\} = \{x \notin \psi_\lambda(X^c)\}$. For any $x \in \mathbb{R}^n$, the function $\lambda \to 1 - G'_x(\lambda)$ defined by

$$1 - G'_x(\lambda) = \Pr\{\Lambda'(x) > \lambda\} = \Pr\{x \in \psi_\lambda(X^c)\}$$

is the pore size distribution of the random closed set X at the point x, with respect to the size criterion ψ_λ. The realizations of the random function $\Lambda'(x)$ are lower semi-continuous on $\mathbb{R}^n$.

G. EXERCISES

G.1. Digitalization for linear size distributions

(a) Suppose the set X in $\mathbb{R}^2$ is digitalized using a rectangular digital mask Z (i.e. a square or hexagonal grid of points (spacing (a)) restricted to a rectangle). Justify the following estimator:

$$Local:\ \Pr\{x \in X \ominus B((k-1)a)\} = P[(k-1)a] \sim \frac{\mathcal{N}(\overbrace{1\ 1\ \ldots\ 1})^{k}}{\underbrace{\mathcal{N}(b\ b\ \ldots\ b)}_{k}}$$

$$P''(ka) \sim \frac{\mathcal{N}(0\ \overbrace{1\ 1\ \ldots\ 1}^{k}\ 0)}{\underbrace{\mathcal{N}(b\ b\ \ldots\ b)}_{k+2}}$$

$$Global:\ A[X \ominus B((k-1)a)] = P[(k-1)a] \sim \mathcal{N}(\overbrace{1\ 1\ \ldots\ 1}^{k})\cdot\mu$$

$$P''(ka) \sim \mathcal{N}(0\ \overbrace{1\ 1\ \ldots\ 1}^{k}\ 0)\mu$$

where $\mathcal{N}(\cdot)$ = number of configurations $\{\cdot\}$

$$b = \text{one or zero}$$

$$\mu = \begin{cases} a^2 & \text{(square grid)} \\ a^2\ \sqrt{3/2} & \text{(hexagonal rid)} \end{cases}$$

and " $\sim$ " means: "is estimated by".

Notice that the denominators of the two local estimates are different. Interpret this difference.

(b) Show that, in the global case, we have:

$$P''(ka) \simeq P[(k+1)a] - 2P(ka) + P[(k-1)a]$$

and interpret the result.

Start from $\mathcal{N}[\overbrace{1\ \ldots\ 1}^{k}] = \mathcal{N}[\overbrace{1\ \ldots\ 11}^{k}] + \mathcal{N}[\overbrace{1\ \ldots\ 10}^{k}].$

The digital structuring elements chosen for P and P'' are equivalent to a second order finite difference.

Although it is natural for technology we have available, this differentiation is not the only possible one (we could also interpolate $P(l)$ locally with a parabola using least squares. In the local case, the estimates satisfy the same relationship when the demoniators $\mathcal{N}(b \ldots b)$ are large enough).

(c) Suppose we apply the local algorithms for calculating P and P'', to the background of a realization of a Boolean disk process (the disks have area fraction $A_A = 0.1$). The curves in Figure X.20 are obtained. Interpret the results.

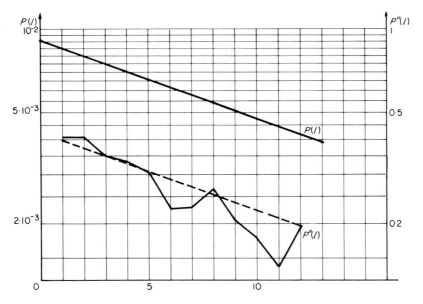

Figure X.20. Moments $P(l)$ and $P''(l)$ for a realization of a Boolean process.

[Theoretically, $P(l)$ and $P''(l)$ are negative exponentials (Ch. XIII), i.e. a straight line with a log scale on the y-axis. And from a practical standpoint, the estimation of $P(l)$ is excellent, allowing calculation of $P'(o)$, although $P''(l)$, which is a second difference, fluctuates enormously. Note the y-axis represents the distribution. Note also the chord joining the two extreme points of $P''(l)$ is parallel to $P(l)$. Clearly, to use information specifically taken from P'' (e.g. bimodality of $P(l)$, angularity of X), we need sample sizes much larger than those for P (from which we take measure and number weighted moments).]

G.2. Physical noise and size distribution

Suppose X is a population of convex particles with variable volumes and specific parameters V_V, S_V and M_V. A physical abrasion process detaches a thin pellicule from the surfaces of the grains, and reduces it to powder. How does the abrasion modify the volume fraction V_V, the specific surface S_V, the mean intercept $E(L)$, and the star $\mathcal{M}(V)$?

(a) [Let V_V' and S_V' be the new volume fraction and specific surface. We represent the abrasion process by an erosion by a ball of radius r. The pellicle crumbles into small cubes of side r. We get:

$$V_V' = V_V \qquad S_V' = 7S_V - 2M_V r]$$

(b) [$E[L] = 4V_V/S_V$; so $E[L'] = 4V_V'/S_V'$. Thus, when $r \to 0$, $E[L'] = \frac{4}{7}E[L]$. The bias is enormous].

(c) [$V_V'\mathcal{M}(V') = (V_V - rS_V)\{E[V-rS]^2/E[V-rS]\}$. Hence, when $r \to 0$ we have $\mathcal{M}(V') \to \mathcal{M}(V)$. There is no bias for the star and, more generally, there is no bias for measure-weighted moments.]

G.3. Linear size distribution for convex particles

Prove that for X a compact convex set, the linear erosion and the geometric covariogram are identical ($P(h) \equiv K(h)$). Show that when X is the union of disjoint convex particles sufficiently far from each other, the linear size distribution can be deduced from the set covariance (global or local).

G.4. Angularities and linear openings

In Section B.2c, we proved that when the boundary of X was smooth, $f(l)$ was linear near the origin. Now consider angular grains, such as polygons. Prove that an angle such as represented in Figure X.21a makes the following additive

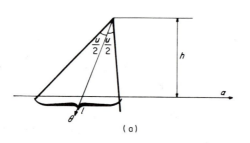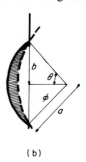

(a)   (b)

Figure X.21.

contribution to $F_\alpha(l)$:

$$h = \frac{l}{2 \sin u} [\cos u - \cos 2(\theta - \alpha)] \quad \text{if } \frac{u}{2} < \theta - \alpha < \pi - \frac{u}{2}$$

and zero if not.

Assuming that there are v angles per unit area, with directions that are uniformly oriented, prove that the resulting size distribution $F(l)$ is:

$$F(l) = \frac{2v}{\pi S_V} \left[1 + \underset{u}{E} \left[(\pi - u) \cot u \right] \right] \cdot l = K \cdot l$$

(E_u: mathematical expectation with respect to the orientations u of the grains).

Interpret the term K [K is the value of $f(l)$ at the origin; the factor between brackets [ ] turns out to be an angularity factor.]

G.5. Angularities and circular openings

We will now analyse the angularities by performing circular openings. Prove that the contribution of the angle u to the opening by a disk of radius r is

$$r^2 \left[\cot \frac{u}{2} - \frac{\pi - u}{2} \right]$$

and that the overall size distribution is:

$$G(r) = A_A v E \left[\cot \frac{u}{2} - \frac{\pi - u}{2} \right] \cdot r^2$$

Note that the term between square brackets looks similar to the one appearing in $F(l)$.

We will now analyse non-angular objects by circular openings. Two local parameters of the boundary ∂X intervene: the radius of curvature ρ, and the angle 2θ through which we see, from the centre of curvature, the zone eliminated by the opening of radius r (see Fig. X.21 b).

Prove that the contribution of such a boundary zone to $G(r)$ is:

— zero when $r < \rho$
— equal to $\rho^2 \arcsin(b/\rho) - b\sqrt{\rho^2 - b^2} - \arcsin(b/r) - b^2 \sqrt{r^2 - b^2}$ when $r \geq \rho$.

Show that a limited expansion of this expression near $r = \rho$, gives

$$\frac{3}{2} (\sin \theta)^3 \rho (r - \rho), \qquad r \geq \rho$$

Interpret the result. [The contribution of the zone to the opening is zero up to $r = \rho$, and a linear function of r for $r \geq \rho$.]

G.6. Size distributions by dodecagonal openings

This exercise is a continuation of Exercise VI-9. Prove that any family of digital dodecagons

$$D[\lambda(y)] = (H_1)^{\oplus i} \oplus (H_i^*)^{\oplus j} \qquad i, j, \text{ integers}$$

generates a size distribution if and only if

$$\lambda(i', j') \geq \lambda(i, j) \Leftrightarrow i' \geq i \quad \text{and} \quad j' \geq j \tag{1}$$

On the other hand, the pseudo-anisotropy required for the dodecagons implies that

$$|i - j| \leq 1 \tag{2}$$

Construct a sequence of digital dodecagons satisfying (1) and (2) with successive areas that increase as slowly as possible.

G.7. Opening size distributions

Interpret the size histograms (densities of hexagonal openings—local approach) for the following two metallographic structures (Fig. X.22):

(a) ferrites (in black) in a phenomenon of phase transformation;
(b) sinter (porous medium in black).

The openings have been performed on the white phases of the photographs.

[In(a), three scales of structures are exhibited. Steps 1–2 correspond to the narrow cracks between the black particles; from step 9 to 22, the medium sized white zones are filled up, as for example to the top right of the photograph. The third scale (steps 31–32) corresponds to the destruction of the large empty zones of the centre.

In (b) the white zone is present at only one scale of thickness. The interesting part is the beginning of the histogram, where we see that the white zone is already opened by a hexagon of size two. So what has happened is that the sintering process has itself eroded the pores with a (presumably) spherical ball of radius r. This value two represents the average radius (weighted in measure) of the disk (obtained by intersecting the sphere with the plane); application of relation (X-61) gives $r \sim 2.2$ steps.]

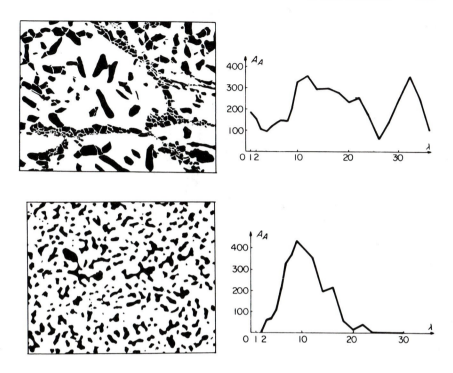

Figure X.22.

G.8. Scaling table for digital openings and closings

Let $B(i)$ denote either the hexagon or the square of size i (i.e. with a digital area $A[B(i)] = 3i^2 + 3i + 1$ for the hexagon, and $(i + 1)^2$ for the square (Ch. VI)). Derive the following scaling table, where X is the set under study, and Z is the measuring mask. The results are given in opening and closing *increments*. Each algorithm is the value of a function of i, and the corresponding abscissa is indicated. These results are strictly digital. For the local version, they are also estimates in $\mathbb{R}^2$, but for the global version they must be multiplied by $a^2\sqrt{3/2}$ (resp. a^2) to become estimates in $\mathbb{R}^2$. For this local version, note also that the openings-closings are expressed per unit area of the *space*, but the openings (or closings) per unit area of the phase X (Fig. X.23).

G.9. Openings of squares by disks (Delfiner)

(a) Suppose that X is the union of separate squares, all with the same length of side a. Let $B(\lambda)$ be a disk of radius λ. Calculate

Function	Algorithm	Abscissa
Global opening Mes $(X_{\lambda B})$	$A[(X \cap Z)_{B(i)}] - A[(X \cap Z)_{B(i+1)}]$	$i + \frac{1}{2}$
Global closing Mes $(X^{\lambda B})$	$A[(X \cap Z)^{B(i+1)}] - A[(X \cap Z)^{B(i)}]$	$i + \frac{1}{2}$
Global opening–closing	id. global opening	$-(i + \frac{1}{2})$
Global opening–closing	id. global closing	$-i + \frac{1}{2}$
Local opening $G_1(\lambda)$(rel. X-22)	$\left\{ \dfrac{A[(X \cap Z)_{B(i)}]}{A[Z \ominus B(2i)]} - \dfrac{A[(X \cap Z)_{B(i+1)}]}{A[Z \ominus B(2i+2)]} \right\} \dfrac{A(Z)}{A(X \cap Z)}$	$i + \frac{1}{2}$
Local closing $G_0(\lambda)$(rel. X-24)	$\left\{ \dfrac{A[(X \cap Z)^{B(i+1)}]}{A[Z \ominus B(2i+2)]} - \dfrac{A[(X \cap Z)^{B(i)}]}{A[Z \ominus B(2i)]} \right\} \dfrac{A(Z)}{A(Z) - A(X \cap Z)}$	$i + \frac{1}{2}$
Local opening-closing $P_B(-\lambda)$(rel. X-21)	$\dfrac{A[(X \cap Z)_{B(i)}]}{A[Z \ominus B(2i)]} - \dfrac{A[(X \cap Z)_{B(i+1)}]}{A[Z \ominus B(2i+2)]}$	$-(i + \frac{1}{2})$
$P_B(\lambda)$ (rel. X-23)	$\dfrac{A[(X \cap Z)^{B(i+1)}]}{A[Z \ominus B(2i+2)]} - \dfrac{A[(X \cap Z)^{B(i)}]}{A[Z \ominus B(2i)]}$	$+i + \frac{1}{2}$

Scaling table

$$G(\lambda) = \frac{A(X) - A(X_B)}{A(X)}$$

and its first two moments.

$$\left[G(\lambda) = \begin{cases} \dfrac{4\lambda^2}{a^2}\left(1 - \dfrac{\pi}{4}\right); & \lambda \leq \dfrac{a}{2} \\ 1; & \lambda > \dfrac{a}{2} \end{cases} \right.$$

$$E[\lambda] = \frac{a}{3}\left[1 + \frac{\pi}{8}\right] = 0.46a;$$

$$\left. \text{Var}(\lambda) = \frac{a^2}{72}\left[1 + \frac{\pi}{4} - \frac{\pi}{8}\right] \sim 7.7 \cdot 10^{-3}a^2 \right]$$

Thus the mean value has a small bias (0.46 instead of 0.50), and the variance is not zero, although the squares are all identical; interpret this.

(b) Suppose that X is now a set of homothetic, separated squares whose sides a are taken at random from a population with distribution function $H(a)$.

Calculate $G(\lambda)$.

$$\left[G(\lambda) = \frac{4\lambda^2(1 - \pi/4)}{\int_0^\infty a^2 H(da)}\left[1 - H(2\lambda)\right] + \int_0^\infty a^2 H(da)\right]$$

Notice that the second term represents the size distribution of the maximum inscribable circles, and the first term that of the residual zones between these circles and the corners of the squares. The two terms respectively correspond to the discontinuous and continuous part of the openings $X_{B(\lambda)}$.

G.10. Size distribution and individual analysis in $\mathbb{R}^2$

This exercise illustrates an extension of relation (VIII-27). Consider an initial individual X_1 in R^2.

We wish to calculate the length distribution $F_1(l)$ of a random chord through X_1. First, fixing the direction α of the chord l, show that:

$$1 - F_1(l; \alpha) = \frac{A_1'(l; \alpha)}{A_1'(0; \alpha)} \tag{1}$$

where $A(l; \alpha) = A[X_1 \ominus B(l, \alpha)]$; $B(l, \alpha)$ = segment (l, α). Note that $A'(l; a) = -D^{(2)}[X_1 \ominus B(l, \alpha)]$ = total diameter in direction α. Show that the average chord $E\alpha(l)$ in direction α is equal to $A(X_1)/D^{(2)}(X_1, \alpha)$.

Now randomize α by a uniform distribution over $[0, \pi]$. Show that:

$$1 - F_1(l) = \frac{1}{U(X_1)} \int_0^\pi A_1'(l; \alpha)\, d\alpha \tag{2}$$

[The probability of having a chord in direction $(\alpha, \alpha + d\alpha)$ is

$$\frac{D^{(2)}(X_1, \alpha)d\alpha}{\int_0^\pi D^{(2)}(X_1, \alpha)d\alpha} = \frac{D^{(2)}(X_1, \alpha)}{U(X_1)}\, d\alpha$$

Thus unconditioning (1) w.r.t. α, we get (2). In particular, the average chord length $E_1(l)$ is:

$$E_1(l) = \frac{\pi A(X_1)}{U(X_1)}$$

which is just Crofton's formula, but extended to non-convex individuals.]

(b) The set X in $\mathbb{R}^2$ is partitioned into I individuals X_i. Show that:

$$F(l) = \frac{\sum U(X_i)F_i(l)}{\sum U(X_i)} \tag{3}$$

Interpret (3) by using the general relation (VII-27).

[The set transformation involved in (1) concerns the *boundary* ∂X_1, and concerns giving to the element $dx \in \partial X_1$, weight $|dx| \cdot |\cos(\theta - \alpha)|$ if dx is followed by a segment $\geq l$ and belonging to X_1, and weight zero otherwise (α is the angle of the outside normal at dx). A'_1 is the integral of this transformation over ∂X_1. Therefore, the individuals in (3) must be weighted by the measures of the corresponding sets i.e. the lengths of the ∂X_i.]

G.11. Size distribution and individual in $\mathbb{R}^3$

(a) Extend the previous analysis to $\mathbb{R}^3$ by showing that:

$$F(l) = \frac{\sum S(X_i) F_i(l)}{\sum S(X_i)}, \qquad S(X_i) = \text{surface area of } X_i \qquad (4)$$

(b) Consider the cross-section $X_1 \cap \Pi(x_1, \omega)$ of a 3-D set X_1 by the random plane $\Pi(x, \omega)$. Assuming that $X_i \cap \Pi(x, \omega)$ has no holes, denote by $F_i(A)$ the histogram of the areas of the connected components of $X_i \cap \Pi(x, \omega)$ when x and ω vary uniformly (as in Chap. VII, Sect. E). Under these conditions show that:

$$F(A) = \frac{\sum M(X_i) \cdot F_i(A)}{\sum M(X_i)} \qquad (M = \text{norm}) \qquad (5)$$

(c) What happens to this result when the particles appearing on the cross-section have holes? The $F_i(A)$ of each 2-D connected component must be weighted by its connectivity number; then (5) remains valid. Note that we need not identify individuals on the section in relations (4) and (5), since the Minkowski functionals are C-additive.]

G.12. Size distribution weighted in volume and in number

Recall the story of the butcher who sells lark pie. Lark meat being expensive, he decides to mix it with horse meat, and advertises "Excellent mixed pie: 50 per cent lark, 50 per cent horse". A few days later, a customer says: "It's curious that your pie does not taste very much of lark". "Strange indeed, Madam, but I guarantee you the proportion: one horse to one lark" . . .

This story sums up everything we want to say about size distribution weighted in volume and in number. However, we offer two more variations on the same theme. The first one, due to W. Feller (1966) deals with time processes, while the second will draw us nearer to image analysis.

(a) A commuter takes the bus each morning. Upon arriving at the bus stop, he inquires about when the last bus passed, in order to measure the time separating two successive buses. He notices that this random time follows a gamma distribution $\theta^2 t \exp(-\theta t)$, with an average $2/\theta = 20\,\text{mins}$. Independently, the bus company times the passing of all buses at this stop, and claims that the random interval between two buses in fact satisfies a Poisson law with an average $1/\theta = 10\,\text{mins}$.

(i) Who is right?

[Both. They weight the same events differently. For the company, all *beginnings* of the inter-bus interval are registered. For the commuter, who has a uniform probability of arriving at the bus stop, each instant dt has equal weight. Therefore, an inter-bus interval t which is given the probability density $f(t)$ by the company, appears to the commuter with a probability density $g(t)$ proportional to $tf(t)$:

$$g(t) = \frac{tf(t)}{K} \quad \text{and} \quad \int_0^\infty g(t)\,dt = 1 \Rightarrow K = \int_0^\infty tf(t)\,dt = E[t]$$

For $f(t) = \theta \exp(-\theta t)$, we get $g(t) = \theta^2 t \exp(-\theta t)$].

(ii) Now take for the density f, the law:

$$f(t) = 0, \quad 0 \le t \le 1; \qquad f(t) = \frac{2}{t^3}, \qquad t > 1$$

[On average, the commuter will not go and complain to the bus company, because he is still waiting for his bus.]

(b) A metallographer wants to estimate the histogram of the areas A of particles, on a polished section. For sampling reasons, he takes only those intersecting the vertices of a square grid (Fig. X.23).

Is the corresponding histogram unbiased? If not, how can it be corrected? What can be done if two vertices belong to the same particle?

[He wishes to estimate a law in number $f(A)$, but his technique gives him an estimate of the law in area $g(A)$. To go from one to the other he must use relation (X-27). Each particle sampled twice must be counted twice.]

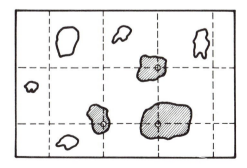

Figure X.23.

G.13. Reconstitution of the size distribution of balls and disks

(a) Using the notations Section E-1, establish

$$N_V^{(1)}[1 - F_1(h)] = N_V^{(3)} \int_h^\infty \frac{\pi}{4}(D^2 - h^2)f_3(D)dD \tag{1}$$

[Notice that $(\pi/4)(D^2 - h^2)$ is the contribution of a ball of diameter D to the traverses $\geq h$; then integrate.]

(b) invert relationship (1) to give $F_3(h)$ in terms of $F_1(h)$. By differentiating with respect to h, show that

$$N_V^{(1)} F_1'(h) = N_V^{(3)} \frac{\pi}{2} h[1 - F_3(h)]$$

Deduce $N_V^{(3)} = (2/\pi)F_1''(o)N_V^{(1)}$, and reconstruct the size distribution of the diameters from that of the traverses.

(c) To reconstruct the size distribution of disks from their traverses, show that

$$N_V^{(2)}[1 - F_2(D)] = \frac{N_V^{(1)}}{\pi} \int_D^\infty \frac{1}{\sqrt{h^2 - D^2}} F_1'(h) dh \tag{2}$$

[Let C_0 be a disk of diameter $D_0 \geq D$. Show that its contribution to the integral of the second member of (2) is:

$$2 \int_D^{D_0} \frac{P''(h)}{\sqrt{h^2 - D^2}} = \pi$$

Notice that this contribution is a constant for all the disks having a diameter $D_0 \geq D$, whose number is $N_V^{(2)}[1 - F_2(D)]$. Derive relation (2).]

(d) Size distribution of spheres from circles.

Show that the appropriate formula is the same as (2) except that the subscripts 1 and 2 are changed respectively to 2 and 3.

G.14. Distribution of the radii of curvature in $\mathbb{R}^2$ (Hass et al., 1967)

(a) Let X be a 2-D regular set (local approach), $N_A^+(\alpha)$ be its specific first contact number in direction α, as defined in Chapter V, B.5, and $\Phi^+(l, \alpha)$ be the distribution function of the marginal intercepts of X in direction $\alpha(\varphi^+(l, \alpha)$ associated density function). In other words, the quantity $N_A^+(\alpha)[1 - \Phi^+(l, \alpha)]\delta a$ represents the total number, per unit area of marginal intercepts (from straight lines of direction belonging to $(\alpha, \alpha + d\alpha)$), whose length is greater than l (Fig. X.24a). The distribution function of the radii of curvatures of X in direction α, is denoted by $\Psi^+(R, \alpha)$.

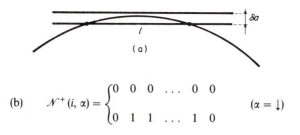

$$\mathcal{N}^+(i, \alpha) = \begin{cases} 0 & 0 & 0 & \ldots & 0 & 0 \\ 0 & 1 & 1 & \ldots & 1 & 0 \end{cases} \qquad (\alpha = \downarrow)$$

(b)

Figure X.24. (a) an example of a marginal intercept, (b) its square digitization.

Show that:

$$N_A^+(\alpha)[1 - \Phi^+(l, \alpha)] = N_A^+(\alpha) \int_{l^2/8\delta a}^{\infty} \left[1 - \frac{l^2}{8\delta a} \right] d\Psi^+(R, \alpha) \qquad (1)$$

Solve this integral equation and express $\Psi^+(R, \alpha)$ as a function of the marginal distribution, i.e.:

$$\Psi^+\left(\frac{l^2}{8\delta a}, \alpha \right) = \Phi^+(l, \alpha) - \frac{l}{2}\varphi(l, \alpha) \qquad (2)$$

Average in α to find the actual distribution $\Psi^+(R)$ of the positive radii of curvature. Is relation (2) still valid if X exhibits angularities?

[Each first contact zone of the boundary which has a positive normal of direction α, makes a contribution to the left member of (1), equal to $\delta a - l^2/8R$ if the radius of curvature $R < l^2/8\delta a$, and to 0 otherwise. Relation (1) results. By differentiating it with respect to l, we obtain

$$\varphi^+(l, \alpha) = \frac{l}{4\delta a} \int_{l^2/8\delta a}^{\infty} \frac{1}{R} \cdot d\Psi^+(R, \alpha)$$

A second differentiation gives, after a change of variables:

$$1 - \Psi^+\left[\frac{l^2}{8\delta a}, \alpha \right] = -\int_l^{\infty} \frac{u^2}{2} d\left[\frac{\varphi^4(u, \alpha)}{u} \right]$$

which yields relation (2) after an integration by parts. The isotropic distribution $\Psi^+(R)$ satisfies the relationship:

$$N_A^+ \cdot \Psi^+(R) = \frac{1}{2\pi} \int_0^{2\pi} N_A^+(\alpha) \cdot \Psi^+(R, \alpha) d\alpha.] \qquad (3)$$

(b) *Decomposition of the specific perimeter*
Calculate the moments of $\Psi(R)$ as functions of $\varphi(l)$ i.e. $\varphi(l)$ $= (1/2\pi) \int_0^{2\pi} N_A^+(\alpha)\varphi(l, \alpha)d\alpha$; decompose the specific perimeter $U_A(X)$ into its

convex and concave parts as follows:

$$U_A(X) = U_A^+(x) + U_A^-(X) = E\left[\int_{R>0} R\,d\alpha\right] + E\left[\int_{R\leq 0} |R|\,d\alpha\right]$$

Express U_A^+ and U_A^-.
  [Relations (2) and (3) imply:

$$E[(R^+)^n] = \frac{n+1}{n(8\delta a)^n} \int_0^\infty l^{2n}\varphi^+(l)\,dl$$

We have

$$U_A^+ = 2\pi N_A^+ \cdot E(R^+) = \frac{\pi N_A}{2\delta a} \int_0^\infty l^2 \varphi^+(l)\,dl$$

Mutatis mutandis all the results of the exercise, in particular U_A^+, are immediately extended to negative radii of curvature.]

(c) *Digitization of* N_A^+, Φ^+ *and* Ψ^+ *(eight-connected graph)*
The Texture Analyser (square grid with a spacing $a = 1$ mm) provides us with the number $\mathcal{N}_\alpha^+(l)$ of events in Figure XI.26b, computed in a zone Z for the four directions of the grid. Estimate the three quantities N_A^+, Φ^+ and Ψ^+ (keep the same symbols for the estimates).
  [We have

$$N_A^+(\alpha) = \frac{1}{A(Z)} \sum_{i=1}^\infty \mathcal{N}(i,\alpha) \quad \text{and} \quad N_A^+ = \frac{1}{4}\sum_{j=0}^3 N_A^+\left(j\frac{\pi}{2}\right).$$

$$\Phi^+(i) = \frac{1}{A(Z)\cdot N_A^+} \cdot \sum_{j=0}^3 \mathcal{N}\left(i, j\frac{\pi}{2}\right)$$

and

$$\Psi^+\left(\frac{i}{8}\right) = \Phi^+(i) - \frac{i}{2}[\Phi^+(i) - \Phi^+(i-1)]$$

where the positive integer i, is a number of millimeters.]

(d) *Digitization of* N_A^- *(eight-connected graph)*
Show that the specific first contact number $N_A^-(\alpha)$ equals the sum of the following numbers of configurations encountered per unit area:

$$\left\{\begin{matrix} & 1 & \\ 1 & 0 & 1 \end{matrix}\right\}; \quad \left\{\begin{matrix} & 1 & 1 & \\ 1 & 0 & 0 & 1 \end{matrix}\right\}; \quad \left\{\begin{matrix} & 1 & 1 & 1 & \\ 1 & 0 & 0 & 0 & 1 \end{matrix}\right\}; \text{ etc.}$$

Deduce that, in any direction α, we have

$$N_A^+(\alpha) - N_A^-(\alpha + \pi) = \frac{1}{A(Z)}\left[\mathcal{N}\begin{pmatrix} 0 & 1 \\ 0 & 0 \end{pmatrix} - \mathcal{N}\begin{pmatrix} 1 & \\ 0 & 1 \end{pmatrix}\right]$$

That is just the digital version of the connectivity number N_A^2 (Ex. VI-4).

XI. Connectivity Criteria

A. INTRODUCTION

The chapter heading, Connectivity Criteria, is used here to mean those set transformations which emphasize a feature of the object associated with its connectivity. There are:

— those which provide a simplified version of the object, which exhibit the same homotopy (i.e. topologically equivalent);
— those which isolate connected components.

It turns out that they can be digitalized only within the framework of the regular model. Thus there exists one set model adapted to each large class of criteria: closed sets for increasing mappings, the regular model for connectivity, and the extended convex right for stereological measurements.

The history of connectivity criteria has, for the last twenty years, seen involvement with the homotopically suspicious and non-digitalizable Euclidean notion of the *skeleton*. Dissatisfaction with its performance has led to short cuts and variants, resulting in the homotopic thinnings, the conditional bisectors, etc. . . . *They* are the positive results of this evolution of ideas. We have devoted the first two sections (B and C) of the chapter to recounting the history of the skeleton. The reader not interested in this review can start immediately with Section D, where the first practical algorithms are presented. However, from critical examination of Sections B and C, one is able to glean the major sign posts of the chapter (digital differentiation between skeletons and median axes, between homotopy and connectivity . . .).

The research on connectivity criteria is based upon two central notions of a thinning and of a sequence of transformations (Section E). The first one is both Euclidean and digital, while the second, although basically digital, has a Euclidean meaning in the two important cases where the set to be transformed belongs to the regular model, and where the structuring element is convex. The concept of a thinning and its variants (thickening, conditional and sequential thinning) introduced here is more general than the usual meaning found in the literature. For example, A Rosenfeld (1975) has called a thinning a digital algorithm which (i) reduces a set X to arcs; (ii) does not modify its homotopy. Strictly speaking, these two conditions are incompatible, and the first one must be weakened. For us, the central notion is the anti-extensivity (the mapping removes a part of X) which is a property of transformations such as skeletons, complements of convex hulls, particle extraction etc. . . . . We have grouped them under the general formalism of a thinning, exactly as we did in Chapter II, with the erosions and the increasing mappings.

This thinning criterion firstly leads to several algorithms already known for the square lattice, such as the homotopic thinning (Sect. F) (See for example S. Castan (1978) S. Levialdi (1971), J. Prewitt (1966), A. Rosenfeld (1975), etc., etc. . . .), and secondly, opens the horizon to new criteria (conditional bisector, ultimate erosion, watersheds, various segmentations . . .). But most important of all, the approach allows one to construct a whole new class of algorithms, by combining thinnings and increasing mappings.

Here, a pedagogical problem arises. One could initiate the learner of a new language with a list of all possible sentences (here, the morphological algorithms), and ask him to learn them by heart. Alternatively, one can provide him with a shorter list, made up only of a basic vocabulary (thinnings, hit and miss transformations) and a few important idiomatic terms. We have chosen the second way. A non-exhaustive, but very basic list of idiomatic terms with regard to the morphology of connectivity is given in Sections G-2 and H. At this juncture, the role of the practical morphologist is to build up his own programs and, of course, to discover new key terms.

By principle, flow charts have not been presented in this book. Take the very simple operation of dilation by a hexagon of size k. There are at least four quite different flow charts for implementing it. One can iterate the dilation of the unit hexagon, or compose three dilations by using three segments of length k oriented at $60°$ from each other. Finally, by working with complementary sets, one could erode by either one of the above two methods. It is important to know that these methods are available, but for presentation, we prefer a more symbolic style, where the flow charts are replaced by set operations. Indeed the sequential formalism defined by relations (XI-27) and (XI-28) corresponds *exactly* to macro-instructions; in addition, it clearly separates the operation itself from the structuring element.

We really treat in detail only the hexagonal digitization, i.e. the simpler and morphologically richer one, and let the reader transpose the algorithms to the square grid: it is *always* easy (a few indications are given in the exercises).

B. THE SKELETON AND MAXIMUM DISKS IN R^n

As we can see on Figure XI.1, the notion of a skeleton of a set X is a very intuitive one. Obviously:

(i) if x is a point of the skeleton, and B_x the largest disk centred at x and contained in X, one cannot find a larger disk (not necessarily centred at x) containing B_x and included in X (one says that B_x is a *maximum disk*);

(ii) the disk B_x hits the boundary ∂X at two or more different places.

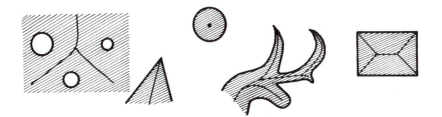

Figure XI.1. Examples of skeletons.

Historically, these two approaches come respectively from Th. Motzkin (1935) and H. Blum (1962). Both can be formulated into a precise definition, they are *almost* identical, but they do not induce the same morphological methods for obtaining them. Therefore, we treat them individually (Sect. B and C). In both cases, we start from the same extensive class of sets, and afterwards improve the morphological properties by specifying certain "good" sub-classes. The main results of Sections B and C are due to G. Matheron (1978) and L. Calabi (1965).

B.1. General properties of the skeleton

The largest class of sets we consider in this book is $\mathscr{F}(\mathbb{R}^n)$ (the closed sets), or equally $\mathscr{G}(\mathbb{R}^n)$ (the open sets), since $X \in \mathscr{F} \Leftrightarrow X^c \in \mathscr{G}$. The property (i), based on inclusions, suggests the use of $\mathscr{G}$ rather than $\mathscr{F}$, and in so doing, we eliminate the sets with zero thickness (such as lines) which are their own skeletons. For the sake of simplicity, we also assume that $X \neq \varnothing$ and that the maximum disks are bounded, i.e. that X contains no half space, or in other words the convex hull $C(X^c)$ is $\mathbb{R}^n$:

$$C(X^c) = \mathbb{R}^n \tag{XI-1}$$

we denote by $\mathscr{G}'(\mathbb{R}^n)$ the corresponding class of X's.

Definition XI-1: Suppose $X \in \mathcal{G}'$. Then the skeleton of X, denoted $S(X)$, is the set of the centres of the maximum open balls $\rho \mathring{B}$ contained in X. Similarly, we denote as $s_\rho(X)$ the set of centres of the maximum open balls of radius $\rho > 0$ ($\mathring{B}$ and $\overline{B}$ are respectively the open and the closed unit ball).

Such maximum balls do in fact exist, since every limit of open balls contained in X is an open ball. Then by Zorn's lemma, every open ball contained in X is also contained in a maximum open ball.

The $s_\rho(X)$ generate a *partition* of the skeleton $S(X)$, since two concentric balls having different radii ρ and ρ' cannot both be maximizing; thus:

$$S(X) = \bigcup_{\rho > 0} s_\rho(X) \tag{XI-2}$$

The following important theorem, due to Ch. Lantuéjoul (1977), allows $S(X)$ to be expressed in terms of erosions and openings of X.

THEOREM XI-1. *The skeleton $S(X)$ is characterized by the relationships*:

$$S(X) = \bigcup_{\rho > 0} \bigcap_{\mu > 0} [(X \ominus \rho\mathring{B})/(X \ominus \rho\mathring{B})_{\mu\overline{B}}] \tag{XI-3}$$

where $\rho\mathring{B}$ is the open ball of radius ρ.

Proof: From the following three equivalences:

$$\begin{cases} \rho\mathring{B}_x \subset X \Leftrightarrow x \in X \ominus \rho\mathring{B} & (X \ominus \rho\mathring{B} \text{ is } \textit{closed}, \text{ see Ex. VI-4}) \\ (\rho + \alpha)\mathring{B}_y \subset X \Leftrightarrow \alpha\overline{B}_y \subset X \ominus \rho\mathring{B} \\ \rho\mathring{B}_x \subset (\rho + \alpha)\mathring{B}_y \Leftrightarrow x \in \alpha\overline{B}_y \end{cases} \tag{XI-4}$$

we see that the open ball $\rho\,\mathring{B}_x$ is maximum in X if and only if the point x itself is a maximum *closed* ball contained in the closed set $X \ominus \rho\mathring{B}$, i.e. if x belongs to $X \ominus \rho\mathring{B}$, but does not belong to any of the morphological openings of $X \ominus \rho\mathring{B}$ by the closed balls $\mu\overline{B}$. Hence relation (XI-3). In practice, one often writes relation (XI-3) in the less rigourous, but simpler following form:

$$S(X) = \bigcup_{\rho > 0} s_\rho(X) = \bigcup_{\rho > 0} [(X \ominus \rho\mathring{B})/(X \ominus \rho\mathring{B})_{d\rho\overline{B}}] \tag{XI-5}$$

The meaning of relation (XI-5) is clear. The part of each eroded set $X \ominus \rho\mathring{B}$ which belongs to the skeleton is made up of the angular points and the lines without thickness of $X \ominus \rho\mathring{B}$, parts which disappear in the difference $(X \ominus \rho\mathring{B})$ $/(X \rho B)_{d\rho\overline{B}}$ (see Fig. XI.2a). Now, starting from the skeleton $S(X)$ it is possible to recover X, since X is the union of the maximum open balls that it contains; so we have:

$$X = \bigcup_{\rho > 0} s_\rho(X) \oplus \rho\mathring{B} \tag{XI-6}$$

Figure XI.2. (a) The skeleton is the union of angular points and of lines of the eroded sets. (b) The skeleton is not closed. The limit point B of the triangles does not induce a limit dendrite on the skeleton.

In other words, the datum of set X is equivalent to that of its skeleton $S(X)$, together with the maximum radius ρ associated with each point of $S(X)$. This maximum radius function is the restriction to $S(X)$ of the *quench function* (Calabi's denomination) defined over $\mathbb{R}^n$ (see Sect. C.1).

B.2. Erosion, dilation, opening, and the skeleton

If we apply the system of equivalences (XI-4) to the two open sets X and $X \ominus \rho_0 \overline{B}$, we see that $s_\rho(X \ominus \rho_0 \overline{B}) = s_{\rho + \rho_0}(X)$. Thus, the skeleton of $X \ominus \rho_0 \overline{B}$ is:

$$S(X \ominus \rho_0 \overline{B}) = \bigcup_{\rho > \rho_0} s_\rho(X) \tag{XI-7}$$

The expression of $X \ominus \rho \overline{B}$ itself immediately results and we have:

$$X \ominus \rho_0 \overline{B} = \bigcup_{\rho > \rho_0} s_\rho(X) \oplus (\rho - \rho_0)\mathring{B}$$

i.e. the skeleton of the eroded set $X \ominus \rho_0 \overline{B}$ is the restriction of $S(X)$ to the points labelled by a quench function with $\rho > \rho_0$. The set $X \ominus \rho_0 \overline{B}$ is obtained by replacing each maximum disk of X of radius $\rho > \rho_0$ by the concentric disk of radius $\rho - \rho_0$.

In order to calculate the open dilate $X \oplus \rho_0 \mathring{B}$, we can start from relation (XI-6), which implies

$$X \oplus \rho_0 \mathring{B} = \bigcup_{\rho > 0} s_\rho(X) \oplus (\rho + \rho_0)\mathring{B} \tag{XI-8}$$

Therefore, the skeleton of $X \oplus \rho_0 \mathring{B}$ is the same as the skeleton of X, and the associated quench function is that of X plus the constant ρ_0. Similarly, for the morphological opening $X_{\rho_0 \overline{B}} = (X \ominus \rho_0 \overline{B}) \oplus \rho_0 \overline{B} = (X \ominus \rho_0 \overline{B}) \oplus \rho_0 \mathring{B}$, of X by the closed ball $\rho_0 \overline{B}$, we find:

$$X_{\rho_0 \overline{B}} = \bigcup_{\rho > \rho_0} s_\rho(X) \oplus \rho \mathring{B}$$

So to obtain $X_{\rho_0 \overline{B}}$ we perform a synthesis of X, but only for the values of ρ larger than ρ_0. Note that in the three cases of erosion, dilation and opening we always use the same skeleton, but change the quench function (with the convention that a point of $S(X)$ negatively labelled is removed). This remark suggests the set transformation that is developed in Exercise XI-8.

B.3. Topological properties of the skeleton $S(X)$

Firstly, note that the skeleton $S(X)$ is *not* necessarily closed (Fig. XI.2b). Secondly, we can describe the degree of fineness of the skeleton by successively asking the following four questions:

is the interior of $S(X)$ empty?
is $S(X)$ negligible with respect to Lebesgue measure?
is the interior of the adherence (topological closure) $\overline{S(X)}$ empty?
is $\overline{S(X)}$ negligible with respect to Lebesgue measure?

The answer to the first question is yes, and to the last one, no (counter-examples are possible). The two intermediary questions are as yet unresolved.

Let us turn now to connectivity and continuity, and state the following three theorems:

THEOREM XI-2. *If $X \in \mathcal{G}'$ is connected, then the adherence $\overline{S(X)}$ of its skeleton is also connected. (Matheron, 1978).* (How to know whether $S(X)$ is connected is unresolved.)

In spite of the rather disappointing connectivity properties of $S(X)$, L. Calabi (1965) succeeded in proving the following beautiful result concerning the partition of the space due to the skeleton:

THEOREM XI-3. *In $\mathbb{R}^2$, every connected component Y_i of $[S(X)]^c$ contains one and only one connected component X_i^c of X^c, and every X_i^c can be obtained in this way. Moreover, if $X_i^c \subset Y_i$ the two are of the same homotopy type (see Fig. XI.4c).*

THEOREM XI-4. *The mapping $X \to \overline{S(X)}$ is a lower semi-continuous mapping of $\mathcal{G}'$ onto $\mathcal{F}$. (More simply: every point x of $\overline{S(X)}$ is limit of a sequence of points $\{x_i\}$, with $x_i \in S(X_i)$). Similarly, for every $\rho > 0$, the mapping $X \to S(X \ominus \rho \overline{B})$ is l.s.c. (Matheron, 1978).*

Note that the mapping $X \to \overline{S(X)}$ is not u.s.c. As a counter-example, take for X_i a regular polygon with i sides, and inscribed in the unit disk $\mathring{B}$ in $\mathbb{R}^2$. The skeleton is made up of i radii. When $i \to \infty$, X_i tends towards $\mathring{B}$, and its skeleton also tends towards $\mathring{B}$, although the skeleton $S(\mathring{B})$ of the limit is reduced to the centre of $\mathring{B}$.

Note also that if we had adopted an approach based on closed sets, Theorems XI-3 and XI-4 would have no equivalent. Indeed, define the skeleton $S'(X)$ of $X \in \mathcal{X}'$ to be the set of the centres of the maximum closed balls included in X, then the counter-example of Figure XI-3a shows that the mapping $X \to S'(X)$ is neither upper nor lower semi-continuous, and that of Figure XI.3b shows the lack of connectivity. However, the skeletons $S'(\overline{X})$ and $S(\mathring{X})$ are extremely similar since we have $S(\mathring{X}) = S'(\overline{X}) \cap \mathring{X}$ (one goes from one skeleton to the other by removing its boundary points).

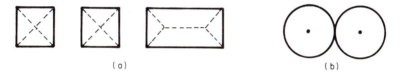

Figure XI.3. (a) X is the union of two closed squares of side a, distance h apart. The skeleton is made up of the four diagonals. For $h = 0$, the two squares become a single rectangle whose skeleton does not contain, nor is included in, the limit of the diagonals. (b) non-connectivity of $S(X)$ for X a connected closed set.

B.4. Exoskeleton and convexity

We have studied the skeleton of X. What about that of the background of X^c? Obviously, if X satisfies relation (XI-1), X^c does not. Therefore, to study $S(X^c)$, we must remove the condition (XI-1) and work on the space $\mathcal{G}(R^n)$ itself. Then, the three topological theorems remain true, but other results must be

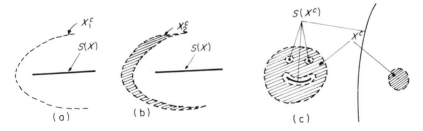

Figure XI.4. (a) and (b) X_1 and X_2 have the same skeleton, and the same quench function on it. (c) The skeleton of X partitions the space into as many connected components as there are in X^c.

completely modified. In particular, the datum of $S(X)$ and the quench function $\rho(x)$ on $S(X)$ no longer determine X (see Fig. XI.4a and b). Indeed, X_1 and X_2 are identical if and only if they have the same skeleton, the same quench function on it, and if X_1^c and X_2^c have the *same convex hull*. (This third condition was trivially fulfilled for the X's satisfying relation (XI-1)). Now consider a set $X \in \mathscr{G}$ such that X^c is convex. Then, and only then, is the skeleton of X in $\mathbb{R}^n$ the empty set $\varnothing$ (Motzkin's theorem).

B.5. The regular model and connectivity

We have gone about as far as we can with an analysis based on the very broad class $\mathscr{G}'$. The main drawback we encountered was the non closure of $S(X)$. Indeed, if we assume that $S(X)$ is topologically closed, the following two strong results, due to L. Calabi (1965), are obtained in the 2-D case:

THEOREM XI-5. *Assuming $S(X)$ is closed, every point $x \in S(X)$ possesses arbitrarily small neighbourhoods which are connected, locally connected, and without loops.*

THEOREM XI-6. *Assuming $S(X)$ is closed, every bounded connected component X_i of X contains one and only one connected component S_i of $S(X)$, and every bounded component of $S(X)$ is so obtained. Moreover, if $S_i \subset X_i$, then the two are homotopically equivalent.*

Calabi's Theorem XI-6 is the central result vis a vis connectivity, but there still remains the question: for which X's is the skeleton a closed set? J. A. Riley (1965) succeeded in proving in $\mathbb{R}^2$ the following result:

THEOREM XI-7. *If ∂X is a locally finite union of curves C_i which meet, if at all, at their end points, and which possess continuously differentiable curvatures, and locally finite critical points for the curvature, then the skeleton $S(\mathring{X})$ is a closed set.*

It is easy to see that for X belonging to the regular model $\mathscr{R}(\mathscr{K})$, in $\mathbb{R}^2$, such conditions are fulfilled. Anyway, the digitalization constraint with regards to homotopy (Theorem VII-2) demands that $\overline{X} \in \mathscr{R}(\mathscr{K})$, so it is wise to restrict ourselves to the $\{\mathring{X}, X \in \mathscr{R}(\mathscr{K})\}$ if we want to use skeletons for homotopy characterization. Then Theorems XI-6 and XI-7 are satisfied (but the question of the largest class such that $S(X) = \overline{S}(X)$, remains open).

C. MEDIAL AXIS AND DERIVATIONS OF THE SKELETON

We now go back to the general class $\mathscr{G}'$, and we approach the notion of a "skeleton" using H. Blum's grass fires.

C.1. Parametrization of X using erosion

Let X be a given set of the class $\mathscr{G}'$. With each point $x \in \mathbb{R}^n$, we associate the non-negative *quench function* $\rho(x; X)$ (or briefly $\rho(x)$):

$$\left.\begin{array}{ll} \rho(x; X) = \text{Sup}\{\rho : \rho\mathring{B}_x \subset X\} & \text{if} \quad x \in X \\ \rho(x; X) = 0 & \text{if} \quad x \notin X \end{array}\right\} \tag{XI-9}$$

ρ is just the distance from the point x to the complementary set X^c, in the sense of Exercise III-2, i.e.:

$$\rho(x; X) = d(x, X^c) = \inf_{y \in X^c} d(x, y)$$

We have the equivalence $\rho_1 \overline{B}_{x_1} \subset \rho_2 \overline{B}_{x_2} \Leftrightarrow \rho_2 - \rho_1 > d(x_1, x_2)$. Moreover, the continuity of the norm implies:

$$|\rho(x) - \rho(y)| \le d(x, y) \qquad x, y \in \mathbb{R}^n \tag{VI-10}$$

We also see that, for any $x \in X$, the relation

$$\rho(y) - \rho(x) < d(x, y) \qquad \forall y \in \mathbb{R}^n, y \ne x \tag{XI-11}$$

simply means that the open ball $\rho(x)B_x$ is maximum in X. Therefore relation (XI-11) *characterizes* the points x of the skeleton $S(X)$. Consider now a point $x \in X$ with parameter $\rho(x)$. Since $\rho(x)\mathring{B}_x$ is included in X, but not $\rho(x)\overline{B}_x$, then ∂X and $\partial \mathring{B}_x$ have at least one common point x_0 (see Fig. XI.5). For every point y belonging to the segment $[x_0, x]$, the open ball of centre y and of radius $d(x_0, y) = \rho(x_0) - d(x_0, y)$ is the largest one centred at y and included in X. In other words

$$\rho(y) = \rho(x) - d(x, y) \tag{XI-12}$$

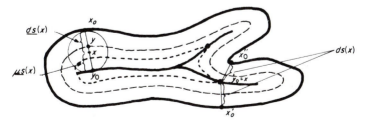

Figure XI.5. Parametrization of $\mathbb{R}^2$ from the family of the eroded sets $\{X \ominus \rho\mathring{B}\}$.

Conversely, is relation (XI-12) is satisfied for $y \neq x$, the half straight line $\{x + \lambda(y - x), \lambda \geq 0\}$ hits ∂X at the point $x_0 \in \rho(x)\overline{B}_x$ and the relation (XI-12) is satisfied for every point $y' \in [x, x_0]$ (if not we would have $\rho(y)B(y) \subset X$). We shall call *downstream* of x all the points y satisfying relation (XI-12), and denote it by $ds(x)$.

Starting from the point x, we call *upstream* of x, i.e. $us(x)$, the closed set of points y satisfying the relationship:

$$\rho(y) = \rho(x) + d(x, y) \qquad x \in X \qquad \text{(XI-13)}$$

If x belongs to the skeleton $S(X)$, $us(x)$ is just x itself, and conversely, because of relation (XI-11). If $x \notin S(X)$, then $ds(x)$ is the segment $[x_0, x]$ and $us(x)$ is the segment $[x, y_0]$ which lengthens $[x_0, x]$, and whose extremity y_0 is the centre of the unique maximum ball containing $\rho(x)\mathring{B}_x$. This remark suggests a definition of the edge of x ($ed(x)$) to be the union of the upstream and the downstream of x:

$$ed(x) = ds(x) \cup us(x)$$

As we just saw, if $x \in X - S(X)$, $ed(x)$ is the unique closed segment in X, going from x_0 to y_0 via x; along the edge, the function $\rho(y)$ increases from x_0 to y_0 with a constant slope equal to one. Concerning the behaviour of the quench function ρ in X, notice that relation (XI-10) implies that $\rho(x; X)$ is continuous in X. G. Matheron (1978) proved the following stronger result:

THEOREM XI-8. *The function* $\rho(x; X)$ *is differentiable at every point* x *of* X *external to the adherence of the skeleton* $(x \in X/\overline{S}(X))$, *and grad* $\rho(x)$ *is the unit vector of* $ed(x)$, *us-oriented.*

C.2. Definition of the medial axis

The downstream set associated with $x \in X$, is non-empty and closed in X (rel. XI-12 is still valid in the limit, since ρ and d are continuous functions). Therefore $ds(x)$ is made up of one or more segments of line starting from x. We say that the point $x \in X$ is a *single point* when $ds(x)$ is a single segment, and that x is a *multiple point* otherwise, and call *medial axis* $Me(X)$ the set of multiple points of X. The reader probably knows the classical, and very suggestive interpretation of the medial axis in terms of a "grass fire".

Think of X as a meadow. At time $t = 0$ a fire is lit all along ∂X. The fire invades X with unit speed; for example, in Figure XI-6, point m will be reached by the fire initiated at m_0 after a time $t = |m_0 - m|$, where m_0 is the nearest point of ∂X to m. The downstream of m is the segment $[m_0, m]$. But some other points, such as p, are reached at the same time, by more than one fire front. The downstream of p is $ds(p) = [p_0, p] \cup [p_1, p]$ and p belongs to the medial axis.

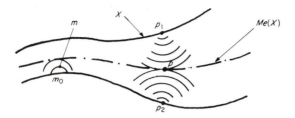

Figure XI.6. The medial axis of X is the set of points of X simultaneously reached by grass fires initiated from at least two different points of ∂X.

For $X \in \mathcal{G}'$, it is not true that the medial axis $Me(X)$ is identical to the skeleton $S(X)$. For example, in Figure XII.4a, the end point of a skeleton (a half straight line) belongs to $S(X)$ but not to $Me(X)$. In fact $Me(X)$ is always contained in $S(X)$. Assume that the ball $\rho_x \overline{B}$ hits the boundary of X at two distinct points x_0 and x_1. Then, any open ball containing $\rho_x \mathring{B}$ is centred at a point y belonging to both half lines $\{x_0 + \lambda(x - x_0); \lambda \geq 0\}$ and $\{x_1 + \lambda(x - x_1); \lambda \geq 0\}$. Therefore $y = x$ and $\rho_x \mathring{B}$ is maximum; i.e. $x \in X$ and so

$$Me(X) \subset S(X) \qquad\qquad (XI\text{-}13)$$

These two sets are extremely close to each other, since $\overline{Me}(X) = \overline{S}(x)$ (Matheron, 1978). Moreover, in $\mathbb{R}^2$, $Me(X)$ is negligible with regard to Lebesgue measure. In conclusion, the medial axis offers little new, but rather gives a different way of looking at the skeleton. There are three interesting restrictions of the skeleton that we discuss now.

C.3. The conditional medial axis

This first restriction of $Me(X)$ is extremely simple. Instead of initiating the grass fire from all the points of the boundary ∂X, choose a sub part $\partial' X \subset \partial X$, and consider the set $Me'(X)$ of points of X reached by more than one front. Obviously this conditional medial axis $Me'(X)$ is included in $Me(X)$, and it is able to exhibit certain features of X better than $Me(X)$. F. Meyer (1979) illustrates this point by taking three cytoplasms of cells without their nuclei (Fig. XI.7), and for the $\partial' X_1$, their outside boundaries.

C.4. The conditional bisector

Subsets of the medial axis can be defined using parametrizations. For example, we have already seen that the skeleton of $X \ominus \rho_0 B$ corresponds to the part of the skeleton $S(X)$ where the quench function $\rho(X)$ is higher than ρ_0. Another example is given by the conditional bisector. This notion appears for the first

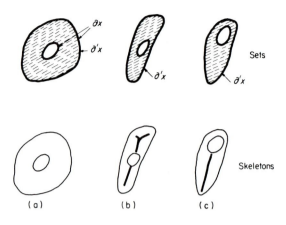

Figure XI.7. Conditional medial axes with respect to the outside boundaries $\partial'X$. They reveal both the shape of the cytoplams and the location of the nuclei (the conditional medial axis is empty in (a), disconnected in (b) and connected in (c)).

time in H. Blum (1971) and has been formalized by F. Meyer (1979) as follows in $\mathbb{R}^2$:

$$S^{\alpha}(X) = \left\{ x\text{:tg} \cdot \lim_{t \to 0} \frac{d[x, X \ominus (\rho(x) + t)B]}{t} > \alpha \right\}$$

Morphologically speaking, α is the speed at which one leaves a point of the medial axis to go to the next one, as the grass fire progresses (Fig. XI.8), and $S^{\alpha}(X)$ is the subset of the medial axis for which this speed is higher than α. Indeed, the largest class of sets for which the above definition is valid is not delineated yet. In practice, we will consider $S(\mathring{X})$ as a finite union of simple arcs; along each of them, we assume that the inverse quench function $\rho^{-1}(x)$ possesses a derivative. Then $S(\mathring{X})$ is closed, and $S(\mathring{X}) = \text{Me}(\mathring{X})$, except for a finite number of end points that we disregard. Then $S^{\alpha}(X)$ is the subset of $S(X)$

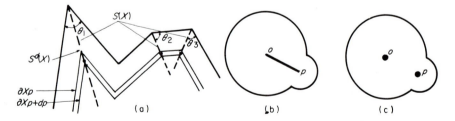

Figure XI.8. (a) conditional bisector of an angular figure. For $\alpha = 2$ the second and third bisectors are suppressed. (b), (c) A small bubble on the contour generates a parasite bone on the skeleton (b), but not on the conditional bisector (c). The two centres always remain because they correspond to an infinite speed.

where $(\rho^{-1})' > \alpha \geq 1$ (derivative w.r. to the curviline co-ordinate of the arc). For $\alpha = 1$ one recovers the whole skeleton $S(X)$. Notice that if set X is a dihedron (Fig. XI.8a) of angle $2\theta\,(0 \leq \theta \leq \pi/2)$, its conditional bisector equals its bisector for $\alpha > 1/\sin\theta$ and is the empty set for $\alpha \leq 1/\sin\theta$. Therefore, one can use it for measuring the angles in two dimensions. The conditional bisector is a rather robust transformation. Indeed it is invariant under similarities (since it depends on a parameter α of dimension zero!), and fairly insensitive to slight modifications of the contour of X (Fig. 8b and c). The class of sets for which the quench function is differentiable over the median axis is not however theoretically known, but it must be rather regular. In practice, F. Meyer (1979) has used the notion mainly in quantitative cytology, where it seems to be a very effective tool.

To find an algorithm which gives the conditional bisector, we must assume that $\overline{X}$ belongs to the regular model of parameter r. Start from the eroded set $X \ominus \rho\mathring{B}$. Interpret ρ as a time, and consider the next instant $\rho + d\rho$ with $d\rho < r$, when the grass fire arrives at $X \ominus (\rho + d\rho)\mathring{B}$ (Fig. XI.9a). The points of $X \ominus \rho B$ which have a speed greater than 1 are those that cannot be reached after a dilation of $X \ominus (\rho + d\rho)\mathring{B}$ by $d\rho \cdot \overline{B}$. They make up that part of the skeleton with a quench function equal to ρ (this interpretation is the direct illustration of algorithm (XI-5)). Similarly, the points of the α-conditional bisector are those which are not accessible by a dilation of $X \ominus (\rho + d\rho)\mathring{B}$ by $\alpha d\rho\overline{B}$. Summing over all the values of ρ, we get

$$S^{\alpha}(X) = \bigcup_{\rho \geq 0} \left[(X \ominus \rho\mathring{B})/(X \ominus \rho\mathring{B})_{d\rho\overline{B}} \oplus (\alpha - 1)d\rho\overline{B} \right] \qquad \text{(XI-14)}$$

for which relation (XI-5) is a particular case.

Figure XI.9. (a) Two elementary steps of the grass fire, (b) points with a fire speed greater than 1 (i.e. skeleton), (c) points with a speed greater than α (i.e. α = conditional bisector).

C.5. The skiz

In 1966, J. Prewitt used the skeleton of the background of isolated particles (called by her an "exoskeleton") in order to characterize their *zones of influence*. This is done by partitioning the space into cells with one particle in each (Fig. XI.10a). Going further in the same way leads to the notion of a *skiz* (skeleton

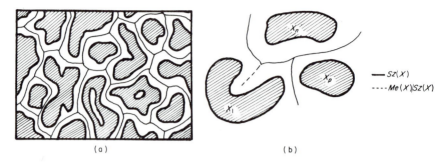

Figure XI.10. (a) example of skiz. (b) difference between skiz and medial axis. The skiz eliminates the bones of Me(X).

by zones of influence), due to Ch. Lantuéjoul (1978). The skiz is a subset of the median axis defined as follows: Suppose the background of X is made up of a locally finite union of N disjoint closed sets K_n:

$$X^c = \bigcup_{n=1}^{N} K_n; \qquad n \neq p \Rightarrow X_n \cap X_p = \varnothing \qquad (XI\text{-}15)$$

Let $\mathcal{G}''$ be the class of sets X satisfying this relation (XI-15). Obviously $\mathcal{G}'' \subset \mathcal{G}'$. We thus restrict the general class of sets $\mathcal{G}'$ to a subclass which is better adapted to the purpose we have in mind. Let $X \in \mathcal{G}''$, and consider a point $x \in \text{Me}(X)$ with quench function $\rho(x)$. The maximal ball $\rho(x)\overline{B}_x$, included in $\overline{X}$, hits X^c at more than one point. By definition, we say that x belongs to the skiz $Sz(X)$ of X if and only if the contact zone $\rho \overline{B}_x \cap X^c$ belongs to more than one of the K_n's (Fig. XI.10b):

$$Sz(X) = \{x \in \text{Me}(X); \; \exists n \neq p : \rho(x)\overline{B}_x \cap K_n \neq \rho(X)\overline{B}_x \cap K_p \neq \varnothing\} \quad (XI\text{-}16)$$

In other words, we partition the space into the skiz, and into N cells $T(K_n)$ associated with every K_n; $T(K_n)$ is the *zone of influence* of K_n, i.e. the set of the points closer to K_n than to any other $K_p (p \neq n)$:

$$T(K_n) = \{x : d(x, K_n) < d(x, K_p), p \neq n\} \qquad (XI\text{-}17)$$

with

$$\bigcup_{1}^{N} T(K_n) = [Sz(X)]^c \qquad (XI\text{-}18)$$

Amongst all the derivatives of the skeleton, the skiz is the one which exhibits the most interesting topological properties. Indeed:

THEOREM XI-8. *For $X \in \mathcal{G}''$, the skiz of X is a closed set.*

The distance d is continuous; therefore, according to (XI-17), the $T(K_n)$'s are open, and according to (XI-18), $Sz(X)$ is closed.

THEOREM XI-9. (*Maisonneuve, 1977, in Ch. Lantuéjoul, 1978*). *For* $X \in \mathcal{G}''$, *the skiz of* X *is a locally finite union of simple arcs.*

THEOREM XI-10. (*Ch. Lantuéjoul, 1978*). *Let* $\{X_i^c\}$ *be a sequence in* $\mathcal{G}''$ *where each element is the union of N disjoint closed terms* $K_{n, i}$, *and* $K_{n, i}$ *converges towards a* $K_n (K_{n, i} \to K_n$ *and* $n \neq p \Rightarrow K_n \cap K_p = \varnothing$). *The skiz is a continuous mapping of* $\mathcal{G}''$ *into* $\mathcal{F} : Sz(X_i) \to Sz(X)$, *where* $X^c = \bigcup_{n=1}^{N} K_n$.

C.6. Are the skeletons morphological mappings?

All the skeletons studied above are invariant under displacements and under scale change, and are semi-continuous. But unfortunately, the local knowledge condition is not always satisfied. A small angle of X, located extremely far away from the actual measuring mask Z, induces a bone of the skeleton which may be very long and appear inside Z. We can only have access to the points of $S(X)$ (and of $\text{Me}(X)$, $S^\theta(X)$, $Sz(X) \dots$) whose quench value ρ is small enough, and such that $Z \ominus \rho B \neq \varnothing$. For these ρ's, we know from relations (XI-2) and (XI-3) that $s_\rho(X)$ is included in $X \ominus \rho B$, i.e. lies in $Z \ominus \rho B$. In practice, this condition on ρ is interpreted in two different ways:

(i) X is bounded and contained in Z (cytology for example).
(ii) X is not necessarily bounded but $X \ominus \rho B$ is empty for ρ greater than a fixed value ρ_0 and $Z \ominus \rho_0 B \neq \varnothing$. Figure XI.10a shows an example of a set exhibiting such a maximal thickness ρ_0.

D. DIGITALIZATION OF THE SKELETON

D.1. Troubles when changing the distance

In order to transpose the skeleton into digital terms, we must firstly stay with $\mathbb{R}^2$, but replace the shape of the disk by that of the square, or the hexagon, or the octagon. In fact, we know from Exercise III-1 that the metrics based upon these different shapes lead to equivalent topologies. Of course skeletons associated with maximal hexagons will be different from those associated with maximal disks (the former depend on the direction chosen), but according to Exercise III-1, they exhibit the same topological properties, and the properties studied above (semi-continuity, connectivity . . .) are topological. The second step of digitalization is then to approximate the continuous similarities of the square or the hexagon, by using a grid of points.

Let us examine this more closely, on the example of the square distance defined in $\mathbb{R}^2$ by

$$d(x, x') = \max(|x_1 - x_1'|, |x_2 - x_2'|) \, ((x_1, x_2) \text{ are the co-ordinates of } x).$$
$$\text{(XI-19)}$$

The digital version of this is given by relation (VI-32). As a first example, take for X the complement of a segment in $\mathbb{R}^2$ (Fig. XI.11a); $S(X)$ is empty (there is no centre of a maximal square at a finite distance), but $\mathrm{Me}(X)$, which usually should be contained in $S(X)$, is the dashed portion of the space. Not only is its area non-zero, but it is infinite. This "skeleton" then is quite fat! The same remarks can be made with regard to the square skiz in Figure XI.11b. All the points of the dashed zone are at the same distance (measured by (XI-19)) from the segments X_1 and X_2. Finally, we see a connected set, namely the open rectangle minus the two median segments, whose skeleton is nevertheless made up of two disjoint parts. The justification of the skeleton (as a picture of the homotopy having a zero thickness) completely breaks down! What has happened?

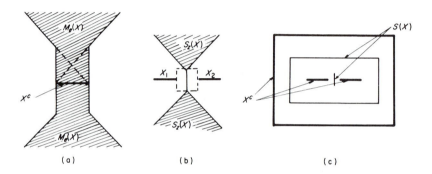

(a)   (b)   (c)

Figure XI.11. Strange properties of the square skeleton. $\mathrm{Me}(X)$ in (a) and $Sz(X)$ in (b) are not included in $S(X)$, and their areas are infinite. In (c), $S(X)$ is disconnected, although X is connected and bounded.

In the proof of relation (XI-13) we assumed that two normals, drawn from two different points of B crossed each other. If ∂B has flat portions, such as the sides of a square, the assumption fails, and $M(X)$ is no longer contained in $S(X)$. Then all the properties involving this relation (XI-13) become invalid (in particular all the topological theorems, apart from those involving lower semi-continuity; Th. XI-4). The properties that we lose are not exclusively topological, but partly metric; we must demand that the curvature of ∂B be everywhere strictly positive. As a (poor) compensation, the contents of sections B1 and B2 remain true and are hopefully digitalizable. For the square (or octagonal or hexagonal) distance, the median axis $\mathrm{Me}(X)$ becomes a concept radically different from $S(X)$, and is no longer an "axis", since if the distance between x_1 and x_2 is unique, there is not a unique minimum path for joining x_1 to x_2, but an infinity of such paths.

D.2. Digital algorithms for the skeleton

According to the preceding discussion, we must restrict ourselves to the digital versions of $S(X)$ involving erosions and openings, and not digitalize the median "axis" associated with a polygonal distance. For example, for the hexagonal grid with smallest hexagon H, the relationships

$$s_n(X) = (X \ominus nH)/(X \ominus nH)_H \quad \text{and}$$

$$S(X) = \bigcup_n s_n(X) \qquad (n = 0, 1, 2, \ldots) \qquad \text{(XI-20)}$$

directly transpose relations (XI-5) and (XI-2). And for the reconstruction of X from $S(X)$ and the quench function, relation (XI-6) becomes:

$$X = \bigcup_n (s_n(X) \oplus nH) \qquad \text{(XI-21)}$$

where X is a set of points in $\mathbb{Z}^2$ such that the convex hull $C(X^c)$ of X^c equals $\mathbb{Z}^2$ itself. Figure XI.12 illustrates, step by step, the implementation of such a skeleton.

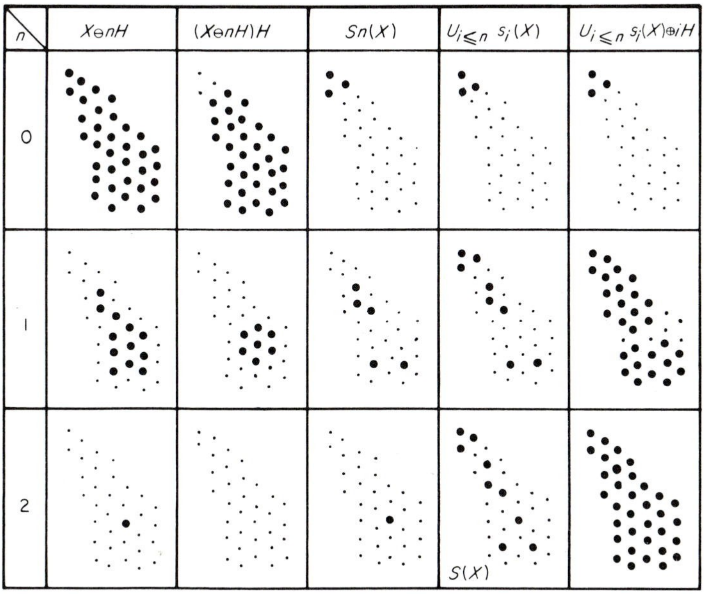

Figure XI.12. Hexagonal skeleton: step by step implementation of relation XI-20 and XI-21.

Similarly the digital conditional bisector is given by the following algorithm, which corresponds to relation (XI-14):

$$S^{\theta}(X) = \bigcup_{n \geq 0} \left[(X \ominus nH)/(X \ominus nH)_H \oplus (\theta - 1)H \right] \qquad \theta \quad \text{integer} \geq 1)$$

$$\text{(XI-22)}$$

Are the skeleton and conditional bisector digitalizable mappings? Let $X \in \mathcal{G}'$ in $\mathbb{R}^2$. Use for X^c the covering representation (Ch. VI, C.1b). Then the complement $\mathcal{T}^c(X, k)$ in $\mathbb{R}^2$ of $\mathcal{T}(X, k)$ defined by relation (VI-45) are increasing open sets which tend towards X as $k \to \infty$. Consider the hexagonal skeleton of the digital version $\mathcal{S}(\mathcal{T}^c(X, k))$. It is made up of a certain number of points $\hat{S}(X, k)$ of $\mathbb{Z}^2$ which can also be considered as points of $\mathbb{R}^2$. When $k \to \infty$, the lower semi-continuity of the skeleton implies that:

$$\mathcal{T}^c(X, k) \uparrow X \quad \text{and} \quad \hat{S}(X, k) \to \hat{S}(X, \infty) \subset S(X) \qquad \text{(XI-23)}$$

In words, when the grid becomes infinitely fine, the limit of the digital skeletons is included in the Euclidean hexagonal skeleton of the limit set X. The result has limited applications, and says nothing about connectivity: anyway, the Euclidean skeleton $S(X)$ does not itself have good connectivity properties. . . . However, note that the reconstruction algorithm (XI-6) is digitalizable by using relation (XI-21). Finally, the digital skeletons are mainly convenient for coding images, owing to relation (XI-21). If we really want to construct homotopic "skeletons" it seems preferable to approach them as *thinning.*

E. THINNINGS AND THICKENINGS

E.1. Definition and properties

We now change our approach to that of digital "skeletons" via *homotopic thinnings*, which is a particular case of the general transformation of a *sequential thinning.* Consider the set $X \in \mathcal{P}(\mathbb{R}^n)$ and the hit or miss transform $X \circledast T$ by the compact structuring element $T = (T_1, T_2)$, i.e.:

$$X \circledast T = (X \ominus \check{T}_1)/(X \oplus \check{T}_2) = (X \ominus \check{T}_1) \cap (X^c \ominus \check{T}_2) \qquad T_1, T_2 \in \mathcal{K} \qquad \text{(XI-24)}$$

By definition, we *thin* X by T when we subtract $X \circledast T$ from X, and we *thicken* X by T when we add $X \circledast T$ to X: denoting these two transformations by $X \bigcirc T$ and $X \odot T$ respectively, we can write:

$$X \bigcirc T = X/X \circledast T \qquad X \odot T = X \cup (X \circledast T) \qquad \text{(XI-25)}$$

Obviously, these operations are of no interest if T_1 and T_2 are not disjoint. In particular the origin cannot belong to both T_1 and T_2 (Fig. XI.13). Therefore, the erosion and the dilation involved in (XI-24) are not both centred, but shifted w.r. to each other (see the list of T's in Fig. XI-14 for example). If T_1, or T_2, is the empty set, $X \circledast T$ is reduced to an erosion, since $X \ominus \varnothing = \mathbb{R}^2, \forall X$. As an example, take T_1 equal to the origin $(T_1 = \{0\})$, and $T_2 = \varnothing$, then $X \circledast T = \mathring{X}$ and $X \bigcirc T = \partial X$. The topological status of the thinnings and the thickenings will be made precise by putting $X \circledast T = (\mathring{X} \ominus T_1)/(\overline{X} \oplus T_2)$. Then:

$$X \in \mathscr{F}(\mathbb{R}^n); \quad T_1, T_2 \in \mathscr{K}(\mathbb{R}^n) \Rightarrow X \bigcirc T \in \mathscr{F}(\mathbb{R}^n)$$
$$X \in \mathscr{G}(\mathbb{R}^n); \quad T_1, T_2 \in \mathscr{K}(\mathbb{R}^n) \Rightarrow X \odot T \in \mathscr{G}(\mathbb{R}^n)$$

The semi-continuity of the erosion implies that for $X \in \mathscr{F}$ and (T_1, T_2) compact sets, the thinning $X \bigcirc T$ is a u.s.c. mapping of $\mathscr{F} \times (\mathscr{K} \times \mathscr{K}) \to \mathscr{F}$. Similarly, for X an open set $(X \in \mathscr{G})$ the thickening is a l.s.c. mapping of $\mathscr{G} \times \mathscr{K} \times \mathscr{K} \to \mathscr{G}$. Therefore thinnings and thickenings satisfy the four principles of Mathematical Morphology.

We just defined two operations in $\mathbb{R}^n$. Indeed relations (XI-24) and (XI-25) remain valid in $\mathbb{Z}^n$ (by putting $X = \mathring{X} = \overline{X}$), and all the algebraic properties we will now develop, hold for both spaces.

(α) $(X \bigcirc T)^c = X^c \bigcirc T^*$ where $T^* = (T_2, T_1)$

The two operations are dual to each other w.r. to the complementation. So for any property of $X \bigcirc T$, there is a corresponding dual property of $X \odot T$.

(β) $T_1 \cap T_2 \neq \varnothing \Rightarrow X \bigcirc T \equiv X \odot T \equiv X$.

(γ) $\{0\} \in T_2 \Rightarrow X \bigcirc T \equiv X, \forall T_1; \{0\} \in T_1 \Rightarrow X \odot T \equiv X, \forall T_2$.

To avoid the two trivial transformations β and γ, we always assume in what follows that the two parts of the structuring element are disjoint and that the origin does not belong to T_2 (resp. to T_1) when we thin (resp. we thicken).

(δ) When T_1 (resp. T_2) is reduced to just the origin, it can be replaced by the empty set when thinning (resp. thickening):

$$T_1 = \{0\} \Rightarrow X \bigcirc T = X \cap [X \oplus T_2];$$
$$T_2 = \{0\} \Rightarrow X \odot T = X \cup [X \ominus T_1]$$

(ε) Let $T' \subset T$ denote the two inclusions $T_1' \subset T_1$ and $T_2' \subset T_2$. Then:

$$T' \subset T \Rightarrow X \bigcirc T' \subset X \bigcirc T \subset X \quad \text{and} \quad X \odot T' \supset X \odot T \supset X$$

The more comprehensive is the structuring element T, the less it removes points of X. Note that the thinning is an increasing mapping with respect to T but not to X.

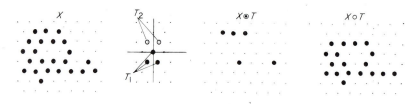

Figure XI.13. Example of a thinning; when T_2 contains the origin, $X = X \odot T$.

Symbol	Structuring element	Thinning	Thickening	Hit or Miss
L		homotopic skeleton	Conditional Segmentation	
M		homot. skeleton (rarely used)		
D		homotopic marking	Pseudo-convex hull (D*)	
C			Hexagonal Convex hull	
E		Skeleton clipping (cond.) skiz		End points
I			Homotopic clipping	Isolated points
F				Triple points
F′				Triple points
R		Erosion (R*)	Dilation Cond: ultimate erosion partly recons.	
H		Boundary		Erosion-dilation (H*) Cond: part. recons. ultimate erosion
K		Sizing by circumscribed hexagons	Ferret's diameter	

(η) It can be deduced from property ε that if ρ is the Hausdorff distance between the origin $\{O\}$ and T_2, then

$$X \ominus \rho B \subset X \bigcirc T \subset X$$

[take $T_1' = \{O\}$ and $T_2' = \{y\}$, where y is an arbitrary point of T_2]. In particular, in $\mathbb{Z}^2$, in the hexagonal case, if T_1 and T_2 belong to the unit hexagon H (centred at the origin), we have:

$$X/\partial X \subset X \bigcirc T \subset X \subset X \odot T \subset X \cup \partial X^c$$

i.e. the thinning of X by T suppresses only points of the boundary ∂X, and its thickening by T adds only points of the boundary $\partial X^c = (X \oplus H)/X$. The thinnings and thickenings associated with such T's are the finest possible, and are discussed in what follows. By reference to Golay's work of 1969, we call the *Golay alphabet* the class of structuring elements on the hexagonal grid such that $T_1 \cup T_2 \subset H$. For the sake of clarity, it is worthwhile to assign a different symbol to the most frequently used of such structuring elements (Fig. XI-14).

(ζ) Extending the two notions, we can define *conditional thickening* $Y \odot T$; X of Y by T, given X, as the restriction of $Y \odot T$ to X:

$$Y \odot T; \quad X = (Y \odot T) \cap X \qquad \text{(XI-26)}$$

and by duality, *conditional thinning* $Y \bigcirc T$; X as the union of $Y \bigcirc T$ with X:

$$Y \bigcirc T; \quad X = (Y \bigcirc T) \cup X$$
$$(\text{and } (Y \odot T; X)^c = Y^c \bigcirc T^*; \quad X^c) \qquad \text{(XI-27)}$$

Warning: X intervenes via an intersection in $Y \odot T$; X but via a union in $Y \bigcirc T$; X! More generally, Definitions XI-26 and XI-27 remain valid if we replace $\odot$ and $\bigcirc$ by $\oplus$ and $\ominus$ respectively, or by any extensive (resp. anti-extensive) morphological mapping.

E.2. Sequential operations

From now, we stay in two dimensions for the sake of digitalization.

Let $X \in \mathscr{F}(\mathbb{R}^2)$ and consider a sequence $\{T^i\}$ of structuring elements. Define the thinning of X by the sequence $\{T^i\}$ as the composition product of X by successive operators T^i, and denote it by $X \bigcirc \{T^i\}$:

$$X \bigcirc \{T^i\} = (\ldots ((X \bigcirc T^1) \bigcirc T^2) \ldots \bigcirc T^i \ldots) \qquad \text{(XI-28)}$$

Figure XI.14. Main structuring elements of the Golay alphabet involved in sequential analysis. These elements are combinations of the 14 possible neighbourhoods of X shown in Figure VI.20 (e.g. L is the union of configurations 3, 4 and 5 of Fig. VI.20 etc . . .).

The ith intermediary result of the process is denoted by $\Phi_i(X \bigcirc T)$. By construction the closed set $\Phi_i(X \bigcirc T) \downarrow X \bigcirc \{T^i\}$, therefore, the latter is a closed set and represents the limit in $\mathscr{F}$, of $\Phi_i(X \bigcirc T)$. However, given $\{T^i\}$, the mapping $\mathscr{F} \to \mathscr{F}$ defined by $X \bigcirc \{T^i\}$ is not necessarily semi-continuous. So we prefer to change the approach and to adopt the digital one. From now, the sets Y, X and T will *systematically* belong to $\mathscr{P}(\mathbb{Z}^2)$, and the gap from $\mathbb{Z}^2$ to $\mathbb{R}^2$ will be bridged in the two most important cases (Sect. F.1 and G.1) via the introduction of the regular model.

By replacing in (XI-28) X by $(Y; X)$ and/or $\bigcirc$ by $\odot$, or $\ominus$, or $\oplus$, (conditional) thinnings, thickenings, erosions and dilations by a sequence can be defined. For example, the conditional dilation by the unit hexagon is written

$$Y \oplus \{H\}; \quad X = [\ldots \{[(Y \oplus H) \cap X] \oplus H\} \cap X \ldots \quad \text{(XI-29)}$$

Sequential thinnings and thickenings satisfy, by extension, the five properties α to ε of Section G.1, by replacing T and T' by $\{T\}$ and $\{T'\}$. The sequences $\{T^i\}$ may be or may not be finite. For example; erosions, dilations, openings and closings by convex sets are easily representable in terms of finite sequences based on the structuring element H of Figure XI-14. But usually, sequences are infinite. Nevertheless, for X bounded, $X \overset{\circ}{\ominus} \{T^i\}$ and $(Y; X) \overset{\circ}{\underset{\oplus}{\ominus}} \{T^i\}$ are reached after a finite number of steps, whether $\{T^i\}$ is finite or not, $\forall Y$.

The sequence $\{T^i\}$ is often generated by successive digital rotations of a basic T. When T^{i+1} is obtained from T^i by the hexagonal rotation of $60°$ (such as $L(2)$ from $L(1)$ in Fig. XI.15), the sequence is said to be *standard*, and the label i is removed in the notation $\{T^i\}$. In particular, all the infinite sequences based on *one* letter of the Golay alphabet, and where all six orientations appear (and reappear after a finite number of terms) are equivalent in the sense that they lead to homotopic results (see Fig. XI.19), and often the same result (convex hull for example). Therefore, we will make a systematic study of the "words" generated by the standard sequences associated with the main letters of the Golay alphabet.

Figure XI.15. Standard rotations.

F. HOMOTOPIC THINNINGS AND THICKENINGS

F.1. Digitalization, thinning and homotopy

We have failed in digitalizing the skeleton because we have been too demanding. It is too much to simultaneously ask the skeleton:

(i) not to modify the homotopy type of X,
(ii) to be reduced to one line thickness,
(iii) to be a median axis,
(iv) to be based on a digital distance.

The examples of Figures XI.11 and XI.12, clearly show that these four prerequisites are contradictory. Digitalization forces us to drop one of them, the choice depending on morphological considerations. In Section D.2, we have cast aside the demand (i). We now keep demands (i) and (iv), drop (iii), and consider No. (ii) as a consequence to be checked; call *homotopic thinnings* the class of digital thinnings defined by the two demands (i) and (iv) (here the hexagonal case is treated in detail; for the square and the 8-connected cases, refer to Ex. XI-4).

The above approach is digital, but what about its Euclidean meaning? Let $X \in \mathcal{K}(\mathbb{R}^2)$ be a compact set in $\mathbb{R}^2$, $(\mathcal{T}(X), 0)$ its planar graph representation with respect to an hexagonal lattice of spacing a, and $\mathcal{S}[\mathcal{T}(X)]$ the digital image of $\mathcal{T}(X)$ (see Fig. VI.29). A homotopic thinning $\hat{\Psi}$ acting on $\mathcal{S}[\mathcal{T}(X)]$ has a Euclidean interpretation if the inverse image $\mathcal{S}^{-1}\hat{\Psi}\mathcal{S}(X)$ of the thinning $\hat{\Psi}$ is a planar graph having the homotopic type of X itself. According to Theorem VII-2, this happens if and only if X belongs to the regular model $\mathcal{R}(\mathcal{K})$ with a parameter $r > a$. For each representative X of the class $\mathcal{R}(\mathcal{K})$, the homotopies of X (in $\mathbb{R}^2$) and of its digital version $\mathcal{S}[\mathcal{T}(X)]$ (in $\mathbb{Z}^2$) are equivalent, and one can homotopically thin X itself via its digital version. Therefore we assume the regular model to hold throughout the present Section F, and work only on the corresponding digital versions (denoted X instead of $\mathcal{S}[\mathcal{T}(X)]$ for simplicity). We established in Chapter VI that the only Golay letters that allow homotopic thinnings are L, M and D (see Fig. VI.21). We will now analyse the associated standard sequences.

F.2. Sequences based on $L = \begin{pmatrix} 0 & 0 \\ \cdot & 1 & \cdot \\ 1 & 1 \end{pmatrix}$

(a) *Thinning* $X \bigcirc \{L\}$.

The study of the structuring element $L = \begin{pmatrix} 0 & 0 \\ \cdot & 1 & \cdot \\ 1 & 1 \end{pmatrix}$ (i.e. $L_1 = \begin{pmatrix} \cdot & \cdot \\ \cdot & 1 & \cdot \\ 1 & 1 \end{pmatrix}$ and

$L_2 = \begin{pmatrix} 0 & 0 \\ \cdot & \cdot & \cdot \\ \cdot & \cdot \end{pmatrix}$) is due to Ch. Lantuéjoul (1976, 1978), who used it to calculate

skizs, and was not long in noticing that the result could be very different from a median axis (1978). Associate the standard sequence with L. Is $X \bigcirc \{L\}$ a homotopic thinning? Now condition (i) requires that $X \bigcirc \{L\}$, or the portion of $X \bigcirc \{L\}$ in which we are interested, is reached after a finite number of steps. Indeed:

— if X is bounded (global case) and assumed to be included in the measuring mask Z, $X \bigcirc \{L\}$ is included in Z, thus reached by a finite decreasing sequence of sets X_i.

— if X^c is bounded (global case), such that $X^c \subset Z$, after $6k$ steps of the iterative algorithm we know the part $X \bigcirc \{L\}$ lying inside the mask $X^c \oplus kH$.

— if X and X^c are unbounded (local case), the first 6 steps of $\{L\}$ require knowledge of $X \ominus H$ which is accessible in $Z \ominus H$ only, etc. . . . . Let k be the largest integer such that $Z \ominus kH \neq \varnothing$ and $6n$ be the first step such that $X_{6n} = X_{6(n+1)} = X \ominus \{L\}$. If $k \geq n$, the limit $Z \bigcirc \{L\}$ is known without error in the mask $Z \ominus nH$; if $n > k$, $Z \bigcirc \{L\}$ is not accessible.

Now for condition (ii). We can analyse it by recording the local configurations of the object (point x plus its six neighbours) which are invariant under H.M.T.'s with respect to L and its rotations. By construction, L suppresses only the configurations 2, 3 and 4 of Figure VI.20. Its action stops when the thinned set possesses only the 11 remaining configurations, i.e. indicated in Figure XI.16 (modulo a rotation). We see that stripes of thickness 2 are impossible, but that $X \bigcirc \{L\}$ eroded by the unit hexagon H is not always empty.

Figure XI.16. Configurations involved in $X \bigcirc \{L\}$.

(b) *Special points and their morphological use*

The neighbourhoods of the special points of a thinning come directly from the inventory of Figure XI.16. The *end points* of a thinning are characterized by the Golay letter E, its multiple points by the letters F and F' (modulo a rotation, in both cases). In many applications, triple points serve as an indicator of specific patterns. Illustrations are given below, in Examples E-2 and E-3. In this section we will shown how the end points allow us to perform *prunings*. The thinning of the daisy X represented in Figure XI.17a led to the skeleton $X_1 = X \bigcirc \{L\}$ (Fig. X.17b) with a nuisance branching, or barb, on the left, that we want to suppress. Noticing that it is shorter than the stalk and the petals of the skeleton, we can eliminate it by a finite thinning using the structuring element E. Associate with E the sequence $\{E^i\}$ obtained by stopping the standard sequence $\{E\}$ after two cycles of six elementary thinnings, and construct the set $X_2 = X_1 \bigcirc \{E^i\}$ (Fig. X.17c). The parasite branch is effectively supressed, but all the others have been shortened by two points. In order to restitute them integrally, we can first extract all the end points of X_2, i.e. construct $X_3 = \bigcup_j (X_2 \circledast E_j)$ (E_j = hexagonal rotation of E). Then we perform on X_3 the sequence $\{H^i\}$ of two dilations conditionally to X_1, and add the result to X_2. We obtain X_4:

$$X_4 = X_2 \cup (X_3 \oplus \{H^i\}; X_1) \tag{XI-30}$$

which is the pruned version of the thinning X_1 (note that $\{E^i\}$ does not generate a homotopic thinning since it eliminates eventual isolated points). If the sequence $\{E^i\}$ and $\{H^i\}$ comprise four cycles instead of two, the pruning becomes more severe and leads to X_5, and if these sequences were infinite, they would remove all X_1 except the small hexagonal ring at the top of the stalk.

Indeed by applying the complete sequence $\{E\}$, we progressively nibble at everything that is accessible from the end points; if $X \bigcirc \{L\}$ does not have of some closed curves, to thin it by $\{E\}$ completely extinguishes it:

$$[X \bigcirc \{L\}] \bigcirc \{E\} = \varnothing$$

But the progressive pruning from the end points stops as soon as we encounter a loop (see the passage from (b) to (c) in Figure XI.18). In other words, the product $[X \bigcirc \{L\}] \bigcirc \{E\}$ is a substitute for the skiz $S^z(X)$.

(c) *Soundness of the method*

Can we consider the hexagonal graphs associated with $X \bigcirc \{L\}$ and $[X \bigcirc \{L\}] \bigcirc \{E\}$ as digital approximations of the skeleton and of the skiz respectively? An easy counterexample shows that the answer is no. The set to be thinned is the background of the four points shown in Figure XI.19. In (a) the thinning sequence turns clockwise and produces two parallel half lines oriented at 60°, which is far away from the perpendicular bisector. . . . Now

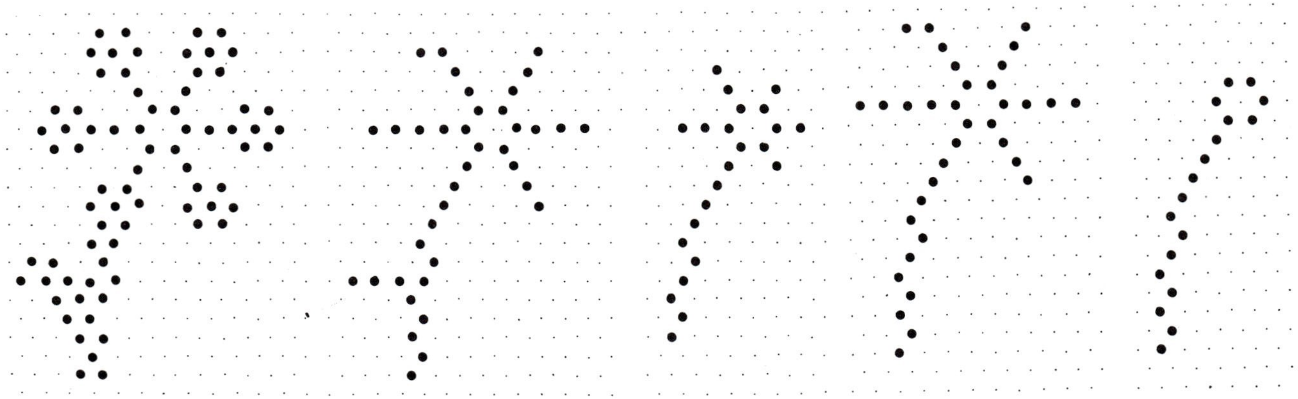

(a) X

(b) $X_1 = X_0 \{L\}$

(c) $X_2 = X_1 \, O \, \{E'\}$

(d) X_4

(e) X_5

Figure XI.17. Prunings on the thinning of a daisy.

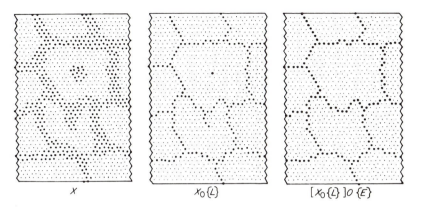

Figure XI.18. Digital skiz defined by the product $[X\bigcirc\{L\}]\bigcirc\{E\}$ of the thining sequence $\{L\}$ followed by the pruning $\{E\}$.

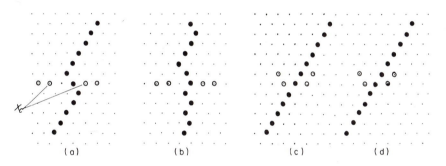

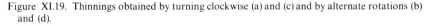

Figure XI.19. Thinnings obtained by turning clockwise (a) and (c) and by alternate rotations (b) and (d).

change the thinning sequence and take $L(1)$, $L(2)$ to $L(6)$, then $L(6)$, $L(5)$, ... to $L(1)$, and again $L(1)$ to $L(6)$ etc, alternating the sense of each cycle. Applying this sequence to the same figure results in Figure XI.19b. The thinning looks like a perpendicular bisector, and should tend towards the Euclidean one as the spacing $\rightarrow 0$. To gain some idea of the robustness of this alternated sequence, we will slightly modify the two segments X^c by turning them (Figs (c) and (d)). The thinning of X by the alternated sequence, shown in (d), has the orientation 60°, exactly as the thinning produced by the standard sequence (Fig. (c)). This counter-example illustrates very well, the general limits of a pure digital approach: when a Euclidean criterion is not digitalizable, the associated digital algorithms are either unstable and fluctuating (here thinnings for the skeleton), or wrong (graph counting for connectivity numbers, Ch. XIII, B.6).

Does it mean that the thinning algorithms are useless? Not exactly, however the class of sets for which they are correct is difficult to delineate. For sets made of unions of long stripes–grain boundaries in Figure XI.24a, the thinnings are good approximations to the associated Euclidean skeletons or skiz. But what is a *long* stripe? Finally, one piece of practical advice we can give corresponds to the division we made here between digital skeletons, and thinnings. If the user wants something which is a good approximation to the medial axis, then we suggest that he forgets about connectivity and uses conditional bisectors. If he wants a line descriptor of the homotopy structure, then he should ignore the median properties and take the thinning by $\{L\}$, or by other similar sequences.

F.3. Sequences based on $M = \begin{pmatrix} & 1 & \cdot & \\ 1 & 1 & 0 \\ & 1 & \cdot & \end{pmatrix}$

The second class of thinnings is associated with sequences derived from the structuring element M of Figure XI.14. We will only give it brief exposure since it is less powerful than L. During the thinning process resulting in $X \bigcirc \{M\}$, M tears out small convex pieces of X, resulting in a jagged (but homotopic!) thinning, often comprising small packs of points. In Figure XI.20 we compare $X \bigcirc \{L\}$ with $X \bigcirc \{M\}$ on a very simple X. The theoretical $\mathbb{R}^2$ skeleton based on the hexagonal distance is just the centre of X; $X \bigcirc \{L\}$ deviates from this exact shape but not too much, but the "star fish" $X \bigcirc \{M\}$ is not even convex.

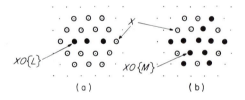

Figure XI.20. Comparison of thinnings by $\{L\}$ and $\{M\}$.

The only case when $\{L\}$ fails however not $\{M\}$ is when X^c has isolated points: their zones of influence cannot be obtained from L. One can overcome the trouble by using $\{M\}$ or, better still, by replacing each isolated point by a doublet (without modifying the homotopy), and then by using $\{L\}$.

F.4. Sequences based on

$$D = \begin{pmatrix} & 0 & \cdot & \\ 0 & 1 & 1 \\ & 0 & \cdot & \end{pmatrix} \text{ and } D^* = \begin{pmatrix} & 1 & \cdot & \\ 1 & 0 & 0 \\ & 1 & \cdot & \end{pmatrix}$$

Obviously, to thin X by D or to thicken X^c by D^* are two equivalent operations:

$$X \bigcirc \{D\} = [X^c \odot \{D^*\}]^c \tag{XI-31}$$

Therefore we can interpret $X \bigcirc \{D\}$ via the thickening of X^c. The structuring element D* is very close to the C used for the hexagonal convex hull (Ex. XI-1); indeed, $X^c \odot \{D^*\}$ is the nearest set to $X^c \odot \{C\}$ which maintains the homotopy of X^c. More precisely, in $X \bigcirc \{D\}$:

(i) each connected component of X is transformed independently of the others;

(ii) each bounded, simply connected component X_j of X becomes a point (the hexagonal convex hull of X_j^c would be the whole space);

(iii) each hole of X_j approaches its own convex hull, but the process stops just before a possible modification of the homotopy;

(iv) (i) and (iii) imply that a shape homotopic to a circular crown is transformed into a loop (i.e. closed contour).

All these possibilities are synthesized in Figure XI.21. In practice, the property ii) is the most often used, serving to label objects in particle extraction problems. The thinnings equivalent to $X \bigcirc \{D\}$ on square or octagonal graphs have been studied by S. Castan (1977). We can order the results given by the three structuring elements L, M and D in the following way. Usually, $X \bigcirc \{L\}$ is made up of lines, with small packs of points where two lines intersect; $X \bigcirc \{M\}$ is less completely thinned and exhibits larger packs of points, and $X \bigcirc \{D\}$ may not be thinned at all, if X is the background of a compact set.

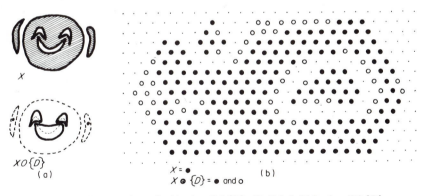

Figure XI.21. (a) schematic thinning $X\bigcirc\{D\}$; (b) digital thickening $X\odot\{D\}$.

G. CONNECTED COMPONENTS ANALYSIS

G.1. Particle marking

The set $X \in \mathcal{P}(\mathbb{R}^n)$ is made up of the union of J disjoint compact connected components X_j $(1 \le j \le J)$ (the particles). We want to extract those X_j which are hit by a given set Y (the marker) i.e. the set

$$\psi(Y;X) = \bigcup_j \{X_j : X_j \cap Y \neq \varnothing\} \quad \text{with } X = \bigcup_j X_j \qquad \text{(XI-32)}$$

and we want to express the mapping $\psi(Y;X)$ in terms of digitalizable operations of the morphology. The problem is in general impossible, but has a solution if $X \in \mathscr{R}(\mathscr{K})$ (henceforth in this Section G, we assume the regular model hypothesis) and we can state:

THEOREM XI-11. *Let X belong to the regular model $\mathscr{R}(\mathscr{K})$ of radius $r > 1$ in $\mathbb{R}^n$, then:*

$$\Psi(Y;X) = Y \oplus \{B\}; X \qquad \text{(XI-33)}$$

Proof: The radius r of $\mathscr{R}(\mathscr{K})$ is defined by relation (V-22). We firstly show that induction (XI-33) transforms each X_j independently of the other. Since X belongs to $\mathscr{R}(\mathscr{K})$ the number J of particles X_j is finite. Put $X_j^0 = X_j \cap Y$ and $X_j^i = (X_j^{i-1} \oplus B) \cap X_j$. Assume that $X^{i-1} = \bigcup_j X_j^{i-1}$ then:

$$X^i = \left[\left(\bigcup_j X_j^{i-1}\right) \oplus B\right] \cap X = \bigcup_j [(X_i^{i-1} \oplus B) \cap X]$$

$$= \bigcup_j [(X_j^{i-1} \oplus B) \cap X_j] = \bigcup_j X_j^i$$

since the inclusion $X_j^{i-1} \subset X_j$ and the regularity of X imply that, for $j' \neq j$, $(X_j^{i-1} \oplus B) \cap X_{j'} = \varnothing$. (Use the same proof for the first step of the induction.) Now consider a grain X_j marked by Y, i.e. there exists a $y \in X_j \cap Y$. Let z be one point of the projection of y onto ∂X_j. The distance $d(y, z)$ is bounded (X_j compact), hence a finite number n_1 of induction steps is needed to reach z from y, so $z \in X_j^{n_1}$. Similarly, since the length of ∂X_j is bounded (X_j regular), after a finite number of iterations n_2, all points of ∂X_j are reached, and $\partial X_j \subset X_j^{n_1 + n_2}$. Finally, every point of X_j is at a finite distance from its projection on ∂X_j and therefore reached after n_3 iterations starting from ∂X_j. So for $i_0 > n_1 + n_2 + n_3$ we find $X_j^{i_0} = X_j^{i_0 - 1} = X_j$ and the process of filling up X_j stops. Thus $X^{i_0} = \psi(Y;X)$.   Q.E.D.

COROLLARY. *The mapping ψ from $\mathscr{R}(\mathscr{K}) \times \mathscr{F}$ into $\mathscr{R}(\mathscr{K})$, the result of finite iterations of increasing and semi-continuous mappings (XI-33), is digitalizable for the covering representation (Ch. VI, C.1).*

In practice then, this means that we can directly transpose relation (XI-32) into digital terms; X and Y become digital sets, B is replaced by the unit hexagon H of the grid, and according to relation (XI-29), we have

$$\psi(Y;X) = Y \oplus \{H\}; X \qquad \text{(XI-34)}$$

Alternatively, an equivalent, but six times slower, digital implementation is given by the conditional thickening of Y, given X, by the sequence $\{R\}$:

$$\psi(Y; X) = Y \odot \{R\}; \ X \quad R = (0, 1) \tag{XI-35}$$

G.2. Applications of particle marking

Very often, the set Y is itself the result of a transformation of X, exhibiting particular properties of the particles we want to extract. Here are a few examples of possible information carried by Y. Others are given in section H.

(a) *Particles lying inside the mask*
Y is the set of points of X which hit the boundary $\partial Z = Z/(Z \ominus H)$ of the measuring mask (see Fig. XI.23).

(b) *Marking by presence, by convexity and by size*
We now use algorithm (XI-23) to remove the three types of artefacts presented in Chapter I (Fig. I.9). The initial specimen, a cervical smear, has been illuminated successively by two fluorescent lights, producing two very distinct images of the nuclei Y_1 and of the cytoplasms (X). By so doing we generated two different sets that are defined within the same mask. We wish to eliminate two types of artefacts, as well as the uninteresting small cells. The rule of the game is to associate a specific criterion with each type of object to be eliminated. The application of the criterion often "damages" all the particles present within the mask, however some of them (the ones we wish to recognize) are characterized by a typical deformation. We then return to the initial sets and remove the unwanted artefacts. In the present case, the features to be filtered out are the following:

  (i) the connected components of the cytoplasm set without a nucleus inside,
 (ii) the nuclei spoiled by the smear technique, making them non convex,
(iii) the nuclei with a maximal diameter smaller than 10μ (the grid spacing equals 1μ).

To filter out the first type of feature, perform the conditional dilation of Y_1, given X, by H; i.e. $X' = Y_1 \oplus \{H\}$; X. The resulting set of cytoplasms X' is the grey one in Figure XI.22b. For the second implementation, we assume that a nucleus is not spoiled if an hexagonal closing of 5μ leaves it unaltered (convex sets are invariant under closings). The set difference $Y_2 = Y_1/(Y_1)_{5H}$ between the initial set of nuclei and its closing in the white part of the photograph in Figure XI.22b. Once again, we use conditional dilation to extract the cells of X' marked by Y_2; i.e. $X'' = Y_2 \oplus \{H\}$; X'. The last step (iii) eliminates the nuclei of X'/X'' with maximal diameter smaller than 10μ by a similar procedure; the marker Y_3 is now the nuclei eroded by the hexagon of size 10μ. The final result,

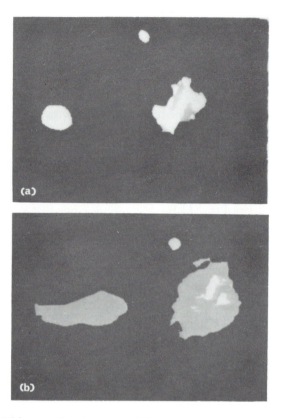

Figure XI.22. (a) A hexagonal closing leaves only the convex nuclei unchanged. (b) set X' and its marker $Y_2 = Y_1 - (Y_1)_{5H}$.

after all the artefacts have been removed, is shown in Figure I.9b, where the cytoplasm and the nucleus have been superimposed.

In conclusion, we first note that here the order of the three steps was independent of the final result. This is not always true when recognizing shapes by mathematical morphology. Secondly, note that we have not used an algorithm that proceeds particle by particle. Erosions, closings, intersections, etc. . . . are carried out in an *identical manner on all the points* of the frame. These transformations have progressively emphasized the contrast between figure and background. The final image, cleaned, is now ready for a cytologic examination that proceeds cell by cell. The method of observation must eventually swing from working with the whole frame, to a particle by particle analysis. This is made possible by using a new marker Y, namely the "first point", which labels and extracts the connected components.

(c) *Labeling and extraction of particles*
Imagine that the moment has come when we have eliminated the various artefacts (types (i), (ii) or others), and that particles $\{X_j\}$, which we want to analyse individually by set transformations and measurements, remain inside the mask. The simplest labeling of the X_j's is performed by scanning $X = \bigcup X_j$ and stopping at the *first point* $y_1 \in X$ met by the scanning. Put $Y_1 = y_1$; then the quantity $Y_1 \oplus \{H\}$; X is just the first particle, say X_1, of X hit by a horizontal test line Δ sweeping the image from north to south. Put $X_2' = X/X_1$; the induction loop is now obvious, and gives:

$$X_j = Y_j \oplus \{H\}; \ X_j' \quad \text{with} \quad \begin{cases} X_j' = X_{j-1}'/X_{j-1} \\ \\ Y_j = F\ P\ (X_j') = \ \text{first point of } X_j' \end{cases} \quad \text{(XI-36)}$$

Algorithm (XI-36) not only *labels* the particles with respect to the sweeping order relation, but also *extracts* them one by one. Exactly as for the dilation (XI-34), the digital particle extraction (XI-36) only has a Euclidean interpretation for the regular model.

(d) *The ultimate erosion*
Let $X \in \mathscr{F}\ (\mathbb{R}^2)$ be a closed set in $\mathbb{R}^2$, and $X_\rho = X \ominus \rho B$ be the erosion of X by the compact disk of radius ρ. Define the ultimate erosion $Y\ (\rho)$ of label ρ, to be the connected components of X_ρ which vanish from any larger erosion $X_\mu\ (\mu > \rho)$:

$$Y_\rho = \bigcap_{\mu > 0}\ [X_\rho/\psi\ (X_{\rho+\mu};\ X_\rho)] \quad (\psi \text{ defined by XI-32}) \quad \text{(XI-37)}$$

As ρ varies, the ultimate erosions $Y(\rho)$ generate a set family depending on the positive parameter ρ. Denote by $Y = \bigcup_\rho Y(\rho)$ the union of the $Y(\rho)$'s. The $Y(\rho)$'s partition Y. The union Y (and a fortiori the $Y(\rho)$'s) maintains neither the homotopy type of X, nor its connectivity (see Figs XI.28, 29). If X is a compact, convex set, then Y is reduced to only one $Y(\rho)$ which is also compact, convex. But if X is made up of two compact convex sets X_1 and X_2 which overlap, Y is in general the union of two disjoint convex particles, the "centres" of X_1 and of X_2. This lack of conservation of connectivity is exploited to disconnect particles which overlap—when they are approximately convex (e.g. nuclei in biology, see Fig. XI.29). Note that $Y(\rho)$ is not necessarily reduced to small packs of points; for X a circular crown, $Y = Y(\rho)$ is the median circle of the crown, but if the hole of the crown is moved slightly, $Y(\rho)$ suddenly reduces down to one point: the ultimate erosion is neither continuous, nor increasing!

Digitalization of the mapping (XI-37) is valid for $X \in \mathscr{R}\ (\mathscr{K})$ (all the connected components of X must appear on the digital version). If X and Y

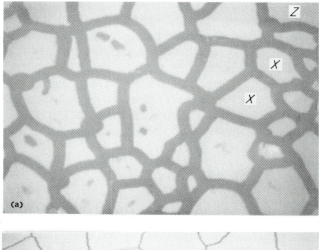

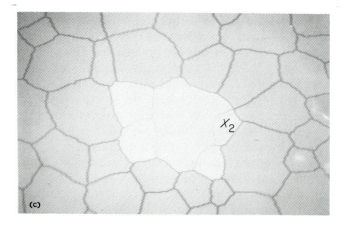

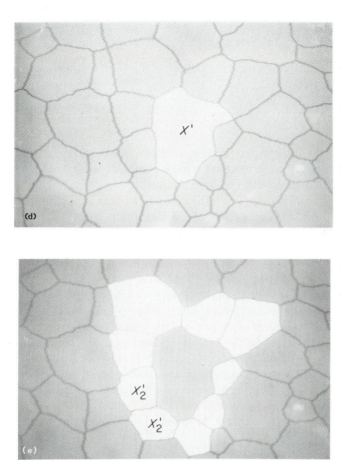

Figure XI.23. Steps involved in the neighbour analysis. (a) polished section X of a polycrystalline ceramic, (in white), seen in a rectangular mask Z; (b) skiz of X^c; k maximum width of the stripe between two neighbour grains (measured by the number of cycles of thickenings necessary for obtaining $X \odot \{L^*\}$: $X_1 = [(X \odot \{L^*\}) \odot \{E^*\}] \cap (Z \ominus kH))$; (c) elimination of the grains whose neighbours are not included in Z;

$$A = \partial(Z \ominus (k-1)H) \oplus \{H\}; X_1$$
$$A' = (A \oplus H) \oplus \{H\}; X_1$$
$$X_2 = X_1/A'$$

(d) datum of the first point Y of the set X_2. Extraction of the grain X' by the first point algorithm: $X' = Y \oplus \{H\}; X_2$. Measurement of the number n of connected components of X'_1 $= (X' \oplus 2H) \cap X_1$ ($n-1$ is the number of edges and of the neighbours of X'); (e) set of the neighbours of X': $X'_2 = (X'_1 \oplus \{H\}; X_1)/X'$ (each neighbour is then individually studied by the procedure of step d).

now denote digital sets, and if $X_i = X \ominus iH$, the digital algorithm associated with (XI-37) is as follows:

$$Y_i = X_i/(X_{i+1}; X_i) \oplus \{H\} \qquad Y = \bigcup_i Y_i$$

H. EXAMPLES

We want to use the few examples which follow, to give an idea of the combination of morphological criteria occurring in practice, as well as to provide the reader with the most frequently used connectivity algorithms. Here digital morphology is used as a language: there is the classical distinction into the three levels, namely *letters*, *words*, and *sentences*. The letters are all possible H.M.T.'s whose support lies in the unit hexagon H. (This Golay alphabet is not the only possible one, but is the most useful pedagogically). The *words* are the thinnings and thickenings produced from the letters (including finite sequences, e.g. the erosions). The *sentences* are a succession of words in a given order. The opening and the closing are sentences of two words; the cell extraction illustrated in Figure XI.23, requires three words. One can also imagine words made up of sequences, each term of which is already a sequence, etc. . . . Anything is possible, but the experimenter must always keep in mind the morphological meaning and constraints of each word (action based on the convexity, difference between homotopy and connectivity, between size and erosion, etc. . . .). The presentation of the examples which is to follow is limited strictly to image processing, without developing the physical implication of the results; they can be found in the references.

H.1. Neighbour analysis

This study, due to Ch. Lantuéjoul (1978), is involved with analysing the dependence between grains in polycrystalline ceramics. This dependence is closely linked with the mechanical properties of the structure and allows partial prediction of the evolution of the material as physical conditions vary (M. Blanc, 1978). We work in two dimensions, and consider a polished Section X of a polycrystalline ceramic, partitioned into grains X_i' (Fig. XI.23). Among a number of possible parameters, we concentrate upon the statistics of the edge numbers $v_0(X_i')$, $v_1(X_i')$ and $v_2(X_i')$ of the X_i''s and of their first and second neighbours.

Physically speaking, the grain boundaries have zero width. They are revealed by a chemical etching which must be strong enough to completely cover the boundaries. As a result, the neighbours of a given grain, say X', are not clearly defined, except if we thin the boundary set into a skiz. Then the neighbours of X' can be defined as being the only particles marked by $X' \oplus H$. Suppose we follow this morphological sentence step by step: mutatis mutandis, the same morphological sentence allows the extraction of the second order neighbours of X', say X'_3 and to compute the term v_2 for each of them. From these data, one can now estimate the joint probability of the pairs (v_0, v_1) and (v_0, v_2). The results will be valid if

(i) the statistical Miles–Lantuéjoul correction (VIII-14) is applied, in order to compensate for oversampling the large pairs of neighbouring grains,
(ii) configuration H of Figure (XI.14) hardly ever occurs. In the present case, thousands of grains were measured without once finding these critical occurrences, but we have met other structures where configuration H does occur.

Remark: We presented this study, not only because of its interest in terms of the above algorithm, but also as a typical example of the method followed in the local approach. The concepts of homotopy and connectivity are essentially *global* ones. Therefore, the two thickenings necessary to produce X_1 are global algorithms. However, the erosion by $k\,H$ is the edge correction of the local case. It is followed by a second *global* edge correction (suppression of $X_2 \cup X_3$) which allows us to perform the individual analysis of the X's under global conditions. Finally, it is only at the step where measurements are taken that we return to the local framework, by applying Miles–Lantuéjoul's correction.

H.2. Defect lines

The unidirectional solidification of eutectic systems may lead to anisotropic structures made up of regularly spaced dendrites. Figure XI.24 shows a transverse cross-section of such a structure. The physical quality of the material (traction and torsion modules) is negatively correlated with the crystallization defects and in particular with the defect lines at the end of the dendrites. In Figure XI.24b we see the result of an automatic extraction of such lines by the sequence of set transformations shown in Figure XI.25.

Finally, thinning and clipping of X_6 results in the defect lines X_7. We can see on Figure XI.24b the lines X_7 superimposed on X_1, for an actual case studied. In spite of its sophistication the transformation $X_7 = \psi(X_1)$ turns out to be experimentally robust, and has been used routinely without trouble (note that we could also have extracted the flexure zones by starting from the triple points of X_2).

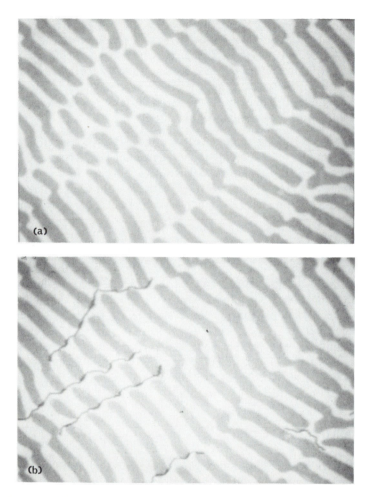

Figure XI.24. (a) transverse section of lamellae, (b) extraction of the defect lines.

H.3. Discrimination between cells and artefacts

Since 1975, F. Meyer has established a library of the major morphological transformations used in quantitative cytology. The most useful consequences of this work are the standard series of transformations for automatic screening of cervical smears (1978, 1979a), developed in cooperation with J. S. Ploem's group, and in particular with A. Van Driel (1979b). The following example comes from one of these programs, applied to cytological material doubly stained by Acriflavine–Feulgen and SITS. The present discrimination is solely based on the nuclei images obtained by fluorescence; these provide "masks" of

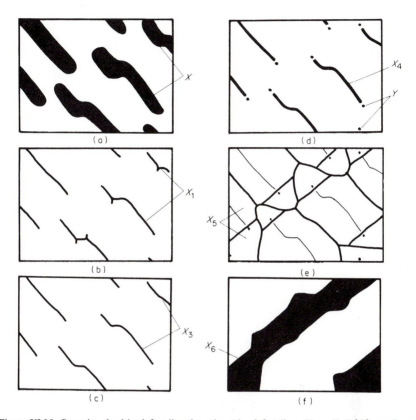

Figure XI.25. Steps involved in defect line detection (the defect lines $X_7 = X_6 \bigcirc \{L\}$ are those visualized in Figure XI.24b). (a) X = initial set, in black (n = mean width of X); (b) thinning of set $X : X_1 = X \bigcirc \{L\}$. (Note that the triple points correspond to the flexure zones.) (c) pruning of $X_1 . X_2 = X_1 \bigcirc \{E^i\}$ where $\{E_i\}$ = n cycles of rotations of E. $X_3 = X_2 \cup ([\bigcup_j (X_2 \circledast E_j)] \oplus \{H\};$ X_1); (d) extraction of the end points. $Y = \bigcup^{j=1}_{6} [X_3 \circledast E(j)]$ (end points). $X_4 = [X_3/(Y \oplus H)] \cup Y$ (disconnection of the end points); (e) skiz of X_4^c i.e. $X_5 = (X_4 \bigcirc \{M^*\}) \bigcirc \{E^*\}$; (f) particle extraction (the small final dilation will connect the particles) $X_6 = (Y \oplus \{H\}; X_5) \oplus H$.

the nuclei, i.e. sets which exactly cover them (Fig. XI.26a). For the sake of pedagogy, we describe step by step the discrimination process on a schematic example

X_1: initial set, corresponds to the union of the white and grey particles of Figure XI.26a.

$X_2 = S^2(X_1) = \bigcup_{n \geq 0} [(X_1 \ominus nH)/(X \oplus nH)_H]$ (Cond. bisector XI-22).

$X_3 = X_1/X_2$ (x grey, particles minus holes)

$X_4 = X_3 \bigcirc \{L\} \bigcirc \{E\}$ = skiz of X_3 (in solid lines)

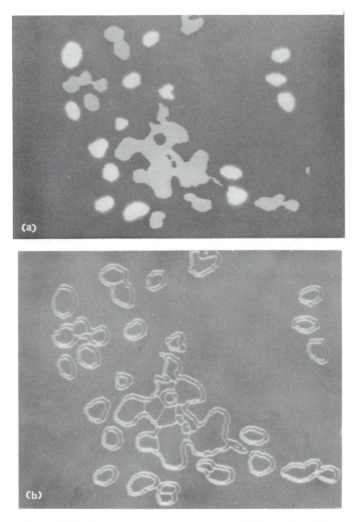

Figure XI.26. (a) cells in white, artefacts in grey; (b) the double skeleton.

$X_5 = (X_1 \oplus H)/X_1$ (stripes in grey)

$X_6 = X_5 \bigcirc \{L\}$ (thinning of X_5, in solid lines)

$X_7 = X_4 \cup X_6$ (union of the two thinnings)

$X_8 = \bigcup_{j=1}^{6} [X_7 \circledast F(j)] \cup [X_7 \circledast F'(j)]$ (triple points of the thinnings)

$X_9 = [X_8 \oplus \{H\}; (X_1 \oplus H)] \cap X_1$ (filling up those particles whose
thinnings exhibit triple points in grey).

In the actual case of Figure XI.26a, the white corresponds to X_1/X_9 ("good" cells) and the grey to X_9; in (b) the double stripes around each particle are X_7. The triple points of the thinning X_4 mark the clusters of overlapping cells; those of X_6 mark the nuclei damaged by narrow gulfs. It is interesting to note that the image is treated globally and not particle by particle. It is only after the discrimination that the connected components of X_1/X_9 are labelled and successively extracted (by the first point technique) for further investigation.

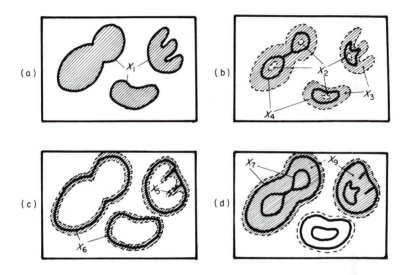

Figure XI.27. Step by step approach of the discrimination process.

H.4. Cluster fast segmentation

It often happens that one wants to analyse separately particles which overlap. A simple example is given in Figure XI.28a, where we wish to know the DNA content of three nuclei (absorption image in Feulgen). The associated fluorescent image is shown in Figure XI.28b, which results, after transformation, in three masks which can be separately superimposed onto (a). The steps involved in this set transformation (due to F. Meyer, 1979) are shown in Figure XI.29.

This type of transformation is made up of two distinct phases, namely: to find markers for the zones to be separated, and to draw the boundaries. Here, the markers are the ultimate eroded sets $Y(i)$. One could replace them by finer markers, for example by the conditional bisector. Indeed, when two convex sets overlap too much, they may give only one ultimate erosion, but may

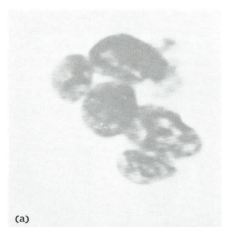

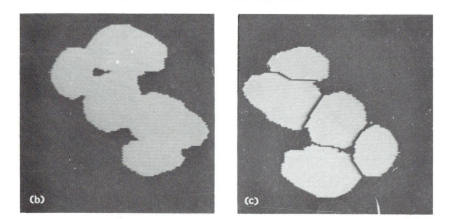

Figure XI.28. (a) nuclei under study (Feulgen), (b) their "masks" obtained by fluorescence, (c) division of the "mask".

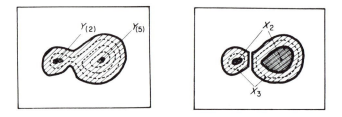

Figure XI.29. Steps involved in the cluster fast separation.

contribute two pieces to the conditional bisector.

$$X = \text{set to be split up; } X_i = X \ominus iH$$

$$Y_i = X_i/(X_{i+1} \oplus \{H\}; X_i) \text{ ultimate eroded sets (rel. (XI-37))}$$

$$m = \max\{i : X_i \neq \varnothing\}$$

$$X_2 = \bigcup_{i=0}^{m} Y(i) \oplus (m-i)H$$

$$X_3 = [X_2 \odot \{L\}] \cap X \text{ (id. Fig. XI.28c)}$$

Now, if we draw the boundaries by taking the exoskeleton of the ultimate erosions, we get lines which are shifted toward the largest cells, as illustrated in Figure XI.30a. Here we overcame this bias by dilating each ultimate component proportionally to the size of the corresponding cell (algorithm X_2), before performing the exoskeleton $X_2 \odot \{L\}$. Afterwards, we keep only the part of the exoskeleton lying in the initial set X. This latter procedure also is criticizable. We implicitly assumed that the grains were nearly convex. If not, there is absolutely no reason why the branches of the skeleton inside X should pass through the points of concavity of ∂X! This drawback can be corrected by the much more subtle algorithm of the watersheds, where the skeleton is progressively constructed *and* intersected by successive sets nested in each other, the last set being X (in other words, the present algorithm is a coarse watershed with only one term in the nested sequence, namely X itself).

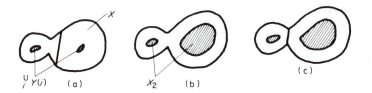

Figure XI.30. (a) segmentation by the exoskeleton of the ultimate erosions. (b), (c) correction of the bias.

H.5. Segmentation by watersheds

The next chapter (Sect. E) will justify the use of such a geographical type term. In brief, the idea is to consider the successive erosions of X as horizontal cross sections of a relief, from the top to the ground level. The watersheds will be the domains of attraction of rain falling over the region (Fig. XI.31). Segmentation by watersheds was proposed for the first time by Ch. Lantuéjoul (1978), and

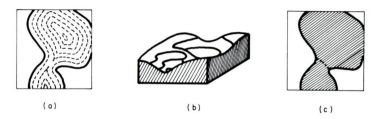

Figure XI.31. Watersheds. (a) set X and its successives erosions (b) associated reliefs (c) the two watersheds of X.

was later improved jointly with S. Beucher (1979). It has been used extensively by Coster *et al.* (1980) to study sintering processes in metallurgy. The segmentation is provided by the mth value $Y(m)$ of an iterative algorithm, where m is, as in the previous example, the range of the ultimate erosion, and where the induction is initiated by the two terms $X(1) = X \ominus (m-1)H$; $Y(o) = X \ominus mH$. The general terms $X(i+1)$ and $Y(i)$ come from the pair $X(i)$, $Y(i-1)$ as follows:

$$\left.\begin{aligned}
&X(i) = X \ominus (m-i)H \\
&X^1(i) = Y(i-1) \odot \{L\}; X(i) \quad \text{(skeletonization of } Y^c(i) \text{ inside } X(i)) \\
&X^2(i) = X^1(i) \odot \{E\}; X(i-1) \quad \text{(skizing by pruning the dendrites} \\
&\qquad\qquad\qquad\qquad\qquad\qquad\qquad \text{finishing inside } X(i-1)) \\
&X^3(i) = X^2(i) \oplus \{H\}; X(i) \quad (X(i)/X^3(i) = \text{ultimate eroded com-} \\
&\qquad\qquad\qquad\qquad\qquad\qquad\quad \text{ponents with label } m-i) \\
&Y(i) = X^2(i) \cup [X(i)/X^3(i)]; \quad X(i+1) = X \ominus (m-i-1)H
\end{aligned}\right\} \quad \text{(XI-38)}$$

Briefly then, at each iteration we thicken $Y(i-1)$ inside $X \ominus (m-i)H$ by $\{L\}$, which preserves connectivity. This thickening, added to the new ultimate eroded components appearing in $X \ominus (m-i)H$ provides the next term $Y(i)$. The limit $Y(m)$ is the required segmentation (Fig. XI.32).

I. EXERCISES

I.1. Convex hulls

(a) In $\mathbb{R}^2$, associate with any compact set X the smallest compact hexagon Y with sides parallel to $0°$, $60°$, $-60°$ that contains X. Is this mapping digitalizable?

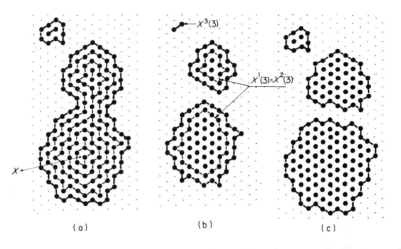

Figure XI.32. (a) initial set X and its successive erosions (here $m = 4$), (b) step $i = 3$; $Y(3) = X^2(3) \cup X^3(3)$, (c) final segmentation. (Compare alg. (XI-38) and (XII-57)).

[The mapping is increasing and upper semi-continuous, hence digitaliz-able for the covering representation. Digital algorithm: take the intersec-tion of the half planes containing the digital X over the three basic directions.]

(b) X is now digital; associate with $C = \begin{pmatrix} & 1 & \cdot \\ 1 & 0 & \cdot \\ & 1 & \cdot \end{pmatrix}$ its standard sequence.

Prove that for X connected $X \odot \{C\}$ is the hexagonal (digital) convex hull of X. Show that if X is bounded $X \odot \{C\}$ is reached after a finite number of steps. What happens if we change the order of the terms in $\{C\}$ in such a way that each direction reappears after a finite time?

(c) Apply algorithm $X \odot \{C\}$ to the following sets of Figure XI.33.

Show that if X is made up of particles X^j whose hexagonal convex hulls are disjoint, then and only then $X \odot \{C\} = \bigcup_j (X_j \odot \{C\})$, and so all the

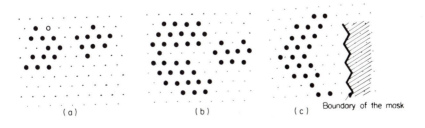

Boundary of the mask

Figure XI.33.

particles can be processed in parallel. Show that if X is included in the mask Z, but not its convex hull, then $[X \odot \{C\}] \cap Z = X \odot \{C\}; Z$.

(d) Define in $\mathbb{R}^2$ the dodecagonal convex hull as the intersection of the half planes containing X, over directions $k\pi/6$, $(0 \le k < 12)$. As in question (a) the mapping is digitalizable, but its digital version on the hexagonal grid cannot be reduced to sequences of the Golay alphabet. Define by C' the structuring element of the sub grid equivalent to C on the basic grid; i.e.

$$C' = \begin{pmatrix} \cdot & 1 & & \cdot \\ 1 & 1 & 1 & 1 \\ \cdot & 0 & & \cdot \end{pmatrix}$$; the group of six hexagonal rotations around the

white points generate the standard sequence $\{C'\}$. Prove that the dodecagonal hull is given by the algorithm $[X \odot \{C\}] \cap [X \odot \{C'\}]$.

I.2. Curve smoothing by pseudo-closing

Let $X \in \mathscr{P}(\mathbb{Z}^2)$. Its hexagonal closing of size n is $X^{nH} = (X \oplus nH) \ominus nH$. Modify this algorithm by replacing $\oplus H$ by $\odot L^*$, $\ominus H$ by $\odot \check{L}$ and the $2n$ iterations by the sequences $\{L^{*i}\}$ and $\{\check{L}^i\}$ of n complete rotations of L^* and $\check{L}$ respectively (each sequence is made up of six n terms):

$$Y(X) = [X \odot \{L^{*i}\}] \odot \{\check{L}^i\}$$

$$\text{with} \quad L^* = \begin{pmatrix} \cdot & 1 & 1 & \cdot \\ & 0 & & \\ 0 & & 0 & \end{pmatrix} \text{ and } \check{L} = \begin{pmatrix} \cdot & 1 & 1 & \cdot \\ & 1 & & \\ 0 & & 0 & \end{pmatrix} \quad (1)$$

(a) Is this transformation extensive? increasing? idempotent? Does it modify the homotopy?

(b) Apply algorithm (1) to the connected curve S which divides the plane into two disjoint parts S_1 and S_2 (Fig. XI.34). Take for X the curve S itself, and then the above part S_1. Are the sets $Y(S)$ and $\partial Y(S_1)$ identical? Is $Y(S)$ a set with a zero thickness?

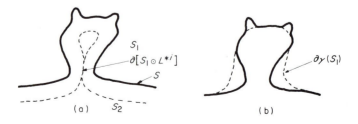

(a) S_2 (b)

Figure XI.34.

I.3. Sizing by the circumscribed hexagon,

$$\text{Ferret's diameter and } K = \begin{pmatrix} & 0 & & 0 & \\ \cdot & & 1 & & \cdot \\ & \cdot & & \cdot & \end{pmatrix} \quad \text{(S. Beucher)}$$

(a) Let $X \in \mathscr{P}(\mathbb{Z}^2)$ be the union of disjoint convex particles X_j. Define the finite sequence $\{K^{3n}\}$ by taking n terms identical to K, followed by n terms

$$K(1) = \begin{pmatrix} & \cdot & & 0 & \\ \cdot & & 1 & & 0 \\ & \cdot & & \cdot & \end{pmatrix}, \text{ and followed by } n \text{ terms } K(2) = \begin{pmatrix} & \cdot & & \cdot & \\ \cdot & & 1 & & 0 \\ & \cdot & & 0 & \end{pmatrix}. \text{ Show}$$

that the particles X_j which are eliminated in $X \bigcirc \{K^{3n}\}$ are those whose circumscribed hexagons are smaller than nH.

(b) We now consider $K^* = \begin{pmatrix} & 1 & & 1 & \\ \cdot & & 0 & & \cdot \\ & \cdot & & \cdot & \end{pmatrix}$, and the infinite sequence K^{*i} all of

whose terms are identical to K^* (without rotation). Transform the particle X_j into $Y_j = X_j \bigcirc \{K^{*i}\}$, show that the number of intercepts of Y_j in directions $+60°$ and $-60°$ are just the corresponding Ferret diameters (i.e. the apparent contour without counting holes and concavities). Is it possible to work directly on X, as in question a)? Why?

I.4. Thinnings and thickenings on the square grid

Re-write all the structuring elements of Figure XI-14 in terms of the square grid, by treating two cases of square and of octagonal graphs separately. Can the elementary square be reduced to four points? Why?

[The equivalent of L, for example, is the well known Levialdi's algorithm for

the skeleton. L becomes $L' = \begin{pmatrix} 0 & 0 & 0 \\ \cdot & 1 & \cdot \\ 1 & 1 & 1 \end{pmatrix}$. The second term of the sequence is

$$L'_4 = \begin{pmatrix} \cdot & 0 & \cdot \\ 1 & 1 & 0 \\ 1 & 1 & \cdot \end{pmatrix} \text{ for the square graph and } L'_8 = \begin{pmatrix} \cdot & 0 & 0 \\ 1 & 1 & 0 \\ \cdot & 1 & \cdot \end{pmatrix} \text{ for the}$$

8-connected graph. The sequence $\{L'\}$ has the successive rotations of L' as its even terms and the rotations of L'_4 (resp. L'_8) as its odd terms.]

I.5. Pruning and homotopy

Define a pruning which preserves the homotopy of the thinning $Y = X \bigcirc \{L\}$.
   [Take the induction $Y^o = Y$, $Y^i = (Y^{i-1} \bigcirc E) \cup (Y^{i-1} \circledast I)$ (notation of Fig. XI.14). Note that neither of the two terms $\bigcirc E$ and $\circledast I$ considered separately, preserves homotopy.]

I.6. Connected digital skeleton (by erosions)

Let H be the elementary hexagon of the grid and let X be a set morphologically closed with respect to H (i.e. $X = X^H$). Prove that the set S^* defined by:

$$S^* = \bigcup_i^{i_0} [Y_i/(Y_i)_H] \quad \text{with} \quad \begin{cases} Y_0 = X \\ \\ Y_i = \left[[Y_{i-1}/(Y_{i-1})_H] \oplus H \right] \cup (X \ominus iH) \end{cases}$$

where $i_0 = \text{Sup}\,\{i : X \ominus iH \neq \varnothing\}$ has the same homotopy as X itself, contains the digital skeleton relation (XI-20), but is not of zero thickness and satisfies only the relationship $A(S^* \ominus 2H) = 0$.

I.7. Homotopy characterization

Consider a bounded connected component $X_{j,1}$ possessing internal pores $Y_{j,1}$ themselves having internal particles $X_{j,2}$ etc. . . . up to $n(j)$ (Fig. XI.35). Let X_j denote the union $X_{j,1} \cup \ldots X_{j,n(j)}$, Y_j the union $Y_{j,1} \cup \ldots Y_{j,n(j)}$, and consider the set X made up of the union of p such X_j's, with the $X_j \cup Y_j$'s disjoint from each other. Given X, how can the $X_{j,n(j)}$'s be extracted?

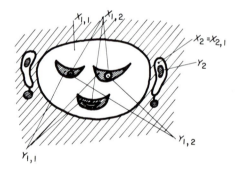

Figure XI.35.

[Take the mask Z large enough to have $X \subset Z \ominus H$. Construct a set $W = Z/\partial Z \oplus \{H\}; X$ whose connected components are the unions $X_j \cup Y_j$'s. Extract them individually via a first point algorithm acting on W and work with $X_j \cup Y_j$. The set $\partial(X_j \cup Y_j) \oplus \{H\}; (X_i \cup X_j)$ equals $X_{j,1}$. Subtract it from $X_j \cup Y_j$ and iterate: the result is the sequence $X_{j,1}; Y_{j,1}; X_{j,2}; Y_{j,2} \cdots \varnothing$. This algorithm is repeatedly used in practice as a means for filling in holes in particles.]

I.8. The quench function transformation
(S. Beucher, Ch. Lantuéjoul)

Let $f(\rho)$ be a bounded function defined on $\mathbb{R}^+$. We assume that for $f(\rho) < 0$, the quantity $f(\rho)\,\mathring{B}$ is the empty set $\varnothing$ and that $X \oplus \varnothing = \varnothing$. Start from relation (XI-6) and define the set transformation $Y = \bigcup_{\rho > 0} s_\rho(X) \oplus f(\rho)\mathring{B}$ of $\mathscr{G}'$ into $\mathscr{G}$. What are the transforms of:

(a) a regular hexagon by, $f(\rho) = \rho$ for $\rho \le \rho_0$ and $f(\rho) = \rho_0$ for $\rho > \rho_0$,
(b) a rectangle for $f(\rho) = \frac{1}{2}\rho$ and $f(\rho) = 2\rho$?

I.9. Non-extensive iterative transforms
(S. Beucher, Ch. Lantuéjoul)

In this exercise, we would like to show the importance of the extensivity property of the thickening, by analysing what happens when it is left out.

(a) Let ψ be a mapping of $\mathscr{P}(\mathbb{R}^2)$ (or of $\mathscr{P}(\mathbb{Z}^2)$) into itself, and $\psi^{(i)}$ the i-th iteration of ψ (e.g. $\psi^{(2)}(X) = \psi[\psi(X)]$). As i increases, $\psi^{(i)}(X)$ can decrease down toward $\varnothing$; or invade the whole space; or be cyclic; or be idempotent; or tend towards a limit; or translate the initial set after n iterations, etc. . . .
Try to give an example of each type of evolution.
(b) *Conway's game*: Start from the two configurations $X = X^{(0)}$ and $Y = Y^{(0)}$ of Figure XI.36. The set $X^{(i)}$ is the locus of the centres x of the hexagons H_x such that exactly two points of the six of $H_x/\{x\}$, belong to $X^{(i-1)}$. Express this mapping in terms of digital erosions. Compute the sets $X^{(7)}$ and $Y^{(7)}$. Check that $Y^{(2i)} = Y^{(6)}$ and $Y^{(2i+1)} = Y^{(7)}$ (cyclic process). Note the total lack of continuity of the mapping; X and Y differ by one point but the transforms rapidly diverge (Fig. XI.36).
(c) *Fractal set*: $X^{(0)}$ is made up of k simply connected particles lying inside the mask Z, and $X^{(i)} = [X^{(i-1)} \ominus \{L\}]^c$. What can one say about the behaviour of $X^{(i)}$ after one hundred iterations. What is the connectivity number of iteration no. 14,593? and that of iteration no. 14,594?

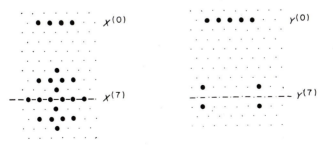

Figure XI.36.

I.10. Trend analysis and length measurements

The techniques for estimating a length or a perimeter from Crofton's formula (VIII-14) assume all the directions to be equiprobable. They are no longer valid following a digital treatment such as a thinning for example. In this case we must base the algorithms on the actual major directions of the grid. To resolve this question, we need a classical result from universal kriging (G. Matheron, 1969); it is quoted here without proof. Let $f(x) = a_0 + a_1 x + \varepsilon(x)$, in R^1, where the residual $\varepsilon(x)$ is a random function with stationary increments, and semi-variogram $\frac{1}{2}E[\varepsilon(x+h) - \varepsilon(x)]^2 = \omega|h|$. On the segment L, $f(x)$ is known at the points x_i and in particular at the two extremities x_0 and x_n of L. Then the straight line $D_{x_0 x_n}$ joining $f(x_0)$ to $f(x_n)$ is the best estimate of the trend $a_0 + a_1 x$ at the middle m of L where best means unbiased, and minimizing the variance of $f - D_{x_0 x_n}$ at the point m. Note that it is not necessary to know $f(m)$ to estimate the trend at the point m; the inside points play no role in the estimation (so the sampling can be irregular). This model of Matheron interpreted in $\mathbb{R}^2$ (Fig. XI.37a) is rather convenient for images produced by a scanner, whose noise is auto-correlated principally in one direction. It allows us to find algorithms for interpolation and for length measurement, which have a Euclidean meaning. Although it is particularly useful for connecting skeletons and conditional bisectors by linear interpolation, we only use it here for trend analysis. The result does not depend on the choice of the axes, and of course $f(x)$ is not assumed to be differentiable.

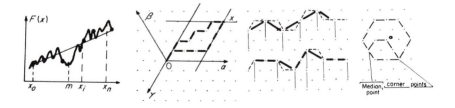

Figure XI.37.

(a) Length of a segment: Consider a point $x \in \mathbb{Z}^2$; let X be the hexagonal convex hull of $\{o; x\}$, and I_α, I_β and I_γ the intercept numbers of X in the basic directions. Interpreted on $\mathbb{R}^2$, the grid becomes a hexagonal lattice with a spacing a; and $|\overrightarrow{ox}|$ denotes the Euclidean length of $\overrightarrow{ox}$. Prove that

$$|\overrightarrow{ox}| = \frac{a}{\sqrt{2}}[(I_\alpha(X) - 1)^2 + (I_\beta(X) - 1)^2 + (I_\gamma(X) - 1)^2]^{1/2} \qquad (1)$$

[The Euclidean co-ordinates of x in the system (α, β) are x_1, x_2. We have $\overrightarrow{ox} = x_1 + jx_2 = (x_2 - x_1)j - x_1 j^2 = (x_1 - x_2) - x_2 j^2$ with 1, j, j^2 the cube roots of 1. So, $2|\overrightarrow{ox}|^2 = 2(x_1^2 + x_2^2 - x_1 x_2) = (x_2 - x_1)^2 + x_1^2 + x_2^2$, which is relation (1) in terms of $\mathbb{Z}^2$. Note that it is not necessary to define a digital segment ox, but only a convex hull, which is less ambiguous. There exist several paths from 0 to x and included in X, which have the projections $I_\alpha(X)$, $I_\beta(X)$ and $I_\gamma(X)$].

(b) Dodecagonal perimeter: The point $x \in \mathbb{Z}$ is now on the boundary of the hexagon $2H$ (Fig. XI.37d). Prove that the only two possible lengths $|ox|$ are $2a$ and $a\sqrt{3}$ (for the median and corner points respectively). Derive an algorithm for measuring Euclidean lengths from digital arcs (homotopic thinnings by $\{L\}$ for example).

[For the first part, see Figure XI.36 and apply relation (1). For the second part, consider an arc Y (Fig. XI.37c), made up of the chain $x_1 x_2 \ldots x_n$. There are two possible partitions of Y into successive groups of the three points. In both cases apply Matheron's algorithm to every group in order to estimate the local trend, and apply relation (1). The total Euclidean length $L(Y)$ of the smoothed arc is given by averaging the results of the two partitions; i.e.

$L(Y) = a(p + \sqrt{3/2}\, p' + 1)$ where p and p' are the numbers of configurations

(·—·—·) and (·—·⟋) respectively (modulo a rotation and a symmetry).]

(c) Trend Analysis: Consider a digital arc (for example a digital boundary). Show that the technique used in b) gives to each point x_i (extremities excepted) a Euclidean slope from among 12 possible directions. Similarly prove that in a hexagonal lattice, the number of possible directions of a segment that joins two points of Hausdorff distance k apart, is equal to the digital perimeter of the hexagon of size k. In particular, show that if the arc Y is the boundary of a set X closed by the hexagon $2H$, one can partition Y into groups of five successive points. Calculate Matheron's algorithm for the trend by using five point neighbourhoods, resulting in 24 classes of directions.

XII. Morphology for Grey-tone Functions

A. FROM SETS TO FUNCTIONS

This chapter is essentially a microcosm of mathematical morphology. All the notions developed elsewhere in the book as set descriptors reappear here as tools for handling functions. Nevertheless, the decision to consider only those function transformations which generalize the set operations of morphology, is quite a restriction. Since none of the set operations are linear, notions such as convolution, or Fourier filtering, will not be included in the scope of the generalization.

After having established a few fundamental results in Section B (the conditions given there allows us to say when a "pile" of sets will generate a function; also we can specify how one function tends towards another one) what follows is based on a stereological approach. Firstly we study transformations where each cross-section is treated independently of the others (Sect. C), then the transformations involving two cross-sections (Sect. D), and finally those requiring knowledge of two consecutive sections (connectivity criteria, Sect. E). The experimental section F deals with the first problem that a practitioner will face, namely translation in terms of sets of the picture under study. The last two sections are devoted to models. Section G proposes a generalization of the convex model, the convex ring and the stereological parameters. Section H extends the approach to random functions; two specific models, which generalize the Boolean model and the Poisson lines of Chapter XIII, are investigated.

Mathematical morphology for grey tone functions has to be investigated from three different points of view. Each transformation acts on the functions themselves, and also on their horizontal cross sections, and finally has a geometrical interpretation. Therefore a transformation is completely known when we are able to express it in terms of each approach, and to derive the corresponding properties. We shall fulfil such a program for all the notions presented below.

Except two mathematical results due to G. Matheron (1969) (compacity of the class of Umbrae, definition of a random function via its umbra), the literature on mathematical morphology for functions dates principally from the second part of the 70's. Thinned of its redundancies, reduces to the following major contributions. J. Serra (1975) started by extending the Hit or Miss transformation and the size distributions from sets to functions. The description of a function by its watersheds, and the associated algorithms, are due to Ch. Lantuéjoul (1977). Later S. Beucher and Ch. Lantuéjoul (1979) formalized the concept by using geodesic distances (this approach is not taken here). In the meantime, F. Meyer (1977) developed contrast descriptors based on residuals of openings (top hat transformation). A new wave of generalizations was introduced by S. Sternberg (1978) who systematically considered the functions as sets in R^3. This allowed him to define a number of morphological transformations for functions and more generally, to design sequences of morphological treatments. Rosenfeld (1977) proposed a generalization of the connectivity to functions, from which he derived a thinning algorithm (in collaboration with Dyer (1977)). The first survey of the question appears in V. Goetcharian's thesis (1979), where the author contributes several original notions such as the lower skeleton, convexity criteria, and so on. For the random version, the reader is referred to the study by R. Adler (1980) with regard to the upcrossings of a random function, and to D. Jeulin's work about roughners of metallic surfaces (1981).

An indirect input to the content of this chapter may be found in Fuzzy set theory (L. A. Zadeh, 1965; A. Kaufman, 1973). To say that a set is fuzzy means that instead of having an indicator function $k(x)$, it is replaced by a membership function taking intermediate values between 0 and 1. In other words, a fuzzy set is no longer a set but a picture bounded by 1. Here the sup and inf are of a special interest, describing the behaviour of a picture w.r. to the t-axis (affinity, addition, rel. (XII-25) etc. . . .), rather than the spatial structure that we develop.

B. GENERAL RESULTS

B.1. Semi-continuous functions and their thresholdings

Following the general line of the book, we are mainly concerned with the Euclidean $\mathbb{R}^2$ framework. Here the notion of an upper semi-continuous function (u.s.c.) defined on the plane corresponds to that of a closed set.

Remember that function $f(x)$, $x \in R^2$ is u.s.c. when for every x and every $t > f(x)$ there exists a neighbourhood V_x of x such that $f(y) > t$ for every $y \in V_x$. According to a classical topological result (Choquet, 1966), $f(x)$ is u.s.c. iff its horizontal cross sections are closed sets:

$$f(x) \text{ u.s.c.} \Leftrightarrow X_t(f) = \{x : f(x) > t\} \quad \text{a closed set } (-\infty < t < +\infty)$$

(Fig. XII.1a). When the context is not ambiguous, the u.s.c. functions are simply called "function"; their class is denoted $\mathscr{F}u$. We call a *picture* a positive u.s.c. function $f(x)$ bounded by the value m:

$$f(x) \text{ picture} \Leftrightarrow 0 \leq f(x) \leq m \qquad \text{(XII-1)}$$

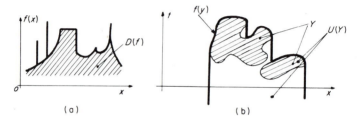

Figure XII.1. (a) umbra of a u.s.c. function (it looks like paint that trickled upwards). (b) umbra $U(Y)$ associated with the set Y.

The class of pictures is denoted $\mathscr{P}i$. Photographs and images are modelled by pictures, but we also need functions to perform a subtraction, or for certain parametrizations (for example, build a function from the set $X \in \mathscr{F}(\mathbb{R}^2)$ by adding its quench function to the negative of the quench function of X^c).

When the point x spans the Euclidean plane $\mathbb{R}^2$, the *support* of function $f(x)$ is the set of points x where $f(x) \neq -\infty$ (resp. $f(x) \neq 0$ for pictures). The class of functions (resp. pictures) with compact support is denoted by $\mathscr{K}u$ (resp. $\mathscr{K}i$).

By thresholding a function $f(x)$ at successive levels, we can associate a family of sets with the function. But by acting on those sets using methods of mathematical morphology, we can generate a new set family. Under what conditions do these new sets represent a function? The following theorem characterizes legitimate set families.

THEOREM XII-1. *Let f be a u.s.c. function defined in $\mathbb{R}^2$, and*

$$X_t(f) = \{x : f(x) \geq t\} \qquad -\infty \leq t \leq +\infty \qquad \text{(XII-2)}$$

is the set family generated by thresholding f at levels t. Then, the X_i's are closed and monotonically decreasing, i.e.

$$t' < t \Rightarrow X_{t'} \supset X_t \quad \text{and} \quad X_t = \operatorname*{Lim}_{t' \uparrow t} \downarrow X_{t'} = \bigcap_{t' < t} X_{t'} \qquad \text{(XII-3)}$$

and we have

$$f(x) = \text{Sup}\{t : x \in X_t\} \tag{XII-4}$$

Conversely, a family $X_t \, (-\infty \leq t \leq +\infty)$ of closed sets generates a u.s.c. function $f(x)$ if and only if conditions (XII-3) are satisfied; $f(x)$ is then defined by (XII-4) and satisfies (XII-2).

Proof: Relation (XII-2) obviously implies relation (XII-3). Moreover, it shows that $\text{Sup}\{t : x \in X_t\} \leq f(x)$ (remember that $\text{Sup}(\varphi(x))$ is the smallest upper bound $M > \varphi(x)$). But $x \in X_{f(x)}$, and $\text{Sup}\{t : x \in X_t\} \geq f(x)$, so relation (XII-4) results. Conversely, let X_t be a family of closed sets satisfying (XII-3). Monotonicity allows the function $f(x)$ defined by relation (XII-4) to be associated with the X_t's. To prove that (XII-2) is satisfied, consider the set family $Y_t = \{x : f(x) \geq t\}$. If $X \in X_t$, we have $f(x) \geq t$ by hypothesis; hence $X_t \subset Y_t$. Conversely if $x \in Y_t, f(x) \geq t$, and according to relation (XII-4), $x \in X_{t'}$ for every $t' < t$; hence $x \in \bigcap_{t' > t} X_{t'} = X_t$.   Q.E.D.

Remarks:
  (i) In proving the converse, the set family *must* be continuously *monotonic*. If not, relations (XII-2) and (XII-3) are incompatible. As a counter-example, take the following family:

$$X_t = \text{closed unit disk}, \, t < 1; \qquad X_t = \varnothing, \qquad t \geq 1$$

The X_t's are closed and decreasing. On one hand, relation (XII-2) implies that for every $t < 1$ the origin $O \in X_t$, hence $f(0) \geq 1$. On the other hand $0 \notin X_1$, since $X_1 = \varnothing$ and according to (XII-4) $f(0) < 1$, which is contradictory. Simply speaking, we could say that the condition "X_t is closed", closes $f(x)$ laterally, while monotonic continuity closes $f(x)$ upwards.
  (ii) In fact the condition "X is closed" is not really obligatory. A more general version of this theorem that avoids giving X_t topological status, can be found in G. Matheron (1975).
  (iii) If $f(x)$ is a random function, $X_t(f)$ becomes a random set, and the probability $\text{Pr}\{x \in X_t(f)\}$ is just the probability that $f(x) \geq t$. More generally, the spatial law associated with n points $x_1, \ldots, x_n$ is given by

$$\text{Pr}\{f(x_1) \geq t_1, \ldots, f(x_n) \geq t_n\} = \text{Pr}\{x \in X_{t_1} \cap X_{t_2} \cdots \cap X_{t_n}\}$$

where n thresholds $t_1 \ldots t_n$ intervene simultaneously. When $n = 2$ we are reduced to the probabilities and the moments of order two studied in Chapter IX, i.e. the classical covariance. But this type of generalization is different from that proposed in the next section, since there one cross-section only is involved.

(iv) Mutatis mutandis the same proof holds for the *lower* s.c. function by replacing the X_t's with the *open* decreasing sets X_t^- such that

$$X_t^- (f) = \{x : f(x) > t\} \qquad \text{(XII-5)}$$

and by replacing condition (XII-3) with lower semi-continuity:

$$X_t^- = \lim_{t' \downarrow t} \uparrow X_{t'}^- = \bigcup_{t' > t} X_{t'}^- = \bigcup_{t' > t} X_{t'} \qquad \text{(XII-6)}$$

B.2. Umbrae

The notion of an umbra (Sternberg 1978) is the link between the functions $f \in \mathscr{F}_n(\mathbb{R}^2)$ and the closed sets $Y \in \mathscr{F}(\mathbb{R}^3)$ of the space. Points of $\mathbb{R}^3$ are parametrized by their projection x on $\mathbb{R}^2$ and their altitude t on an axis perpendicular to $\mathbb{R}^2$. The umbra $U(Y)$, $Y \in \mathscr{F}(\mathbb{R}^3)$ is the dilate of Y by the positive axis $[0, +\infty]$ of the t's, i.e. since $(0, -\infty)$ is the transposed of $[0, +\infty]$:

$$U(Y) = Y \oplus [0, -\infty] = \{(x, t') : (x, t) \in Y; \ t' \le t\} \qquad \text{(XII-7)}$$

Note that $U(Y)$ is morphologically open by the negative t-axis $[0, -\infty]$ and closed by the positive t-axis $[0, +\infty]$. The class $\mathscr{U}$ of the umbrae of $\mathscr{F}(\mathbb{R}^3)$ plays a fundamental role in the morphological study of the functions $\mathscr{F}_u$. It is stable for intersections and for finite unions and is compact in $\mathscr{F}(\mathbb{R}^3)$ (Matheron, 1969). Each umbra U induces a unique function $f(U)$, whose value at point x is the sup of the t's such that $(x, t) \in U$. Indeed, the horizontal section $X_t = U \cap \Pi(t)$ of U by the horizontal planes $\Pi(t)$ at altitude t satisfy the conditions (XII-3), and generate a function $f(U)$. Conversely, given a function $f \in \mathscr{F}_u$, the set:

$$U(f) = \{(x, t) : f(x) \ge t\} \qquad x \in \mathbb{R}^2, \qquad t \in \mathbb{R} \qquad \text{(XII-8)}$$

is by construction an umbra, and we have:

$$U(f) = \{(x, t) : x \in X_t(f)\} \qquad X_t(f) = \{x : (x, t) \in U(f)\} \qquad \text{(XII-9)}$$

(In analysis, $U(f)$ is often called the graph of function f; we prefer to avoid this term which could be confused with its use in graph theory (Ch. VI, for example), and because the same concept of umbra applies to sets and to function). Finally we have the following implications:

$$X \in \mathscr{F}(\mathbb{R}^3) \Rightarrow U(X) \in \mathscr{U}(\mathbb{R}^2 \times \mathbb{R}) \rightleftharpoons f(U) \in \mathscr{F}u \qquad \text{(XII-10)}$$

There is a unique umbra and a unique u.s.c. function associated with each closed set X of $\mathbb{R}^3$; conversely, each u.s.c. function f corresponds to a unique umbra $U(f)$, but to the infinity of the closed sets of $\mathbb{R}^3$ which possess the same umbra $U(f)$.

B.3. Topology

The simplest way to provide functions (and pictures) with a topology suitable for the morphological treatment is to use the topology induced by the Hit or Miss topology on their umbrae. The induced properties can be easily derived from the general theorems of Chapter III. Without going into all the details, we will quote the criteria of convergence.

THEOREM XII-2. *A sequence* $\{f_i\}$ *of u.s.c. functions converges towards the u.s.c. function f, if and only if the two following conditions are fulfilled:*

(a) *for every point* $x \in \mathbb{R}^2$, *there exists a sequence* $\{x_i\} \to x$ *such that* $f_i(x_i) \to f(x)$

(b) *if a subsequence* $\{x_{i_k}\} \to x$, *then the subsequence* $\{f_{i_k}(x_{i_k})\}$ *satisfies* $\overline{\lim} \{f_{i_k}(x_{i_k})\} \leq f(x)$

   Moreover, if $\underline{\text{Lim}}\ f_i$ *and* $\overline{\text{Lim}}\ f_i$ *are respectively the intersection and the union of the limits of all the subsequences of* $\{f_i\}$ *which converge, then*

(a') *the function* $\underline{\text{Lim}}\ f_i$ *is the sup of the u.s.c. functions satisfying* (a);

(b') *the function* $\overline{\text{Lim}}\ f_i$ *is u.s.c. and is the inf of the u.s.c. functions satisfying* (b).

Remarks:

 (i) The second part of the theorem is a criterion for semi-continuity for semi-continuous mappings, as defined in III-24.

 (ii) If, for every t, the sequence $\{X_t(f_i)\}$ is (semi) continuous w.r. to Hit or Miss topology in $\mathbb{R}^2$, then $\{f_i\}$ is also (semi) continuous, but the converse is false.

 (iii) Note that for this topology, the sup of two u.s.c. functions is a continuous operation, but their inf, their sum and their product are only semi-continuous.

C. TRANSFORMATIONS BASED ON ONE CROSS-SECTION

The content of this section dates from 1975 (Serra), it was the first attempt to extend mathematical morphology to functions.

   Throughout the section, each different level of the transformed function is obtained from only one level of the original $f(x)$.

C.1. Elementary transformations

(a) *Translation* f_h *of f by the vector h in* $\mathbb{R}^2$

$$f_h(x) = f(x - h) \Rightarrow X_t(f_h) = [X_t(f)]_h$$

(b) *Affinity λf of f* with respect to the t-axis:

$$X_t(\lambda f) = \{x : \lambda f(x) \geq t\} = \{x : f(x) \geq t/\lambda\} = X_{t/\lambda}(f) \qquad \text{(XII-11)}$$

(affinity is stable for pictures if and only if $\lambda \geq 0$).

(c) *Similarity $f(\lambda x)$ of f* with respect to the plane $\mathbb{R}^2$, $(\lambda \geq 0)$:

$$X_t[f(\lambda x)] = \lambda X_t[f(x)] \qquad \text{(XII-12)}$$

For $\lambda > 1$, affinity stretches $f(x)$ upwards without modifying its base, whereas similarity enlarges its sections, without modifying its altitudes (see Fig. XII.3a).

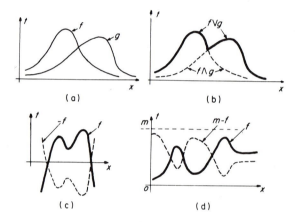

Figure XII.2. (a), (b) sup $f \vee g$ and inf $f \wedge g$ of functions f and g. (c), (d) the two modes of complementation for functions (b) and for pictures (c).

(d) *Addition $f(x) + \mu$ of a relative constant μ to f*:

$$X_t[f(x) + \mu] = X_{t-\mu}(f) \qquad \text{(XII-13)}$$

(the class of pictures is stable under addition if and only if $0 < \mu < \infty$).

(e) *Complementation*

— for functions: defined by symmetry with respect to the plane $t = 0$.

$$[X_t^-(-f)]^c = \{x : -f \leq t\} = X_{-t}(f) \qquad \text{(XII-14)}$$

— for pictures: defined by symmetry with respect to the median plane $m/2$:

$$[X_t^-(m-f)]^c = \{x : m-f \leq t\} = \{x : f \geq m-t\} = X_{m-t}(f) \qquad \text{(XII-15)}$$

In photography for example, this latter operation corresponds to taking the negative of a photograph (see Fig. XII.16). Note that in both cases, the sets $\{X_t^-\}$ associated with the l.s.c. functions $-f$ and pictures $m-f$ are open. Two function (resp. picture) transformations ψ and ψ^* are said to be dual of each other w.r. to the complementation when:

$$\left.\begin{array}{ll} \psi^*(f) = -\psi(-f) & \text{for function} \\ \psi^*(f) = m - \psi(m-f) & \text{for pictures} \end{array}\right\} \quad \text{(XII-16)}$$

C.2. Erosions and dilations by 2D-sets

Suppose we transform f into g ($f, g \in \mathscr{F} u$), with sections $Y_t(g)$. If we demand that for each $t, Y_t(g)$ comes exclusively from $X_t(f)$, the set mapping $Y_t = \Psi_t(X_t)$ must be increasing in X and decreasing in t. We will assume for the moment that Ψ is independent of t, and put aside the case when $\Psi = \Psi(t)$ for Exercise XII.I9. Then (Ch. II, Sect. F) Ψ is composed of unions of erosions, or equivalently of intersections of dilations. Indeed the two simplest prototypes, i.e. erosion and dilation, of each X_t by the same compact set B satisfy the implications:

$$X_t \subset X_{t'} \Rightarrow X_t \oplus B \subset X_{t'} \oplus B; \qquad X_t \downarrow X_{t'} \Rightarrow X_t \oplus B \downarrow X_{t'} \oplus B$$

(and ditto when $\oplus$ is replaced by $\ominus$), and therefore generate a function f', which will be denoted by $f' = f \oplus B$ (resp. $f \ominus B$).

The geometrical interpretation of $f \oplus B$ and $f \ominus B$ is extremely easy. Denote the sup (resp. the inf) of $f(x)$ and $g(x)$ by $f \vee g$ (resp. $f \wedge g$), and notice that

$$\left.\begin{array}{l} X_t(f \vee g) = \{x : f(x) \text{ or } g(x) \geq t\} = X_t(f) \cup X_t(g) \\ X_t(f \wedge g) = \{x : f(x) \text{ and } g(x) \geq t\} = X_t(f) \cap X_t(g) \end{array}\right\}$$

By taking for g all the translates f_x of f as x runs over $\check{B}$, we obtain:

$$\left.\begin{array}{l} X_t(f \oplus B) = \{x : \exists y \in \check{B}_x, f(y) \geq t\} \\ X_t(f \ominus B) = \{x : \forall y \in \check{B}_x, f(y) \geq t\} \end{array}\right\} \quad \text{(XII-17)}$$

or, equivalently

$$\left.\begin{array}{l} (f \oplus B)(x) = \sup\{f(y); y \in \check{B}_x\} \\ (f \ominus B)(x) = \inf\{f(y); y \in \check{B}_x\} \end{array}\right\} \quad \text{(XII-18)}$$

As we can see on Figure XII.3, $f \ominus B$ reduces the peaks and enlarges the valleys, and vice versa for $f \oplus B$.

Note that both functions and pictures are stable with respect to Minkowski operations. However, their behaviour differs under complementation since relations (XII-11), (XII-12) and (XII-14) imply:

$$\left.\begin{array}{ll} [X_{-t}(f \oplus B)]^c = X_t^-(-f \ominus B) & \text{for functions} \\ [X_{m-t}(f \oplus B)]^c = X_t^-[(m-f) \ominus B] & \text{for pictures} \end{array}\right\} \quad \text{(XII-19)}$$

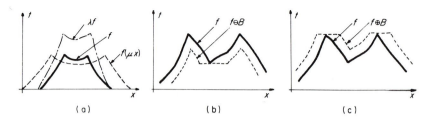

Figure XII.3. (a) For $\lambda > 0$, affinity stretches without enlarging, similarity enlarges without stretching. (b) erosion, (c) dilation, of f by B.

The equivalent expression in terms of functions is immediate:

$$\left.\begin{array}{l} -(f \ominus B) = -f \oplus B \qquad \text{for functions} \\[4pt] m - (f \ominus B) = (m - f) \oplus B \qquad \text{for pictures} \end{array}\right\} \qquad \text{(XII-20)}$$

The basic algebraic properties of the Minkowski operations are given in Exercises XII-1. According to Theorem XII-2 (remark (ii)) the mappings $f \oplus B$ and $f \ominus B$ of $\mathcal{U}(\mathcal{F}) \times \mathcal{K}$ into $\mathcal{U}(\mathcal{F})$ are upper semi-continuous, even when dilation is continuous. The three major prototypes of erosion and dilation are obtained by taking B to be a pair of points, a segment, and a disk. There is no particular difficulty encountered in transposing over all the morphological results already obtained for the set case. Consider for example, the chord distribution (B a segment) and the law of first contact (B a disk). To make things simpler, assume that the support S of f is bounded and choose a point x uniformly in $S \oplus B$ (Ch. VIII, 1st situation) in order to generate the randomness. Under these conditions:

(a) Denote by $P_t(l, \alpha)$ the probability that segment $(l, \alpha)_x$ is included in X_t. The chord distribution $P''_t(l, \alpha) - P'_t(0, \alpha)$ (derivatives are with respect to l) is just the proportion of arrows flying at level t and direction α, that pierce through the umbra $U(f)$ and travel along for a length l.

(b) Denote by $Q_t(\rho B)$ the probability that the disk ρB_x is included in the pores of X_t. The law of first contact $Q_t(\rho B)/Q_t(0)$ is the conditional probability that the first point x' around x, where $f(x') \geq t$, is at a distance greater than ρ, given $f(x) < t$.

C.3. Size distributions

Generally speaking, any set transformation mapping $\mathcal{U}(\mathcal{F})$ into itself, which satisfies the morphological principles and Matheron's axioms for sizing (Ch. X, Sect. A.2) defines a size distribution on function f, via its umbra $U(f)$. If we now restrict ourselves to the size criteria accessible from *one* cross-section, essentially we find the three following types of size distribution.

(a) *Thresholding*
Define

$$\Psi(X_t) = X_t \quad \text{for } t \leq \lambda \qquad (\lambda > 0)$$
$$\Psi(X_t) = \varnothing \quad \text{for } t > \lambda$$

The transformation cuts off all those points of $U(f)$ with an altitude higher than $\lambda > 0$ (for pictures only).

(b) *Connectivity*
Consider the connected components of $X_t(f)$, and remove those whose areas (or perimeters, or diameters of the inscribable disks, and so on . . .) are smaller than λ (for functions and pictures).

(c) *Openings and closings*
For B a compact convex set, the family $(f \ominus \lambda \check{B}) \oplus \lambda B = f_{\lambda B}$ is the size distribution of f by openings (for functions and pictures). The geometrical interpretation of the opening $f_{\lambda B}$ is given by the two reciprocal formulae

$$X_t(f_{\lambda B}) = \{x : \exists B \in x, \text{ and } \forall z \in B; f(z) \geq t\} \Big\rbrace$$
$$f_{\lambda B}(x) = \text{Sup}\{\inf\{f(z), z \in \lambda B_y\} : y \in \lambda \check{B}_x\} \Big\rbrace \qquad \text{(XII-21)}$$

At the point $x, f_{\lambda B}(x)$ has the highest value of the infimums of f taken over all the B's containing x. Equivalently, $x \in X_t(f_{\lambda B})$ if and only if $U(f)$ can be penetrated by a vertical cylinder $C(\lambda B)$ of section λB, which hits the point x, and such that all the altitudes of f within $C(\lambda B)$ are higher than t. We leave the reader to derive by duality similar interpretations for closings, and to enlarge the approach by replacing the set B with a function $g(u)$. Figure XII.4 illustrates the changes of shapes that result from openings, and shows how it cuts down the peaks and the saddles. Note that the deformation is the same whatever the altitude of the peak or the saddle; it depends only on their "sharpness". Any size distribution on f, which is u.s.c. and translation invariant, is a union of openings. This operation turns out to be the prototype of sizing in

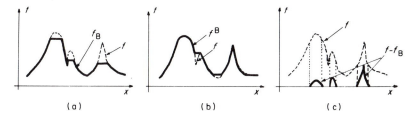

(a)   (b)   (c)

Figure XII.4. (a), (b) opening and closing of a function by a compact convex set, (c) difference $f - f_B$ between f and its opening.

exactly the same way as dilation was the prototype of an increasing mapping. Moreover, if we want sizings compatible with the group of similarities, we must restrict ourselves to opening by homothetic convex sets λB only.

Opening can be interpreted as a 2-D filtering process, but in a rather special sense. Consider the Fourier expansion of a function f (in one dimension for simplicity's sake). Let $\{a_i\}$ $(a_i \geq 0)$ be the energies associated with the frequencies. In Fourier space, a filter $\{\varphi_i\}$ $(0 \leq \varphi_i \leq 1)$ acts by replacing $\{a_i\}$ with $\{\varphi_i a_i\}$. If we iterate the filtering we obtain $\{\varphi_i^2 a_i\}$, which is identical to $\{\varphi_i a_i\}$ if and only if $\varphi_i = 1$ or 0. If so, we can see that the sizing axioms are satisfied for $\{a_i\}$, i.e. satisfied in the *Fourier* plane. Such is the case, for example, for the low pass filters $(\varphi_i = 0$ for $i \geq i_0, \varphi_i = 1$ for $i < i_0)$. On the other hand, the opening acts in the plane of definition of f itself. It happens for example that f is always above $f_{\lambda B}$ (first axiom), but can be below its filtered version obtained by a low pass filter. The second difference between the two techniques is that the opening is *not* linear, but only satisfies the relationship $f \leq g \Leftrightarrow f_B \leq g_B \forall B \in \mathcal{K}$. The third, and main difference is not logical but morphological. The residual $f - f_B$ extracts the sharp peaks, ridges and saddles of f. Then a further morphological treatment can easily separate them from each other. Nothing equivalent exists with Fourier transformation. Finally the opening is essentially a digital technique, whose numerical implementation in $\mathbb{R}^2$ is simpler than the fast Fourier transform. All the mappings presented in this Section C are digitalizable using the covering representation, defined on the umbra $U(f)$, and therefore are particularly robust. The digital algorithms are formally identical to the corresponding Euclidean ones, and the transforms exhibit the same number of grey levels as the original pictures.

D. TRANSFORMATIONS BASED ON TWO CROSS-SECTIONS

D.1. Sums and differences

We now assume that two cross-sctions can be obtained simultaneously, and the results combined. The new major pieces of information then accessible are addition and subtraction of the cross-sections. Since the difference between two pictures may not be a picture, a distinction must be made here between functions and pictures. In fact,

$$f, g \in \mathcal{P}i \Rightarrow f - g \in \mathcal{P}i \quad \text{if and only if} \quad X_t(g) \subset X_t(f),$$
$$0 \leq t \leq m \tag{XII-22}$$

This relationship is automatically satisfied, in particular when g is the transform of f by an *anti-extensive* and u.s.c. mapping, such as an opening, or an erosion by a structuring element containing the origin.

THEOREM XII-3. *For f and $g \in \mathscr{F}u$, the sum $f + g \in \mathscr{F}u$ and its sections satisfy the formula*:

$$X_{t_0}(f+g) = \bigcup_t [X_t(f) \cap X_{t_0-t}(g)] \qquad f, g \in \mathscr{F}u \qquad \text{(XII-23)}$$

Proof: We have:

$$X_{t_0}(f+g) = \{x : f + g \geq t_0\} = \bigcup_t \{x : f = t; g \geq t_0 - t\}$$

The last equation is also true if we replace $f = t$ by $f \geq t$ on the r.h.s.

Indeed, let $x \in X_t(f)$, i.e. $f(x) = t + t'_0$, with $t'_0 \geq 0$. If $x \notin X_{t_0-t}(g)$, it does not contribute to $\bigcup_t \{x : f = t; g \geq t_0 - t\}$. If $x \in X_{t_0-t}(g)$, then according to (XII-23), $x \in X_{t_0+t'_0}(f+g)$, which is contained in $X_{t_0}(f+g)$.   Q.E.D.

COROLLARY 1: *For f and $-g \in \mathscr{F}u$, the difference $f - g$ is given by the formula*

$$X_{t_0}(f-g) = \bigcup_t [X_t(f)/X^-_{t-t_0}(g)] \qquad f, -g \in \mathscr{F}u \qquad \text{(XII-24)}$$

In the same way, for f and g lower semi-continuous functions, we have

$$X^-_{t_0}(f+g) = \bigcap_t [X^-_t(f) \cup X^-_{t-t_0}(g)] \qquad -f, -g \in \mathscr{F}u$$

Both relations come from (XII-23) by applying the complementation formula (XII-11). Notice the formal similarity between relation (XII-23) and the convolution product.

COROLLARY 2: *Relations (XII-23) and (XII-24) remain valid for pictures assuming $f - g \geq 0$ in the latter relation, and we can restrict the range of variation of t to be a bounded zone, namely*:

$$X_{t_0}(f+g) = \bigcup_{t=t_0-m}^{m} [X_t(f) \cap X_{t_0-t}(g)] \qquad f, g \in \mathscr{P}i \qquad \text{(XII-25)}$$

$$X_{t_0}(f-g) = \bigcup_{t=0}^{m-t_0} [X_{t+t_0}(f)/X^-_t(g)] \qquad f, -g \in \mathscr{P}i \qquad \text{(XII-26)}$$

(*When summing up pictures, negative values of t for which $X_t(f) = \mathbb{R}^2$ always intervene, resulting in a picture with range of greys $(0, 2m)$.*)

D.2. Anamorphoses

When dealing with a picture, one often changes its scale of greys, by analogic or digital means (contrast of a photograph, correction of a T.V. camera, cut-off of high values). The most usual transformations of this type are $f(x) \rightarrow af(x) + b(a, b > 0)$; $\text{Log } f(x)$; $[f(x)]^2$; $\sqrt{f(x)}$; $f(x)$ for $f(x) \leq \lambda$, λ for $f(x) \geq \lambda$, and their various combinations. It is essential that we know the interferences between these grey scale changes and the morphological operations.

Define an anamorphosis $\Psi(f)$ to be an *increasing* and *continuous* mapping of $\mathscr{P}$ onto $\mathscr{F}u$. Indeed if f is u.s.c. $\Psi(f)$ is also u.s.c.; the inverse mapping Ψ^{-1} exists, and is increasing. Two simple rules govern the interaction of an anamorphosis Ψ with the morphological operations, namely

$$\text{Sup}\{\Psi(f(x)); x \in Z\} = \Psi(\text{sup}\{f(x); x \in Z\}) \qquad \text{(XII-27)}$$

i.e. the sup of the transform $\Psi(f(x))$ over a domain Z occurs at the same point x_0 where $f(x)$ is sup, and is the transform $\Psi(f(x_0))$ of $f(x)$ at this point.

$$X_t(\Psi(f)) = \{x : \Psi(f) \geq t\} = \{x : f \geq \Psi^{-1}(t)\} = X_{\Psi^{-1}(t)}(f) \text{ (XII-28)}$$

i.e. an upcrossing of $\Psi(f)$ at level t is just the upcrossing of f at level $\Psi^{-1}(t)$.

By applying these two rules, we can re-interpret the morphological transformations encountered in this and the previous sections. As an example, take the anamorphosis $\text{Log}(f)$ which takes the set of *pictures* into the functions. First of all, we see from relations (XII-15) and (XII-27) that all the operations which combine erosions and dilations commute with the anamorphosis. For example $(\log f)_B = \log(f_B)$, i.e. the opening of $\log f$ by B is equal to the log of the opening of f by B. Secondly, all the formulae concerning sums and differences of pictures are transposed into products and ratios, and relation (XII-25) becomes, via relation (XII-28)

$$X_{\text{exp. } t_0}(f.g) = \bigcup_{t_0-m}^{m} [X_{\text{exp. } t}(f) \cap X_{\text{exp. } (t_0-t)}(g)] \qquad \text{(XII-29)}$$

In other words, the algorithm for the product is exactly that for the sum, up to a change in the labelling of the cross-sections.

D.3. Examples of differences

In the following three examples $f \in \mathscr{P}i$ is a picture and B is the unit disk.

(a) *The law of first contact*
The quantity $f - (f \ominus \rho B)$ is a picture whose value at the point x equals the sharpest drop in altitude around x within a distance ρ (*mutatis mutandis* for $(f \oplus \rho B) - f$). In terms of sets, $X_t[f - (f \ominus \rho B)]$ is the set of points x whose altitude decreases more than t within a distance ρ.

(b) *Top hat transformation*
Now consider $f - f_{\rho B}$. This quantity is again a picture, made up of the residual from the opening by ρB and f (Fig. XII.4c). It extracts topographical peaks, isthmuses and capes. More precisely, take all the disks ρB_y containing the

point x. With each of them associate the sharpest drop from x to any point of ρB_y. The largest of these sharpest drops is $f - f_{\rho B}(x)$. In the reciprocal representation $X_t(f - f_{\rho B})$ is the set of points x such that in every ρB_y containing x, the maximum drop is equal to at least t. Using relation (XII-25) we obtain the relation

$$\left.\begin{array}{ll} X_{t_0}(f - f_{\rho B}) = \displaystyle\bigcup_{t=0}^{m - t_0} [X_{t + t_0}(f)/X_t^-(f_{\rho B})] & \text{continuous case} \\[3mm] X_{i_0}(f - f_{\rho B}) = \displaystyle\bigcup_{i=0}^{i_{max} - i_0} [X_{i + i_0}(f)/X_{i + 1}(f_{\rho B})] & \text{digital case} \end{array}\right\} \qquad \text{(XII-30)}$$

which represents the original algorithm given by F. Meyer, when he proposed the top-hat transformation (1977). Since F. Meyer's pioneering work, devoted to the extraction of chromatin in cell nuclei, thousands of experiments have proved the power of algorithm (XII-30) for contrast extraction in pictures. Not only does it lead to a size distribution involving the contrast of the image (by varying t_0 given ρ, or varying ρ given t_0, in (XII-30)) but it also provides one of the best algorithms for segmenting images, since it is insensitive to low variations of grey tones (see Fig. XII.5).

(c) *"Gradients"*
The subsection heading is in inverted commas because there is a distinction between the meaning of gradient in physics, and its use in some of the algorithms proposed below. We assume that f is differentiable everywhere in its domain of definition, except on a set of regular curves S of $\mathbb{R}^2$; but each partial derivative has a limit on both sides of S. The difference between the two values of the partial derivative is called their jump when crossing S. Under such conditions, it can be shown (L. Schwartz, 1967, Ch. II.2) that the derivative of f with respect to the co-ordinate h of direction α, is

$$\frac{\partial f}{\partial h}(x) = \left\{\frac{\partial f}{\partial h}\right\}(x) + \sigma \cos(\theta - \alpha)\delta_x(S) \qquad \text{(XII-31)}$$

where the function $\{\partial f / \partial h\}$ is the derivative of the continuous term, and σ is the jump at the point $x \in S$, θ is normal to S at the point x, orientated towards positive jumps, and $\delta_x(S)$ is the Dirac measure at the point x (Fig. XII.6b). It turns out that $\partial f / \partial h$ is a measure, whose second term is non-zero only when the derivative is taken at a point $x \in S$. This model is fairly irregular, since it tolerates vertical cliffs, such as in Figure XII.6a or on the photograph in Figure XII.5b, but not isolated peaks or walls with a zero thickness. Because of these restrictions on the irregularities allowed, $\partial f / \partial h$ turns out to be a measure, and so is numerically accessible following a small smoothing.

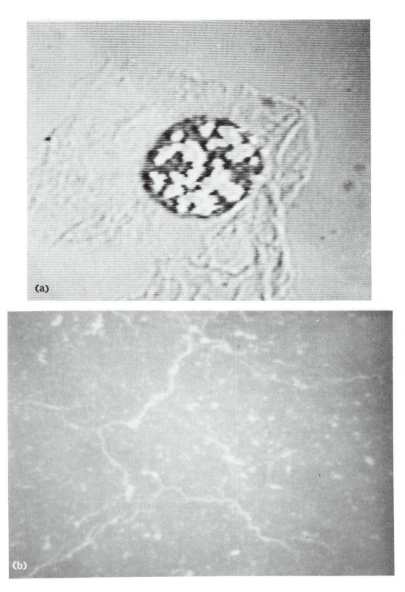

By definition the *gradient* of f at the point x is the vector (or vectorial measure) $\{\partial f/\partial x_1, \partial f/\partial x_2\}$ associated with the differential df by

$$df = \frac{\partial f}{\partial x_1} dx_1 + \frac{\partial f}{\partial x_2} dx_2 \qquad \text{(XII-32)}$$

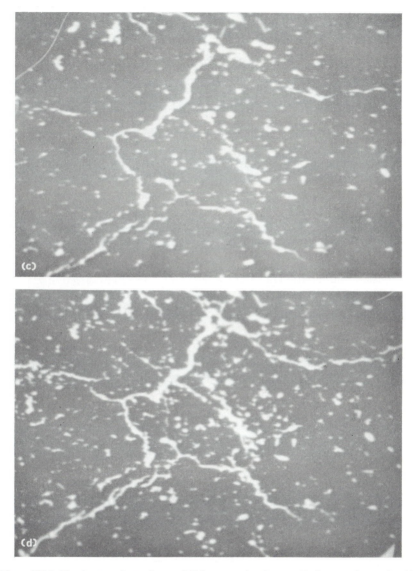

Figure XII.5. Top-hat transformation used (i) for extracting the zone Z of coarse chromatine (a) in a nucleus. (Afterwards, one can measure the DNA content in Z, or study the size and the shape of the coarse chromatine by performing openings and skeletons on Z and on its complement.) (ii) in a specimen of soil (b), for extracting the fissures (d) (which are non-accessible by a simpler thresholding the specimen (c)).

and is independent of the orthogonal system (x_1, x_2) (in spite of the formalism of (XII-32)).

Figure XII.6. (a) Type of discontinuity tolerated by the model (vertical cliff) (b) crossing a cliff.

In algebraic topology, one derives the notion of a *ridge*, (or a *crest-line*) from the gradient of a regular function. Let ρ be the modulus of $\overrightarrow{\text{grad}}\ f$. Consider a section $X_t(f)$ and the variation of ρ along the boundary ∂X_t. Those points $y \in \partial X_t$ where this variation reaches a maximum are by definition the points of the ridge of $f(x)$ at level t. Similarly, one defines also the ruts (or thalweigs) of f as to be the ridges of $-f$. Remark that ridges and ruts are *local* notions, exactly as the gradient is.

Consider the variation of f along a vector $\overrightarrow{dh}$ with a *fixed* direction α. From (XII-32) we have

$$\frac{df}{dh} = \frac{\partial f}{\partial x_1} \cos \alpha + \frac{\partial f}{\partial x_2} \sin \alpha = \langle \overrightarrow{\text{grad}\ f}, \overrightarrow{u}\ (\alpha) \rangle = \rho \cos (\theta - \alpha) \qquad \text{(XII-33)}$$

where θ and ρ are respectively the direction and the module of $\overrightarrow{\text{grad}}\ f$ at a point x. This formula bridges the gap between knowing f itself and knowing it from intercepts and suggests an algorithm for estimating the module ρ of $\overrightarrow{\text{grad}}\ f$. Let

$$g_1(x) = \text{Sup}\{f(x), f(x + dh)\} \quad \text{and} \quad g_2(x) = \inf\{f(x), f(x + dh)\}$$

Obviously

$$\frac{(g_1 - g_2)}{dh} = \left|\frac{df}{dh}\right| = \rho|\cos(\theta - \alpha)| \qquad \text{(XII-34)}$$

and by averaging over α

$$\rho(x) = \frac{1}{4} \int_0^{2\pi} \frac{g_1(x) - g_2(x)}{dh} d\alpha \qquad \text{(XII-35)}$$

In digital analysis, this formula is approximated for example on the hexagonal grid, by

$$\rho^*(x) = \frac{1}{24} \sum_{i=1}^{6} |f(x) - f(x_i)|; \qquad x_i \text{ the six neighbours of } x. \qquad \text{(XII-36)}$$

The range of grey levels of ρ^* is six times that of f. If such a detail seems excessive for a particular application, ρ can be replaced by another and simpler gradient ρ_1 defined from circular erosions:

$$\rho_1(x) = \lim_{\lambda \to 0} \frac{(f \oplus \lambda B) - (f \ominus \lambda B)}{2\lambda} \qquad (B \text{ closed unit disk}) \quad \text{(XII-37)}$$

proposed by S. Beucher (1978). For the function models adopted in the present section ρ and ρ_1 coincide, but not for more general $f \in \mathscr{F}u$ (if they exhibit vertical walls without thickness for example).

Both gradients ρ and ρ_1 provide pictures proportional to the *degree* of "verticality" at the point x. If now we want an indicator of the "horizontality" at x, we must perform elementary transformations of $f(x)$ along the t-axis. The two simplest ones are:

$$\rho_3 = \lim_{\lambda \to 0} \left[\frac{f(x+\lambda) - f(x)}{\lambda} \right] \qquad \lambda \geq 0$$

$$\rho_4 = \lim_{\lambda \to 1} \left[\frac{f(\lambda x) - f(x)}{\lambda} \right] \qquad \lambda \geq 1$$

The readers interested in digitizing the gradients can consult W. Pratt's book (1978).

D.4. Erosions and dilations by functions

The extension of Minkowski operations to functions and pictures is due to S. Sternberg who developed from this starting point, a comprehensive set of morphological generalizations (openings, rolling ball, . . .) and of new notions (perceptual graphs) (1978) (1980). It takes the *set* Minkowski operations performed on umbrae and transposes the results in terms of functions.

The special role played by the third dimension (t-axis) leads us to introduce the symmetry by reflection. Given a set Y, the reflected set $\hat{Y} = \{(x, t);\ (x_1 - t) \in Y\}$ is symmetrical to Y with respect to the horizontal plane of coordinates (Fig. XII.7). This new transformation interacts with the elementary operations of transposition $Y \to \check{Y}$, of complementation $Y \to Y^c$ and of umbriz-

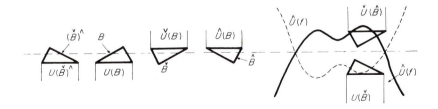

Figure XII.7. Set B, its reflected $\hat{B}$, its transposed $\check{B}$, and the corresponding umbrae.

ation $Y \rightarrow U(Y)$, according to a few algebraic rules, such that:

$$(\hat{B})^{\wedge} = B; \quad (\check{B})^{\wedge} = (\hat{B})^{\vee}; [\hat{U}(B)]^{\vee} = U[\hat{B}^{\vee}];$$
$$\hat{U}(f) = [U(-f)]^{c} \qquad \text{(XII-38)}$$

Concerning the dilation and the erosion, the equivalences:

$$\check{B}_{x,t} \Uparrow U(Y) \Leftrightarrow \check{U}(B_{x,t}) \Uparrow U(Y)$$

and

$$(\hat{B}_{x,t})^{\vee} \subset U(Y) \Leftrightarrow U(\check{B}_{x,t})^{\wedge} \subset U(Y) \Leftrightarrow [\hat{U}(B_{x,t})]^{\vee} \subset U(Y)$$

imply that

$$U(Y) \oplus B = U(Y) \oplus U(B); \qquad U(Y) \ominus \hat{B} = U(Y) \ominus \hat{U}(B) \quad \text{(XII-39)}$$

Suppose now that Y itself is an umbra; therefore it characterizes the function f given by the relation $U(f) = Y$. Then we define the Minkowski sum $f \oplus B$ of the function f by the structuring element $B \in \mathcal{K}(\mathbb{R}^2 \times \mathbb{R})$ via the umbrae, by putting:

$$U(f \oplus B) = U(f) \oplus B \qquad f \in \mathcal{F}u, B \in \mathcal{K}(\mathbb{R}^3) \qquad \text{(XII-40)}$$

It is not possible to introduce the Minkowski subtraction simply by replacing $\oplus$ by $\ominus$ in relation (XII-40), since $U(f) \ominus U(B)$ is reduced to the point at $-\infty$, whatever f and B are. Preferably, we will derive it by duality, by starting from algorithm (XII-16) (for functions). This algorithm, applied to the present case, implies that

$$\hat{U}(f \ominus B) = [U(-f) \oplus B]^{c}$$

The right-hand side of this expression delineates the set of points (x, t) such that $\check{B}_{x,t}$ misses $U(-f)$, by reflection the condition becomes $(\check{B}_{x,t})^{\wedge} \subset U(f)$ and finally

$$U(f \ominus B) = U(f) \ominus \hat{B} \qquad f \in \mathcal{F}_u, B \in \mathcal{K}(\mathbb{R}^3) \qquad \text{(XII-41)}$$

The functions $f \oplus B$ and $f \ominus B$ belongs to $\mathcal{F}_u$. As a particular case, if we take for B a 2-D set lying in the horizontal plane $t = 0$, relations (XII-40) and (XII-41) lead to the dilation $f \oplus B$ and the erosion $f \ominus B$ of a set by a function, as presented in Section C.2. Indeed, in this case, the umbra $U(B)$ is just the half cylinder of top B (Fig. XII.8b).

The two definitions (XII-40) and (XII-41) can immediately be extended to operations for functions. By taking for $B = U(g)$ the umbra of function g, we obtain:

$$\left.\begin{array}{l} f \oplus g \rightarrow U(f \oplus g) = U(f) \oplus B = U(f) \oplus U(B) = U(f) \oplus U(g) \\ f \ominus g \rightarrow U(f \ominus g) = U(f) \ominus \hat{B} = U(f) \ominus \hat{U}(B) = U(f) \ominus \hat{U}(g) \\ \text{with} \quad -(f \ominus g) = (-f) \oplus g \end{array}\right\} \quad \text{(XII-42)}$$

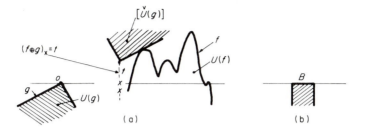

Figure XII.8. (a) Dilations of umbrae. (b) Flat structuring element and its umbra.

For f, $g \in \mathscr{F}_u$, and g with a compact support, $f \oplus g$ and $f \ominus g$ belong to $\mathscr{F}_u$. The dilation is continuous, increasing with respect to f and to g and extensive with respect to f when g contains the origin. The erosion is u.s.c., increasing with respect to f and decreasing with respect to g (i.e. $f \le f' \Rightarrow f \ominus g \le f' \ominus g$; $g \le g' \Rightarrow f \ominus g \ge f \ominus g'$), and extensive with respect to f when g contains 0. All the classical properties of distributivity and iterativity for sets (Ch. II) are still valid for the Minkowski operations on functions.

Relations (XII-42) provide not only the definitions of $f \oplus g$ and $f \ominus g$, but also the geometrical interpretations in terms of umbrae; their translation in terms of sup and inf results and gives:

$$(f \oplus g)_x = \{ \sup t : \check{U}(g)_{x,t} \Uparrow U(f) \} = \{ \inf t : \check{U}(g)_{x,t} \subset [U(f)]^c \}$$

$$(f \ominus g)_x = \{ \sup t : [\hat{U}(g)_{x,t}]^{\vee} \subset U(f) \} = \{ \inf t : [\hat{U}(g)_{x,t}]^{\vee} \Uparrow [U(f)]^c \}$$

(XII-43)

(see Fig. XII.8a).

Alternatively, relation (XII-43) can be written in algebraic terms as follows:

$$(f \oplus g)_x = \operatorname*{Sup}_{y \in \mathbb{R}^2} [f(y) + g(x - y)]$$

$$(f \ominus g)_x = \inf_{y \in \mathbb{R}^2} [f(y) - g(x - y)]$$

(XII-44)

where by convention, $f(x) = g(x) = -\infty$ for $x \notin$ support of f (resp. g). In practice, these relations lead to rather simple algorithms. For example, the dilation of f by the cube–octahedron of Exercise VI-12, with a unit spacing between planes, is the function:

$$\operatorname{Sup}\{f(a) + 1; f(b) + 1; f(c) + 1; f(d) + 1; f(e); f(g); f(f); f(h); f(o)\}$$

According to relation (XII-44), the sections X_{t_0} of $f \oplus g$ and $f \ominus g$ satisfy the relationships:

$$X_{t_0}(f \oplus g) = \{ x; \exists y : f(y) + g(x - y) \ge t_0 \}$$ (XII-45)

$$X_{t_0}(f \ominus g) = \{ x; \forall y : f(y) - g(x - y) \ge t_0 \}$$ (XII-46)

The expression of $X_{t_0}(f \oplus g)$ as a function of the $X_t(f)$'s and $X_t(g)$'s comes from the addition relation (XII-23) applied to relation (XII-45), which gives:

$$\left\{ x : \bigcup_t [X_t(f) \cap X_{t_0 - t}(\check{g}_x)] \neq \varnothing \right\} = \bigcup_t \left\{ x : X_t(f) \cap \hat{X}_{t_0 - t}(g_x) \neq \varnothing \right\}$$

and finally:

$$X_{t_0}(f \oplus g) = \bigcup_t [X_t(f) \oplus X_{t_0 - t}(g)] \qquad \text{(XII-47)}$$

By duality, we get for the erosion:

$$X_{t_0}(f \ominus g) = \bigcap_t [X_t(f) \ominus X_{t - t_0}(g)]$$

We now are able to construct openings and closings by composition of erosions and dilations. The symbol $\check{g}$ (transposed of g) denotes the function $\check{g}(x) = g(-x)$. We define the opening f_g of f by g from their umbrae as follows:

$$U(f_g) = [U(f) \ominus \check{U}(g)] \oplus U(g) \qquad \text{(XII-48)}$$

The umbra of f_g is the part of the domain of the umbra $U(f)$ spanned by all the translates of $U(g)$ which are included in $U(f)$. By noticing that $\check{U}(g) = \hat{U}(\check{g})$, we can interpret relation (XII-48) in terms of functions, and:

$$f_g = (f \ominus \check{g}) \oplus g$$

We shall introduce the notion of a closing f^g by duality by writing

$$f^g = -(-f)_g \Leftrightarrow f^g = (f \oplus \check{g}) - g$$

The relation between umbrae derives immediately, and we have:

$$U(f^g) = [U(f) \oplus U(\check{g})] \ominus \check{U}(\check{g}) \qquad \text{(XII-49)}$$

The complement of the umbra of $U(f^g)$ is the zone spanned by all the translates of $\hat{U}(g)$ which are included in $[U(f)]^c$. The difference $f - f_g$ defines the *rolling ball transform* of f by g (S. Sternberg, 1980), a notion which generalizes the top hat transform to structuring elements other than the half cylinder. The theory of size distribution for sets extends integrally to functions via their umbrae and relations (XII-48) and (XII-49). Finally, the approach followed in this section has allowed us to transpose a very comprehensive class of operations from sets to functions.

From the same basis, we could also generalize gradients, ultimate erosions, conditional bisectors, etc. Its major limit lies in the fact that the set difference of two umbrae is not an umbra (nor the transposed or the reflected of an umbra), and notions such as the Hit or Miss transformation or the thinnings escape its range of possibilities.

E. CONNECTIVITY CRITERIA

E.1. Homotopy for functions

The connectivity criteria are notions which depend on taking two possible successive cross sections (such as X_t and X_{t+dt}). Nevertheless this material is worthy of a new section because we wish to talk about segmenting functions and also introduce new topological tools.

(a) *Definitions concerning the relief*
The notions of summit and sink of functions are the basis from which we call topological classes equivalent. We will define the summit of f in terms of cross-sections as follows: a connected component $M_t(f)$ of the cross-section $X_t(f)$ is a summit if for every $(x_0, t) \in M$ and for every $t' > t$ we have $(x_0, t') \notin X_{t'}$, i.e.

$$M_t(f) \text{ summit} \Leftrightarrow (x_0, t) \in M_t \Rightarrow (x_0, t') \notin X_{t'}, \forall t' > t \qquad \text{(XII-50)}$$

The summits of f are the sinks of $-f$ and conversely. Note that a summit or a sink set cannot be reduced to a point (it can be a lake for example) and that both sets belong to the boundary of the umbra $U(f)$. In the literature, the expression "local maximum" and "local minimum" are sometimes used for summit and sink. We have preferred to adopt a different term, and to give a global meaning to the words "maximum" and "minimum" (the maximum of the Himalaya is Mount Everest). By themselves, summits and sinks do not carry enough topological information for what we have in mind. For example, they do not describe the situation represented in Figure XII.11c with a volcano, a lake in its crater, and an island in the lake. Therefore, we need to associate with each summit and sink some knowledge about their surroundings. This we will do via the notions of *watersheds* and *hills* (Ch. Lantuejoul, 1978).

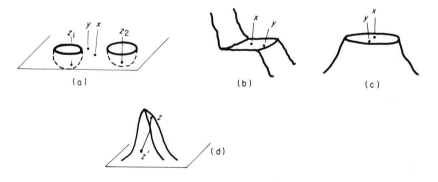

Figure XIII.9. Examples of flat zones, where x is higher than y.

Intuitively, the notion of a watershed is clear. Imagine that it rains over the region, the water streams down, reaches a sink, and stops there. The domain of attraction of this sink is called its watershed basin. In fact, presented in this way, the concept is ambiguous.

Consider Figure XII.9, which exhibits an infinite flat summit (Fig. XI.9a), and also flat steps. To which basin do the flat regions belong? None of them, all of them? We shall remove the ambiguity by adopting an order relation based on two conditions.

First, we will define the lines of swiftest descent as follows. Let x be a point at the surface of the umbra outside of a sink:

(i) If x does not belong to a plateau, consider a point y on the surface, whose horizontal projection describes a small disk centred on the projection of x. As y turns, the slope of vector x-y reaches some negative minimum values, corresponding to the positions $y_1, y_2 \ldots$ of point y. Then the small vectors $xy_1, xy_2 \ldots$ define the elementary arcs of swiftest descent downstream x.

(ii) Point x belongs to a plateau. Let $C(x)$ be the largest closed disk centred at x and included in the plateau. $C(x)$ hits the downstream edges of the plateau at points $y_1, y_2 \ldots$. Then the elementary arcs of the swiftest descent are the vectors $xy_1, xy_2 \ldots$ (Fig. XII.9b).

Definition XII-1. The *lines of swiftest* descent are generated by connecting the elementary arcs which have point in common. They necessarily end at the sinks of function f. The *downstream of point* x is made of all the portions of lines of swiftest descent starting from x. This downstream is generally constituted of a single arc, but may exhibit several arcs (when x is located on a ridge, for example).

Definition XII-2. The *watershed* of a sink M^* of the function $f(x)$, is the set of points $(x, f(x))$ of the surface $\partial[U(f)]$, whose lines of swiftest descent $l(x)$ only reach M^*. Those points of $\partial[U(f)]$ which belong to no watershed are called the *divides*.

Remarks
(i) Denote by W_t the set of points x which are projections of the divides above or at level t. The W_t's are monotonically decreasing as t increases and generate the u.s.c. divide function $d(x)$, i.e. the vertical shadow of the divide lines when lit from above. The support of $d(x)$, i.e. the projection of the divide lines, is a closed set, made up of locally finite unions of simple

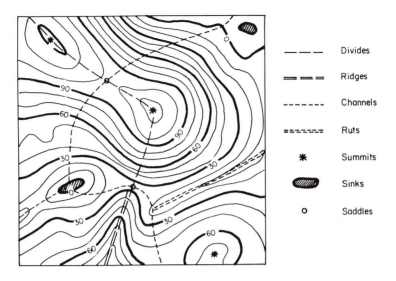

— — —	Divides
≡ ≡ ≡	Ridges
- - - - - -	Channels
⋍⋍⋍⋍⋍⋍	Ruts
✳	Summits
⬬	Sinks
o	Saddles

Figure XII.10. Summits and sinks, divides and channels, watersheds and hills, ridges and ruts of a function.

arcs (properties of the skiz, Ch. XI, C.5). This closed set is the boundary of the (open) projections of the watersheds.

(ii) By duality with respect to complementation of the function f, the notions of watershed and divide become *hills* and *channels* respectively. The umbra of the channels generates a u.s.c. function $c(x)$.

(iii) The divides and the channels of a constant function are the empty set. Conventionally in the case of Figure XII.11a (only one summit, whose support does not contain the point at the infinity of $\mathbb{R}^2$) $c(x)$ is the point at

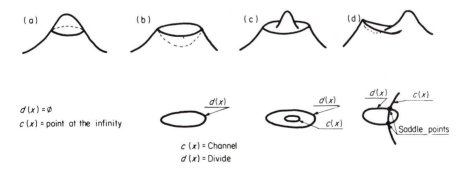

Figure XII.11. Functions and their homotopy characteristics.

the infinity and $d(x)$ does not exist, by duality the unique sink is characterized by $d(x) = $ point at the infinity, and $c(x)$ missing.
(iv) Channels and divides are made of loops drawn in the space, which may be infinite.

(b) *Homotopy*

We are now in a position to define the basic topological equivalence class for functions:

Definition XII-3. Two u.s.c. functions f and f' are said to be homotopic when both supports of $d(x)$ and $c(x)$ of f are jointly homotopic to both supports of $d'(x)$ and $c'(x)$ of f'. (i.e. homotopic in the product space $\mathbb{R}^2 \times \mathbb{R}^2$).

Remarks
  (i) The adverb "jointly" in the definition means that the supports of $d(x)$ and of $c(x)$ create two tessellations of the plane. If a cell of the c-tessellation hits, or is included in a cell of the d-tessellation, the same intersection or inclusion relationships also exist between the corresponding cells of d' and c'.
 (ii) When the divides and the channels of f are homotopics to those of f', then their summits, their sinks, their watersheds and hills are also homotopic. A crater with a hill in it in the relief f corresponds to a crater with a hill in it in f', and so on . . . regardless of the exact location of the hill in the crater, or of their relative altitudes.
(iii) Consider the situation where we want to transform a function f into homotopic f' by working section by section. Clearly, if for every t, $X_t(f)$ and $X_t(f')$ are homotopic sets, f and f' also are. But the condition of homotopy on the X_t's is much more demanding; in particular it says that the altitudes of corresponding summits and sinks are the same. In spite of this rigidity, we only work with these constraints in this chapter. Note that the condition of the X_t' constitute a direct generalization of the homotopy structure for sets.
(iv) Consider a closed set X in $\mathbb{R}^2$, and take the negative of quench functions of X (remember that the quench function is identically zero outside of X). The sinks of the function so generated are the ultimate erosions of X, and the supports $d(x)$ of the divides are nothing but the skiz of X.
 (v) In 1977, A. Rosenfeld proposed a definition for connectivity of functions, by stating that points $(x, f(x))$ and $(y, f(y))$ are connected if there exists a path on the surface $\partial[D(f)]$ all of whose points have altitudes smaller than or equal to both $f(x)$ and $f(y)$. In collaboration with Ch. Dyer (1979) he used this notion to build skeletons for images. But there is something not

right about the approach. Firstly, this connectivity is not transitive: take z_1 and z_2 in two different holes (Fig. XII.9a) and x in a flat zone between them; z_1 and x are connected, as are x and z, but not z_1 and z_2. In fact this definition does not really differentiate the watershed: when a portion of the divide line is quite vertical, one can easily cross it via a descending path zz', where the two extremities belong to two different basins (see Fig. XII.9d, which represents an open book seen from the back). Now, even if this definition were correctly formalized, it would still be criticizable from a morphological point of view. We do not see why the notion of a homotopy should be reduced to the divides and the dual aspect ignored. On the contrary, we feel that a crestline of mountains between two disconnected valleys (divide) is as meaningful a feature of relief as an isolated peak (channel). Definition XIII-3 results.

(c) *Digitalization*

The actual extraction of relief features (summits, divides, . . .) is based on sequential algorithms. A similar problem exists here as for sets (Ch. XI, E.2). Certainly *one* step of thinning can be an u.s.c. mapping of $\mathcal{F}u$ into itself. But semi-continuity, which is central to the Euclidean approach, is not preserved after an infinite sequence of thinnings. Then, a real digital approach is better than a sham continuous symbolism. Indeed, by intersecting a Euclidean function f, with a parallelepipedic or a hexagonal prismatic grid in $\mathbb{R}^3$ one can generate a digital function whose divides and channels projections onto $\mathbb{R}^2$ are digital 2-D graphs. If these graphs, interpreted in $\mathbb{R}^2$, are homotopic to the projections of divides and channels of f itself, then we do not lose the key pieces of information used in the connectivity criteria; thus the digital approach is valid. It can be proved that this will happen if the Euclidean umbra $U(f)$ is open and closed with respect to a small vertical cylinder with circular base of radius r_0 and height h_0 (proof is similar to that of Theorem VII-2), and if the spacing of the grid is small enough relative to r_0 and h_0. The corresponding class of functions will be called the *regular model* $\mathcal{R}u$ for functions ($\mathcal{R}u \subset \mathcal{F}u$).

An example of a digital algorithm is the detection of the summits and sinks of f at level i, denoted by M_i and M_i^* respectively. For simplicity, we keep the same symbol f for the function and for its digital version with respect to a prismatic grid. Translated into digital terms, relation (XII-50) becomes

$$M_i(f) = X_i(f)/[X_{i+1}(f)\oplus\{H\};X_i(f)] \quad -\infty < i < +\infty \quad \text{(XII-51)}$$

(the meaning of the notation on the right-hand side is explained in Ch. XI, E.1). The summits M_i at level i are the parts of the cross-sections X_i which cannot be recovered from the marker X_{i-1}. As the label i varies, the M_i's do not generate a function but only a family of sets. However, if necessary, one can always introduce a summit function $m(x)$ whose sections $X_i(m)$ are defined by the relationship

$$X_i(m) = \bigcup_{j \leq i} M_i(f) \qquad \text{(XII-52)}$$

(see example in Figure XII.19a). The function $m(x)$ is just the umbra of the summits M_i. By duality, an algorithm similar to relation (XII-51) that extracts sinks M_i^* of function f is:

$$M_i^*(f) = X_i^c(f)/[X_{i-1}^c \oplus \{H\}; X_i^c] - \infty < i < +\infty \qquad \text{(XII-53)}$$

E.2. Thinnings and thickenings

Here, we cannot directly transpose the formalism developed in Chapter XI, E.1, since the set $U(f)\circledast T$ is not necessarily an umbra. Therefore, we will define the Hit or Miss transform of function f w.r. to $T = (T_1, T_2)$ as follows:

$$U(f\circledast T) = U[(U(f) \ominus (\check{T}_1)\hat{\,})\cap(U(f)\oplus\check{T}_2)^c]$$

In words, the umbra $U(f\circledast T)$ of the Hit-or-Miss transform $f\circledast T$ is the umbra of the set of those points x of R^3 such that $(T_1)_x \subset U(f)$ and that $(T_2)_x \subset [U(f)]^c$. The definition of thickening $f \odot T$ results, by putting

$$U(f\odot T) = U(f) \cup U(f\circledast T).$$

Applying the duality relation XII-16, we obtain the thinning $f \bigcirc T$, and

$$U - (f\bigcirc T) = U(-f\odot T^*) \quad \text{with } T^* = (T_2, T_1)$$

or equivalently

$$\hat{U}(f \bigcirc T) = [U(-f)\odot T^*]^c$$

The definition of a thickening can be translated in terms of an algebraic algorithm. Let g_1 and g_2 denote the two functions characterized by the umbrae $\hat{U}(T_1)$ and $U(T_2)$. By using relation XII-44, we obtain the following rule: when

$$\sup_{z\in(T_2)_x} [f(z)+g_2(z-x)] < f(x) \leq \inf_{y\in(T_1)_x} [f(y)-g_1(y-x)]$$

then

$$(f\odot g)(x) = \inf_{y\in(T_1)_x} [f(y)-g_1(y-x)]$$

when not $(f\odot g)(x) = f(x)$.

(Similar rule for the thinning by taking the sup instead of the inf). We will no longer develop this general formalism, and prefer to focus the approach on the simpler case of the *flat* structuring elements, acting on *pictures* in a *digital* framework. The corresponding operations will be called lower thinnings and thickenings.

(a) *Lower thinnings*

Given the flat structuring element $T = (T_1, T_2)$, each section $X_i(f)$ of f is thinned and the ith section of the lower thinned picture $f \bigcirc T$ is defined by:

$$X_i(f \bigcirc T) = \bigcup_{j=i}^{m} [X_j(f) \bigcirc T] \qquad (m = \text{maximum})$$
$$1 \leq i \leq m \qquad \text{(XII-54)}$$

As i decreases, the union on the right hand side increases, and the $X_i(f \bigcirc T)$'s generate a picture. By duality the thinning (XII-54) induces a lower thickening $(m - f) \odot T^*$ on the complementary picture $m - f$; indeed, relation (XII-15) enable algorithm (XII-54) to be written in the equivalent form:

$$X_{m-i}[(m-f) \odot T^*] = \bigcap_{j=i}^{m} [X_{m-j}(m-f) \odot T^*]$$
$$0 \leq i \leq m-1 \qquad \text{(XII-55)}$$

To thin level i of function f is equivalent to thickening level $m - i$ of function $m - f$. Algorithm (XII-55) applied to f itself, gives:

$$X_i(f \odot T) = \bigcap_{j=0}^{i} [X_j(f) \odot T] 1 \leq i \leq m \qquad \text{(XII-56)}$$

Lower thinnings possess the following property: if the set thinning $X \bigcirc T$ preserves homotopy of the set X, then the picture thinning $f \bigcirc T$ is homotopic to f, in the sense of Definition XII-3. Set thinning is an anti-extensive mapping and the passage $X_i \rightarrow X_i \bigcirc T$ removes a part, say Z_i, of X_i. By taking the union $\bigcup(X_j \bigcirc T)$ in relation (XII-54), where $X_i \bigcirc T$ is joined by sets which are all included in $X_i \bigcirc T$, we can only modify the set Z_i to $Z_i' \subset Z_i$. If after removing from X_i each point $z_i \in Z_i$ the homotopy is not modified, then $X_i(f \bigcirc T) = X_i/Z_i'$ has the same homotopy as $X_i(f)$. Therefore, according to Definition XII-3, remark iii), f, and $f \bigcirc T$ are homotropic. Similarly, f and $f \odot T$ are homotropic.

This topological property makes lower thinnings and thickenings suitable tools for constructing "skeletons" of pictures. It will be sufficient to take for the structuring element one of the patterns L, M, or D, of Table XI-14. For example, consider picture (Fig. XII.12a) and the structuring element

$L = \begin{pmatrix} 0 & 0 \\ \cdot & 1 & \cdot \\ 1 & 1 \end{pmatrix}$ The three levels of $f \bigcirc L$ are shown in Figure XII.12b. We see that

the point x, removed by the operation $X_2 \bigcirc L$ is added by the union $(X_2 \bigcirc L) \cup (X_3 \bigcirc L)$. Now, iterate the transformation by turning L through

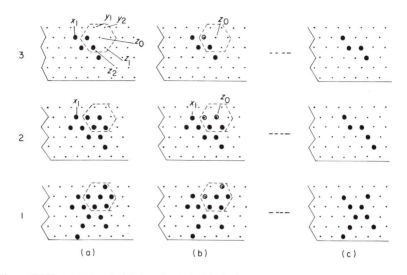

Figure XII.12. (a) picture f with three levels 1, 2, 3; (b) lower thinning $f \bigcirc L$; (c) lower sequential thinning (i.e. skeleton) $f \bigcirc \{L\}$.

$60°$, and so on . . . and generate a sequential transformation of f: the result is the skeleton $f \bigcirc \{L\}$ of f shown in Figure XII.12c.

The approach is general to any sequence (other skeletons, clippings, etc. . . .). The interpretation of the operation $f \bigcirc T$ in terms of functions is clear. Let T_{z_0} be the structuring element centred at z_0. If for any $z \in (T_1)_{z_0} f(z) \geq f(z_0)$ and for any $y \in (T_2)_{z_0} f(y) < f(z_0)$, then replace $f(z_0)$ by the sup of the $f(y)$'s. If not, do not modify $f(z_0)$. In the case of Figure XII.12, we have $f(z_0) = 2$, $f(z_1)$ and $f(z_2) \geq f(z_0)$ and $f(y_1)$ and $f(y_2) < f(z_0)$. Therefore we replace $f(z_0)$ by sup $(f(y_1), f(y_2)) = 1$. A similar interpretation holds for lower thickening, by inverting the roles of sup and inf. Using precisely this sup and inf presentation, V. Goetcharian introduced the lower skeleton (1979). Figure XII.13, extracted from his text, illustrates clearly the modification to the picture.

To a first approximation, the lower thinning is the umbra of the ridge. When the ridges surround a watershed, skeletonization gives each of the points of the watershed the minimum value (see the inside lake between the branches of the chromosome). The other points x (i.e. those which are not associated with a ridge or with an inside watershed) are given the value zero (that of the outside watershed).

What would happen if $f(z_0)$ were replaced by the inf of the $f(y)$'s, instead of the sup? Certainly homotopy would not hold, as shown in Figure XII.14a. The point $f(z_0) = 3$ would be replaced by zero and not by 2. This counter-

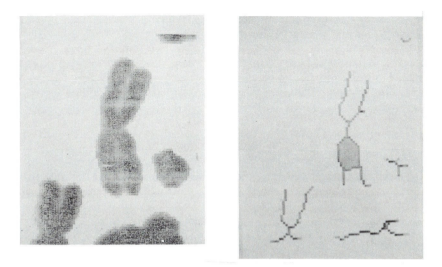

Figure XII.13. Lower skeleton of chromosomes (V. Goetcharian, 1979).

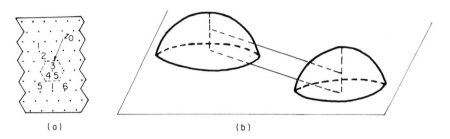

Figure XII.14. (a) counter-example (b) the lower skeleton penetrates inside the umbra $U(f)$.

example also shows that we do not exactly retain the ridges, since the ridge at point z_0 has value 3, and the algorithm (XII-54) will give it the value 2. This phenomenon is illustrated in Figure XII.14e, where $f(x)$ is made up of two domes linked by a thin wall of a level lower than their tops. Skeletonization transforms the domes into their central vertical axes, and extends the wall to now go from axis to axis. If we want the top line of the skeleton to run over the surface of the function (like the great wall of China extending over the landscape) we have to turn to another method, the *upper thinning*.

(b) *Upper thinnings: ridges and ruts detection*
For lower thinnings we started from the tops of the picture and gave preference to transformations in the t-dimension (before turning the structur-

ing element). However here, we completely transform a level before going to the next one, and we start from the lowest level.

The upper thinning does not have an explicit representation, but is given via the intermediary function g, defined through its sections $Y_i = Y_i(g)$ as follows:

$$Y_1 = X_1 \text{ (minimum level)}; \qquad Y_{m+1} = Y_m \bigcirc \{L\} \text{ (maximum level)}$$

$$\left.\begin{array}{l} X_i^1 = Y_{i-1} \bigcirc \{L\}; \; X_i \\ X_i^3 = X_i^1 \ominus \{H\}; \; X_i \\ Y_i = X_1^i / [X_i^3 / X_i] \end{array}\right\} \tag{XII-57}$$

The mechanism which generates the sections Y_i, is illustrated by Figure XII.15.

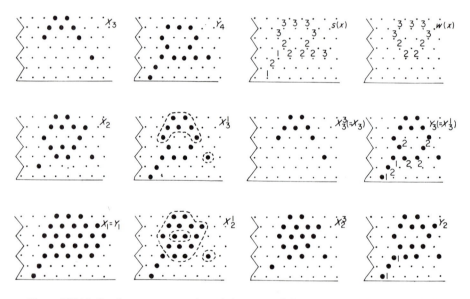

Figure XII.15. Step-by-step construction of the upper skeleton $s(x)$, and derivation of the watershed $w(x)$.

Starting from the lowest level X_1, we skeletonize it by $\{L\}$, conditional on X_2. We obtain a set X_2^1 exhibiting the same dendrites as the skeleton of X^1, provided they grow outside of X_2. Then we go from X_2^1 to Y_2 by inserting in X_2^1 the new holes which appear between levels X_1 and X_2 (i.e. the sinks of the function at level 1) and so on .... One continues the induction up until $n+1$, where m is the maximum level with $X_m \neq \emptyset$. Since $X_{m+1} = \emptyset$, the last step gives the section Y_{m+1} to simply be the skeleton of the section Y_m. The passage from the function Y to the upper skeleton $s(x)$ with section S_i, is as follows.

Define the difference Y_i/X_i $(1 < i \le m + 1)$ as the set of points of $s(x)$ with level *lower* than i; then

$$S_i(s) = \{x : s(x) \ge i\} = Y_{m+1}/[Y_i/X_i] \qquad \text{(XII-58)}$$

This progressive construction of the skeleton $s(x)$ from below, is illustrated in Figure XII.15. Level 1 of $s(x)$ is detected from Y_2, levels 1 and 2 from Y_3 and so on. . . . The relations (XII-57) imply that the S_i's of relation (XII-58) are decreasing as i increases, and that for every x such that $s(x) \ne 0$, we have $s(x) = f(x)$. The upper skeleton is a path drawn on the surface of the umbra of f. It is different from the lower skeleton when one can find $0 < s(x) < f(x)$. However for $s = 0$, the lower skeleton may actually pass *over* the upper one (Fig. XII.16c). By construction, the upper skeleton preserves connectivity of the connected components of X_i^c (i.e. the holes and the background) but not those of X_i. For example the top section of the function of Figure XII.12 is divided into two components by upper skeletonization (Fig. XII.16c). The divides of a picture and of its upper skeleton are homotopic but not their channels. In this sense, the lower skeleton, which preserves the homotopy of pictures, is a better topological descriptor. Anyway, the creation of new summits is rather infrequent, so that one can consider the upper skeleton as a *ridge-following* technique. In fact the algorithm finds a ridge even in a vertical cylinder such as the crab of Figure XII.16a. The problem is that the method has a tendency to follow the geographical ridges somewhat too much. For example, it adds two claws of altitude 1 to the crab (Fig. XII.16b); then the ridge in the zone of the cylinder is changed completely, becoming a prolongation of the claws! Indeed the ridges detected by the upper skeleton have no logical reason to coincide with the true ridges defined from the

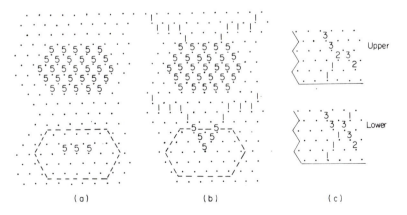

(a)   (b)   (c)

Figure XII.16. Upper skeleton of a crab without (a) and with (b) its claws. (c) upper and lower skeletons of function of Figure XII-12.

gradient in Section D-3c, since the latter are *local* notions and the former *global* ones.

Corresponding to the skeleton by upper thinning there is a dual skeleton by upper thickening. The algorithm, which compares to relation (XII-57), is:

$$\left.\begin{aligned}
Y_m &= X_m & Y_0 &= Y_1 \odot \{L\} \\
X^1_{m-i} &= Y_{m-i+1} \odot \{L\}; \; X_{m-i} \\
X^3_{m-i} &= X^1_{m-i} \oplus \{H\}; \; X_{m-i} \\
Y_{m-i} &= X^1_{m-i} = X^1_{m-1} \cup (X_{m-i}/X^3_{m-i})
\end{aligned}\right\} \qquad \text{(XII-59)}$$

The set X_{m-i}/Y_{m-i} is all those points of ruts with an altitude $m-i$.

(c) *Divides and channels detection*
The original algorithm (Ch. Lantuéjoul 1977) to find the watersheds of f is an extension of relations (XII-57) or (XII-59). It suffices to replace the skeleton by a skiz, i.e. to perform a clipping of the conditional thinning X^1_i. We will not rewrite the complete algorithm, since it has already been given for the thickenings: it is exactly relation (XI-38) for segmenting particles. There the grey tone function was the quench function, but this particular choice makes no difference to the algorithm itself.

Indeed, algorithms (XI-38) or (XII-57) are complicated, and their implementation is tedious, since they work grey level by grey level. Fortunately one can prove (S. Beucher, 1981) that the *sinks* of the upper and lower skeletons coincide; this results in the following practical rule of calculation of ridges and divides:

(1) Compute the lower skeleton $f \bigcirc \{L\}$ of function f.
(2) Detect those points $z \in \mathbb{R}^2$ which do not correspond to minima of $f \bigcirc \{L\}$; then the points $[z, f(z)]$ of the space constitute the lines of ridges and divides.
(3) For extracting the divides, prune the lower skeleton by $\{E\}$, and replace $f \bigcirc \{L\}$ by its pruned version $f \bigcirc \{L\} \bigcirc \{E\}$ in step 2.

F. FROM FUNCTIONS TO SETS: SEGMENTATION BY WATERSHEDS

The classical meaning in image analysis of segmenting a picture, is to partition its support into subsets inside which the picture has a homogeneous texture. We say that the picture has a homogeneous texture in a zone Z when it can be represented as a realization of a stationary random function with a range (of the covariance) that is small with respect to the dimensions of Z. However this definition itself is not particularly operational since there is no suggestion as to how the zones Z might be detected. The question of segmentation is an exceedingly complex one. The problem becomes more precise (and thus accessible) when one knows a physical interpretation for the words "homogeneous textures". They can be cells of a certain type in a tissue, petrographic phases in a mineral, ridge lines in a relief, etc. . . .

When the phases present have almost constant grey levels, and when those greys are quite different from one phase to another, the picture is thresholded with respect to chosen values of t. The thresholds are often visually chosen, which can be risky (As an experiment, give the same "almost" black and white picture to ten operators and compare the areas of the threshold zones).

The next step is to improve the visual thresholding by plotting the density function of f (i.e. the histogram of its greys) (Fig. XII.17a). Minima of the density provide possible thresholds, limited of course by the quality of the histogram. An accurate location of the minimum is not always possible (Fig. XII.17b). In this case the contrast of f can be increased, and/or an additional piece of information can be added to the density function. The contrast improvement techniques are numerous (median filter (Ex. XII-7), contrast algorithm (Ex. XII-8), linear combinations of f, of the module of its gradient and of its Laplacian, deconvolution filters, and so on). The simplest additional piece of information is contained in the grey histogram of gradient f, superimposed on the density distribution of f (Fig. XII.17b). It allows a more accurate location of the thresholds.

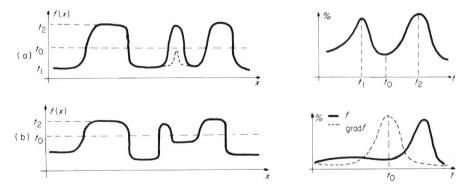

Figure XII.17. (a) automatic threshold from the density function, (b) additional information of the gradient density.

Another quite elaborate technique is to use the shapes and sizes of the objects to be segmented. The openings and closings of the function f are particularly convenient for this purpose. The support of the function $f - f_B$

(top hat transformation) corresponds to the zones with peaks and ridges that are narrow relative to B. Figure XII.5b illustrates the approach. Also, this technique can be combined with the previous one; e.g. if the central peak of f, in Figure XII.17a, is too thin to be detected (dashed lines on the figure) it can be extracted by a top hat transformation. Now also threshold the complete function, and take the union of the two resulting sets, and so on . . . .

Unfortunately, all these methods fail when the same phase exhibits different levels at different places. Then it seems wise to resort to topological features of f, for example by introducing the divides of the grey picture itself, or any morphological transform of it (gradient, combination of f and of its dilated versions, etc. . . .). We will now see three examples of this technique.

F.1. X-ray photographs (Ch. Lantuéjoul, 1978)

The specimen of Figure XII.18a is an X-ray photograph of pores in a radiographic material, and one wants to estimate the number and the size distribution of the pores. The bubbles are surrounded by diffraction haloes which make them appear larger than they actually are, and with fuzzy non-thresholdable contours. After the image has been digitalized on 16 grey levels, its gradient function $s(x)$ is computed from algorithm (XII-36) and the divides of the *gradient function* are calculated on the texture analyser. This results in photograph Figure XII.18b, which gives an over-segmented image. The false boundaries can be eliminated by introducing the sinks M^* of the initial picture $f(x)$, as a new piece of information (Fig. XII.19a). The bubbles of Figure XII.18b which are marked by a connected component of the sinks M^* have to be filled in and the others rejected. The final result of the detection is shown in Figure XII.19b. The weakness of this method does not lie with hyper-segmentation; on the contrary it lies with possible elimination of particles. If the thresholds on the gradient function are crescents and not rings for a particular bubble, then it will be missed by the watershed process.

F.2. Fractures in steel (S. Beucher, 1979)

In this second example, we start from a S.E.M. micrograph of cleavage fractures in steel (Fig. XII.20a). The facets of cleavage show up as heterogeneous grey values which we locate by calculating the divides of the gradient function (Fig. XII.20b). As before, the result is hyper-segmented. However, it is relatively simple to eliminate the parasite boundaries by adding new conditions (for example minimal facet areas and facet widths).

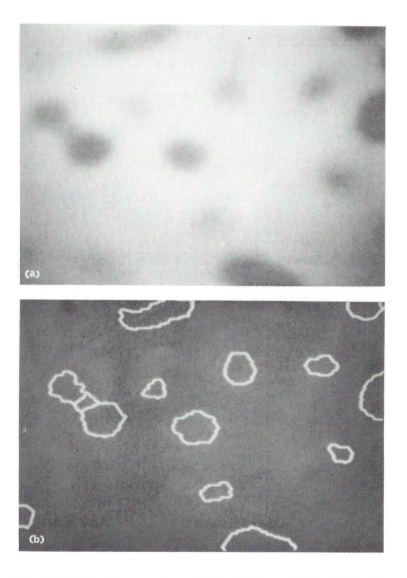

Figure XII.18. (a) bubbles on a radiographic plate. (b) divides of the gradient function.

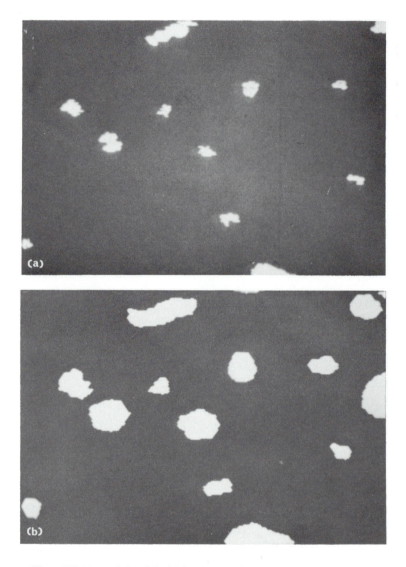

Figure XII.19. (a) sinks of the initial picture f, (b) final result of the detection.

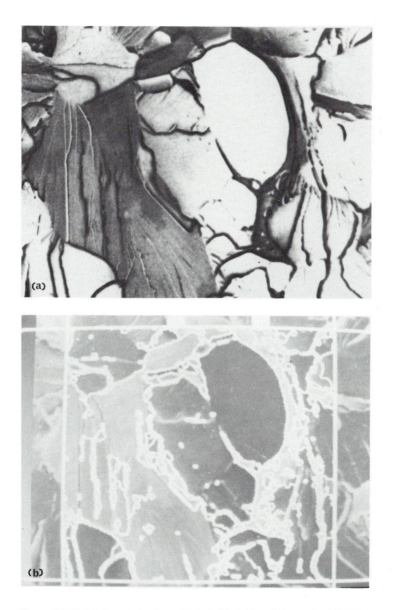

Figure XII.20. (a) cleavage fractures in steel, (b) divides of the gradient function.

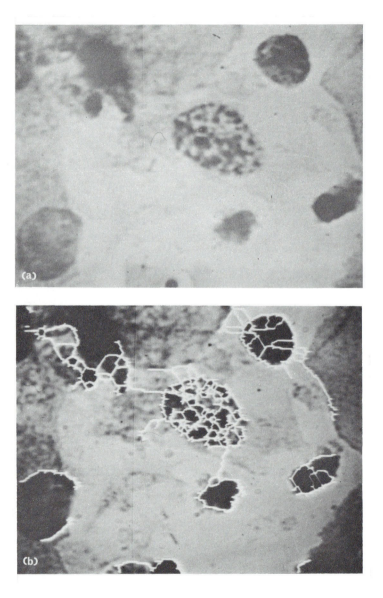

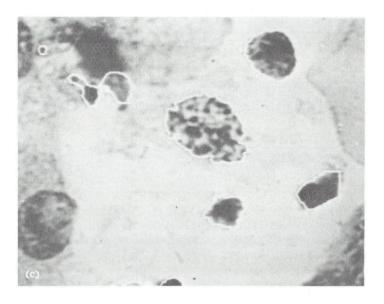

Figure XII.21. (a) cervical smear stained in Papanicolaou, (b) divides of the synthetic grey picture, (c) final result.

F.3. Cervical smears (F. Meyer and A. Van Driel, 1979)

F. Meyer and A. Van Driel studied the closings of contours of cells by divides. The material was stained via the classical Papanicolaou technique (Fig. XII.21a). It is clearly not possible to extract the cells by thresholding (their grey levels are variable) nor by thresholding their gradient. We will segment an image of this type by constructing a "grey-tone" function as follows: take a small number of gradient levels (here two are sufficient) and complete the grey-tone function with a few dilations or thickenings (here two steps).

Then apply the watershed algorithm to this artificial grey tone image with four levels; (Fig. XII.21b). The cells are over segmented, but the contour is easily simplified, for example by keeping only the boundaries between the background and those points which are not accessible from the background. The final segmentation is presented in Figure XII.21c.

Two features of this example are instructive: firstly, the grey tone function can be one hundred per cent synthetic; secondly, the watershed method is an effective technique even when used with a very low number of grey levels.

G. CONVEXITY AND INTEGRAL GEOMETRY

Apart from the notion of convexity number $N(f)$, everything we develop in this section only uses individual sections. Nevertheless for presentation purposes the set parameters provided by the X_t's are summed over t_0.

G.1. Convexity

(a) *Convex model*
A function $f \in \mathscr{F}u$ (resp. $\mathscr{K}u$) is said to be a *s*-convex (resp. *c*-convex) when each of its sections X_0 is convex. (resp. compact convex). The class of all such functions is denoted by $\mathscr{C}(\mathscr{F}u)$ (resp. $\mathscr{C}(\mathscr{K}u)$). A convex function in the classical sense (i.e. $x, y \in U(f)$ implies $[x, y] \in D(f)$) is necessarily *s*-convex, but the converse is not true (Fig. XII.22). Define the *s*-convex hull of $f \in \mathscr{F}u$ as the function generated by the convex hulls of its sections. The convex hull is an l.s.c. mapping of $\mathscr{F}u$ into itself and a continuous mapping of $\mathscr{K}u$ into itself.

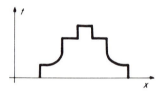

Figure XII.22. Example of a function that is *s*-convex but not convex.

These results remain true for the class $\mathscr{K}i$ of compact pictures. Moreover, four Minkowski functionals can be associated with every $f \in \mathscr{C}(\mathscr{K}i)$. Denote by $W_i(X_t(f))$ ($i = 0, 1, 2$) the Minkowski functionals of $X_t(f)$, and by $\mu(dt)$ the Lebesgue measure on the positive *t*-axis. The first three Minkowski functionals of f are defined by the quantities $\int \mu(dt) W_i(X_t)$ and the fourth $W_z(f)$ is set equal to 1.

(b) *Convex ring*
The approach that has been used already to construct a convex ring on sets (Ch. V) can now be applied to the functions in $\mathscr{C}(\mathscr{K}i)$. With every finite sample $f_i, \ldots, f_j, \ldots, f_n$ of *n* pictures belonging to $\mathscr{C}(\mathscr{K}i)$ associate the picture

$$g(x) = \sup\{f_j(x)\} \qquad f_i, \ldots, f_j, \ldots, f_n \in \mathscr{C}(\mathscr{K}i) \qquad \text{(XII-60)}$$

This approach will generate the convex ring $\mathscr{S}(\mathscr{K}i)$ of compact *c*-convex pictures, which is in turn used for research into certain interesting global parameters, or functionals.

A functional φ on $\mathscr{S}(\mathscr{K}i)$ is said to be sup-additive when

$$\varphi(f_1) + \varphi(f_2) = \varphi(f_1 \wedge f_2) + \varphi(f_1 \vee f_2) \qquad \text{(XII-61)}$$

By iterating this relation, we see that, for g given by (XII-60) we have

$$\varphi(g) = \sum_{j=1}^{n} \varphi(f_j) - \sum_{j_1 < j_2} \varphi(f_{j_1} \wedge f_{j_2}) + \cdots$$
$$+ (-1)^{n-1} \varphi(f_1 \wedge f_2 \ldots \wedge f_n) \qquad \text{(XII-62)}$$

where all the functions involved in the right-hand side are compact c-convex pictures. Therefore, any functional φ on $\mathscr{C}(\mathscr{K}i)$ satisfying relation (XII-61) has an extension to $\mathscr{S}(\mathscr{K}i)$ given by relation (XII-62). In particular, the Minkowski functionals W_i on the class $\mathscr{S}(\mathscr{K}i)$ can be defined by writing relation (XII-62) using W_i of the compact c-convex class. More generally, the following theorem (which generalizes Hadwiger's classical theorem to pictures) provides the general form of "good" functionals.

THEOREM XII-4. *A functional φ on $\mathscr{S}(\mathscr{K}i)$ is invariant under displacements, sup-additive and continuous for the sub-class $\mathscr{C}(\mathscr{K}i)$ if and only if there exist three continous and bounded functions $a_0(t)$, $a_1(t)$, $a_2(t)$ and a bounded constant a_3 such that*

$$\varphi(f) = \int \left[a_0(t) A(X_t) + a_1(t) U(X_t) + a_2(t) N_2(X_t) \right] dt + a_3 N_3(f)$$
$$\text{(XII-63)}$$

Proof: Firstly, we examine the convex case. For $f \in \mathscr{C}(\mathscr{K})$ and $N_3(f) = W_3(f) = 1$, the right hand side of (XII-63) is invariant under displacements, is sup-additive, and $\varphi(f_1) \to \varphi(f)$ as $f_1 \to f$. Conversely, let $\varphi(f)$ be a functional satisfying the condition of the theorem, and let f be a cylinder with convex section X_1 and height t_1. Considering φ as a functional acting on the class of convex cylinders of height t_1, and applying Hadwiger's theorem, we can write

$$\varphi(f) = \varphi(X_1, t_1) = a_0(t_1) A(X_1) + a_1(t_1) U(X_1) + a_2(t_1) + a_3$$

Now take a finite family $f_1, \ldots, f_j, \ldots, f_n$ of cylinders (X_j, t_j) with increasing heights $t_{j+1} > t_j$ and with sections nested inside each other, i.e. $X_{j+1} \subset X_j$. Then sup $\{f_j\}$ generates a c-convex step function g_n, and sup-additivity of the functional φ implies:

$$\varphi(g) = a_0(t_1) A_1 + a_1(t_1) U_1 + \sum_{j=2}^{n} \left[a_0(t_j) - a_0(t_{j-1}) \right] A_j \quad \text{(XII-64)}$$

$$+ \sum_{j=2}^{n} \left[a_1(t_j) - a_1(t_{j-1}) \right] U_j + a_2(t_n) + a_3$$

Any c-convex function g is a uniform limit of c-convex step functions $\{g_n\}$ as $n \to \infty$.

Therefore relation (XII-64) is satisfied in the limit, and relation (XII-63) results (at least for class $\mathscr{C}(\mathscr{K}i)$). Now if $g \in \mathscr{S}(\mathscr{K}i)$, it can be decomposed into p c-convex terms $f_1, \ldots, f_j, \ldots, f_p$. Relation (XII-63), which is satisfied for each f_j, plus the sup-additivity (XII-62) imply the validity of relation (XII-63) for the class $\mathscr{S}(\mathscr{K}i)$.   Q.E.D.

Remark: The theorem puts more emphasis on the role of the W_i of the sections X_t, than on the $W_i(f)$. In fact the main usefulness of this theorem is that it leads us to consider four basic measurements. Two of these are more important, and so are studied in more detail: the first functional is valid on a more general class than the convex ring, while the second has an extremely useful interpretation in local terms. Therefore we leave the convex ring model for the moment.

G.2. Volume of the graph $D(f)$

Every $f \in \mathscr{F}u$ is Lebesgue measurable as the lim sup of measurable functions, namely the areas of the X_t's (rel. (XII-4)): thus

$$V(D) = \int_{\mathbf{R}^2} f(x)\, dx = \int_{-\infty}^{+\infty} A[X_t(f)]\, dt \qquad \text{(XII-65)}$$

This integral is necessarily finite for f a picture having bounded support. More generally, by replacing dt with any bounded measure $\varphi(dt)$ on the right hand side of (XII-65), another overall parameter is defined on the graph D, which still satisfies the four principles of morphology. The possibly infinite quantity (XII-65) may be replaced with a local sum of a slice of f, such as $\int_{t_1}^{t_2} A[X_t(f) \cap Z]\, dt$, which is always finite.

In the case of pictures, it is often instructive to quantify how far f is from a black-and-white function. This is done by calculating the variance of the areas $A(X_t)$'s, as t varies from 0 to m:

$$m^2 \sigma^2(A) = \int_0^m A^2(X_t)\, dt - \left[\int_0^m A(X_t)\, dt\right]^2 \qquad \text{(XII-66)}$$

By using this algorithm after openings and closings, possible correlations between the details and the grey parts of the image can be calculated. A local version of parameters (XII-65) and (XII-66) exists, by replacing $A(X_t)$ with $A_A(X_t)$.

G.3. Gradient of f and perimeters of the X_t's

In this subsection, we will study the morphological meaning of the perimeters of the X_t's in the context of the model already chosen for defining gradients (Sect. D.3c). By integrating relation (XII-35) with respect to x over a domain Z, and by changing the order of integration on the right hand side, we obtain

$$\int_Z \rho(x)\,dx = \frac{1}{4}\int_0^{2\pi} d\alpha \int_0^m \Big\{ A\big[(X_t \cap Z)\oplus\{dh\}\big]$$

$$- A\big[(X_t \cap Z)\ominus\{dh\}\big]\Big\}\frac{dt}{dh} \qquad \text{(XII-67)}$$

$$\int_Z \rho(x)\,dx = \int_0^m U(X_t \cap Z)\,dt \qquad \text{(XII-68)}$$

i.e. the integral over Z of the module of $\overrightarrow{\text{grad}}\, f$ equals the sum of the perimeters of the sections of f. The next step in the approach is to generalize the rose of directions. Let $Z(\theta)$ be the union of the small open sets in Z, where $\overrightarrow{\text{grad}}\, f$ has direction θ. Put $\rho_Z(\theta) = \int_{Z(\theta)} \rho(x)\,dx$. The $Z(\theta)$'s partition Z as θ ranges from 0 to 2π. Therefore, relation (XII-34) becomes, after integration over Z,

$$\int_Z \left|\frac{df}{dh}\right| dx = \int_0^{2\pi} \rho_Z(\theta)\,|\cos(\theta-\alpha)|\,d\theta$$

$$= 2\int_0^m D_\alpha(X_t \cap Z)\,dt = 2d_Z(\alpha) \qquad \text{(XII-69)}$$

where $D_\alpha(X_t \cap Z)$ is the total diameter of the set $X_t \cap Z$ in direction α (Fig. V.6). The deconvolution of relation (XII-69) is similar to that of the rose of directions (Ch. IX, C.2) and we can write

$$d_Z(\alpha)+d_Z''(\alpha) = \rho_Z\left(\alpha+\frac{\pi}{2}\right)+\rho_Z\left(\alpha-\frac{\pi}{2}\right) \qquad \text{(XII-70)}$$

Figure XII.23 illustrates this algorithm. The experimental conditions are identical to those of Chapter IX (same spacing for the α's, same estimation method for d_α''), and the implementation holds on 20 grey levels. The small peak of the rose is due to the slow trend of greys, the two peaks in the large branch correspond to the two directions of dendrites.

G.4. Connectivity numbers

We return to the convex ring $\mathscr{S}(\mathscr{K}i)$ for the last two functionals. Following the notation of Ch. V, $N^{(2)}$ and $N^{(3)}$ denote the connectivity numbers in $\mathbb{R}^2$ and $\mathbb{R}^3$ respectively. Obviously the integral over t of $N^{(2)}(X_t)$ is equal to the sum of the

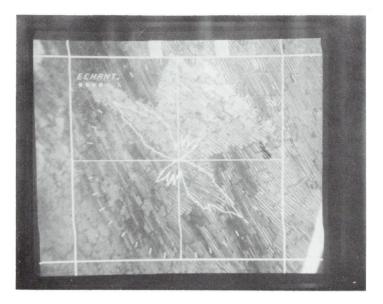

Figure XII.23. Photograph of dendrites and of their rose of directions. The small peak is due to a continuous gradient.

drops in height from summits to saddles of $f(x)$, minus those for $-f(x)$ (modulo the constant m). Unlike the first three, the last Minkowski functional $N^{(3)}$ is not associated with the sections X_t, but with the complete umbra $U(f)$, and is equal to the number of summits plus the number of sinks minus the number of saddles of $f(x)$.

H. RANDOM FUNCTIONS

(We have placed this section in Chapter XII for the sake of consistency. However, pedagogically speaking, its reading should follow that of Chapter XIII.)

H.1. General properties

Probability theory classically deals with the notion of a random function without recourse to the umbra of the function and the computations involved come mainly from the spatial law (i.e. the joint distribution of $f(x)$ at n point $x_1, \ldots x_n$, n finite). This approach is not well adapted to the probabilistic description of features such as ridges, saddles, sinks, etc. . . . nor to the construction of models exhibiting discontinuities (vertical cliffs . . .). The way we propose now, where we consider random functions in terms of their umbrae, allow such descriptions.

Consider the set $\mathcal{U}$ of those closed sets of $\mathbb{R}^3$ which are umbrae of a u.s.c. function, and provide $\mathcal{U}$ with the Hit or Miss topology. Associate with this topology the σ-algebra σ_u generated by the open sets of $\mathcal{U}$. A random umbra U will be defined by the datum of a probability P on the measurable space $(\mathcal{U}, \sigma_u)$. The compacity of $\mathcal{U}$ grants us that such probabilities do exist. Finally, the u.s.c. function f associated with the random umbra U is, by definition, a random function (G. Matheron, 1969).

Most of the features of the random sets have a corresponding feature in terms of random functions (semi Markov property, infinite divisibility, stable random functions, etc. . . .). The reader will easily transpose these results. The most fundamental one is the transcription of the Matheron–Kendall Theorem XIII-4 to random functions. Consider a random function $f(x)$, $x \in R^2$; let $B(x, t)$ be the translate by vector (x, t) of a compact set B centred at the origin. According to the Matheron–Kendall theorem, $f(x)$ is characterized by the probabilities $\text{Pr}\{B(x, t) \subset [U(f)]^c\}$, that $B(x, t)$ misses the umbra $U(f)$, with B spanning the class of compact sets centred at the origin. But $B(x, t)$ misses $U(f)$ if point (x, t) does not belong to $U(f \oplus \check{B})$ (rel. XII-40). Therefore we can state:

THEOREM XII-5. *Every random function $f(x)$ is completely determined by the probabilities $Pr\{x \in X_t(f \oplus B)\}$ that point (x, t) belongs to the umbra $U(f \oplus \check{B})$ as (x, t) spans R^3 and B spans the class of compact sets centred at the origin.*

Similarly we can also take for the B's the finite unions of half cylinders with a compact basis (Prop. 2.3.1 in G. Matheron, 1975). Then Theorem XIII-5 is replaced by the equivalent, but more algebraic following version:

THEOREM XII-6. *The random function $f(x)$ is completely determined by the joint probabilities*

$$\text{Pr}\left\{\sup_{x \in B_1} f(x) < \lambda_1; \quad \sup_{x \in B_2} f(x) < \lambda_2 \ldots; \quad \sup_{x \in B_i} f(x) < \lambda_i\right\}$$

for every finite sequence $B_1 \ldots B_i$ of compact supports in $\mathbb{R}^2$ and of numerical values $\lambda_1 \ldots \lambda_i$.

H.2. Boolean and Poisson models

There is no time to review all the properties of random functions investigated via their umbrae. Instead we will indicate a few results about two basic models, the *Boolean functions* and the *Poisson functions*, corresponding directly to the Boolean set model and to the Poisson lines model (the reader should look at Chapter XIII before reading this section).

(a) *Boolean random functions*

Define a primary function f' to be one realization of a random function whose support is almost surely compact. The translate of f' by the vector x is f'_x. Generate a Poisson point process in $\mathbb{R}^2$ of density θ, and denote the I points of the realization as x_i $(i \in I)$. Centre at each x_i a *different* realization f'_{x_i} of the primary function. Then define the function

$$f(x) = \sup\{i \in I, f'_{x_i}(x)\} \tag{XII-71}$$

to be one realization of the Boolean model. As for Boolean sets, we have to link the functional moments of f to those of f' (assumed to be given). By construction, the random function f is stationary in $\mathbb{R}^2$. Therefore the functional moments

$$Q(B, t) = \Pr\{B(x, t) \subset [U(f)]^c\} = \Pr\{x \notin X_t(f \oplus \check{B})\}$$

do not depend on the point x. Using the proof which gives the functional moments of the Boolean sets (Ch. XIII, B.2) yields the corresponding result for the Boolean functions, i.e.

$$Q(B, t) = \exp\{-\theta \overline{A}[X_t(f' \oplus \check{B})]\} \tag{XII-72}$$

Thus a Boolean function is infinitely divisible for the sup operation. Also, its section by vertical planes is still Boolean (in $\mathbb{R} \times \mathbb{R}$), and the dilate of f by an arbitrary function g with a compact support continues to be a Boolean function. We will illustrate the basic relation (XII-72) by a few examples:

(i) for B the origin we have $\Pr\{f < t\} = \exp\{-\theta \overline{A}[X_t(f')]\}$

(ii) for f' an s-convex function, and for $B = \lambda B_0$, B_0 a *plane* compact convex set, we have

$$Q(B, t) = \Pr\{x \notin X_t(f' \oplus B)\} = \exp\left[-\theta \overline{A}(X_t(f')) \right.$$
$$\left. + \lambda \frac{\overline{U}(X_t(f')) \cdot U(B_0)}{2\pi} + \lambda^2 A(B_0) \right]$$

By taking various shapes (segment, hexagon . . .) and sizes for B, this relation will enable us to test whether a given function can be considered to be a Boolean realization.

(iii) for f' an s-convex function, and $h(t) = \Pr\{X_t(f') \neq \varnothing\}$, the number of summits of f per unit area that are above level t is $\theta \cdot h(t) \exp[-\theta A(X_t(f'))]$.

An instructive use of the Boolean model for functions has been developed by D. and P. Jeulin (1980) in studying the roughness of metallic surfaces.

(b) *Poisson functions*

Suppose that binary lines of density $\theta(dx)$ are generated. Consider one line D_i of the realization. At each point x of D_i centre a given realization of a primary grain f'_i (use the *same* realization for all the points of D_i, but a *different* realization per line). The Poisson function is then defined to be the sup over all the functions f_i centred along the Poisson lines.

The Poisson function f is a stationary random function, it is infinitely divisible for the sup operation and its sections by vertical planes are still Poisson (in $\mathbb{R} \times \mathbb{R}$).

Let B be a compact convex set in $\mathbb{R}^3$, and let B_α be its projection perpendicular to a horizontal direction α. Then the basic relationship of the model is as follows:

$$Q(B, t) = \exp\left\{ -\int_0^\pi \theta(d\alpha)[\overline{L}(f'_\alpha \oplus B_\alpha)] \right\}$$

In particular, by taking θ to be a constant (i.e. $\theta(d\alpha) = \theta\, d\alpha$) and B a plane convex set, one finds

$$Q(B, t) = \Pr\{x \notin X_t(f \oplus B)\} = \exp\{ -\theta[\overline{U}(X_t(f')) + U(B)] \}$$

where U denotes here the perimeter.

I. EXERCISES

I.1. Algebraic properties of the Minkowski operations

(a) For $B \in \mathcal{K}(\mathbb{R}^3)$ and f and g s.c. functions (or pictures), prove the following results:

$$\lambda f \oplus B = \lambda[f \oplus B]$$
$$(f \vee g) \oplus B = (f \oplus B) \vee (g \oplus B)$$
$$(f \wedge g) \oplus B \leq (f \oplus B) \wedge (g \oplus B);$$
$$(f + g) \oplus B \leq (f \oplus B) + (g \oplus B)$$

as well as those obtained by replacing $\oplus$ by $\ominus$, $\vee$ by $\wedge$ and $\geq$ by $\leq$ respectively.

(b) Prove that $f \leq g$ implies:

$$f \ominus B \leq g \ominus B \quad \text{and} \quad f \oplus B \leq g \oplus B, \forall B \in \mathcal{K}.$$

I.2. Sums and differences

Prove that relation (XII, 23 and 24) are simplified when the supports of functions f and g are disjoint; i.e.

$$\varnothing = \{x : f(x) > 0, g(x) > 0\} \Rightarrow \begin{cases} X_{t_0}(f+g) = X_{t_0}(f) \cup X_{t_0}(g) \\ X_0(f-g) = X_{t_0}(f)/X_{t_0}(g) \end{cases}$$

Note that only the first relation holds when f and g are pictures.

I.3. Stars

For a 2-D set X, the star $S(x)$ at the point x is the set of points $y \in X$ directly seen from x (Ch. X, B.1). Generalize this notion to pictures by taking for $S^*(x)$ the volume of the space inside the umbra which is directly visible from horizontal light rays emanating from a vertical neon tube placed at x (Fig. XII.24).

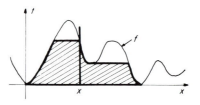

Figure XII.24. Star of point x.

Prove that $E[S^*] = \int P_t(l, \alpha)/P_t(0, \alpha) \, l \, dl \, dx \, dt$, where $P_t(l, x)$ denotes the probability that a segment of length l and direction α is included in X_t.

I.4. S-Convex functions

Let $C(f)$ denote the s-convex hull of the u.s.c. function f. Show that:

$$f = C(f) \Leftrightarrow f \text{ is } s\text{-convex}$$

$$f, g \text{ are } s\text{-convex} \Rightarrow f \wedge g \text{ is } s\text{-convex}$$

I.5. Picture marking (S. Beucher, 1980)

The aim of this exercise is to extend the particle marking algorithm (XI-34) to pictures, and to use the result for extracting the extrema of a digitalized picture.

(a) Let f and g be two pictures satisfying $1 \leq g(x) \leq f(x) < m$; $\forall x$. Picture h is defined from its section as follows:

$$X_t(h) = X_t(g) \oplus \{H\}; X_t(f) \quad \text{(i.e. rel. XI, 34)} \tag{1}$$

Prove that $g(x) \leq h(x) \leq f(x)$. Draw $h(x)$ in the case of Figure XII.25a.

(b) *Application to summits detection.* Given picture g, take $f = g + 1$ and apply algorithm (1) for generating $h(x)$. Put:

$$u(x) = g(x) - [h(x) - 1] \tag{2}$$

What does this new picture represent? Express $X_{t_0}(u)$ as a function of the $X_t(f)$'s. Construct the function $u(x)$ associated with Figure XII.25b.

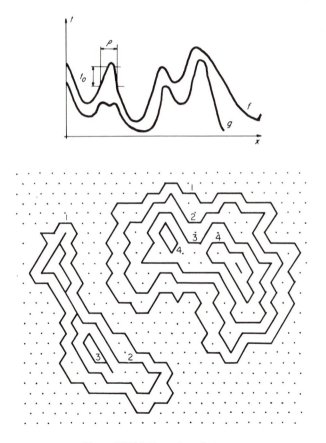

Figure XII.25. Examples of pictures.

(c) *Weighted summits picture.* Show that picture $w(x) = f(x) \wedge u(x)$ satisfies the following property:

$$w(x) = t \quad \text{if} \quad x \quad \text{belongs to a summit at level } t$$
$$w(x) = 0 \quad \text{if} \quad x \quad \text{does not belong to a summit}$$

and calculate the $X_t(w)$'s. Extend questions (b) and (c) to the extraction of sinks.

I.6. "Top-hat" Transformation in $\mathbb{R}^1$ and $\mathbb{R}^2$ (F. Meyer, 1980)

(a) *Top-hats in $\mathbb{R}^1$.* Let f be a 1-D picture, and ρL be the segment of length ρ. Using relation (XII-30), calculate the X_t's of picture

$$h_\rho(x) = f - (f \ominus \rho L) \oplus \rho L = f - f_{\rho L} \tag{1}$$

Interpret, and draw it for f given by Figure XII.25a.
$[X_{t_0}(h)$ is just the projection of the summits of f having a width $\leq \rho$ and a height $\leq t_0]$

(b) *Top-hats in $\mathbb{R}^2$.* Let f now be a picture in $\mathbb{R}^2$, and ρB denote the regular hexagon of size ρ. Define the hexagonal top-hat transform by $h'_\rho = f - f_{\rho B}$ and draw $h_i(x)$, $i = 1, 2, 3$, in the case of Figure XII.25b. Notice that $h_i(x)$ extracts not only the peaks but also the narrow ridges, saddles and capes of f. Find an algorithm which filters only the peaks and leaves the other summit features. Apply it to Figure XII.25b, and compare the results.

[Let L_α, L_β, L_γ be the unit segments in the three basic directions of the hexagonal grid. Replace the hexagonal opening by ρB by the union of the three openings by ρL_α, ρL_β and ρL_γ. This results in the new picture $h''_\rho = f - \sup\{f_{\rho L_\alpha}, f_{\rho L_\beta}, f_{\rho L_\gamma}\}$. We have $h'_\rho(x) \geq h''_\rho(x)$ since B is opened by each of the three segments L; the difference $h'_\rho - h''_\rho$ represents the long summit features, and h''_ρ the round ones.]

I.7. Picture cleanings (S. Beucher, 1980)

In this exercise we compare methods for picture cleaning based on opening–closing and on median filtering (the latter for digital pictures only).

(a) *Opening–closing for pictures.* Suppose F is a picture in $\mathbb{R}^1$, which presents an impulse noise. Pulses are both positive and negative, but with a constant width (Fig. XII.26a). For L greater than the noise width, the linear opening f_L smoothes off the positive impulses, and the linear closing f^L smoothes off the negative ones. One can symmetrize the process by taking the picture

$$r(x) = f^L(x) + f_L(x) - f(x) \tag{1}$$

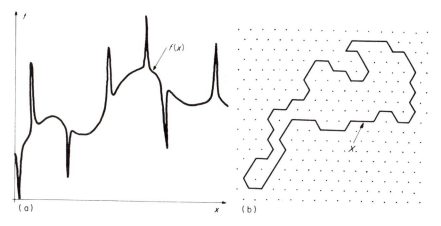

Figure XII.26.

Plot $r(x)$ in the case of Figure XII.26a, give the expression of the $X_t(r)$'s. Show that relation (1) holds in $\mathbb{R}^2$, by replacing segment L by a ball, or a hexagon.

(b) *Opening–closing for sets.* f is now the indicator function $k(x)$ of a closed set X. Express the $X_t(r)$'s as a Boolean function of X, and check that $\{X_t(r)\}$ is a sequence associated with the indicator function of a set Y. As an application, draw X_B, X^B and Y, for B the unit hexagon and for X given by Figure XII.26b.

(c) *Median filtering for pictures.* Consider a point x_0 on a digitalized picture f. Let $f_0 = f(x_0)$ and $f_1, \ldots, f_6$ be the grey tones of the six hexagonal neighbours of x_0. The median filtering med (f) is the operation in which we replace f_0 by the median value of $(f_0, f_1, \ldots, f_6)$. Draw the median-filtered transform of Figure XII.27a.

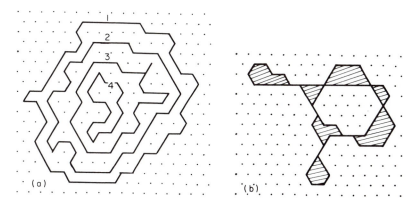

Figure XII.27.

(d) *Median filtering for sets.* Show that, applied to the indicator function $k(x)$ of a set X, median filtering results in the indicator function of a new set Y. Draw the transform Y of X, given in Figure XII.26b. Compare with the opening–closing algorithm. Show that $Y = \bigcup_k (X \ominus T_k)$, where the T_k's are Golay configurations of the neighbourhood. Find these configurations. Prove that the set transformation $X \to Y$ is increasing, but neither extensive, nor anti-extensive, nor idempotent.

(e) When f is a picture, show that the increasing property proved in (d) implies that

$$X_i[\mathrm{med}(f)] \subset X_j[\mathrm{med}(f)] \qquad \forall j \le i$$

Derive that $\{X_i[\mathrm{med}(f)]\}$ generate a picture, and that we have:

$$X_i[\mathrm{med}(f)] = \bigcup_k [X_i(f) \ominus T_k]$$

with the above Golay configurations (i.e. the median filtering commutes with the operation of thresholding).

(f) *Comparison between median filtering and opening–closing filtering. Lemma:* prove that if the digital set X is open by the elementary hexagon H (i.e. $X = X_H$) the neighbourhood of every point $x \in X$ has at least four points which belong to X.

Now consider a digital picture f. It is said to be regular when all its sections are both open and closed by H:

$$\forall i : X_i(f) = [X_i(f)]_H = [X_i(f)]^H$$

Prove that if f is regular, then $\mathrm{med}(f) \equiv f$. What about the converse (see XII.27b)?

I.8. Picture enhancement (Kramer and Bruckner, 1975)

Consider a pixel x_0 and its six neighbours. Find the maximum and minimum grey values among the seven values. If $f(x_0)$ is closer to the minimum, change it to that of the minimum. If not, change to the maximum. This transformation, which sharpens the transition between object and background is called contrast enhancement.

If $g(x)$ denotes the enhanced image, show that

$$g(x) = f(x) \oplus H \quad \text{if} \ (f \oplus H) - f < f - (f \ominus H)$$
$$g(x) = f(x) \ominus H \quad \text{otherwise}$$

What happens if we replace H by nH, with $n \to \infty$?

I.9. Minkowski operations for functions

In this exercise, we generalize the Minkowski operations to functions by an approach different from that of Section D.4, the emphasis being here on similarity.

(a) Prove that the relations

$$X_t(f \oplus g) = X_t(f) \oplus X_t(g)$$

$$X_t(f \ominus g) = X_t(f) \ominus X_{-t}(g) \qquad f, g \in \mathcal{F}_u \qquad (1)$$

actually generate two functions $f \oplus g, f \ominus g \in \mathcal{F}_u$, satisfying the principle of duality (i.e. $-f \oplus g = -(f \ominus g)$). Show that:

$$X_t(f \oplus g) = \{x : \exists y : g(x-y) \geq t \Rightarrow f(y) \geq t\}$$

$$X_t(f \ominus g) = \{x : \forall y : g(x-y) \geq t \Rightarrow f(y) \geq t\}$$

and

$$(f \oplus g)_x = \mathrm{Sup}\,\{f(y) : (y, f(y)) \in D[g(x-y)]\}$$

$$(f \ominus g)_x = \mathrm{Inf}\,\{f(y) : (y, f(y)) \in \check{D}[g(x-y)]\}$$

Interpret geometrically. How can we choose g to make algorithm (1) equivalent to dilations and erosions by a 2-D set B?

[Imagine the umbra $U(g)$ to be a drilling centred at x and piercing the graph $D(f)$; $f \oplus g$ is taken to be the level of the highest point of the core. For $f \ominus g$, the drilling is turned upside down (i.e. $U(g) \to \check{U}(g)$), and *detects the lowest point of the space* beyond the topographic surface f. Figure XII.28 shows the relationship between $f \oplus g$ and the shape of g. The less g

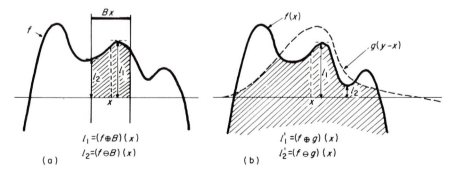

Figure XII.28. Comparison between the dilation and the erosion of f, (a) by a 2-D set B and (b) by a function g.

spreads out as t decreases, the less it affects the low values of $f(x)$. In particular, if g is an infinite vertical cylinder with horizontal cross section B, the transformations (1) coincide with the dilations and erosions of Sect. C-2.]

(b) Show that

(i) $(f \ominus g) \ominus k = f \ominus (g \oplus k)$

(ii) A family of functions f_λ depending on a positive parameter λ satisfies $f_\lambda \oplus f_\mu = f_{\lambda+\mu}$, $\lambda, \mu \geq 0$ if and only if $f_\lambda = f(\lambda x)$ for $f \in \mathscr{C}(\mathscr{K}u)$ i.e. if and only if f_λ is similar to the convex function $f(x)$, with a ratio λ.

(iii) The dilation $f \oplus g$ is an extensive operation with respect to f iff g contains the neutral element, i.e. here the t-axis (positive and negative values).

PART IV. THE RANDOM MODELS

XIII. Random Closed Sets

A. INTRODUCTION

This last part of the book deals with *random* set models. We have already devoted the eighth chapter to notions of randomization, but we did not question whether or not the object under study could be interpreted as a realization of a random set.

We have defined many criteria, e.g. erosions of various types, size distributions, skeletons, etc., so that when we study a given object, each may enlighten us on different aspects of its structure. It would be extremely helpful to condense these elementary pieces of information into a single structural synthesis. In this chapter we shall try to achieve a first stage by constructing a comprehensive range of random sets, and then in the next chapter, confront them with the actual physical phenomena. Indeed, our aim is rather an ambitious one; we do not claim that our analysis is anything more than a limited attempt.

The quantitative description of random sets has to be carried out at two different levels. First of all, we have to properly define them, i.e. provide them with adequate axiomatics, and then derive their main mathematical properties (characterization of random sets by their Choquet functional, infinite divisibility, etc., Section G). However a purely mathematical approach is not sufficient, and it must be complemented with a more forthright description of random sets, which effectively gives recipes for construction of random sets possessing desirable morphological properties. We shall start at this level, to give the reader examples before venturing into the rather heavy, but necessary, mathematical definitions and proofs. Sections B to F and the major part of the exercises essentially cover the basic random set models used in morphology. There is a real danger of transforming the material into a handbook of properties. So our strategy has been to not be exhaustive, but rather to focus on the description of key points such as the infinite divisibility. From an initial scan, we can distinguish three families of models; the Boolean model and its derivatives (Sections B and C); the tesselations of Euclidean space (Sections D and E), and the point process (Section F). Our presentation of the first family is fairly complete. In fact, apart from a few exceptions, notably B. Matern (1960),

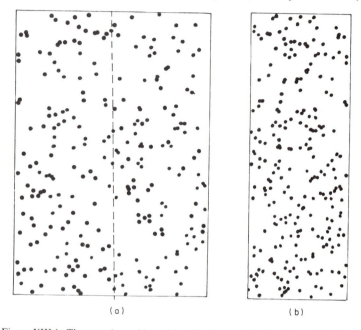

(a)   (b)

Figure XIII.1. The superimposition of two Boolean realizations is again Boolean.

R. E. Miles (1976), D. Stoyan (1978) and, to a lesser extent H. R. Solomon (1953), H. Giger (1968), N. Cressie (1979), the main studies on the subject are due to the Fontainebleau School, (P. Delfiner (1973), J. Serra (1969), (1974), F. Conrad (1972), D. Jeulin (1979)) and especially G. Matheron (1967), (1975) who conceived the family, found its general moment $Q(l)$, and derived from it the two notions of infinite divisibility and semi-Markov random sets.

A tesselation of the space here means a partition of it, in which we do not affect the boundaries to any class. For example, a set of lines or curves splits the plane into topologically open zones. The lines themselves form a random closed set, and the zones are called the classes, or the "polygons" of the tesselation. We have concentrated on two types of tesselations, namely those coming from Poisson flats (called Poisson tesselations), and those known as the metallic grain models. The main names attached to the former are G. Matheron and R. E. Miles, and to the metallic grain tesselations, J. L. Meijering and E. N. Gilbert. The literature devoted to the final family,

Figure XIII.2. Two examples of phenomena suitably modelled by Boolean sets. (a) (above) ferrite crystals in an iron sinter, (b) (below) portion of the forest of Fontainebleau. (See also micrographs of clay Figure V.31 (× 3.000) and oogoniae Figure IX-13(a), which are both Boolean. In the second example, the primary grain is made of two disjoint cells.)

namely point processes, is incomparably more abundant than for the other two. Therefore we decided to choose and analyse five typical prototypes of this family. Of course this is far from a complete inventory, however the majority of the characteristic features met in practice are covered.

The central types from which most of the other models are derived are the Boolean model, the Poisson flats and the Poisson point process. These three models are all infinitely divisible; and it is this basic property which finally implies all the other properties that they satisfy. "Infinitely divisible" means the following: if one picks out two realizations of the model, and superimposes them, taking the set union of phases X, then the result is again a model of the same family. For example, look at Figure XIII.1. Figure XIII.1a shows a realization of a circular Boolean model. Figure XIII.1b has been obtained by reflecting the left side of Figure XIII.1a onto its right side, producing a Boolean realization with double density.

In fact, for the Poisson point process and the isotropic Poisson flat process, the property is even stronger, since the resulting set, after union, is *similar* to the preceeding ones, i.e. identical to it, up to a scale factor. The models which satisfy this self similarity property are said to be *stable under union* (see Section E).

Moreover, the Boolean models and the Poisson flats are also stable under cross sections. If they are defined in $\mathbb{R}^n$ and if one considers the restriction of $\mathbb{R}^n$ to $\mathbb{R}^3$ or to $\mathbb{R}^2$, the sections of the realizations are again Boolean (or Poisson flat) realizations in $\mathbb{R}^3$ or in $\mathbb{R}^2$. Such a characteristic leads us to define and study them independently of the dimensions of the Euclidean space.

From a theoretical point of view, which we do not develop further here, one can interpret their kinships by noting that the three models are all Poisson points in convenient abstract spaces (K. Krickberg 1974, G. Matheron 1975). Other theoretical results, which are not developed here, concern ergodic theorems for random sets (R. Cowan 1978, D. Stoyan 1979) and central limit theorems for random sets (N. Cressie 1978). From a more physical point of view, we note that infinite divisibility plays a role for the random sets similar to that of summation for random variables giving a Gaussian law. (Curiously, the Boolean models often appear as the limit of infinite unions of *other* random sets (Ex. XIII, 11).)

B. THE BOOLEAN MODEL

B.1. Definition of the Boolean model

The definition of the Poisson point process in $\mathbb{R}^n$ is well known (see Section F). This random set of points is characterized by the following two properties:

(a) if B and B' are two sets such that $B \cap B' = \varnothing$ the numbers $N(B)$ and $N(B')$ of points falling in B and B' are two independent random variables;

(b) the elementary volume dv contains one point with probability $\theta(dv)$ and no points with probability $1 - \theta(dv)$. The measure θ is called the density of the process. Here we will take $\theta = \text{constant}$, because it leads to more geometrically interpretable results. The two properties imply that for any bounded B the random integer $N(B)$ follows the Poisson law with parameter

$$\Phi = \int_B \theta(dv)$$

$$\Pr\{N(B) = n\} = \frac{\Phi^n}{n!} \exp(-\Phi)$$

(XIII-1)

Suppose we take a realization of a Poisson process of constant density θ, and consider each point as the germ of a crystalline growth. If two crystals meet each other, we suppose that they are not disturbed in their growth, which stops independently for each component. Let us transpose this description in terms of random sets. The I points of the Poisson realization are at the points x_i ($i \in I$), in $\mathbb{R}^n$. The primary grain is a nonstationary random set X'; we pick out various realizations X'_i of X' from its space of definition, and implant each X'_i at the corresponding point x_i. The different X'_i are thus independent of each other. We shall call the realization X of a Boolean model the *union* of the various X'_i after implantation at the points x_i:

$$X = \bigcup_{i \in I} X'_i$$

(XIII-2)

The Boolean model is extremely flexible; it is a first step, where one admits only negligible interactions between the particles X'_i.

B.2. A fundamental property of the Boolean model

In this and the following two sections we will always work in the space $\mathbb{R}^n$, and make no assumption about the convexity of the primary grain. Centre this primary grain X' at the origin. Then the random set X' is characterized by the family of functionals $\eta(B)$

$$\eta(B) = \Pr\{B \Uparrow X'\}$$

(XIII-3)

If X' is translated from the origin to the point z, it has the new functional $\eta_z(B)$, which is easily deduced from (XIII-3)

$$\eta_z(B) = \Pr\{B \Uparrow X'_z\} = \Pr\{B_{-z} \Uparrow X'\} = \eta(B_{-z})$$

(XIII-4)

We now consider the Boolean model X itself, namely the union of all the X'_i, and compute the probability of B being included in the pores of X^c.

According to property (a) of the Poisson process, each element of volume dz of the space makes its contribution independently of the others. In dz, centred at z, two mutually exclusive "favourable" events can occur:

(1) no germ in dz: probability $1 - \theta dz$;
(2) one germ in dz, but the grain X'_z does not reach B: probability θdz $[1 - \eta(B_{-z})]$.

By composition of these two probabilities and integration of z over $\mathbb{R}^n$, we find

$$Q(B) = \Pr\{B \subset X^c\} = \exp\left[-\theta \int_{\mathbb{R}^n} \eta(B_{-z}) dz \right] \qquad \text{(XIII-5)}$$

To interpret the geometrical meaning of the integral in (XIII-5), let $k(z)$ be the indicator function of the random set $X' \oplus \check{B}$, that is

$$k(z) = 1 \quad \text{when} \quad X' \cap B_z \neq \varnothing$$
$$k(z) = 0 \quad \text{when} \quad X' \cap B_z = \varnothing;$$

then Mes $(X' \oplus \check{B}) = \int_{\mathbb{R}^n} k(z) dz$

By taking the mathematical expectation, and inverting it with integration, we find

$$E[\,\text{Mes}\,(X' \oplus \check{B})] = \int_{\mathbb{R}^n} \eta(B_{-z}) dz$$

Thus formula (XIII-5) may be written as:

$$Q(B) = \Pr\{B \subset X^c\} = \exp\{-\theta E[\text{Mes}\, X' \oplus \check{B}]\} \qquad \text{(XIII-6)}$$

which is the *fundamental* formula for the Boolean model. It links the functionals of X to those of the primary grains. In other words, the proportion $1 - Q(B)$ of the dilated set $X \oplus \check{B}$ (final state) is related to the measure $E[\text{Mes}\, X' \oplus \check{B}]$ of the dilated primary grain (primary state) via an exponentiation. Conversely, if we start with a function $Q(B)$ that satisfies all the mathematical conditions of a Choquet capacity, then it completely determines the random model (see Th. XIII-4). Unfortunately, this theoretical result is mainly of heuristic value since it does not provide us with a way to actually calculate the probability $P(B)$ that a given compact set B is contained in the random set X. For some cases (for example X' a convex set and B a segment) we are able to determine $P(B)$, but not in general.

B.3. The distribution of $N_0(B)$, the number of X' hitting B

We will now calculate the distribution of the number of primary grains X'_i that hit B. Suppose that X' takes only a discrete number of values and regroup the

X_i' in classes of given size and shape, indexed by $j \in J$. In the class j, all the primary grains are translates of one another. Let $x_{i/j}$ be their centres. The points $x_{i/j}$ contribute a subset of the x_i randomly chosen. So they too are a realization of a Poisson point process, density θ_j. Let us fix the class j; the structuring element B hits one X_j' centred at z if and only if $B_z \uparrow X_j'$; i.e., if and only if z falls in the set X_j' dilated by B. The number $N_\theta(B)$ of such z has a Poisson distribution with parameter $\theta_j E[\text{Mes}(X_j' \oplus \check{B})]$. To include all the classes, we sum up all the independent Poisson laws over j. The total is thus given by the Poisson distribution with parameter $\sum_j \theta_j E[\text{Mes} X' \oplus \check{B})]$, a quantity which is precisely $\theta E[\text{Mes}(X' \oplus \check{B})]$. Hence:

$$N_\theta(B) = \text{Poisson}\{\theta E[\text{Mes}(X' \oplus \check{B})]\} \qquad (\text{XIII-7})$$

The formula (XIII-7) generalizes the previous fundamental property (XIII-6), which corresponds to $N_\theta(B) = 0$ in (XIII-7). However in practical work, equation (XIII-6) is more useful, since one cannot always separate out the different X_j', but only test the fact that B hits none of them.

D. Stoyan (1978) enlarged the preceding result by proving that the number of hittings remains a Poisson variable, even when the density $\theta(x)$ is not constant over the space.

B.4. Applications of the fundamental property

Direct use of formula (XIII-6) leads to the computation of the usual functionals. Firstly notice that the right hand side of the expression is constant when B varies only by translation: for θ a constant, the Boolean model is therefore *stationary*.

(a) *Porosity*
Reduce the structuring element B to one point. Then (XIII-6) expresses the probability q that a given point belongs to the pores:

$$q = \exp\{-\theta E[\text{Mes} X']\} \qquad (\text{XIII-8})$$

(b) *Covariance*
If $B = \{x, x+h\}$ is now the two extremities of vector h, we have

$$E[\text{Mes}(X' \oplus \check{B})] = E[2\,\text{Mes}\,X' - \text{Mes}(X' \cap X'_{-h})] = 2K(O) - K(h)$$

so the covariance of the pores $C_0(h)$ is given by

$$C_0(h) = Q(B) = \exp\{-\theta[2K(O) - K(h)]\}$$

We notice that the porosity $q = C_0(O)$ equal $\exp[-\theta K(O)]$ and

$$C_0(h) = q^2 \exp[\theta K(h)] \qquad (\text{XIII-9})$$

For the covariance $C_1(h)$ of the grains, we see that from

$$\Pr\{x \in X\} = \Pr\{x; x + h \in X\} + \Pr\{x \in X; x + h \in X^c\}$$
$$C_1(h) = 1 - 2q + q^2 \exp[\theta K(h)] \qquad \text{(XIII-10)}$$

(c) *Specific volume and surface*
Assume the primary grain X' to be sufficiently regular, and to a.s. belong to the convex ring $\mathscr{F}(\mathscr{K})$. Then the corresponding Boolean set X is a locally finite union of compact covex sets and a.s. belongs to the extension $\mathscr{F}(\mathscr{F})$ of the convex ring (Ch. V, c.4). Its $n + 1$ Minkowski measures exist and possess densities (Matheron 1975 Th. 4-7-2), i.e. the specific parameters. We, have calculated the first of them (rel. VIII-8) while the second one can easily be derived from the covariance and leads to the specific surface s. Applying Cauchy's formula (V-10) to the restriction $Z \cap X^*$ of any realization X^* of X, to a measuring mask Z, together with a mathematical expectation, we find

$$s = \frac{E[n W_1^{(n)}(Z \cap X^*)]}{\text{Mes } Z} = \frac{-1}{b_{n-1}} \int_{\Omega_n} C'_{1,\omega}(o) d\omega$$

$$= -\frac{\theta_q}{b_{n-1}} \int_{\Omega_n} K'_\omega(o) d\omega$$

$$s = \theta q E[S(X')] = \theta \exp\{-\theta E[\text{Mes } X']\} \cdot E[S(X')] \qquad \text{(XIII-11)}$$

According to stereological notation, s is written S_V in $\mathbb{R}^3$, U_A in $\mathbb{R}^2$ and N_L in $\mathbb{R}^1$; relation (XIII-11) becomes:

$$\text{in } \mathbb{R}^3 : S_V = \theta E[S(X')] \cdot \exp\{-\theta E[V(X')]\}$$
$$\text{in } \mathbb{R}^2 : U_A = \theta E[U(X')] \cdot \exp\{-\theta E[A(X')]\} \qquad \text{(XIII-12)}$$
$$\text{in } \mathbb{R}^1 : N_L = \theta E[N(X')] \cdot \exp\{-\theta E[L(X')]\}$$

These results are true whether the primary grain X' is convex or not.

One could derive expressions for the other Minkowski densities, but they are very complicated and we prefer to restrict ourselves to the 2-D and 3-D cases with convex X's. (Moreover the W_k's, with $k \geq 2$, pose a difficult problem with regard to digitalization, as we will see).

(d) *The law of first contact*
Suppose a random point x in the pores is chosen uniformly, and let R be its smallest distance to any grain; R is called the first contact distance. What is the distribution function $F(r)$ of the random variable R? We remark that $1 - F(r)$ represents the conditional probability of a ball of radius r centred at x, being

contained in the pores, given that x is in the pores. We can write:

$$1 - F(r) = \frac{Q(B_r)}{Q(O)} \qquad \text{(XIII-13)}$$

where B_r is the ball of radius r.

(e) *Boolean models induced on subspaces*

Obviously, when a primary grain is compact, a Boolean set in $\mathbb{R}^n$ induces another Boolean model on any hyperplane of dimension k ($1 \leq k \leq n$). However equally obviously, knowledge of the induced model is not sufficient for complete knowledge of the initial one. For example, if one adds to the initial Boolean sets a set of Poisson points in $\mathbb{R}^n$, the induced model is not changed, although the $Q(B)$ of the initial ones are markedly perturbed. Basically, the only properties that we can recover from induced sections are those which come from $Q(B)$'s where B is itself a subset of the section. Formally speaking, the problem of characterizing the primary grain from those induced on sections, is identical with the random sectioning of individuals that has already been treated in Chapter VIII, E.4 and F.1. For example, if $\theta_i(\omega)(i = 1, 2)$ denotes the induced densities on a line Δ_ω or a plane Π_ω, and θ_1 the original density, formulae (VIII, 43) and (VIII, 44) become:

$$\theta_3 = \frac{\theta_2(\omega)}{E[L(X'|A\omega)]} = \frac{\theta_1(\omega)}{E[A(X'|\Pi\omega)]} \qquad \text{(XIII-14)}$$

(the derivation is the same as in Ch. VIII, F.1). For the passage $R^2 \Rightarrow R^1$, we obtain:

$$\theta_1(\alpha) = \theta_2 E[L(X'|\Delta\alpha)]$$

Similarly, formulae (VIII, 37) to (VIII, 40) remain valid, with X', $X' \cap \Pi$, $X' \cap \Delta$ for the 3-D, 2-D and 1-D cases respectively. As in Chapter VIII, isotropy and convexity assumptions are not needed here, they do simplify the formulation of the result (see Ex. XIII-4). On the contrary, the results of Section 5 which follow, demand convexity of X'.

B.5. The Boolean model with compact convex primary grain

(a) *Linear size distribution of the pores*

X' is now assumed to be (almost surely) convex, and B is the segment of length l and direction α. Using the notations of Chapter X, we denote this particular $Q(B)$ by $Q_\alpha(l)$. For X' convex

$$\text{Mes}(X' \oplus \check{B}) = \text{Mes}(X') + l D_\alpha$$

where D_α is the projection of the contour X' in direction α. Thus, $Q_\alpha(l)$ is equal to

$$Q_\alpha(l) = q\exp[\theta K'_\alpha(O)l]$$

The distribution function of the intercepts of pores $F_\alpha(l)$ satisfies

$$1 - F_\alpha(l) = \frac{Q'_\alpha(l)}{Q'_\alpha(O)} = \exp[\theta K'_\alpha(O)l] \qquad \text{(XIII-15)}$$

Thus the linear size distribution (in number) is an exponential. This interesting property is a particular case of the general notion of a semi-Markov process (see Sect. E). If x is a point in the pore, and D a straight line containing x of an arbitrary direction, any event which happens on one side, say to the left of x, is independent of any event on the right. The Boolean model with convex grains is obviously semi-Markovian. After cutting a grain, a straight line cannot cut it again because of convexity, and events above the cut are independent of those below. So the semi-Markovian structure is present, resulting in an exponential size distribution.

Relation (XIII-15) also shows that after averaging over α, we cannot directly find the mean perimeter of the primary grain. However, if we assume that the primary grains have no preferred orientation, then their functional $\eta(B)$ no longer depends on the orientation of B, and (XIII, 15) becomes:

$$1 - F(l) = \exp\left\{-\frac{\theta l}{\pi} E[U(X')]\right\} \qquad \text{(XIII-16)}$$

The slope at the origin of $F(l)$ provides us with an estimator of $\pi^{-1}\theta E[U(X')]$, but does not allow us to separate the two factors "density θ" and "average perimeter $E[U]$".

In order to illustrate relation (XIII-16), we can estimate the function on the simulation photographed in Figure XIII.1a. The original figure is a square with an area of $0.5\,m^2$ and containing 500 Poisson points. The two functions $1 - F(l)$ and $-f(l) = -F'(l)$ are given in Exercise X-1. Formally, the density function $f(l)$ carries the same information as $F(l)$, but due to the finite size of the simulation, it considerably amplifies fluctuations.

(b) *Linear size distribution of the grains*
When X' is convex, the semi-Markov property allows us to compute the distribution function $F_1(l)$ of the grain traverse lengths.

First, we can easily deduce from the covariance (XIII-9), the conditional probability

$$\Pr\{x + h \in X \mid x \in X^c\} = 1 - \exp\{-\theta[K(O) - K(h)]\} \qquad \text{(XIII-17)}$$

which is the probability of having $x + h$ in the grains when x is in the pores.

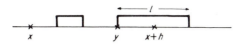

Figure XIII.3.

Second, we denote by y the beginning of the grain which contains $x + h$ (Fig. XIII.3).

According to the semi-Markov property, the probability (XIII-17) is the sum of all products like:

$$\mathrm{Pr}\{y \in X^c \,|\, x \in X^c\} \cdot \mathrm{Pr}\{l \geq x + h - y\}$$

as y varies from x to $x + h$. In other words:

$$1 - \exp\{-\theta[K(O) - K(h)]\} = \theta K'_\alpha(O) \int_0^h \exp\{-\theta[K(O) - K(y)]\}$$
$$\times [1 - F_1(h - y)]dy$$

Thus the size distribution F_1 is given by a convolution equation which can be solved by taking Laplace transforms of both sides. Let $R(\lambda)$ and $\Phi_1(\lambda)$ denote the Laplace transforms of $\exp\{-\theta[K(O) - K(h)]\}$ and of the density function F'_1 associated with F_1, i.e.:

$$[F'_1(h)] = \Phi_1(\lambda); \qquad [\exp\{-\theta[K(O) - K(h)]\}] = R(\lambda)$$

Then we find:

$$1 - \Phi_1(\lambda) = \frac{1 - \lambda R(\lambda)}{\theta K'_\alpha(O)R(\lambda)} \qquad \text{(XIII-18)}$$

giving us F_1 after inversion (for an application see Ex. no. XIII.5).

(c) *Law of first contact*

Due to the a.s. convexity of X', we can now use Steiner's formula to expand relation (XIII-13):

$$\text{in} \quad \mathbb{R}^3\!: 1 - F(r) = \exp\left\{-\theta_3\left[\frac{r}{2\pi}E[S(X')]\right.\right.$$
$$\left.\left. + r^2 E[M(X')] + \frac{4}{3}\pi r^3\right]\right\} \qquad \text{(XIII-19)}$$

$$\text{in} \quad \mathbb{R}^2\!: 1 - F(r) = \exp\left\{-\theta_2[rE[U(X')] + \pi r^2]\right\} \qquad \text{(XIII-20)}$$

i.e. exponential polynomials in r. Note that there is no assumption here about uniform *orientation* of the primary grain. It turns out that the law of first contact is a good description of porous media in general, and recently

G. Matheron (1979) has proved that it is connected with the permeability characteristics of the structure.

B.6. Connectivity number of the 2-D Boolean model

Up to now, we have ignored derivations of connectivity numbers, for the very good reason that this computation presents us with a forbidding problem of digitalization that makes them useless in experimental work. To give some idea how and why, we will treat the 2-D and the 3-D cases of isotropic compact convex primary grains. The 2-D analysis of this section which goes back to J. Serra (1974), formula (XIII-21), has been independently treated by R. Miles (1976).

In $\mathbb{R}^2$, the first two specific parameters A_A and U_A are digitalizable, since the realizations of X are locally finite unions of compact convex sets (Th. VII.3). But unfortunately, the 2-D connectivity number N_A cannot be approached in the same way. Let us calculate it in three different ways.

(a) *First contact points*

Now N_A is the difference between the two convexity numbers N_A^+ and N_A^- (Ch. V.B5). But N_A^+ is the proportion of the first contact points per unit area in a given direction, say α. Each X_i', centred at x_i is convex, and therefore provides only one element of perimeter oriented in direction α. The probability that this element is not covered by another grain is q, the porosity of X. Since there are θ_2 primary grains per unit area, we find:

$$N_A^+ = q\theta_2 = \theta_2 \exp\{-\theta_2 E[A(X')]\}$$

and N_A^- is due to the first contact points of X^c, which can occur only on angular points where the boundaries of two primary grains overlap. More precisely, denoting by θ the value of the angle at one angular point, and by n the specific number of such points, we may write:

$$N_A^- = n\frac{E(\theta)}{2\pi}$$

Note that each angle θ appears with the same probability as $\pi - \theta$, implying that $E(\theta) = \pi/2$. To calculate n consider an elementary vector $(x, x+dx)$ where the two extremities x and $x+dx$ belong to a $\delta X_i'$. Another boundary of primary grain, say $\delta X_j'$ crosses $(x, x+dx)$ with a probability $-2C_0'(O)$ (configurations 01 and 10). The total length per unit area of such elementary vectors is $\theta_2 E[U(X')]$; thus applying relation XIII-9, we find:

$$2n = -\theta_2 E[U(X')]2C'(o) = \frac{2}{\pi}\theta_2^2 E^2[U(X')]\exp\{-\theta_2 E[A(X')]\}$$

(2n because each angular point is counted twice in this evaluation). So finally, the connectivity number N_A is given by:

$$N_A = N_A^+ - N_A^- = \theta_2 \exp\{-\theta_2 E[A(X')]\}$$
$$\times \left\{1 - \frac{\theta_2 E^2[U(X')]}{4\pi}\right\} \qquad \text{(XIII-21)}$$

(b) *Euler's relation for digitalization on a square grid*
We now consider the plane to be digitalized by a square grid of points with a spacing a. To apply Euler's relation for the square planar graphs (Ex. VI, 4) reduces to taking for the connectivity number N_A the following limit:

$$N'_A = -\lim_{a \to 0} \frac{Q\begin{pmatrix} 1 & 1 \\ 1 & 1 \end{pmatrix} - Q\begin{pmatrix} 1 \\ 1 \end{pmatrix} - Q\begin{pmatrix} 1 & 1 \end{pmatrix} + q}{a^2} \qquad \text{(XIII-22)}$$

where the symbols between brackets denote the square or the segment of "size" (a). Since dilation is continuous for the convex ring, one can substitute the full square and segment for their respective vertices. The division by a^2 is due to the local version of the algorithm, and we take the limit in order to see what happens when the digitalization becomes infinitely precise. The different probabilities Q, are given by the fundamental relation (XIII-6), where $E[\text{Mes}(X' \oplus \check{B})]$ is provided by Steiner's formula. Finally, we obtain:

$$N'_A = \theta_2 \exp\{-\theta E[A(X')]\} \left\{1 - \frac{\theta_2}{\pi^2} E^2[U(X')]\right\} \qquad \text{(XIII-23)}$$

(c) *Euler's relation for triangular digitalization*
If the grid is made up of equilateral triangles of side (a), the limiting connectivity number when $a \to 0$, is given by:

$$N'_A = -\lim_{a \to 0} \frac{Q\begin{pmatrix} 1 & 1 \\ & 1 \end{pmatrix} + Q\begin{pmatrix} & 1 \\ 1 & 1 \end{pmatrix} - Q\begin{pmatrix} & 1 \\ 1 & \end{pmatrix} - Q\begin{pmatrix} 1 \\ & 1 \end{pmatrix} - Q\begin{pmatrix} 1 & 1 \end{pmatrix} + q}{a^2 \sqrt{\frac{3}{2}}}$$
$$\text{(XIII-24)}$$

which leads to:

$$N''_A = \theta_2 \exp\{-\theta_2 E[A(X')]\} \left\{1 - \theta_2 \frac{\sqrt{3}}{2\pi^2} E^2[U(X')]\right\} \qquad \text{(XIII-25)}$$

The comparison between the (experimentally inaccessible) right-hand side of the expression for N_A and the two digital versions N'_A and N''_A is interesting. The bias is far too large to be ignored. For example, take for X' a disk with a fixed radius r and express the three connectivity numbers as a function of the porosity $q = \exp(-\theta_2 \pi r^2)$, i.e.:

$$\frac{N_A}{\theta_2} = q(1 + \log q); \qquad \frac{N'_A}{\theta_2} = q\left(1 + \frac{4}{\pi}\log q\right);$$

$$\frac{N''_A}{\theta_2} = q\left(1 + \frac{2\sqrt{3}}{\pi}\log q\right) \qquad\qquad \text{(XIII-26)}$$

For a very standard value of q, say 0.5, we get:

$$\frac{N_A}{\theta_2} = 0.153 \qquad \frac{N'_A}{\theta_2} = 0.059 \qquad \frac{N''_A}{\theta_2} = 0.118$$

which is not encouraging (as usual in quantitative measurements, the square grid is particularly disastrous). We plotted the three functions (XIII-26) in Figure XIII.4. Extrapolating beyond the disk to other convex shapes, we learn from these graphs that it is wise to try and estimate N_A only when the porosity q is rather large (although it is wiser still, not to estimate N_A at all, as we propose below).

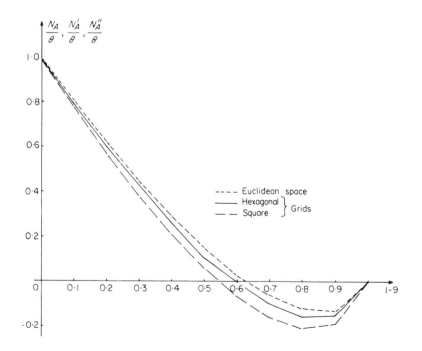

Figure XIII.4. Digitalization causes a grid dependent bias in the measurement of N_A.

B.7. Specific parameters for the 3-D Boolean model

The specific volume and surface have already been calculated in Section B.4; from this it can be seen that they do not pose digitalization problems. Once again assuming convexity and isotropy of X', we can immediately find the specific norm M_v from the 2-D connectivity numbers of the plane sections of X. From formulae (XIII, 14), (XIII, 21) and from Exercise XIII, 4, we obtain:

$$M_v = 2\pi N_A = \theta_3 \exp\{-\theta_3 E[V(X')]\}$$
$$\times \left\{ E[M(X')] - \frac{\pi^2}{32} E^2[S(X')] \right\} \qquad \text{(XIII-27)}$$

The expression for the 3-D connectivity number N_v has been calculated by R. Miles (1976), in a similar way to the above derivation for $\mathbb{R}^2$. In $\mathbb{R}^3$, the calculus is more complicated, since three terms contribute to N_v: one due to the regular portions of δX, one due to the lines along which two X'_i intersect, and a third due to the points where three X'_i intersect; this results in:

$$N_v = \theta_3 \exp\{-\theta_3 E[V(X')]\}$$
$$\times \left\{ 1 - \frac{\theta_3}{4\pi} E[M(X')] \cdot E[S(X')] + \frac{\pi^2 \theta_3^2}{96} E^3[S(X')] \right\} \qquad \text{(XIII-28)}$$

Exercise (XIII-6) is devoted to the calculation of the digital version N'_v of N_v, for a cubic grid. The result, worse still than in $\mathbb{R}^2$, leads us to the question:

B.8. How can one test for a Boolean model?

(a) *Overview*
If the primary grain X' is not convex, we simply cannot proceed with testing the model except when X' is virtually completely defined by external conditions. For example in Chapter IX, Section E.1, the elementary structure X' which governs the ovogoniae division in the ovary, is a pair of balls of given radius r, and distance apart of constant value $d > 2r$. Here the model is almost completely specified. The test consists of checking the Boolean location of the X', using relation (XIII-6) and deriving the numerical estimates of θ and d.

When X' is a compact convex set, we must estimate its first two (in $\mathbb{R}^2$) or first three (in $\mathbb{R}^3$) Minkowski functionals, as well as the density θ. The robustness considerations developed just above, means that we must refuse the estimation of numbers (N_A, N_v, M_v) and also base the experiments on erosion area measurements which are always digitalizable (Th. VII-6).

(i) In $\mathbb{R}^2$: the technique consists of eroding the pores, on one hand by a segment length l, and on the other hand by the non-degenerate convex sets λB

that are similar to a basic given shape B (square, hexagon, octagon). According to the fundamental property, we must obtain:

$$
\left.
\begin{aligned}
Q(l) &= \exp\left\{-\theta_2\left[E(A(X')) + \frac{l}{\pi}E(U(X'))\right]\right\} \\
Q(\lambda B) &= \exp\left\{-\theta_2\left[E(A(X')) + \frac{\lambda}{2\pi}U(B)\cdot E[U(X')]\right.\right. \\
&\qquad\left.\left. + \lambda^2 A(B)\right]\right\}
\end{aligned}
\right\} \quad \text{(XIII-29)}
$$

(here isotropy of X' is assumed; if not, one can obviously introduce orientation parameters of B in the formulae). The linear and parabolic interpolations of $\log Q(l)$ and $\log Q(\lambda B)$ respectively, by least squares for example, first of all allows us to judge whether the data correctly fit the two types of equation (XIII-29), and if so, then to identify the coefficients. As a supplementary test, each of the relations in XIII-29 independently give an estimator of $E(A(X'))$ and $E(U(X'))$, so they can be compared. The quality of this method comes from the fact that the three key parameters are derived from two robust experimental *curves*, and not just from three image transformations. Afterwards, one can always estimate the specific Minkowski parameters of X by putting $E^*(A)$, $E^*(U)$ and θ_2^* into relations (XIII-11), (XIII-12) and (XIII-21), (the latter with the usual caveat).

(ii) In $\mathbb{R}^3$: the structure under study is known through plane sections, each of which we investigate as a 2-D phenomenon. The only difference is that now the segment and the plane convex set are interpreted as 3-D geometric figures in Steiner's formula (rel. V-15). We have:

$$
\left.
\begin{aligned}
Q(l) &= \exp\left\{-\theta_3\left[E(V(X')) + \frac{l}{4}E(S(X'))\right]\right\} \\
Q(\lambda B) &= \exp\left\{-\theta_3\left[E(V(X')) + \frac{\lambda}{8}U(B)E(S(X'))\right.\right. \\
&\qquad\left.\left. + \frac{\lambda^2}{2\pi}A(B)E(M(X'))\right]\right\}
\end{aligned}
\right\} \quad \text{(XIII-30)}
$$

Since the volume of B is zero, the last term in Steiner's formula disappears, and with it, the possibility of estimating θ_3. We have reached the limits of a 2-D analysis. To go further, we can either work on *thick* sections, or add extra specifications onto the primary grain (see next Section B.9). Using the thickness of the section is a bit risky, since the pores under study must effectively be transparent and the grains opaque. If this condition holds, and if we can cut a section of a given thickness d and a given direction $\vec{\omega}$ in the initial

X, then the normal projection of the slice generates a new set Y. To say that a given two dimensional set B is included in the pores Y^c of the projected image is equivalent to saying that the prism $B \oplus \check{\omega} d$ is included in the pores of the 3-D material X. It follows that if the projection Y is Boolean, then the original structure X is also Boolean, since the fundamental relation (XIII-6), applied to Y with a structuring element B, is just the erosion of X by $B \oplus \check{\omega} d$. Thus the two relations of (XIII-29), *holding on Y*, can be interpreted in terms of erosions on X, and we have:

$$\left.\begin{aligned}
Q(l) &= \exp\left\{-\theta_3\left[E(V(X')) + \frac{(l+d)}{4}E(S(X')) + \frac{ld}{4\pi}E(M(X'))\right]\right\} \\
Q(\lambda) &= \exp\left\{-\theta_3\left[E(V(X')) + \frac{\lambda}{8}(U(B)+d(n-2))E(S(X'))\right.\right. \\
&\quad \left.\left. + \frac{\lambda^2}{4\pi}(2A(B)+dU(B))E(M(X')) + \lambda^3\,dA(B)\right]\right\}
\end{aligned}\right\} \quad \text{(XIII-31)}$$

(In the second relation (XIII-31) we particularized B as a regular polygon with n sides. For finding (XIII-31) from (XIII-6) use relations (IV-26) and (IV.33).) As we see, $\log Q(\lambda B)$ is now a polygon of degree 3 in λ, whose coefficients lead directly to estimation of the four key parameters of the model. The first relation of (XIII-31) can be used as a check on the estimators obtained from $Q(\lambda B)$. Note that in these two equations the left hand side is concerned with probabilities on the projection Y, and the right-hand side with the 3-D structure X.

The method proposed in this section is just a first step in solving the estimation problems. Trouble with ergodicity, calculation of estimation variances of the Minkowski functionals, etc., have not been resolved. Moreover, a primary grain is not completely determined from knowledge of the average values of its Minkowski functionals. Consider for example a

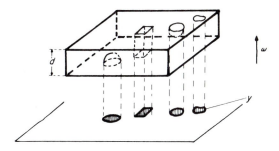

Figure XIII.5. Thick section of a Boolean model.

random primary grain which is similar to a fixed compact convex set X_0', with a random proportionality coefficient k, whose distribution function is $F(k)$. How can $F(k)$ be estimated after "Booleanization"? (see some results on this in D. Stoyan (1978)). Our goal here was a limited one; we merely tried to focus attention on the experimental data we must handle so that the most important parameters could be estimated.

(b) *The Boolean models where θ_3 is accessible from sections*
When it is impossible to make thick sections, one can alternatively try to fit particular models. Therefore, we now develop three more specific models where the 3-D Minkowski functionals of X' are accessible from 3-D and even 1-D sections (another, with cylindric primary grains, is considered in Jeulin (1979)).

(i) Non random primary grain. The two parameters θ_3 and $V(X')$ are derived immediately from the pore covariance $C_0(h)$ (rel. XIII-9), since:

$$\left. \begin{array}{c} \log C_0(O) = \log q = \theta_3 V(X') \\[2mm] \displaystyle\int_{\mathbb{R}^3} \log \frac{C_0(h)}{q^2}\, dh = \theta_3 V^2(X') \\[2mm] \left(= 4\pi \displaystyle\int_0^\infty r^2 \log \frac{C_0(r)}{q^2}\, dr, \text{ under isotropy} \right) \end{array} \right\} \qquad \text{(XIII-32)}$$

The relations in (XIII-32) hold for any compact set X', whether or not it is convex. Now if X' is convex the system (XIII-30) determines all the Minkowski functionals of X' and therefore the estimation problems of the model are solved.

How can the hypothesis of a deterministic X' be tested from sections? We will stay with X' convex, and see whether the variance $\text{var}[V(X')]$ $= \mathcal{M}[V^2(X')] - \mathcal{M}^2[V(X')]$, of $V(X')$ weighted in volume, is equal to zero. The approach is similar to that used in Chapter X, Section D-3. Take for B the group of three points $\{x; x_1; x_2\}$, where $x_1 - x = h_1$ and $x_2 - x = h_2$. We have:

$$Q(B) = \exp\{-\theta_3[3K(O) - K(h_1) - K(h_2) - K(h_1 - h_2) + T(h_1 h_2)]\}$$

where $T(h_1 h_2)$ is the average volume of the primary grain eroded by the triangle $(x\ x_1\ x_2)$. We know (rel. X-32) that:

$$\mathcal{M}[V^2(X')] = \frac{\theta_3}{\log q} \int_{\mathbb{R}^3} \int_{\mathbb{R}^3} T(h_1 h_2)\, dh_1\, dh_2$$

Also,

$$\frac{1}{\log q} \int_{\mathbb{R}^3} \frac{\log C_0(h)}{q^2}\, dh = \mathcal{M}(V)$$

Hence var $(V(X'))$ can now be estimated.

(ii) Spherical primary grain. Denote the distribution function of the radius of X' by $F_3(r)$. The model is completely determined by the covariance $C_0(h)$. Indeed, by noticing that $K(r)$ is in the present case, the average volume of X' eroded by a segment of length r, the problem reduces to recovering the distribution of sphere diameters from their linear intercepts (Ch. X, Section E); we have:

$$\frac{1}{2\pi r}\left[log\ C_0(2r) \right]^{(ii)} = \theta_3 \left[1 - F_3(r) \right] \tag{XIII-33}$$

and, in particular, when $r \rightarrow 0$:

$$\frac{1}{\pi}\left[\log C_0(2r) \right]^{(iii)}_{r=0} = \theta_3$$

Using the left-hand side to estimate θ_3 would result in a large estimation variance, so it seems wiser to open the model with a small ball of radius r_0, i.e. essentially replace θ_3 by $\theta_3[1 - F_3(r_0)]$ given by (XIII-33).

(iii) Poisson polyhedron primary grain. The Boolean set with Poisson polyhedra as primary grains, seems to be rather appropriate when the sizes of the X''s cover a large range. It was first proposed by P. Delfiner (1970), and used to model petroleum-associated media. More recently, J. C. Celeski and D. Jeulin successfully developed it as a model for mineralization associated with various coking coals (1977). Curiously, in both cases the primary "grains" represent the *pores* of the medium, as if the overlapping condition of the Boolean model could more easily be satisfied by air bubbles than by solid materials. A Poisson polyhedron is a random set which has the advantage that it depends on only one parameter, which is called the linear density, and denoted here by ρ. Its covariogram is the negative exponential $K(h) = (6/\pi^4\rho^3)\exp(-\pi\rho h)$. Once again, the two basic quantities used for estimation are $Q(l)$, the result of eroding the pores by a line segment and the covariance $C(h)$. Putting $\theta' = 6\theta/\pi^4\rho^3$, we obtain:

$$\left. \begin{aligned} Q(l) &= q \exp(-\pi\rho\theta' l) \quad \text{with } q = \exp(-\theta') \\ C(h) &= q^2 \exp(\theta' e^{-\pi\rho h}) = q^2 \exp[\theta K(h)] \end{aligned} \right\} \tag{XIII-34}$$

The two parameters θ and ρ which characterize the models can now be estimated from (XIII-34).

(c) *An example*

The Celeski–Jeulin study on cokes will provide us with an example of estimation. The photograph in Figure XIII.6 is meant to be representative of a certain type of coke. Twenty-five similar fields were measured per section, on each of twenty-one polished sections of this type of coke. The purpose of the

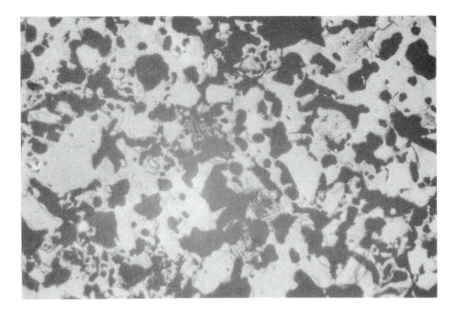

Figure XIII.6. Micrograph from a polished section of formcoke. The carbon appears as white and the pores as dark. The latter will be considered to be the grains when modelled by a Boolean set.

study is to find condensed but rather exhaustive descriptors of the structure (in order to follow how the coke changes when its components and its cooking conditions are modified). The following procedure describes a typical approach for estimating Boolean models.

(a) Using the texture analyser or a similar image analyser, threshold the image into two complementary sets, compute estimates $Q^*(l)$ of the functional $Q(l)$ for the two complementary phases of the structure under study (or more when multiphased structures). The edge effect must be corrected for by suitable formulae (e.g. rel. (II-11)). Plot the quantities $\log Q^*(l)$ versus l, as in Figure XIII.7a.

(b) Measure $Q(l)$ in different directions to detect possible anisotropies. In the present case, the structure is isotropic (i.e. $Q(B)$ does not depend on the orientation of B), and this has been tested for B's that are pairs of points, segments and regular hexagons. When the structure is not isotropic, one must take the rotation average of $Q(B)$ for each B in order to apply relation (XIII-29). At the same time the anisotropies can be described by applying techniques such as the rose of directions.

(c) By the use of geostatistics G. Matheron (1965) calculated the number of

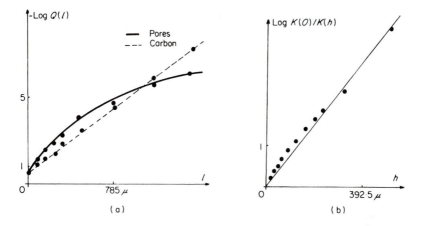

Figure XIII.7. (a) log of the linear erosion estimates for the coal and the pores, (b) log of the geometric covariogram estimate, for the pores primary grain.

sections and the number of fields per section, needed to obtain a given estimation variance for the $Q(B)$'s.

(d) The $Q(l)$ of the coke, in step (a), exhibits a negative exponential behaviour. One might suspect a Boolean structure with convex primary grains. Check this conjecture by estimating $Q(\lambda B)$, of relation (XIII-29), for B a square or a hexagon (here the measurements were performed with B a hexagon, giving a parabolic log $Q^*(\lambda B)$, which corroborates the statement).

(e) If we want to be rigorous, the next step is to check the semi-Markov and the infinite divisibility properties (Section G). The semi-Markov structure has already been shown, but not proved, by the exponential $Q^*(l)$ of the coke. To go further, it would be necessary to test whether relation (XIII-94) is true, for various shapes of B and B'. One checks the infinite divisibility by storing two disjoint fields of analysis and forming the union of the pore sets. On this union, compute log $Q^*(l)$ and log $Q^*(\lambda B)$ and check that they are still a straight line and a parabola respectively (but with different parameters of course). This control was not carried out in the present case, and we assume that it would have led to positive results. If so, according to Matheron's characterization Theorem XIII-9, the structure can be fitted with a union of Boolean sets with convex primary grains, and of cylinders. If there are cylinders, the ranges of the covariance, i.e. the quantities $\int_{\mathbf{R}^2} |h|^n C(|h|) d|h|$ are infinite for $n > 1$ (Ex. XIII-17). It is obviously not the case here (see Fig. XIII.7b). Therefore, the Boolean model with convex primary grains is acceptable.

(f) It remains to estimate the characteristics of the primary grain X'. The next measurement, now, is the covariance. By plotting the covariogram

$\log \theta K^*(h)$, estimated by the algorithm $\log[\log C^*(h) - 2 \log q^*]$, we obtain the curve (XIII.7b) which is virtually a straight line. The primary grain is probably a Poisson polyhedron (although the property is not characteristic). We can confirm this assumption by noticing that for a Poisson polyhedron X':

$$E[V(X')] = \frac{6}{\pi^4}(\rho)^{-3}; \quad E[S(X')] = \frac{24}{\pi^3}(\rho)^{-2};$$

$$E[M(X')] = 3(\rho)^{-1} \tag{XIII-35}$$

the linear density ρ is estimated from the experimental data by a least squares straight line fit to the points $\log[K(o)/K(h)]^*$, and the parameter θ' (whence θ follows immediately) via two possible routes:

— from the experimental porosity, and relation $q = \exp(-\theta')$ and $\theta' = 6\theta/\pi^4\rho^3$
— from the slope of $-\log^* Q(l)$, and relation (XIII-34).

The estimates θ^* and ρ^* and relation (XIII-35) provide estimates of $\theta E(V)$, $\theta E(S)$, $\theta E(M)$ that can be compared with those given by relation (XIII-29) (fitted on the square or hexagonal erosion measurements $Q^*(\lambda B)$). The numerical results are the following:

$$\rho^* = 2.55 \, \text{mm}^{-1}, \theta^* = 141 \, \text{mm}^{-3}(\text{from } q^*) \quad \text{and}$$
$$\theta^* = 144 \, \text{mm}^{-3} \, (\text{slope of } Q(l))$$

We use the second estimate of θ, since it is based on many more measurements than the other. Then:

Parameters calculated from	$\theta E(V)$	$\theta E(S)$	$\theta E(M)$
ρ^* and θ^{**}	0.545	17.14 mm^{-1}	169 mm^{-2}
$Q^*(\lambda B)$ B hexagon	0.539	13.0 mm^{-1}	95 mm^{-2}

The Poisson polyhedron model is acceptable in practice. As usual, accuracy decreases with the dimension of the parameter.

C. MODELS DERIVED FROM BOOLEAN SETS

The Boolean model is the basis for a considerable number of random sets. Some are obtained by taking the density θ to be a realization of a random function, others by iterating (inserting a second model whose centres belong to the grains of the first, for example), others by restricting the Boolean realizations to remain inside Poisson polyhedra, and so on (see Ex. XIII-12). We hope that the derivation of the three-phased Boolean model and the Dead Leaves model, which follow, will suggest further ideas to the reader.

C.1. Three phased textures

There are many ways of building up models for multiphased textures. Depending upon whether or not we wish to emphasize the dependence between two phases, we could use one or the other of the following models, drawn from metallurgy and petrology.

(a) *Sinter textures model*
This random model (A. Greco *et al.*, 1979), was developed in order to describe the morphology of sinter textures. It often happens in metallography that one type of crystal, say X_2, locally replaces all others, and these can only survive in places left by X_2 (i.e. in X_2^c). This leads to a three phased texture with:

$$X_2; X_3 = X_1 \cap X_2^c; X_4 = X_1^c \cap X_2^c = X_3^c \cap X_2^c$$

Assume that X_1 and X_2 are two independent sets. (If the corresponding primary grains are convex, the assumption can be tested very simply: the intercept histograms of X_2^c and of X_3 should be negative exponentials (tests for underlying Boolean structures) and two successive intercepts on X_2 and X_3 should be uncorrelated (tests for independence of the Boolean sets)).

We saw in Chapter IX, Exercise 10 that the second order analysis of a three phased partition of the space is completely determined when we know the covariances of the three phases. In the present case, the independence hypothesis simplifies the covariance calculation, and in terms of $C_1(h)$, $C_2(h)$, we have:

$$
\begin{aligned}
C_2(h) &= C_2(h) \quad \text{(unchanged)} \\
C_3(h) &= C_1(h)\left[1 - 2q_2 + C_2(h)\right] \\
C_4(h) &= \left[1 - 2q_1 + C_1(h)\right]\left[1 - 2q_2 + C_2(h)\right]
\end{aligned}
\qquad \text{(XIII-36)}
$$

From the last two covariances, when left hand sides can be determined experimentally, we can recover $C_1(h)$ and $C_2(h)$. Then from C_1 and C_2, one is able to obtain the characteristics of the respective primary grains by applying relation (XIII-10).

An example of this model is given in Figure XIII.8, where we put side by side a photograph of the sinter texture and a simulation of the model. The experiment is performed on a series of polished sections of sinters, where

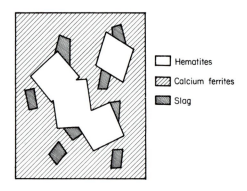

Figure XIII.8. (a) Plate-like calcium ferrite (grey) in contact with haematite (white) or with slag (dark); (b) (below) representation of the structure by a three phased Boolean model.

measurements were made on about 400 fields of the size of photograph XIII.8a. First the Boolean assumption must be tested. We computed the linear erosion of the calcium ferrite matrix, and that of the complement of the haematite. The results (Fig. XIII.9a) leave no doubt about the Boolean structure. The second step is now to recover the primary grain distribution of the haematite and of the pores. On the section, the grains do not exhibit any preferential orientation (indeed the physical process for producing the sinter is isotropic), the crystal sections seem polygonol with angles not far from 90°,

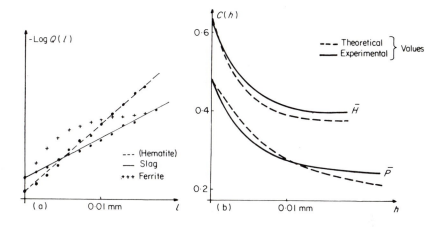

Figure XIII.9. (a) experimental linear erosion of the complement of the haematite, of the slag and of the ferrite: only the first two come from Boolean models, (b) experimental and theoretical covariances of the complement of haematite and the complement of the slag.

and their shapes vary between a maximal and minimal limit. Therefore, we will consider the haematite crystal to be a rectangular parallelepiped with sides a_1, b_1 and c_1, where a_1, b_1 and c_1 are independent random variables, uniformly distributed with a common mean m_1. For the pores, we adopt the same model, but with mean parameter m_2. Fix a_1, b_1 and c_1, and average the haematite covariogram over the rotations in R^3, i.e.:

$$K_1(r) = a_1 b_1 c_1 - \frac{r}{2}(a_1 b_1 + b_1 c_1 + c_1 a_1)$$
$$+ \frac{2}{3\pi} r^2 (a_1 + b_1 + c_1) - \frac{r^3}{4\pi} \quad r \leq \inf(a_1, b_1, c_1)$$

Now averaging over a_1, b_1 and c_1, we obtain:

$$E[K_1(r)] = m_1^3 - \frac{3}{2}m_1^2 r + \frac{2}{\pi}m_1 r^2 - \frac{r^3}{4\pi}$$

and in particular,

$$m_1 = -\frac{3}{2}\frac{E[K_1(O)]}{E[K_1'(O)]}$$

The most robust estimators of $E[K(O)]$ and $E[K'(O)]$ are given by:

$$\log P_1(l) = \theta E[K_1'(O)] l + \log q \quad \text{and} \quad \log C_1(O) = -E[K_1(O)].$$

We apply the same treatment to the pores to finally obtain:

	m (*micron*)	θ (*microns*$^{-3}$)
Haematite	19.2	0.105
Pores	4.3	3.911

It remains then to go through a verification stage, where we check whether the parallelepiped was an appropriate choice for the primary grain. We do so by comparing the theoretical covariances with the experimental ones (Fig. XIII.9b); the agreement seems acceptable.

(b) *Clay soils*
This second model differs from the first in its construction and hence in its results. However, since the statistical treatment is the same, we forego certain details. In cooperation with J. C. Fies, we studied the organization of clay in synthetic soils, and tried to model the process with a three phased Boolean model. The soil was made up of (1) calibrated quartz grains (size fraction 125μ -175μ), (2) clay, (3) pores. The morphological analysis was performed on two series of 40 photographs each, the series coming from soils with 10 per cent and 20 per cent clay material. In both cases the clay partly surrounds the quartz grains, and partly spreads out in the pores, looking like small spots. The same model is applicable to both series, so we will only present the series corresponding to 20 per cent clay. A Boolean model is chosen whose primary grain is:

— with probability p, a ball of quartz of radius r_1 surrounded by a spherical crown of clay of thickness $r_2 - r_1$ (which we call contour clay);
— with probability $1 - p$, a ball of clay of radius r'_2 (which we call isolated clay).

We Booleanize and rule that when quartz and clay overlap, the quartz obliterates the clay. Denote the quartz by (1) and the clay by (2) and the pores by (3) in the resulting texture. The photographs of Figure XIII.10 show the two materials, clay and quartz, which we want to model.
The volumic proportions of the three phases are:

$$V_{V_1} = p_1 = 1 - \exp(-\tfrac{4}{3}\pi\theta\,\mathrm{pr}_1^3)$$

$$V_{V_2} = p_2 = 1 - p_1 - p_3$$

$$V_{V_3} = p_3 = 1 - \exp\{-\tfrac{4}{3}\pi\theta\,[\mathrm{pr}_2^3 + (1-p)r_2'^3]\}$$

For a second order analysis, we shall consider the covariances C_1 and C_3 and the cross covariances $C_{1,3}$ between the quartz and the complement to the pores.

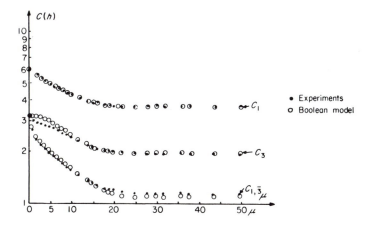

Figure XIII.10(a) (above). Thin section of clay soil. The quartz is in white, the pores in grey and the clay is dark (20%). (b) (below): the experimental and theoretical values of the three covariances (XIII-37).

They all depend on the volume $K(h, r_1, r_2)$ of the intersection of a ball of radius r_1 with a ball of radius r_2, whose centres are separated by a distance h. We thus obtain:

$$
\left.
\begin{aligned}
C_1(h) &= 2p_1 - 1 + (1 - p_1) \exp\left[-\theta K(h, r_1, r_1)\right] \\
C_3(h) &= p_3^2 \exp\left\{-\theta\left[pK(h, r_2, r_2) + (1 - p)K(h, r_2', r_2')\right]\right\} \\
C_{1,\bar{3}}(h) &= (1 - p_1)p_3 \exp\left[p\theta K(h, r_1, r_2)\right]
\end{aligned}
\right\} \quad \text{(XIII-37)}
$$

The mathematical expression for $K(h, r_1, r_2)$ is given in Exercise IX-8 (namely the difference $K - K_X - K_Y$, according to the notation of the exercise). Plotting the linear intercept distributions demonstrates the acceptability of the model, and the comparison between theoretical and experimental covariances is given in Figure XIII.8b. It is worthwhile emphasizing two points. First, notice that angular grains (the quartz) can be adequately modelled by balls (cov. C_1). On the other hand, the behaviour of the experimental covariance $C_{1,\bar{3}}$ near the origin decreases more quickly than the model suggests. The difference arises from the irregularity of the clay crowns surrounding the quartz. Of course in such a case the parameters must be estimated from C_1 and C_3 only; this provides the following results:

r_1	$r_2 - r_1$	r_2'	p	θ
$150\,\mu m$	$60\,\mu m$	$39\,\mu m$	0.02	$210\,\text{mm}^{-3}$

It is also interesting to note the stereological aspect of the result: r_1, r_2 and r_2', radii of 3-D structures, are recovered from 2-D measurements.

C.2. The Dead Leaves model

This further variation of the Boolean model that we now describe is due to G. Matheron (1968). It has the dual advantage that it provides us with a tesselation of the space as well as a model for non overlapping particles.

(a) Dead leaves tesselation

When one looks at a cloudy sky, one only sees the lowest clouds, which hide those above. Standing at the beginning of a forest, one only sees the first few trees, the others being hidden behind. The dead leaves model is just a quantitative description of this type of superimposition. Although the basic relation (XIII-38) is independent of the dimension of the space, we will take our realizations to be in $\mathbb{R}^2$ (Fig. XIII.11), and develop the model in this space. Let X' be a random primary grain in $\mathbb{R}^2$. Place the origin of time t in the far

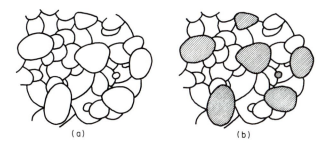

(a)   (b)

Figure XIII.11. A realization of the Dead Leaves model. The remaining boundaries make up a closed set X.

past at $-\infty$. From the origin to the present (i.e. $t = 0$) take realizations of Boolean sets independent of each other with identically distributed primary grains X', at time instants given by a constant spatial density θdt. The grains appearing between $-t$ and $-t + dt$ *hide* the portions of former grains. At time zero, the plane is completely covered (since the origin of t is $-\infty$), giving a *random tesselation* of the space.

According to the Matheron–Kendall theorem, the tesselation is determined by knowing the probability $Q(B)$ that B misses the visible boundaries X of the tesselation (i.e. is included in a connected component of X^c), for all compact B. The probability that, at a time $-t$, a primary grain covers B is $\theta dt\, E[\mathrm{Mes}(X' \ominus \check{B})]$; moreover the union of the grains appearing from $-t$ to zero is a Boolean set with a density $\theta' = \theta t$, and B belongs to the pores with a probability $\exp\{-\theta t\, E[\mathrm{Mes}(X' \oplus \check{B})]\}$. The grain which covers B can appear at any time t with the same probability, thus $Q(B)$ is given by the integral:

$$Q(B) = -\theta E[\mathrm{Mes}(X' \ominus \check{B})] \int_{-\infty}^{0} \exp\{-\theta t\, E[\mathrm{Mes}(X' \oplus \check{B})]\}\, dt$$

i.e.

$$Q(B) = \frac{E[\mathrm{Mes}(X' \ominus \check{B})]}{E[\mathrm{Mes}(X' \oplus \check{B})]} \qquad \text{(XIII-38)}$$

We will use this as the basic formula to describe the model. From it, we immediately derive the probability $C(h)$ that two points x and $x + h$ belong to a same class of the tesselation. Taking $B = \{x, x + h\}$, we get:

$$C(h) = \frac{K(h)}{2K(0) - K(h)} \qquad \{K(h) = E[\mathrm{Mes}(X' \cap X'_h)]\}$$

The *linear size distribution* is also immediate from the basic formula. For the sake of simplicity, we restrict ourselves to convex X' (the general case is no more difficult, but the formalism is heavier). Denote by $P_0(h)$ the probability that the segment $B = [x, x+h]$ belongs to a single class of X^c (i.e. the tesselation boundaries are excluded), where h is the vector of module r and direction α

$$E[\text{Mes}(X' \ominus \check{B})] = K(h)$$
$$E[\text{Mes}(X' \oplus \check{B})] = K(o) - rK'_\alpha(o)$$

(all the derivatives here are taken with respect to r, for α fixed) thus:

$$P_0(h) = \frac{K(h)}{K(0) - rK'(o)} \qquad \text{(XIII-39)}$$

Denoting by F_0 the distribution function in number of the intercepts of X^c:

$$1 - F_0(h) = \frac{P'_0(h)}{P'_0(o)} = \frac{K(o)}{2K'(o)} \left\{ \frac{K'(h)}{K(o) - rK'(o)} \right.$$
$$\left. + \frac{K'(o) \cdot K(h)}{[K(o) - rK'(o)]^2} \right\} \qquad \text{(XIII-40)}$$

whose integral over h (α fixed) from 0 to ∞ is the average intercept m_0:

$$m_0 = -\frac{1}{P'_0(o)} = -\frac{K(o)}{2K'(o)} \qquad \text{(XIII-41)}$$

Note that if $F(h)$ is the distribution function of the intercepts included by the primary grain X' on a straight line, and m the corresponding mean intercept, then:

$$1 - F(h) = \frac{K'(h)}{K'(O)} \quad \text{and} \quad m = -\frac{K(O)}{K'(O)} = 2m_0$$

Thus, in comparison to that of the primary grain, the average intercept in the tesselation is reduced by 50 per cent.

From relation (XIII-41), we can also derive the *specific perimeter* U_A of the tesselation, i.e.:

$$U_A = \frac{1}{2\pi} \int_0^{2\pi} P'_0(O) d\alpha = -\frac{2}{\pi} \frac{E[A(X')]}{E[U(X')]}$$

(the 3-D tesselation gives $S_V = -\frac{1}{2}(E[V(X')]/E[S(X')])$). For the 2-D connectivity number N_A, see Ex. XIII-8.

(b) *Relief grains*
For this example, the assumption of convexity of X' is essential. However, a 2-D isotropic analysis is used only for simplicity, and can be extended to R^3.

Imagine that the observer is now able to decide whether or not a given class of the final tesselation corresponds to an *entire* primary grain that has not been partially obscured by another (Fig. XIII-11b). The union of all these classes generates a random set X made up of disjoint, compact convex grains; we can give such a grain the generic name of X_1.

Denote by A' and U' the random area and perimeter of the primary grain X', let $F(dA', dU')$ be the law of these two variables, and $\bar{\omega}(B; A', U')$ be the probability that $B \subset X'$, where X' is a primary grain with functionals A' and U' and located at the origin of R^2. The probability that B is contained in an X' which locates itself in dx during the time interval $(-t, -t + dt)$, has an area A' and a perimeter U', and is missed by the latter grains, is:

$$\theta\,dt\,dx\bar{\omega}(B_{-x}; A', U')F(dA', dU')\exp\left\{-\theta t\left[A' + E(A')\right.\right.$$

$$\left.\left. + \frac{U'E(U')}{2\pi}\right]\right\} \qquad \text{(XIII-42)}$$

By integrating over t and x, we get the probability $Q_1(B)$ that a connected B is contained in a single relief grain X_1:

$$Q_1(B) = \int_{\mathbb{R}^2} dx \int \frac{\bar{\omega}(B_{-x}; A', U')\, F(dA', dU')}{A' + E(A') + (U'E(U')/2\pi)} \qquad \text{(XIII-43)}$$

(Note: when B is not connected, $Q_1(B)$ is *not* the probability that $B \subset X_1$.)

Another interesting problem is to find the specific number N_A, and the distribution $F_1(dA, dU)$ of the areas A_1 and the perimeters U_1, of the relief grains X_1. The probability that a primary grain locates itself in dx, during $(-t, -t + dt)$, admits an area A' and a perimeter U' and is disjoint from the latter grains, is just (XIII-42), with $\bar{\omega} = 1$ (i.e. B_{-x} is reduced to the point x). After integrating over time, we obtain:

$$N_A = \int \frac{F(dA', dU')}{E(A') + A' + [U'E(U')/2\pi]} \qquad \text{(XIII-44)}$$

and similarly:

$$N_A F_1(dA_1, dU_1) = \frac{F(dA', dU')}{E(A') + A' + [E(U') \cdot U'/2\pi]} \qquad \text{(XIII-45)}$$

We see on the denominator the bias affecting the size distribution of the areas and perimeters; i.e. large values appear with a lower frequency in the relief grains X_1 than in the primary grains X' (see more precise results in Ex. XIII-9).

D. POISSON TESSELATION

D.1. Definition

The best way to introduce Poisson tesselation is probably to show a realization of this model, e.g. Figure XIII.12. As we see, the Poisson tesselation in $\mathbb{R}^2$ is simply obtained by cutting the plane by lines that are random in both orientation and location. This type of definition however, needs to be better formalized if we want to avoid Bertrand type paradoxes. We propose two approaches. In Exercise XIII-13, we go through a step-by-step construction and discussion of the model in $\mathbb{R}^2$. In this section, we decide to stay in $\mathbb{R}^n$, since the tesselations induced on subspaces by a Poisson tesselation in $\mathbb{R}^n$ are again Poisson.

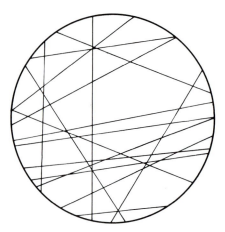

Figure XIII.12. A Poisson lines realization.

In a certain sense, Poisson tesselations go back to Crofton; Crofton's famous theorem can be interpreted in terms of the conditional expectation of the intercept length when a Poisson line hits a given compact convex set. In the twentieth-century literature, one can find a few results in J. Deltheil (1926) and M. G. Kendall and P. A. Moran (1962). However, the main properties have been proved by R. E. Miles (Miles' thesis 1962, or 1969), and G. Matheron (1971, 1972, 1975). Among others, one owes to R. E. Miles the general laws regarding induced tesselations, the functionals $Q(B)$ for B a disk in $\mathbb{R}^2$ or a ball in $\mathbb{R}^3$ and the general formula linking distributions in number and in measure (XIII-51). His work is limited by the absence of the concept of a set transformation (in particular erosions and openings), which probably stopped him short of discovering the property of erosion invariance. It was elegantly proved via a geometric argument by G. Matheron (1971), generalizing to n dimensions the notion of lack of memory of the Poisson process in $\mathbb{R}^1$. This property governs all erosion and size distribution analyses.

Let us define an $n-1$ dimensional flat in $\mathbb{R}^n$ to be the set of points defined by the scalar product:

$$\sum_{i=1}^{n} \omega_i x_i = r \qquad (\text{XIII-46})$$

where the unit vector $\omega = (\omega_1, \ldots, \omega_n)$ is a point on the half unit sphere of $\mathbb{R}^n$, say $\frac{1}{2}\Omega$ (in order to avoid counting the same flat twice for (ω, r) and $(-\omega, -r)$). Start from the straight line Δ of direction ω going through the origin. Let $d\omega$ be an elementary solid angle centred on ω, and realize on Δ a Poisson point processes of density $\lambda_n d\omega$ (λ_n is a positive constant). At each point of the process implant an $n-1$ dimensional flat normal to ω. Consider the union of the various flats as Δ sweeps the half sphere $\frac{1}{2}\Omega$, each direction possessing a realization of the point process that is independent of the others. The union of all these realizations generates a tessellation of the space whose classes are the *Poisson polyhedra Y*. The hyperplanes themselves form a *Poisson flat set X*.

D.2. Intersections with a compact convex set; induced tessellations

Let B be a given compact convex set, $N(B)$ the random number of flats which hit B, and $D(\omega)$ denote the width of B in direction ω. For ω fixed, the number of intersections is a Poisson random variable with parameter $\lambda_n D(\omega) d\omega$. As ω varies, the contributions from each direction are independent. Therefore, upon summing, we see that $N(B)$ follows a Poisson distribution with parameter:

$$\lambda_n \int_{\frac{1}{2}\Omega} D(\omega)\, d\omega = \lambda_n M^{(n)}(B) = n\lambda_n W^{(n)}_{n-1}(B)$$

(see Ch. IV, C.1a). Thus, $N(B)$ depends only on λ_n and on the norm $M^{(n)}$ of B. In particular, the probability $Q(B)$ that B is included in a polyhedron of X^c is:

$$Q(B) = \exp[-\lambda_n M^{(n)}(B)] \qquad (\text{XIII-47})$$

We see that $Q(B)$ does not depend on the location of B: the tessellation is stationary. Now for the induced tessellation, R. E. Miles (1962) showed that the restriction of X to an $n-k$ dimensional hyperplane $E_{n-k} \subset \mathbb{R}^n$ is made up of Poisson flats of $\mathbb{R}^{n-k}$. Let λ_{n-k} be their density. For any compact convex set $B \subset E_{n-k}$, the probability (XIII-47) is the same, whether B is considered in $\mathbb{R}^n$ or in $\mathbb{R}^{n-k}$. So we can write:

$$\lambda_n M^{(n)}(B) = \lambda_{n-k} M^{(n-k)}(B)$$

i.e. (Hadwiger, 1957):

$$\lambda_{n-k} = \frac{b_{n-1}}{b_{n-k-1}} \lambda_n \qquad (b_i = i\text{-volume of the unit ball}). \qquad \text{(XIII-48)}$$

In particular for $\mathbb{R}^3$ and $\mathbb{R}^2$:

$$\lambda_3 = \frac{2}{\pi} \lambda_2 = \frac{\lambda_1}{\pi} \quad \text{and} \quad \lambda_2 = \frac{\lambda_1}{2} \qquad \text{(XIII-49)}$$

For example, if in $\mathbb{R}^2$, B is respectively the segment of length l, the disk of radius r, and the convex set of perimeter U, we have:

$$Q(B) = \exp(-2\lambda_2 l); \qquad Q(B) = \exp(-2\lambda_2 \pi r);$$
$$Q(B) = \exp(-\lambda_2 U) \qquad \text{(XIII-50)}$$

and in $\mathbb{R}^3$ if B is the ball of radius r:

$$Q(B) = \exp(-4\pi\lambda_3 r)$$

Now, if we interpret the Poisson lines involved in relation (XIII-50) as the result of a plane section through Poisson planes of $\mathbb{R}^3$, we must replace λ_2 by $(\pi/2)\lambda_3$ in the three equalities (XIII-50) in order to find the probabilities that a segment, a disk and a planar convex set are embedded in a polyhedron of the original 3-D Poisson tesselation (with parameter λ_3).

D.3. Laws in measure and laws in number

In several previous sections, the distinction between laws in number and in measure has been made (see for example Ch. X, Sect. D.2 and Ex. X-12). Indeed there are two very different ways to look at the Poisson polyhedra. From a first point of view, we could consider them as individuals that form a statistical population. Then we could give the same weight to each polyhedron Y (recall that in this case, expectation is denoted by the symbol $E(\cdot)$). Alternatively, without loss of generality, we could focus our attention on a polyhedron Y_0 which contains a given point, the origin O, since the tesselation is stationary. This means that the random polyhedron is implicitly weighted according to its n-volume, since the origin had more chances of belonging to a large Y than to a small Y. This generates a law in measure. Using an ergodic theorem, R. E. Miles (1962) established the following property: if Ψ is a characteristic of a polyhedron Y and if V is its n-volume, the corresponding laws in number $F(d\Psi, dV)$ and laws in measure $G(d\Psi, dV)$, are linked by the relationships:

$$G(d\Psi, dV) = \frac{V}{E(V)} F(d\Psi, dV);$$

$$F(d\Psi, dV) = \frac{V^{-1}}{\mathcal{M}(V^{-1})} G(d\Psi, dV) \qquad \text{(XIII-51)}$$

($\mathcal{M}(\cdot)$ = moment weighted in measure). Exercise X-12a was in fact an application of this relation. The times between which a bus passes a given point can be interpreted as a 1-D section of Poisson polyhedra. We saw that the density distribution of the intercepts, in number, was the exponential $\lambda_1 \exp(-\lambda_1 x)$ although the associated law in length (i.e. in measure) was the Gamma law with a density $(\lambda_1)^2 x \exp(-\lambda_1 x)$. Among other differences, the mean intercept doubles; i.e. goes from $1/\lambda_1$ (law in number) to $2/\lambda_1$ (law in measure).

Another application of relation XIII-51 is to the erosion formula (XIII-47) in which $Q(B)$ is the probability that B is included in a Y, with fixed set B, which implies a law in measure. The translation to the law in number is the following:

$$Q(B) = \frac{E[\mathrm{Mes}\,(Y \ominus \check{B})]}{E[\mathrm{Mes}\,Y]} = \frac{E[V(Y \ominus \check{B})]}{E[V(Y)]} \qquad \text{(XIII-52)}$$

That is $Q(B)$ is just the mean n-volume of the eroded set $Y \ominus \check{B}$, divided by the mean volume of Y, both quantities being averaged with respect to the law in number.

Experimentally speaking, if the space is swept by a plane ($\mathbb{R}^3$) or a line ($\mathbb{R}^2$) with a given direction, and if each Poisson polyhedron (polygon) is labelled by its first contact point with the sweeping plane (line), then a classification in number is obtained. For a classification in measure, associate with each vertex of a regular grid, the polyhedron (or polygon) which contains it.

D.4. Conditional invariance

The property presented now is basic to Poisson polyhedra. Take the Poisson polyhedron Y_0 and two compact convex sets B and B', such that all three contain the origin O. According to relation XIII-47, the probability of the event $B \oplus B' \subset Y$ is given by:

$$Q(B \oplus B') = \exp\left[-\lambda_n M^{(n)}(B \oplus B')\right]$$

Now remember that the norm is additive with respect to Minkowski addition (rel. IV-19):

$$Q(B \oplus B') = \exp\left\{-\lambda_n [M^{(n)}(B) + M^{(n)}(B')]\right\}$$
$$= Q(B) \cdot Q(B') \qquad \text{(XIII-53)}$$

Conversely, G. Matheron (1972) proved that *only* the Poisson flats satisfy relation XIII-53, which thus characterizes them.

What is the probabilistic meaning of this relation? The event $B \oplus B' \subset Y_0$ is equivalent to $B' \subset Y_0 \ominus \check{B}$ (Ex. II-9). Therefore, we can write the relation XIII-53 in the following form:

$$\Pr(B' \subset Y_0 \ominus \check{B}) = \Pr(B \subset Y_0) \cdot \Pr(B' \subset Y_0)$$

or, equivalently, by using conditional probabilities:

$$\Pr(B' \subset Y_0 \ominus B \,|\, 0 \in Y_0 \ominus \check{B}) = \Pr(B' \subset Y_0) \qquad \text{(XIII-54)}$$

i.e. conditional on the eroded set $Y_0 \ominus B$ being non empty, it has the same probability characteristics as Y_0 itself (from the measure point of view). Or equivalently, an erosion by B does not change the population of Y's (from the number point of view). Thus the smallest polyhedra vanish but the larger ones are reduced in such a way that they replace and exactly compensate for their disappearances.

Now what does lack of memory in a Markov process mean? Suppose that events occur in time, labelled by points in Figure XIII.13a; consider the law $f(t)$ of the time separating two successive events. If our clock does not start when an event occurs, but at a given time t_0, the law $f(t)$ is unchanged; then the system has no memory of its past. Now replace the time axis by the space $\mathbb{R}^n$, the Poisson points by Poisson flats, and the time t_0 by an arbitrary compact convex set in $\mathbb{R}^n$, and Matheron's theorem becomes a direct geometrical generalization of the lack of memory property.

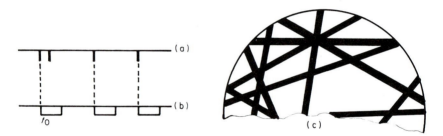

Figure XIII.13. (a), (b) in $\mathbb{R}^1$ the lack of memory is equivalent to conditional invariance under erosion. (c) statistically speaking the polygons of Figure 13(c) and of Figure 12 are identical.

D.5. Specific parameters

The results given in this section were originally derived by R. E. Miles in a general approach (1962). We present them here with simpler proofs.

Let us denote by $N_v^{(n)}$ the specific number of the polyhedra (number per unit of volume); $N_v^{(n)}$ satisfies the relation:

$$E(V^{(n)}) = \frac{1}{N_v^{(n)}} \qquad \text{(XIII-55)}$$

since we are dealing with a tesselation (i.e. partition) of the space.

$N_v^{(n)}$ is also equal to the expectation of the number of vertices per unit of volume. Let $S_v^{(n)}$ denote the *specific area* of X, or equivalently the mathematical expectation of the Lebesgue measure (in $\mathbb{R}^{n-1}$) of the faces per unit of volume.

The two specific parameters $N_V^{(n)}$ and $N_V^{(n-1)}$ are linked. Let Π be one of the flats of X; the other flats induce on it a new Poisson flat process in $\mathbb{R}^{n-1}$, of specific number $N_V^{(n-1)}$, whose density λ_{n-1} is given by (XIII-48). Let us consider a volume V of $\mathbb{R}^n$. The average area of the flats intersected by V is equal to $VS_V^{(n)}$. These portions of hyperplanes contain $(V/n)S_V^{(n)}N_V^{(n-1)}$ vertices of X (each vertex belongs to n hyperplanes). But these vertices, considered as vertices in $\mathbb{R}^n$ are $N_V^{(n)}\cdot V$ in number. Hence:

$$nN_V^{(n)} = S_V^{(n)}\cdot N_V^{(n-1)} \tag{XIII-56}$$

To calculate $S_V^{(n)}$, consider the ball B of radius r in $\mathbb{R}^n$, of volume $b_n r^n$. The average number of flats intersected by B is $\lambda_n M^n(B) = nb_n\lambda_n r$. The average measure of each of these sections is $b_n r^n/2r$. Then $S_V^{(n)}$ is given by:

$$S_V^{(n)} = \frac{nb_n\lambda_n r}{b_n\cdot r^n}\cdot\frac{b_n r^n}{2r} = \frac{n}{2}b_n\lambda_n \tag{XIII-57}$$

Now notice that using relation (XIII-48), relation (XIII-56) and (XIII-57) lead to the recurrence relation:

$$N_V^{(n)} = \frac{b_n\lambda_n}{2}N_V^{(n-1)} = \frac{b_n}{b_{n-1}}\lambda_2 N_V^{(n-1)} \tag{XIII-58}$$

For $n=1$, we have $N_V^{(1)} = \lambda_1$. Consequently:

$$N_V^{(n)} = b_n(\lambda_2)^n = b_n\left(\frac{b_{n-1}}{2}\right)^n(\lambda_n)^n \tag{XIII-59}$$

In the case of the most common spaces, we obtain

$$\text{in } \mathbb{R}^2: N_A = \pi\lambda_2^2 \qquad U_A = \pi\lambda_2$$

$$\text{in } \mathbb{R}^3: N_V = \frac{\pi^4}{6}\lambda_3^3 \qquad S_V = 2\pi\lambda_3$$

Now taking a law in number point of view, we immediately derive from (XIII-57) and (XIII-59), the average volume and the average area of the polygon Y:

$$E[V^{(n)}] = \frac{1}{b_n(\lambda_2)^n}, E[S^{(n)}] = \frac{2S_V^{(n)}}{N_V^{(n)}} = \frac{2n}{b_{n-1}(\lambda_2^{n-1})} \tag{XIII-60}$$

In particular in $\mathbb{R}^2$: $E[A] = \dfrac{1}{\pi\lambda_2^2}$; $\qquad E[U] = \dfrac{2}{\lambda_2}$

and in $\mathbb{R}^3$:

$$E[V] = \frac{6}{\pi^4\lambda_3^3}; \qquad E[S] = \frac{24}{\pi^3}\cdot\frac{1}{\lambda_3^2}$$

D.6. Size distributions by openings

Let B be a compact convex set containing the origin O. $\{\mu B\}$, $\mu > 0$, are the homothetics of B. The size distribution of the random polyhedron Y_0, with respect to the $\{\mu B\}$ is just the probability $P(-\mu)$ that a fixed point (the origin) belongs to $X_{\mu B}^c$:

$$P_B(-\mu) = \Pr\{O \in (Y_0 \ominus \mu\check{B}) \oplus \mu B\} = \Pr\{O \in X_{\mu B}^c\}$$

(a law in measure point of view)   (XIII-61)

Equivalently, $P_B(-\mu)$ is also equal to:

$$P_B(-\mu) = \frac{E[V(Y_{0,\mu B})]}{E[V(Y_0)]}$$

(a law in number point of view)   (XIII-62)

(for the definition of $P_B(-\mu)$, see Ch. X, Sect. C.1). It is the second expression for $P(-\mu)$ that we will use in calculations. The denominator $E[V(Y_0)]$ is given by relation XIII-60, and as for the numerator, notice that the set $(Y_0 \ominus \mu\check{B}) \oplus \mu B$ is empty, with zero volume, when $Y_0 \ominus \mu\check{B}$ is empty. When $Y_0 \ominus \mu\check{B} \neq \varnothing$, then $Y_{0,\mu B}$ has the same probability characteristics as $Y_0 \oplus \mu B$ (conditional invariance). Therefore, we can write:

$$E[V(Y_{0,\mu B})] = \exp\{-\mu\lambda_n M^{(n)}(B)\} E[V(Y_0 \oplus \mu B)]$$

and the general form of the size distribution, in terms of openings, is given by:

$$P_B(-\mu) = \frac{E[V(Y_0 \oplus \mu B)]}{E[V(Y_0)]} \cdot \exp[-\mu\lambda_n M^{(n)}(B)]$$ (XIII-63)

For example, we find:

— in $\mathbb{R}^1$, for linear openings:

$$P_B(-\mu) = \lambda_1 \mu \exp(-\lambda_1 \mu)$$

— in $\mathbb{R}^2$ for openings by disks:

$$P_B(-\mu) = (1 + 2\pi\lambda_2\mu + \pi^2\lambda_2^2\mu^2)\exp(-2\pi\lambda_2\mu)$$

— in $\mathbb{R}^3$ for the openings by balls:

$$P_B(-\mu) = \left(1 + 4\pi\lambda_3\mu + \frac{\pi^4}{2}\lambda_3^2\mu^2 + \frac{2}{9}\pi^5\lambda_3^3\mu^3\right)\exp(-4\pi\lambda_3\mu)$$

The first two formulae still hold in $\mathbb{R}^2$ and $\mathbb{R}^3$, upon changing the densities λ, according to relation (XIII-49).

D.7. Other results and a conclusion

We now give, without proofs, a few other minor results (minor, because seeing we are fortunate enough to have the characteristic property of conditional invariance at our disposal, we should base all our statistical inferences on the $Q(B)$'s and not on the results which follow):

In $\mathbb{R}^3$:

$$E(V) = \frac{6}{\pi 4} \cdot \frac{1}{\lambda_3^3}; \qquad E(V^2) = \frac{48}{\pi^6 \lambda_3^6}; \qquad E(V^3) = \frac{1344}{\pi^9 \cdot \lambda_3^9}$$

$$\mathcal{M}(V) = \frac{8}{\pi^2} \cdot \frac{1}{\lambda_3^3}; \qquad \mathcal{M}(V^2) = \frac{224}{\pi^4 \lambda_3^6}$$

$$E(S) = \frac{24}{\pi^3 \lambda_3^2}; \qquad E(S^2) = \frac{240}{\pi^4 \lambda_3^4}; \qquad \mathcal{M}(S) = \frac{16}{\pi} \cdot \frac{1}{\lambda_3^2}$$

In $\mathbb{R}^2$:

$$E(A) = \frac{1}{\pi \lambda_2^2}; \qquad E(A^2) = \frac{1}{2 \cdot \lambda_2^4}; \qquad E(A^3) = \frac{4\pi}{7 \lambda^6}$$

$$\mathcal{M}(A) = \frac{\pi}{2 \cdot \lambda_2^2}; \qquad \mathcal{M}(A^2) = \frac{4\pi^2}{7 \lambda_2^4};$$

$$E(U) = \frac{2}{\lambda_2}; \qquad E(U^2) = \left(\frac{\pi^2}{2} + 2\right)\frac{1}{\lambda_2^2}; \qquad \mathcal{M}(U) = \frac{\pi^2}{2\lambda_2}$$

Moments of the number N of sides of the Poisson polygon are:

$$E(N) = 4; \qquad E(N^2) = \frac{\pi^2}{2} + 12; \qquad \mathcal{M}(N) = \frac{\pi^2}{2}$$

Note in particular the size of the ratios $\mathcal{M}(V)/E(V)$ and $\mathcal{M}(A)/E(A)$ which are respectively equal to 13 and 5. They indicate large relative variances for the Poisson polyhedra and polygons.

In contrast to Boolean sets, Poisson tesselations are rarely met in practice. The model depends upon only one parameter, which makes it rather inflexible. However, indirectly, Poisson tesselations play a fundamental role in set modelling, since they provide a method for generating primary grains, for Boolean models for example. They have been also the basis for a number of filters (Modestino, 1979), for simulations (A. Journel, 1978), for fractal models (B. Mandelbrot, 1976), for random functions with an exponential covariance or a linear variogram (Ex. XIII-18 below) for superimposition of Boolean sets (Conrad, 1972), and for hierarchic tesselations (Ch. IX). We now develop one of these models in some detail, since it has immediate application to the general phenomenon of the milling of rocks.

E. OTHER TESSELATIONS

E.1. Poisson twin flats

(a) *Definition and induced models*
The twin flats model was proposed by Ph. Cauwe (1973) and studied by Cauwe and Matheron (1972). In their joint work, they calculated the basic criteria associated with the model. We start from Poisson planes in $\mathbb{R}^3$ with parameter λ_3, and replace each plane by a doublet of planes, parallel to, and at a distance $h/2$ each side of the initial plane. The distance separating the twin planes is thus h, where we suppose h is a random variable with distribution $F_3(h)$. The values of h are independent from one twin to another.

Clearly, the set induced by the intersection of the model with a plane section or with a straight line, is also a twin flats model with induced densities λ_i satisfying equation (XIII-49). Denote by $F_1(h)$ the distribution of the twin flats distances induced on a line, the z-axis for example. Per unit length, the number of these distances smaller than h is $\lambda_1 F_1(h)$. Consider all twin planes whose normals make an angle between φ and $\varphi + d\varphi$ with the z-axis (Fig. XIII.14). The normals occupy an area equal to $2\pi \sin\varphi\, d\varphi$ on the unit sphere. The corresponding planes induce a density $\lambda_3 \cos\varphi$ on the z-axis, and contribute to

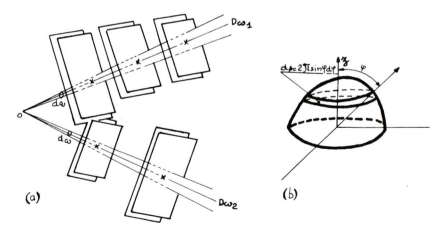

Figure XIII.14. (a) construction of Poisson twin planes, (b) portion of the unit sphere whose normals make an angle between φ and $\varphi + d\varphi$ with the z-axis.

$F_1(h)$ when their twin distance is smaller than $h\cos\varphi$. Thus we have:

$$\lambda_1 F_1(h) = \int_0^{\pi/2} \lambda_3 \cos\varphi \cdot F_3(h\cos\varphi)\cdot 2\pi\sin\varphi\,d\varphi$$

Putting $\cos\varphi = u$, and noting that $\lambda_1 = \pi\lambda_3$ (rel. XIII-69), we obtain:

$$F_1(h) = 2\int_0^1 u F_3(uh)\,du \tag{XIII-64}$$

In the same way, the passage from $\mathbb{R}^3 \to \mathbb{R}^2$ and $\mathbb{R}^2 \to \mathbb{R}^1$, is found to be (respectively):

$$\left.\begin{aligned} F_2(h) &= \frac{4}{\pi}\int_0^1 \frac{u^2}{\sqrt{1-u^2}}F_3(uh)\,du \\ F_1(h) &= \int_0^1 \frac{4}{\sqrt{1-u^2}}F_2(uh)\,du \end{aligned}\right\} \tag{XIII-65}$$

By a change of variable to $uh = v$, the integral Equation (XIII-64) can be solved, giving:

$$F_3(h) = \frac{1}{2h}\frac{d}{dh}[h^2 F_1(h)] \tag{XIII-66}$$

The two equations of relation (XIII-65) need a more sophisticated treatment (similar to that used in Ex. X-13); we finally obtain:

$$\left.\begin{aligned} F_3(h) &= \frac{1}{2h^3}\int_0^h \frac{v}{\sqrt{h^2-v^2}}\frac{d}{dv}[v^3 F_2(v)]\,dv \\ F_2(h) &= \frac{2}{\pi h^2}\int_0^h \frac{v}{\sqrt{h^2-v^2}}\frac{d}{dv}[v^2 F_1(v)]\,dv \end{aligned}\right\} \tag{XIII-67}$$

(b) *Evaluation of $Q(B)$*

The probability $Q(B)$ that a compact convex set B is included in a single member of the partition has quite a simple form. Therefore we will calculate it in $\mathbb{R}^n$. For each direction ω and each linear element $(x, x+dx)$ $(x > 0)$ of the generator axis Ox (see Fig. XIII.14), one of the three following mutually exclusive events may occur, which will result in the event in question.

— no doublet in direction ω, prob.: $(1 - \lambda_n \omega dx)$
— one doublet which is outside of B, prob.: $\lambda_n d\omega\,dx\,F_n(x)$
— one doublet containing B, prob.: $d\omega\,dx(1 - F_n(D_\omega^{(n)}(B) + x))$

After combining, they give the elementary probability:

$$\exp\{\lambda_n[1 - F_n(D_\omega^{(n)}(B)+1) - 1 + F(x)]\,d\omega\,dx\}$$

For $-D_\omega^{(n)}(B) \le x < 0$, only the first event remains, with a prob. $\exp\{-\lambda_n d\omega\, dx\}$. Integrating over x and over ω, we find:

$$\log Q(B) = -\lambda_n \int_{\frac{1}{2}\Omega} D_\omega^{(n)}(B)d\omega$$

$$+\lambda_n \int_{\frac{1}{2}\Omega} d\omega \int_0^\infty [1 - F_n(x) + D_\omega^{(n)}(B)] - 1 + F_n(x)]dx$$

i.e.:

$$\log Q(B) = -\lambda_n \int_{\frac{1}{2}\Omega} d\omega \int_0^{D_\omega^{(n)}(B)} [2 - F_n(x)]dx \qquad \text{(XIII-68)}$$

In particular the probability $Q(B)$ that a segment of length l is embedded in a single class of the twin model (λ_1, F_1) induced on a straight line is:

$$Q(l) = \exp(-\lambda_1) \int_0^l [2 - F_1(x)]dx \qquad \text{(XIII-69)}$$

The length distribution $F(l)$, of the segments induced on a straight line by the twin planes, is immediate from $Q(l)$:

$$1 - F(l) = \frac{Q'(l)}{Q'(0)} = [1 - \tfrac{1}{2}F_1(l)]\exp\left\{-\lambda_1 \int_0^l [2 - F_1(x)]dx\right\}$$

(mean intercept $m = 1/2\lambda_1$).

In $\mathbb{R}^2$, the probability $Q(C_r)$ associated with the disk C_r of radius r is equal to:

$$Q[C_r] = \exp\left\{-\pi\lambda_2 \int_0^{2r} [2 - F(x)]dx\right\} \qquad \text{(XIII-70)}$$

Taking relation XIII-67 into account, we see that the linear and circular probabilities $Q(l)$ and $Q(C_r)$ are related by the equation:

$$\log Q(l) = \frac{2}{\pi} \int_0^1 \frac{\log Q[C_{ul}]}{\sqrt{1 - u^2}} du \qquad \text{(XIII-71)}$$

Experimentally speaking, the estimators of $Q(l)$ and $Q(C_r)$ are two independent pieces of information. Thus relation (XIII-71) offers a test of the model.

Before ending the section, two other results related to the plane sections, are worth quoting (proofs can be found in G. Matheron, 1972).

(i) Take the twin lines model X in $\mathbb{R}^2$, eroded by the disk C_r. The connectivity number of the eroded set $X \ominus C_r$ is equal to:

$$N_A(r) = N_A(X \ominus C_r) = \pi\lambda_2^2 [2 - F_2(2r)]^2 Q(C_r)$$

Notice that $N_A(r)/N_A(0)$ is just the size distribution of the disks inscribable in the Poisson twin polygons.

(ii) the size distribution according to circular openings, $P_C(-r)$, is given by:

$$P_C(-r) = \left[1 + 2\lambda\pi r(2 - F_2(2r))\right]^2 \exp\left\{-\lambda_2\pi \int_0^{2r} [2 - F(x)]\, dx\right\}$$

(c) *A physical example of the twin planes process*
The context in which Cauwe's study is set was briefly presented in Chapter I, E.2b. Imagine the process of milling a rock as a black box, and model the outcome to be equivalent to partitioning the mineral space into Poisson twin polyhedra.

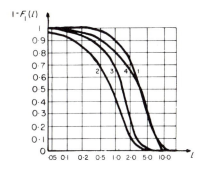

Figure XIII.15. (1) intercept distribution of Poisson polyhedra, (2) quartz; roll crusher, (3) solomite; jaw crusher, (4) galena; roll crusher. Points are experimental, curves 2, 3 and 4 follow eq. XIII-69 (Poisson twin planes model).

Cauwe performed thirty milling tests on various rocks under different crushing conditions. Samples of resulting milled products were mixed with a teflon powder, compressed and sintered in order to provide polished sections (for more details, see Ph. Cauwe (Thesis, 1973), or (1976)). Quantitative image analyses were then carried out on a Texture Analyser, programmed to measure linear and hexagonal erosions, i.e. programmed to estimate $Q(l)$ and $Q(C_r)$. The experimental curves were smoothed (by fitting a quadratic equation) before calculating the logarithm and taking derivatives. The first results can be seen on Figure XIII-16a. The simple Poisson plane model shown as curve 1, systematically underestimates small intercepts: this discovery led Cauwe to reject this as a first model and suggested the possibility of Poisson *twin* planes. Indeed, we see from the three curves 2, 3 and 4, that a suitable choice of parameters in equation (XIII-69) gives an excellent fit. This result seems general, at least for the thirty experiments of the study: independent estimation of the two erosion probabilities of relation (XIII-71) also give positive

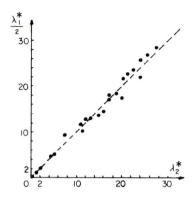

Figure XIII.16. relationship between λ_2 estimated from $Q(l)$ and from $Q(C_r)$.

confirmation of the model. Another experimental confirmation is provided by the computation of λ_2 from both $Q(l)$ and $Q(C_r)$. We see from Figure XIII-16b the high correlation between the two, when calculated for the thirty samples (average value of $\lambda_1^*/2\lambda_2^* = 1.023$, with standard deviation 0.085).

For the distribution $F_3(h)$ of the inter-plane distances (in $\mathbb{R}^3$), Cauwe chose a uniform law:

$$F_3(h) = \frac{h}{d} \qquad O \le h \le d; \qquad F_3(h) = 1 \quad h > d$$

which has the advantage of depending on only one parameter. Here again the results confirm the model. For example, denote by d^* and d^{**} the two estimates of d obtained from $Q^*(l)$ and $Q^*(C_r)$ respectively. The ratio d^*/d^{**}, measured on the thirty tests has a mean value 1.067, with a standard deviation of 0.208. Thus the pair (λ, d) turns out to be a complete morphological summary of the results of milling.

E.2. "Metallic grain" tesselations

(a) Definition
In metallography, a grain is a polyhedron possibly with curved faces. A piece of metal is then generally made up of disjoint monocrystals which exhibit such shapes (C. S. Smith, 1954) and whose union virtually fills the entire space. The photograph of Figure XIII.17a, gives an idea of a 2-D section of such metallic arrangements.

This type of structure suggests a family of tesselation models that we will now develop. A realization of a *metallic grain model* is obtained from the skiz of the realization of a random model of disjoint compact sets in $\mathbb{R}^n$. In the

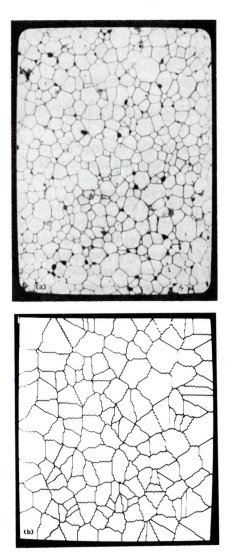

Figure XIII.17. (a) metallic section of iron—it can be modelled by an isovolumic model, (b) Voronoï simulation in $\mathbb{R}^2$.

sequel, this primary set will be denoted by Y. As yet, we have been vague about Y; three particular types of the metallic grain model are of interest: the isovolumic, the Voronoï, and the Johnson–Mehl tesselations. Heuristically speaking, these models are less informative than those already discussed

(Boolean and Poisson). Their mean features do not obey simple laws, and in particular, the sections of a model of a given type does not belong to the same type. Therefore, to avoid making this section a handbook of results, we shall restrict ourselves to their major characteristics, i.e. their Minkowski functionals and their linear size distribution. For more details, we refer the reader to J. L. Meijering (1953), E. N. Gilbert (1961), R. E. Miles (1972) and Ch. Lantuéjoul (1977); these four studies contain all the basic results on the subject. Interesting simulations have been performed by M. Durand (1978) and by H. Stienen (1979).

(b) *Iso-volumic tesselations*
There are two possible ways of obtaining regular tesselations of the space, starting from a periodic cubic grid of points in $\mathbb{R}^3$. The primary set Y may be either the centre of the cubic grid or the centre of the face of the cubic grid. In both cases, the set Y is made up of points the skiz of which partitions the space into either tetrakaïdodecahedra (14 faces) or rhombododecahedra (12 faces) respectively (Fig. II.13).

To test an iso-volumic tesselation, first calculate the volume weighted moments of order one and two ($\mathcal{M}(V)$ and $\mathcal{M}(V^2)$) of the grains, using the method described in Chapter X, Section D-3. If the quantity $\mathcal{M}(V^2) \cdot [\mathcal{M}(V)]^{-2}$ is statistically equivalent to one, then the grains have the same volume, and their number per unit volume is $\theta_3 = [\mathcal{M}(V)]^{-1}$. On the other hand, θ_3 is also obtainable from the specific number of grains induced on lines and on planes (θ_1 and θ_2 respectively; see table Fig. XIII.9), giving a further test of the model. Notice that the expression for θ_3 as function of θ_1 and of θ_2 do not really differ for both iso-volumic models. Indeed, since an actual metal is due to the combination of elementary crystals of various shapes (C. S. Smith, 1954), it is probably hardly worth considering two models.

(c) *Voronoï tesselations*
The Voronoï tesselations are the skizs associated with the points of a Poisson point process (see Fig. XIII.17b). Thus they are necessarily isotropic and depend upon a single parameter θ the Poisson density, which is in fact a magnification factor. However, in order to enlarge the range of possibilities of the model, notice that the 3-D or 2-D section of an n-dimensional Voronoï tesselation are *not* Voronoï; thus the dimension n of the space can be used as a parameter of the more general model. This approach leads to rather tedious calculations, and the only useful result obtained is $P(l)$, the probability that a segment is included in a single class of the partition (Gilbert, 1961). From it, we can derive the linear intercepts distribution, the average star of 2-D or 3-D

sections and the specific surface S_V of the 3-D sections. For example:

$$S_V = 4 \cdot N_L = 4\theta^{1/n} \frac{2\Gamma(2-1/n) \cdot \Gamma^{(2-1/n)}(n/2+1) \cdot \Gamma(n)}{n\Gamma^2(n/2+\frac{1}{2}) \cdot \Gamma(n-\frac{1}{2})} \quad \text{(XIII-72)}$$

In $\mathbb{R}^2$, R. E. Miles (1971) systematic calculations of the Voronoï polygon, led to number weighted averages for the elementary polygon:

area: θ^{-1}; number of sides: $4\theta^{-1/2}$; number of edges = number of vertices = 6; density distribution of the angles:

$$f(u) = 4(-u \cos u + \sin u)\frac{\sin u}{3\pi}; \text{ hence } E(u) = \frac{2\pi}{3} \text{ and}$$

$$\text{var}(u) = \frac{\pi^2}{9} - \frac{5}{6} \quad \text{(XIII-73)}$$

In $\mathbb{R}^3$, A. Haas calculated the result of erosion by a ball $B(r)$ of radius r:

$$Q(B) = \text{Pr}\{B \subset \text{a single class}\} = 4\pi\,\theta_3 \left\{ \int_0^r \exp\left[-\frac{32\pi}{3}\theta_3 r(\rho^2+r^2) \right] d\rho \right.$$

$$\left. + \int_r^\infty \exp\left[-\frac{4\pi}{3\rho}\theta_3(\rho+1)^4 \, d\rho \right] \right\} \quad \text{(XIII-74)}$$

Other results are given in Table (XIII-19). Various generalizations have been proposed to enhance the flexibility of the model. For example, R. E. Miles suggested defining a class of the tesselation as the set of points for which a particular "germ" is the kth nearest (k fixed). The resulting class is again convex (and Voronoï for $k = 1$). However, the most powerful generalization is that proposed by W. A. Johnson and R. F. Mehl in 1939.

(d) *Johnson–Mehl tesselations*
Here the primary set Y is a set of points in $\mathbb{R}^n \times \mathbb{R}^t$, which are constructed as follows. The point process in $\mathbb{R}^n$ starts at time $t = 0$; if the germs appearing during the time $(t, t+dt)$ are not disturbed, then they grow isotropically at a constant speed v (call them "grains"); i.e. they generate balls of radius vt' at time $t + t'$. This rule is altered by two possibilities:

(i) if a germ appears at time t at a point already occupied by a grain, then the germ is removed;
(ii) when two grains touch each other, the growth stops in the contact direction, but continues everywhere else. The growth of a grain is complete when it is stopped in all directions.

For example, in two directions, the boundaries are arcs of hyperbolae. We can get an idea of the overall structure by looking at the Johnson–Mehl simulation, performed by Ch. Lantuéjoul on a Texture Analyser and shown in Figure XIII.18.

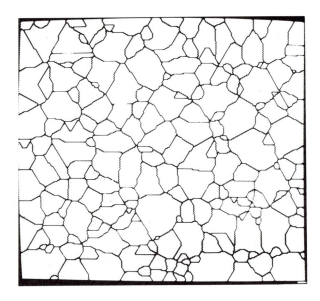

Figure XIII.18. Johnson–Mehl simulations.

Assume that the points (i.e. the germs) appearing in $\mathbb{R}^n$ during $(t, t + dt)$ are Poisson, with parameter $\mu(dt)$ a positive measure on $\mathbb{R}^+$. For $\mu = \theta \delta_{t_0}$, we recover the Voronoï model. Consider the topological closure of those grains generated up to time t; it is obviously a realization of a Boolean model whose primary grains are balls $\{\, B(x)\,\}$. Thus if $Q_t(C)$ denotes the probability that the compact set C is included in the pores of this set, we have:

$$Q_t(C) = \exp\left[-\int_0^t \mathrm{Mes}\,\{C \oplus B(v(t-u))\}\,\mu(du) \right]$$

and in particular $Q_t(\{0\}) = q_n(t) = \exp[-b_n v^n m_n(t)]$ where $m_n(t)$ is then nth convolution moment of the measure μ:

$$m_n(t) = \int_0^t (t-u)^n \mu(du) \qquad i = 0, 1, 2 \ldots \qquad \text{(XIII-75)}$$

We see that as $t \to \infty$, the process tends towards a tesselation if and only if $Q_\infty(\{O\}) = 0$, i.e. $m_n(t) \to \infty$. Assume this condition is fulfilled and calculate the number weighted mean volume $E(V_n)$ of the grains. At time t, the proportion of the space which contains no grains is $q_n(t)$. During $(t, t + dt)$ $\mu(dt)$ germs appear. The final total number θ_n of germs is thus $\int_0^\infty q(t)\mu(dt)$, and:

$$\frac{1}{\theta_n} = E(V_n) = \frac{1}{\int_0^\infty q(t)\mu(dt)} \qquad \text{(XIII-76)}$$

Using a similar approach, Ch. Lantuéjoul calculated the average volume $E[V_n(t)]$ of the grains born up to time t:

$$E[V_n(t)] = \frac{nb_n \int_0^\infty q_n(t + r/v)\, r^{n-1}\, dr}{q(t)} \qquad \text{(XIII-77)}$$

In $\mathbb{R}^3$, R. E. Miles proved that the expression: $kE(V_3)\int_0^\infty [m_2(t)]^v q_3(t)dt$ is equal to the mean face area, mean edge length, mean vertex number of a polyhedron when (k, v) is equal to

$$\left(\frac{64\pi^2 v^5}{3}, 2\right), \left(\frac{64\pi^4 v^7}{5}, 3\right) \text{ and } \left(\frac{102\, 4\pi^5 v^9}{105}, 4\right)$$

respectively. Similarly Ch. Lantuéjoul showed that in two dimensions, the mean perimeter of the grains is

$$16\pi v^3 E[V_2]\int_0^\infty [m_1(t)]^2 q_2(t)\, dt$$

The complete calculation of all the integrals appearing in these formulae is easy when $\mu(dt)$ is a linear combination of Dirac measures or of monomials of the form αt^β, with $\beta > 1$.

For the constant measure μ, Gilbert has also calculated the probability $P(l)$, in $\mathbb{R}^n$.

(e) *Comparison of the various metallic grain models*

In conclusion then, the various metallic grain models leave us feeling a little unsatisfied. They are rather difficult to handle (the general form of the erosion $Q(B)$ is unknown) and to test (how can $\mu(dt)$ be estimated for example?). The major pieces of information are carried by the probability $P(l)$. In particular, we are interested in the stars $\mathcal{M}(L)\mathcal{M}(S)$ and $\mathcal{M}(V)$, obtainable from 2-D sections. In the table of Figure XIII.19 we compare the main parameters of the four grain models in $\mathbb{R}^3$ (for the Johnson–Mehl model, time measure μ is taken to be constant).

	Parameter	Voronoi	Johnson–Mehl	Tetrakai-decahedron (isovolumic)	Rhombodo-decahedron (isovolumic)
for the tesselation	θ_1	$1.45\,\theta_3^{1/3}$	$1.28\,\theta_3^{1/3}$	$1.33\,\theta_3^{1/3}$	$1.33\,\theta_3^{1/3}$
	θ_2	$1.46\,\theta_3^{2/3}$	$1.22\,\theta_3^{2/3}$	$1.34\,\theta_3^{2/3}$	$1.37\,\theta_3^{2/3}$
	$\mathcal{M}(L)$	$0.950\,\theta_3^{-1/3}$	$1.10\,\theta_3^{-1/3}$		
	$\mathcal{M}(S)$	$1.04\,\theta_3^{-2/3}$	$1.50\,\theta_3^{-2/3}$		
	$\mathcal{M}(V)$	$1.24\,\theta_3^{-1}$	$2.20\,\theta_3^{-1}$	θ_3^{-1}	θ_3^{-1}
for one polyhedron	face area	$5.82\,\theta_3^{-2/3}$	$5.14\,\theta_3^{-2/3}$	$5.31\,\theta_3^{-2/3}$	$5.34\,\theta_3^{-2/3}$
	edge length	$17.50\,\theta_3^{-1/3}$	$14.71\,\theta_3^{-1/3}$	$16.04\,\theta_3^{-1/3}$	$15.60\,\theta_3^{-1/3}$
	face number	15.54	> 13.2	14	12
	edge number	40.61	> 33.8	36	24
	vertex number	27.07	22.56	24	14

Figure XIII.19. Comparison of the four metallic grain models in $\mathbb{R}^3$.

F. POINT MODELS

F.1. Introduction

Random point models are by far the most extensively studied class of random sets. The Poisson point process is the oldest stochastic model in geometry; it arises as far back as the last century and every classical book on geometrical probabilities contains a development of it (Detheil, Kendall and Moran, Santalo, etc.).

The point process literature stemmed from Bartlett (1963, 1976). One can find a general overview of the subject in P. A. W. Lewis (ed.) (1972).

More recently, two important subsequent works are due to K. Krikeberg (1973), and to B. D. Ripley (1976, 1977). Their concern is primarily for the moments and the statistical aspects of the analyses.

Of course points are an excellent material for geometrical probability, since they cannot overlap and they are associated with elementary open sets dx. Such good properties allow one to construct a multitude of variants from the initial Poisson point model. In a certain sense, we could say "too many variants", since the proliferation has resulted in a large catalogue of theoretical models, mostly untried and not possible to test from experiments. This probably explains why recent significant work (K. Krickeberg, B. D. Ripley) is deliberately oriented towards moment estimation. However, regardless of the quality of the estimators, there often remains a deeper problem. For example, the Poisson points model (possessing a regionalized density), and the Neymann–Scott clustering model, both constructed in completely different ways, are indistinguishable in terms of their covariances (and probably also from higher order moments). In 1960, B. Matern proposed three methods for generating hard-core models; i.e. processes whose points are separated by a minimal distance $r_0 > 0$, and calculated the covariances for two of them. However, even though there is a theoretical difference between them, there is no practical way of choosing which to fit to experimental data. Theoretical distinctions are likely to be too thin to imply statistically robust differences in the moments.

This gap between models and moment estimators need not always be serious. Indeed, for purposes of prediction the *whole* model is rarely used and often the covariance is sufficient. There are also other (more descriptive and heuristic) sources of interest in a model, besides statistical inference. For example, the morphologist wants to see how his criteria, such as function $H(r)$, or prob. $P_n(B)$ used below, reflects the presence of clusters, or alignments, or other features on the set of points; the difference between one mode of clustering or another is secondary in the initial stages.

For these reasons, we will consider only five models, which show five very contrasting morphological features, namely:

(i) the Poisson point model; i.e. spatial structure of degree zero—serves as a reference for the lack of structure;

(ii) the Neymann–Scott clustering model; i.e. a Boolean model with a cluster of points as the primary grain;

(iii) a hard-core model—models repulsion between points;

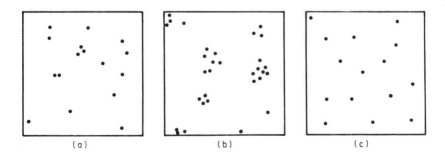

Figure XIII.20. Realizations (a) of a Poisson point process, (b) of a Neyman–Scott cluster process, (c) of a hard core model.

(iv) the Poisson alignment model—exhibits features of dotted lines;
(v) a Brownian motion model—accumulation points represent a limiting form of clustering.

The first three models are well known (Fig. XIII. 20), but the last two seem to be quite new (other random point sets are given in exercises). Before describing them one by one, we would like to briefly develop the necessary probabilistic tools.

F.2. Probabilistic tools

Apart from the problem of astronomy, random point sets are basically used in the description of 2-D phenomena, such as trees in forests, spacial distribution of towns in a region , etc. Therefore, we work within the framework of the 2-D space, although generalizations to $\mathbb{R}^n$ are usually obvious.

There are two possible approaches. First, we could apply the general methods developed for random sets; i.e. calculate the probability $Q(B)$ that a compact set B misses the set. We could also introduce more explicitly the fact that we are dealing only with points, by considering the number of points $Z(B)$ which fall inside B. This second approach turns out to be more fruitful, although it is of course more general (the first turns out to be a particular case for $Z(B) = 0$). We want the number $Z(B)$ to be a random variable, which implies that the random set X of points does not admit accumulation points (see G. Matheron (1969), quoted in Theorem XIII-11, below). We shall call random point processes of this type "germ models".

Mathematically speaking, $Z(B)$ is a measure. Affect Dirac measures to all the points of X contained in B, and take their sum; this gives precisely $Z(B)$. The treatment of random versions of measures is by now well known (see for example Daley and D. Vere-Jones in Stochastic Point Processes (1972)). The task before us then is to find the distribution function of $Z(B)$, i.e. the probabilities $\{P_n(B)\}$ of B containing 0, 1, 2, ... n points. In practice, estimators depend only on $P_0(B)$ and the first two moments $E[Z(B)]$ and $\mathrm{var}[Z(B)]$, the mean and variance of $Z(B)$. One can classify the germ models in two families, stationary or not. The latter, which serve principally as intermediary processes (primary grains for example) are studied in Exercise XIII-22 and we devote the present section to the stationary case. Under stationarity, a general theorem states that:

(i) the average value $E[Z(B)] = m\,\mathrm{mes}(B) = m \cdot A(B)$, where the constant m is known as the intensity or density of the model.
(ii) the variance of $Z(B)$ and the covariance between $Z(B)$ and $Z(B')$ are integrals of the centred covariance measure $C(dx)$. We see here exactly the same formulation as in the regularization of stationary random functions,

developed in Chapter IX, Section B.2e; here $C(dx)$ replaces the point covariance $C(x)\,dx$. More precisely, the expression for the covariance between $Z(B)$ and $Z(B')$ is given by:

$$E[Z(B) - m\,A(B)][Z(B') - m\,A(B')]$$

$$= \int_{\mathbb{R}^2} A[B \cap B'_\mu]C(du) \qquad \text{(XIII-78)}$$

where the measure $C(du)$ acts on the cross product $\mathcal{K}' \times \mathcal{K}'$; $(B, B' \in \mathcal{K}')$. Usually in applications, B and B' are two equal masks distance h apart whose geometric covariogram is $K(h) = A[B \cap B_h]$. Then the right-hand side of (XIII-78), call it $C_B(h)$, turns out to be:

$$C_B(h) = \int_{\mathbb{R}^2} K(h + u)C(du) = K * C \qquad \text{(XIII-79)}$$

$$C_B(0) = \text{Var}[Z(B)] = \int_{\mathbb{R}^2} A(B \cap B_u) \cdot C(du) \qquad \text{(XIII-80)}$$

According to a property that is general to all stationary germ processes (direct proofs of the property will be given for each particular model), the covariance measure $C(dh)$ can be decomposed into:

$$C(dh) = m\delta + C_0(dh) \qquad m > 0 \qquad \text{(XIII-81)}$$

where δ is the Dirac measure, and C_0 a covariance without Dirac measure. Thus XIII-79 may be written:

$$C_B(h) - m\,K(h) = K * C_0 \qquad \text{(XIII-82)}$$

This decomposition is very useful when the range of C is large compared to the size of B (see the forest example below).

When the phenomenon under study is isotropic, K. Krickeberg (1974) and B. D. Ripley (1976) suggest that the covariance be replaced by an equivalent quantity which has a more geometrical interpretation.

Consider a disk B of radius r centred at an arbitrary point x_i of the germ model X and let $mH(r)$ be the expectation of the number of points of X (not including x_i) belonging to the disk. From XIII-80, we have firstly:

$$E[Z(B) \cdot Z(dx_i)] = E[Z(B/dx_i)|x_i \in X]m\,dx_i + E[Z(dx_i)|x_i \in X]m\,dx_i$$

$$= m^2\,H(r)\,dx_i + m\,dx_i$$

and on the other hand:

$$E[Z(B) \cdot Z(dx_i)] = \left[2\pi \int_0^r C(\rho)d\rho + m^2\pi r^2\right][dx_i]$$

Thus:

$$2\pi \int_0^r C(\rho)\, d\rho - m = 2\pi \int_0^r C_0(\rho)\, d\rho = m^2 \left[H(r) - \pi r^2 \right]$$
$$= m^2 H_1(r) \qquad\qquad \text{(XIII-83)}$$

From the decomposition XIII-81 we see that $H(r) - \pi r^2$ starts from zero and is continuous at the origin. Noting that $m\,\pi r^2$ is the unconditional expectation of $Z(B)$, we can interpret $m\left[H(r) - \pi r^2 \right]$ as the average excess of points falling in a B centred at x_i (x_i excluded) over and above the average number $E[Z(B)]$; this excess is due to B being randomly implanted at an x_i. Another interpretation is the following: for a Poisson point process, $C(du)$ $= m\delta$, and $H(r) = \pi r^2$. Therefore the quantity $H(r) - \pi r^2$, for a non Poisson germ model, allows a comparison of the features (clusters, alignments, etc.) of the texture under study with the Poisson reference (B. D. Ripley prefers to estimate and plot the function $H(r)$; here we shall work with $H_1(r) = H(r) - \pi r^2$, for the reasons given above).

The probabilistic technique for calculating the key parameters $P_n(B), C(du)$ and $H^1(r)$ is to proceed via the Laplace transform $\Gamma(\lambda, B)$ of the random variable $Z(B)$:

$$\Gamma(\lambda, B) = E \exp\left[-\lambda Z(B) \right]$$

which makes the convolution of variables very easy, and also gives the first two moments (among others), since:

$$E[Z(B)] = -\left[\log \Gamma(\lambda, B) \right]^{(i)}_{\lambda=0}$$
$$E(Z(B))^2 = \left[\log \Gamma(\lambda, B) \right]^{(ii)}_{\lambda=0} \qquad\qquad \text{(XIII-84)}$$

For discrete distributions p_n, we shall also use the generating function $G(s)$ $= \sum p_n s^n$. Knowing the generating function (or the Laplace transform in the continuous case) is equivalent to knowing the distribution function of the random variable.

F.3. Models

(a) *Poisson points with parameter θ*
The contribution $d\Gamma(\lambda, B)$ from the elementary zone dx of the space, to the total Laplace transform, is:

$$d\Gamma(\lambda, B) = (1 - \theta\, dx)\, e^0 + \theta\, dx \exp\left[-\lambda \cdot 1_B(x) \right]$$
$$\sim \exp\left\{ -\theta[1 - \exp(-\lambda \cdot 1_B(x))] \right\} dx$$

Since the process is independent from one dx to another, $\log \Gamma(\lambda, B)$ is the integral of $\log(d\Gamma(\lambda, B))$:

$$\log \Gamma(\lambda, B) = \theta \int \{1 - \exp[-\lambda \cdot 1_B(x)]\} dx$$

$$= [1 - \exp(-\lambda)] \theta A(B) \qquad \text{(XIII-85)}$$

which is the Laplace transform of a Poisson distribution with parameter $\theta A(B)$; thus:

$$P_n(B) = \frac{[\theta A(B)]^n}{n!} \exp[-\theta A(B)], \text{ and in particular } Q(B) = P_0(B)$$

$$= \exp[-\theta A(B)]$$

Applying the differentiation rules (XIII-84) to relation (XIII-85), we find:

$$E[Z(B)] = \theta A(B) \quad \text{and} \quad \text{var}[Z(B)] = \theta A(B)$$

which implies that $m\,dx = \theta\,dx$ and $C = \theta\delta$ are the mean measure and the covariance measure respectively (to determine C, use the variance $\theta A(B)$ with XIII-80). Note that relation XIII-83 obviously implies that $H_1(r) \equiv 0$.

(b) *Shooting model or Neyman–Scott cluster process*

The shooting model was independently defined and its properties studied by G. Matheron (1969) and by Neyman and Scott (1972). It is simply a Boolean model whose primary grain X' is itself made up of points. More precisely, X' is the union of v points $x_i, \ldots x_v$, where the number v is a random variable with generating function $G(s) = \sum p_v s^v$. Consider the origin as the "bulls-eye"; each point $x_1 \ldots x_v$ is "shot" towards the origin independently of each other, and hits around the bulls-eye with a spatial probability measure $f(dx)$ (measure or density according to the regularity of f).

The Laplace transform from the shooting of one point of the primary grain is $\int \exp[-\lambda \cdot 1_B(y)] F(dy)$. By composition, the Laplace transform from the whole primary grain is thus:

$$\sum_v p_v \left[\int \exp[-\lambda \cdot 1_B(y)] F(dy) \right]^v = G\left[\int \exp[-\lambda \cdot 1_B(y)] F(dy) \right] \qquad \text{(XIII-86)}$$

(As an aside, note that relation XIII-86 is interesting in itself, since it describes a non stationary random point set). To "Booleanize", we must compose (XIII-86) (translated to the point x) with the probabilities of having $0, 1 \ldots$ etc. points in dx; this finally gives:

$$\log \Gamma(\lambda, B) = -\theta \int \left\{ 1 - G\left[\int \exp[-\lambda \cdot 1_B(y-x)] F(dy) \right] \right\} dx \qquad \text{(XIII-87)}$$

By differentiating this expression and putting $\lambda = 0$, we find:

$$\left[\log \Gamma(\lambda, B)\right]^{(i)}_{\lambda = 0} = E[Z(B)] = \theta E(v) A(B)$$

i.e. the mean measure is the Lebesgue measure $\theta E(v)\, dx$. Differentiating twice, and applying rule XIII-84, we have;

$$\mathrm{var}[Z(B)] = \theta E(v) A(B) + \theta E[v(v-1)] \int dx \int\int 1_B(y-x) 1_B(z-x) F(dy) F(dz)$$

To make the notation simpler, assume that $(dx) = f(x)\, dx$ possesses a density f, and let $g(h) = \int f(x) f(x+h) dx$ denote the covariogram of $f(x)$. Then:

$$\mathrm{var}[Z(B)] = \theta E(v) \cdot A(B) + \theta E[v(v-1)] \int A(B \cap B_h) \cdot g(h)\, dh$$

and the covariance measure of the process is $C = \theta E(v) \cdot \delta + \theta E[v(v-1)]g$. Here again, we see a Dirac measure at the origin, but this time completed by a continuous second term, which is responsible for the spatial "clustering" of the process. If $f(x)$ is isotropic, the function $H_1(r)$ becomes:

$$H_1(r) = \frac{E[v(v-1)]}{\theta E^2[v]} 2\pi \int_0^r u g(du) \qquad \text{(XIII-88)}$$

When $r \to \infty$ the quantity $2\pi \int_0^r u g(du) \to \int_{R^2} g(h)\, dh = 1$. The behaviour of $H_1(r)$ near the origin depends on the regularity of the density $f(x)\, dx$. For the most commonly used densities (Gaussian in $\mathbb{R}^2$, uniform distribution over a given disk) $H'_1(0) = 0$ and $H_1(r)$ exhibits a parabolic behaviour; but if $F(dx)$ contains a Dirac term (for example), $H_1(r)$ possesses a jump at the origin $(H_1(0) = 0$, and $H_1(r) \to C_0 > 0$ as $r \to 0)$ (see Fig. XIII.24). We will now obtain the probabilities $\{P_n(B)\}$ from their generating function $T(s)$. Putting

$$\alpha = \int F(dy) 1_B(y-x) = F * 1_B, \quad \text{and} \quad \beta = 1 - \alpha, \quad \text{relation (XIII-87) becomes}$$

$$\log \Gamma(\lambda, B) = -\theta \int \{1 - G[\exp(-\lambda)\alpha + \beta]\}\, dx$$

Notice that by substituting $\exp(-\lambda)$ for s in XIII-87, we have $T(s)$; thus:

$$\log T(s) = -\theta \int [1 - G(\lambda_\alpha + \beta)]\, dx \qquad \text{(XIII-90)}$$

An expansion in α provides us with:

$$\log T(s) = -\theta \int \sum_{k=1}^{\infty} (1 - s^k)\alpha^k \sum_{n=k}^{\infty} \binom{n}{k} \beta^{n-k} p_n\, dx \qquad \text{(XIII-91)}$$

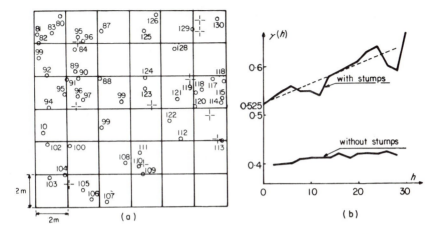

Figure XIII.21. (a) portion of the planting plan:. trees, + stumps, (b) semi-variograms of the trees, after a moving average by a square of $2\,m \times 2\,m$.

Thus $\log T(s)$ is the sum of the log of the generating functions of independent Poisson k-plets, with parameters $\theta p_n \int \alpha^k \beta^{n-k} dx$. Each k-plet is due to a primary grain of n points, k of which hit B. In particular, the probability $Q(B)$ that no point hits B is

$$\log Q(B) = \log T(0) = -\theta \int [1 - G(\beta)]\, dx$$

For example, if the number of points v in the primary grain

— is fixed and equal to v_0, then:

$$\log Q(B) = -\theta \int [1 - (1 - F * 1_{\check{B}})^{v_0}]\, dx$$

— is random with a binomial distribution (N, q), then:

$$\log Q(B) = -\theta \int [1 - q \cdot 1_B * F]^N$$

— is random with a Poisson law λ, then:

$$\log Q(B) = -\theta \int [1 - \exp(-\lambda F * 1_{\check{B}})]\, dx$$

It is also possible to calculate the specific connectivity number N_A of the shooting model (J. Serra 1974). The expression is complicated, so that we confine ourselves to the particular case of point doublets.

(i) Point doublets. This particular model has already received some attention within the framework of cell division in embryology (structures of the type seen in Figure IX.13a). The primary grain is now the union of two points. Thus $G(s) = s^2$ and:

$$\log T(s) = -\theta[1 - s^2] \int \alpha^2(x)\,dx - 2\theta(1 - s) \int \alpha(x)[1 - \alpha(x)]\,dx$$

B is hit by both points of the Poisson doublets, in a number given by a Poisson variable with parameter $\theta \int \alpha^2\,dx$, plus by single points (half doublets) in a number given by a Poisson variable with parameter $2\theta \int \alpha(1 - \alpha)\,dx$. We see immediately that:

$$\int \alpha(x)\,dx = \text{mes } B = A(B)$$

$$\int \alpha^2(x)\,dx = \int K(h)g(h)\,dh$$

(K is the geometric covariogram of B, and $g = f * \check{f}$, i.e. is the probability density of the vector joining the two points of a primary grain.) The integral $\int \alpha^2(x)\,dx$ turns out to be the expectation of the area of the intersection $B \cap B_h$, where $h = x_2 - x_1$ is the random distance between x_1 and x_2, the two points of the primary grain. In particular:

$$\log Q(B) = 2\theta A(B) - \theta \int K(h)g(h)\,dh$$

Now replace each point x of a realization X of the doublet model by a hexagon B with length of side r; i.e. dilate X by B. The resulting specific connectivity number N_A'', in the sense of algorithm (XIII-24), can be calculated, and by comparing it to relation XIII-25, we have a mean of analysing the clusters of the model. Thus:

$$N_A'' = \theta Q(B)\left[3\sqrt{3}r^2 + 2\left(1 - \int_B (f * \check{f})\,dy\right) - \frac{3\sqrt{3}}{2}\theta\left(r - \int_0^r (f * \check{f} * 1_B)\,dz\right)^2\right]$$

(B is the hexagon of size r, centred at the origin).

(ii) An illustrative example. A large number of examples of point models (particularly for analysing forest structures), can be found in the literature. In fact the basic work of B. Matern, and his hard core model, comes from these forestry applications. A few years ago, we were confronted with the problem of how to estimate certain vegetation reserves, which led us to compute variograms on forests, and then to try to model them (C. Millier et al., 1970). In the example presented here, the zone analysed is a rectangular plot

50 m × 1000 m in the Haye's forest, near Nancy (France); the tree population is made up of one hundred year old beeches. In 1960, the Forestry Department decided to clear the zone, and reduce the population by about 70 per cent. Figure XIII.21a shows a recent partial view of the plot. Over the whole zone, the exact location of the stumps is known. Therefore, we can calculate the variogram of either the trees, or the set "trees + stumps". In both cases, a square 2 m × 2 m grid is superimposed onto the plan, and each vertex (i, j) is attributed with the number of trees (respectively trees + stumps) contained in the square centred at (i, j). In mathematical terms, this technique is the result of putting a Dirac measure at each tree-point, and then regularizing these measures by the square $B = 2$ m × 2 m. Rather than covariances, we calculated variograms, since the range of the structure is not reached inside the limits of the increments h (see Ch. IX). Anyhow, the regularized variogram is

$$\gamma_B(h) = (\gamma * K) - \int K(h)\gamma(dh) \text{ where } K(h) = \int 1_B(x) \cdot 1_B(x+h)\,dx$$

the results are the following:

$$\gamma_B^*(h) = 0.525 + 0.004\,h; \qquad h > 0 \text{ (trees + stumps)}$$
$$\gamma_B^{**}(h) = 0.41; \qquad\qquad h > 0 \text{ (trees only)}$$

The scale of regularization (2 m × 2 m) is small with respect to the larger structure which is still linear beyond 30 m and has no appreciable effect on it. It is only the Dirac measure at the origin of the variogram-measure which is smoothed by the 4 m^2 square, appearing as a nugget effect; i.e. a jump at the origin of the experimental curves. While $\gamma^{**}(h)$ exhibits a Poisson structure with a total lack of autocorrelation, $\gamma^*(h)$ *is* structured and its shape indicates clusters, with the possibility also of repulsion phenomena (the two dips in the curve). Curiously, the Forestry Department have destroyed the tree clusters just enough to turn the wood into a Poisson point realization!

The question now is: how can these clusters be modelled? If we start from a Neyman–Scott process, $\gamma_B(h)$ must be of the type $k_1 - k_2\,g\,(k_1, k_2$ constant), where $g(h)$ is the covariogram of the spatial density $f(x)$ of the primary grain. But unfortunately, if we completely change the model, and assume the forest structure to be a Poisson process with a *regionalized* density θ, then it can be shown that once again $\gamma_B(h)$ is of the type $k_1 - k_2 g$, where now g is the variogram of the Poisson density θ. Which of the two possibilities we choose *cannot* be determined by experimentation. The second, the regionalized density, contains the Neyman–Scott process as a particular case. To distinguish between them, it would be necessary to test whether $N(B)$, the number of points hitting B, is a sum of independent n-tuplets. For $n > 2$ such a test is totally unrealistic.

This leads us to conclude that the refined specifications of random set models is too sophisticated for the actual power of discrimination of the statistics. Indeed, these classes of models (cluster, repulsion, alignment . . .) are often used as a heuristic tool for providing us with the general behaviour of the basic properties ($H(r)$ in Fig. XIII.24, for example) rather than for their fine details.

(c) Hard core models

Since B. Matern introduced hard core models into the literature (1960), many variants have been developed (see Snyder (1976) or Ripley (1976) for a survey). One of these variants, by Matern himself, we will now discuss, since its covariance measure clearly exhibits the main features of the hard core type.

By hard core process (or model), we mean a model where the distance between two points cannot go below a given bound R. A simple construction of such a model, is to start for example from a Poisson point realization, and remove each pair of points whose separating distance is $< R$. The mean measure and covariance measure are easily calculable. This time, we shall take the direct approach, rather than via the Laplace transform. In a small open set dx of the plane $\mathbb{R}^2$ one point occurs with the probability Pr $= \theta\, dx \cdot \exp\{-\theta A[C(r)]\}$, *or* no point occurs with probability $1 - \mathrm{Pr}$, where $C(R)$ is the repulsion disk or radius R. Thus the average number of points in dx is:

$$m\, dx = \theta \exp\{-\theta A[C(R)]\}\, dx = \theta q\, dx \qquad \text{(XIII-88)}$$

giving the mean measure of the process. To calculate the covariance, notice firstly that the probability of simultaneously having one point in dx and another in dy, is:

$$\begin{cases} m\, dx\, dy & \text{if } \rho = 0 \\ 0 & \text{if } 0 < \rho < R \\ \theta^2 [q^2 \exp\{\theta K_c(\rho)\}]\, dx\, dy & \text{if } \rho \geq R \end{cases}$$

where $\rho = |x - y|$, and $K_c(\rho)$ is the geometric covariogram of the disk $C(R)$. Denote by $k(\rho)$ the function equal to zero for $0 \leq \rho < R$ and equal to $q^2 \exp[\theta K_c(\rho)]$ for $\rho > R$. Then using the decomposition XIII-81, the variance $\mathrm{var}[Z(B)]$ of the number of points falling in B is:

$$\mathrm{var}[Z(B)] = mA(B) + \theta^2 \int_B \int_B [k(\rho) - q^2]\, dx\, dy$$

$$= mA(B) + \theta^2 \int_{\mathbb{R}^2} K_B(h)[k(|h|) - q^2]\, dh$$

$(K_B(h)$ = geometric covariogram of B). The covariance measure $C(dh)$ results:

$$C(dh) = m\delta(dh) + \theta^2[k(|h|) - q^2]dh$$

The model being isotropic, one can also calculate $H_1(r)$ by applying (XIII-83):

$$H_1(r) = \begin{cases} -\pi r^2 & 0 < r \leq R \\ 2\pi m^2 \displaystyle\int_R^r \exp[(\theta K)_c(\rho)]_\rho \, d\rho - \pi m^2 r^2 & R < r \leq 2R \\ 2\pi m^2 \displaystyle\int_R^{2R} \left\{ \exp[(\theta K)_c(\rho)] - \frac{4}{3} \right\}\rho \, d\rho & (= \text{neg. constant}) \quad r > 2R \end{cases} \qquad \text{(XIII-89)}$$

The curve is plotted in Figure XIII.24c. Its behaviour is typical and common to any hard core model; the anti-cluster structure is seen on the curve as a negative function, which reaches its range at $2R$, the diameter of the empty neighbourhood $C(R)$.

We shall not calculate the probabilities $P_n(B)$ of the model, but simply note that for $B \subset C(R)$ (modulo a translation), $P_0(B) = q \cdot A(B)$.

(d) *Poisson alignments*

Consider the set of intersection points of the isotropic Poisson line process in $\mathbb{R}^2$ (density λ); we call this point model "the Poisson alignment model" (see Fig. XIII.22a). In order to determine its mean measure and its covariance measure, we need to calculate the average number and the variance of the points of the process falling in a compact convex set B.

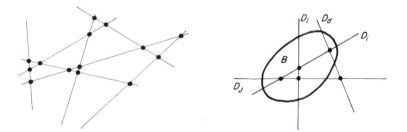

Figure XIII.22. (a) realization of Poisson alignments, (b) types of points falling in a compact convex set B.

With each pair of Poisson lines (D_i, D_j) we associate the indicator variable $N_{ij} = 1_{D_i \cap D_j \cap B}$, which equals 1 when the intersection $D_i \cap D_j$ lies in B, and zero otherwise. Put:

$$N = \sum_{i < j} N_{i,j}$$

If n is the number of Poisson lines hitting B, we have:

$$E(N) = E\left[\frac{n(n-1)}{2}\right]E(N_{ij})$$

The number n is Poisson with parameter $\lambda U(B)$, thus $E[n(n-1)] = \lambda^2 U^2(B)$. We know (Ex. X-10) that the mean length of a traverse $D_i \cap B$ (when $D_i \cap B \neq \varnothing$) is:

$$E[D_i \cap B] = \frac{\pi A(B)}{U(B)}$$

Noting that $E(N_{ij})$ is the probability that a line D_j hitting B also hits the segment $D_i \cap B \subset B$, we have:

$$E[N_{ij}] = \frac{2E[D_i \cap B]}{U(B)}$$

Thus:

$$E(N) = \pi \lambda^2 A(B)$$

so that the mean measure $m(dx)$ is proportional to Lebesgue measure dx:

$$m(dx) = \pi \lambda^2 \, dx$$

To find $\mathrm{var}(Z(B))$, we first calculate $E(N^2)$. The quantity $N^2 = \sum_{i<j} \sum_{l<s} N_{ij} N_{ls}$ is made up of:

— $\dfrac{n(n-1)}{2}$ terms of the type $(N_{ij})^2$

— $n(n-1)(n-2)$ terms of the type $N_{ij} \cdot N_{is}$ $(s \neq j)$

— $\dfrac{n(n-1)(n-2)(n-3)}{4}$ terms of the type $N_{ij} \cdot N_{ls}$ (all indices distinct)

We already know that:

$$E[N_{ij}^2] = E[N_{ij}] = \frac{2\pi A(B)}{U^2(B)}$$

$$\text{and} \quad E[N_{ij} \cdot N_{ls}] = E[N_{ij}] \cdot [N_{ls}] = \frac{4\pi^2 A^2(B)}{U^4(B)}$$

To calculate $E(N_{ij}, N_{is})$, work with the random length $\rho = L(D_i \cap B)$. Let $f(\rho)$ denote the density function of ρ. For a given ρ, the conditional probability that D_j and D_s hit $D_i \cap B$, is $[2\rho/U(B)]^2$. Thus:

$$E[N_{ij}, N_{is}] = \frac{1}{U^2(B)} \int_0^\infty \rho^2 f(\rho)\,d\rho = \int_0^\infty \rho[1 - F(\rho)]\,d\rho$$

Now the function F is linked to the geometric covariogram $K(\rho, \alpha)$ of B (see Ch. X, B.2)

$$1 - F(\rho) = -\frac{1}{U(B)} \int_0^\pi K'_\alpha(\rho, \alpha)\, d\alpha$$

Hence:

$$E[N_{ij}, N_{is}] = \frac{4}{U^3(B)} \int_{\mathbb{R}^2} \frac{1}{\rho} \cdot K(\rho, \alpha) \cdot \rho\, d\rho\, d\alpha$$

So that finally:

$$\mathrm{var}\,(B) = E(N^2) - [E(N)]^2 = \lambda^2 \pi A(B) + 4\lambda^3 \int_{\mathbb{R}^2} \frac{1}{\rho} \cdot K(\rho, \alpha) \cdot \rho\, d\rho\, d\alpha$$

Thus the covariance measure is $C(dh) = \lambda^2 \pi \delta(dh) + 4\lambda^3 (dh/|h|)$.

(Note that $C(dh)$ also represents the "density" of the connectivity number of the Poisson polygons, since at each vertex x the sum of the normal rotations for the four polygons of vertex x is always 2π.)

From $C(dh)$, the function $H_1(r)$ is immediate:

$$H_1(r) = 8\,(\pi\lambda)^{-1} r$$

The difference between this and all the preceding cases is that $H_1(r)$ increases without bound as $r \to \infty$. That is, the alignment process gives rise to clusters of an infinite size. (For the reader who is interested in a more mathematical analysis of this model, the book by G. Matheron (1975, 6-2) contains the generalization of $C(dh)$ to n dimensional space.)

(e) A Brownian motion example in $\mathbb{R}^3$

This final example exhibits a hyperclustered process that is not amenable to the usual forms of analysis. Consider a large ball B of radius R. At time $t = 0$, we release a Poisson number N (where $E(N) = R$), of Brownian particles uniformly distributed over the surface of the sphere B. Let X denote the union of their $\mathbb{R}^3$ traces. The set $X \cap B$, is the part of X which penetrates the ball B; it is the restriction to B of an istropic stationary random set, studied by G. Matheron (1975, 3-4). The probability $Q(K)$ that a compact set $K \subset B$ misses the set X, is given by $\exp[-\Psi(K)]$, where Ψ is the Newtonian capacity (e.g. for K a ball of radius r, $\Psi(K) = r$ and for K a disk of radius r, $\Psi(K) = 2r(\pi)^{-1}$).

Now suppose that $X \cap B$ is sectioned by a plane Π, for example by the equatorial plane of B. The "traces" of X perforate Π in a very strange way: think of a needle pricking virtually the same point an infinite number of times and then suddenly jumping far away and repeating the process (see in Mandelbrot, 1976, simulations of such sets). The number of points induced on

Π is uncountable, and the presence of accumulation points makes an analysis of this random set by covariance methods (measure $C\,(dx)$ or function $H_1\,(r)$), or by probabilities $P_n\,(K)$, $n > 0$ (see Th. 3 below), impossible. The only information that is quickly accessible is the probability $P_0\,(K)$; e.g. for K a disk of radius r:

$$Q(K) = P_0(K) = 1 - \exp\left(-2r/\pi\right)$$

Notice that $1 - Q\,(K)$ is also the area proportion of the dilated set $(X \cap B \cap \Pi) \oplus K$, (the 2-D Brownian points dilated by a disk). As $r \to 0$, $1 - Q(K) \to 2r/\pi$, and from relation (V-27) we see that the fractal dimension of the Brownian points is one (the dimensions of a regular curve in $\mathbb{R}^2$!)

(f) *Conclusion*

Let us now ask how, as experimenters, we see the various types of point sets presented above. In reality, the phenomenon under study is made up of very small 2-D particle particles that we *decide* to treat as points in $\mathbb{R}^2$. From there, our first experimental step is to estimate $Q\,(B)$ for increasing disks B (equivalently hexagons or octagons), and plot $Q\,(B)$ against the size r of B (radius or length of side). If the decreasing behaviour is parabolic at the origin (Fig. XIII.23a), then one can try to fit a germ model. If not, the phenomenon under study looks like a fractal set, with accumulation points, and there our investigation stops; however, the slope of the line at the origin is interpretable as an intensity factor of the infinite clustering around the accumulation points.

Figure XIII.23. (a) a germ model exhibits a parabolic behaviour $Q\,(rB)$ at the origin, (b) for a fractal set of points $Q\,(rB) \to k \cdot r$ when $r \to 0$.

If it seems reasonable to fit a germ model, the experimenter's second step is to look at the covariance. For example, by measuring the variances of the numbers $Z\,(B(d))$ for rectangles $B\,(\alpha)$ elongated in direction α, and varying α, one can detect anisotropy of the structure. If anisotropy is present, the simplest method of analysis is to choose a B (square or regular hexagon) which is small compared to the ranges of the structure, and then to compute the covariances or the variograms of $Z\,(B)$ in different directions (as we did in one direction in the Haye forest example). For $|h|$ larger than the diameter of B, $C_B\,(h)$ (or $\gamma_B\,(h)$)

are virtually identical to the continuous part of the covariance (or variogram) measure, modulo an upwards or downwards shift.

If the structure is isotropic, instead of measuring the covariances it is preferable to estimate the function $H_1(r)$ directly. For the types of point processes described above, $H_1(r)$ exhibits the various behaviours shown in Figure XIII.24.

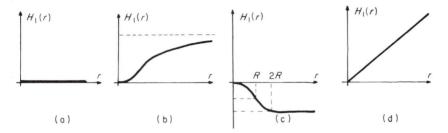

Figure XIII.24. Function $H_1(r)$ as a cluster descriptor (a) Poisson points—no cluster, (b) clusters of shooting model type, (c) hard core model (negative clusters), (d) Poisson alignments (infinite clusters).

Estimators for the function $H_1(r)$, can be found in B. D. Ripley (1976). Here the analysis should stop, as we have already remarked that the second order analysis (by covariance or by $H_1(r)$) is not definitive enough to decide upon a model. To go further would require us to fit all the observed probabilities $\{P_n^*(B)\}$, with those of the models under consideration assuming that it is possible to calculate the theoretical $P_n(B)$, and then to test their fit.

G. RANDOM SETS

Some very worthwhile preliminary reading to this section, is Chapter III, Section D, where the link was made with the Hit or Miss topology and its physical meaning. Section G.1 to follow, is just a random version of the same fundamental approach.

G.1. Random closed sets (RACS)

Previously, random sets (or RACS) were discussed via their realizations (X will denote a RACS as well as its realization). However, just as a random variable, a RACS is mathematically defined from a collection of events, namely the relationships "K misses X", $K \in \mathcal{K}$ and "G hits X", $G \in \mathcal{G}$. These events are subject to the classical axioms of probability theory, which require firstly a σ-algebra, call it σ_f, and then a probability P on the measure space $(\mathcal{F}, \sigma_f)$.

Generate σ_f by taking countable unions of the events, together with their complements (i.e. σ_f is a Borel algebra). In fact, because of the properties of the Hit or Miss topology, the σ-algebra σ_f can be generated just by the relationships "K misses X", i.e. the class $\mathscr{F}^K$ as defined in Chapter III, Section D.1, or equivalently just by the relationships "G hits X", i.e. $\mathscr{F}_G$ (G. Matheron, 1975 p. 27). We still have to define a probability P on $(\mathscr{F}, \sigma_f)$. This can be done by associating with any event V of σ_f, the probability that this relation V is true. The compactness of the space $\mathscr{F}$ ensures that such probabilities on σ_f actually exist.

It is not obvious that when we transform the realization of a RACS, by dilating them for example, we again obtain a RACS. The following theorem (G. Matheron, 1975) resolves the problem, at least for semi-continuous mappings:

THEOREM XIII-1: *If Ω is a topological space, $\mathscr{C}$ its Borel algebra, and Ψ an u.s.c. or l.s.c. mapping from Ω into $\mathscr{F}$, then Ψ is measurable, and defines a RACS if a probability P' is given on $\mathscr{C}$.*

Hence, according to the results of Chapter III, if F and F' are RACS, so are $F \cup F'$, $F \cap F'$, δF, $(\mathring{F})^c$ and $\mathring{F}$. Moreover, if K is a non empty compact set, $F \oplus K$, $F \ominus K$, F_K and F^K are also RACS.

Note that it is possible to provide $\mathscr{P}(\mathbb{R}^n)$, the set of all subsets of $\mathbb{R}^n$, with the σ-algebra generated in $\mathscr{P}(\mathbb{R}^n)$ by:

$$\mathscr{F}_G = \{X : X \in \mathscr{P}(\mathbb{R}^n), \quad X \cap G \neq \varnothing\}$$

But the following equivalence:

$$X \cap G \neq \varnothing \Leftrightarrow \overline{X} \cap G \neq \varnothing \qquad \text{(XIII-90)}$$

shows that $\mathscr{F}_G$ has the same meaning in $\mathscr{F}$ as in $\mathscr{P}(\mathbb{R}^n)$. In other words, it is completely equivalent to study a set X with the σ-algebra σ_f, as it is to study its closure $\overline{X}$ with the same σ_f. Hence:

THEOREM XIII-2: *If $\mathscr{R}_f$ is the equivalence relationship "$X = X'$ if $\overline{X} = \overline{X}'$" in $\mathscr{P}(\mathbb{R}^n)$, the quotient space $\mathscr{P}(\mathbb{R}^n)\,\mathscr{R}_f$ with the σ-algebra generated by $\mathscr{F}_G$ is isomorphic to $\mathscr{F}$ with σ_f.*

This last theorem reduces the class of experimental sets to realistic proportions. It eliminates the artificial problems of the introductory example of Chapter III, where we considered a set which contained only half its boundary. It also shows that it is sufficient to construct random model for *closed* sets only. To conclude this section, we will give a general result on the measurability of RACS (G. Matheron, 1969).

THEOREM XIII-3: *Every RACS is measurable in the following sense: the mapping* $k : \mathbb{R}^n \times \mathscr{F} \to \{0, 1\}$ *defined by setting:*

$$k(x, X) = \begin{cases} 1 \text{ if } x \in X \\ 0 \text{ if } x \notin X \end{cases}$$

is measurable for the product σ-*algebra* $\sigma(\mathscr{G}) \times \sigma_f$.

The theorem implies the following important corollary:

COROLLARY: *Let* μ *be a positive measure on* $(\mathbb{R}^n, \sigma(\mathscr{G}))$. *Then the mapping* $X \to \mu(X) = \int k(x, X) \mu(dx)$ *from* $\mathscr{F}$ *into* R_+ *is a random variable* ≥ 0 *whose expectation is:*

$$E[\mu(X)] = \int P\{x \in X\} \mu(dx) \qquad \text{(XIII-91)}$$

As a consequence, in the random version of all those sets considered in mathematical morphology (either the original closed sets, or the sets resulting from a semi-continuous transformation), the volume (in $\mathbb{R}^3$) or the area (in $\mathbb{R}^2$) does exist. The reader should take heed that the other Minkowski functionals may sometimes not exist. For example, consider our RACS to be a bounded part of the trace of a 3-D particle in a Brownian motion; the length of this RACS is infinite (fractal dimension). The only general cases where we can be sure of the existence of measurable Minkowski parameters are the regular model and the convex ring.

G.2. The functional $T(K)$

It is well known that the probability distribution associated with an ordinary random variable is completely determined if the corresponding distribution function is given. There is a similar relationship for random closed sets. If X is a RACS and P the associated probability on σ_f, define:

$$T(K) = P(X \cap K \neq \varnothing) \qquad \text{(XIII-92)}$$

to be the probability that X hits a given compact set $K \in \mathscr{K}$. That is, we obtain a function T on $\mathscr{K}$ associated with the probability P. Conversely:

THEOREM XIII-4. (*G. Matheron, 1971, and D. G. Kendall, 1974*). *The probability P on* σ_f *is completely determined if the functional* $T(K)$ *on* $\mathscr{K}$ *is given.*

With respect to the RACS, the role of T is thus similar to the distribution function for random variables. It directs us towards the function to calculate when studying a RACS.

The functions defined on $\mathscr{K}$, the space of compact sets, are of great importance in integral geometry (Hadwiger, 1957). Thus, it is interesting to find the necessary and sufficient conditions for a given function T to be associated with a RACS. First, it is required that T be null for the empty set: $T(\varnothing) = 0$, because the empty set does not hit any other set; and also $T \leq 1$. If K_n is a decreasing sequence of compact sets and $K = \cap K_n$ is its intersection, the sequence $X \cap K_n$ is also decreasing in $\mathscr{F}$ and its intersection is $X \cap K$. Thus, the sequence $T(K_n)$ must converge towards $T(K)$ (sequential continuity). This property appears to be equivalent to upper semi-continuity of the function T on $\mathscr{K}$, equipped with the Hausdorff metric. Finally, let $S_n(K_0; K_1, \ldots, K_n)$ denote the probability that X misses the compact set K_0, but hits the other compact sets $K_1, \ldots, K_n$. These functions are obtained by the following recurrence formulae:

$$S_1(K_0; K_1) = T(K_0 \cup K_1) - T(K_0)$$
$$S_n(K_0; K_1, \ldots, K_n) = S_{n-1}(K_0; K_1, \ldots, K_{n-1})$$
$$- S_{n-1}(K_0 \cup K_n; K_1, \ldots, K_{n-1})$$

Clearly, these functions S_n must be ≥ 0 for any integer n and any compact sets $K_0, K_1 \ldots, K_n$.

These three conditions must be satisfied by the function T. We can summarize the two most important of them (sequential continuity and the positivity of the S_n), by saying that T is *an alternating capacity of infinite order*. It turns out that the converse is true, and we may state the following:

THEOREM XIII-5. *A functional T on $\mathscr{K}$ is associated with a random closed set X satisfying $P(X \cap K \neq \varnothing) = T(K)$ for any compact K, if and only if T is an alternating capacity of infinite order satisfying $T(\varnothing) = 0$ and $T \leq 1$.*

This theorem goes back to Choquet (1953). Its greatest interest for us is to determine the class of functions T on $\mathscr{K}$ having a probabilistic interpretation, and to tie together integral geometry and random set theory. It turns out that many properties, classical in integral geometry, have a probabilistic interpretation, and conversely. We now give a few examples.

Consider first the case where the function T is C-additive on the class of the compact convex sets (Hadwiger, 1957) i.e. T satisfies:

$$T(K \cup K') + T(K \cap K') = T(K) + T(K') \qquad \text{(XIII-93)}$$

if K, K' and their union $K \cup K'$ are compact and convex. If the conditions of Choquet's theorem are also satisfied by T, then the associated random closed set X is *almost surely convex*, and the converse is also true: thus, strong additivity of the function T and almost sure convexity of the random set X can be identified as saying the same thing.

(a) *RACS that are infinitely divisible with respect to the union*

A random closed set X is infinitely divisible with respect to the union if, for any integer n, X is equivalent to the union $X'_1 \cup \ldots \cup X'_n$ of n independent and equivalent RACS $X'_1, \ldots, X'_n$. Let us denote by:

$$Q(K) = 1 - T(K) = P(X \cap K = \varnothing)$$

the probability that X does not hit the compact set K. Clearly, X is infinitely divisible if and only if the conditions of Choquet's theorem are satisfied by the functionals:

$$T_n = 1 - Q^{1/n}$$

In order to eliminate inessential complications, let us also assume that there are no fixed points (i.e. no points x such that $P(x \in X) = 1$). Then we have the following theorem (G. Matheron, 1971):

THEOREM XIII-6. *A function T on $\mathscr{K}$ is associated with an infinitely divisible RACS without fixed points, if and only if there exists an alternating capacity of infinite order Ψ satisfying $\Psi(\varnothing) = 0$ and $Q(K) = 1 - T(K) = \exp\{-\Psi(K)\}$ for any compact set K.*

Note that the condition $T \le 1$ is no longer required.

Theorem XIII-6 opens the door to the Boolean sets, but not only to them. Imagine for example, the union in $\mathbb{R}^3$ of a Boolean set and trajectories of Brownian motions. The result is infinitely divisible, but not reducible to Boolean structures. For the random variables, a classical theorem of P. Levy states that an infinitely divisible variable is a sum of independent Gaussian and Poisson variables. It seems that the equivalent of such a theorem does not exist for random sets, although the Boolean sets and the Poisson flats are in fact the major representatives of this type of model.

(b) *RACS stable with respect to the union*

Looking beyond infinite divisibility, there is another more demanding structure, namely that of the RACS *stable* with respect to the union. A RACS belongs to this category if for any integer n, a positive constant λ_n can be found such that the union $X_1 \cup X_2, \ldots \cup X_n$ of n independent RACS is equivalent to the RACS $\lambda_n X$. Obviously a stable RACS is necessarily infinitely divisible, but the converse is false. The following theorem (G. Matheron, 1975) characterizes the stable RACS:

THEOREM XIII-7. *A RACS X without fixed points is stable with respect to the union if and only if its functional $Q(K)$ is equal to $\exp(-\Psi)$ for a capacity of infinite order Ψ satisfying $\Psi(\varnothing) = 0$ and homogeneous of degree $\alpha > 0$, i.e.:*

$$\Psi(\lambda K) = \lambda^\alpha \Psi(K) \qquad (\lambda > 0, K \in \mathscr{K})$$

For instance, in $\mathbb{R}^n$ the isotropic Poisson flats of dimensions $n - k$ ($n > k > 0$), are stable RACS. In their.case, when the compact set K is convex, the associated functional Ψ is just the Minkowski functional W_{n-k}. Another example is provided by Brownian motion in $\mathbb{R}^3$ (Sect. F). The notion of a stable RACS is the correct mathematical concept needed to describe self similarity of sets. We already noticed in Chapter V, Section D, a stable, or self similar RACS is not necessarily fractal and conversely.

G.3. The semi-Markov property

The semi-Markov property (Matheron, 1967) leads to another interpretation for the *C-additive functions* of integral geometry. We say that two compact sets K and K' are *separated* by a compact set C if any segment (x, x') joining a point $x \in K$ to a point $x' \in X'$ hits the compact set C. Then a random closed set X is said to be semi-Markov if, for any compact sets K and K' separated by another compact set C, the RACS $X \cap K$ and $X \cap K'$ are conditionally independent, given $C \cap X = \varnothing$.

It can be shown that X is semi-Markov, if and only if its functional $Q = 1 - T$ satisfies the relationship:

$$Q(K \cup K' \cup C)Q(C) = Q(K \cup C)Q(K' \cup C)$$

where K and K' are separated by C. In particular, if the compact sets K and K', not necessarily convex themselves, have a convex union, then they are separated by their intersection $K \cap K'$, (Prop. IV-8) and thus the semi-Markov property implies:

$$Q(K \cup K')Q(K \cap K') = Q(K)Q(K') \qquad \text{(XIII-94)}$$

In other words, the function $\Psi = -\log Q$ is C-additive. The converse is also true if X is infinitely divisible. We therefore have:

THEOREM XIII-8. *An infinitely divisible random closed set X is semi-Markov if and only if the function on $\mathscr{K}$ defined by:* $\Psi(K) = -\log P(X \cap K = \varnothing)$ *is C-additive on the class of convex compact sets.*

The corresponding class of functionals Ψ (alternating capacities of infinite order, null for $K = \varnothing$ and *C-additive* for convex compact sets) is probably the most interesting class for integral geometry. In the particular case where X is stationary and isotropic (i.e. Ψ invariant under translation and rotation), the characterization theorem of H. Hadwiger (Ch. IV.C1f) shows that $\log Q(K)$ for K convex compact sets, is given by the formula:

$$\log Q(K) = \sum_{i=0}^{n} \beta_i W_i(K)$$

where β_i are non negative coefficients and where $W_i(K)$ is the ith Minkowski functional of K. Remember that the W_i's are proportional, in $\mathbb{R}^2$, to the area, the perimeter and to 1, and in $\mathbb{R}^3$ to the volume, the surface, the norm and to 1. That explains the structure of relations (XIII-29) and (XIII-30) and shows that the Boolean model with primary grains is semi-Markov. One can discover this result intuitively in the following way. Intersect X by the straight line Δ, and pick up one point x in the pores of $X \cap \Delta$. It separates Δ into two half lines Δ_1 and Δ_2. A primary grain X of X' which hits Δ_1 cannot hit Δ_2 since X' is convex; moreover, the various primary grains are independently located. Thus any event on $X \cap \Delta$ is independent of any other one on $X \cap \Delta_2$, given $x \in X^c$.   Q.E.D.

The characterization relation (XIII-94) points out that it is interesting to combine the three properties of stationarity, infinite divisibility and semi-Markov. We have the following result:

THEOREM XIII-9: *Any stationary infinitely divisible and semi-Markov RACS in $\mathbb{R}^n$ is equivalent to the union of n stationary independent RACS $X_1, \ldots, X_n$, where X_1 is a Boolean set with convex primary grains and $X_2 \ldots X_n$ are the unions of cylinders the base of which are Boolean sets with convex grains in $\mathbb{R}^{n-1}, \ldots, \mathbb{R}^1$ respectively* (G. Matheron, 1975).

Note that if the sub-dimensional Boolean sets have points as primary grains, the associated cylinders become Poisson hyperplanes. In other words, the Poisson flats and the Boolean sets turn out to be the prototypes from which any stationary, infinitely divisible and semi-Markov set may be constructed. One might wonder what happens if we replace infinite divisibility by the stability condition in Theorem XIII-9 (other things being equal). The result is that we restrict the possible class of sets, and the only RACS which are stationary, stable and semi-Markov are the Poisson flat networks (planes and lines in $\mathbb{R}^3$, lines in $\mathbb{R}^2$).

We would like to end this section by a remark on the use, and misuse, of Markov processes in Picture Processing. The fundamental difference between the line and the plane, or the space, is that the line is ordered and the others are not. Markov chains and processes are strongly based on this notion of ordering (the future is independent from the past, given conditions x, y, etc.). Of course, one can define order relations in $\mathbb{R}^2$ (order of TV scanning for example), but unfortunately there are too many ways of doing this, and all of them give priority to particular directions of the space. Moreover, when using the Markov chain framework for a phenomenon which in fact evolves in a continuous way, one has to impose a digital spacing a, which results in a loss of information. (If two successive pixels are independent for a spacing a, are they still independent when $a' = a/r$?). Finally, the fact that the linear sections of a picture are Markov does not imply anything about the "Markov" structure of the 2-D phenomenon. One can easily construct counter-examples (Ch. IX, Sect. D) of plane textures where the intersection by any line is Markov which, nevertheless, exhibit splendid clusters of particles in $\mathbb{R}^2$!

On the contrary, the semi-Markov concept does not require any order relation in $\mathbb{R}^2$, and thus is direction free. Defined in the continuous space, it can lend itself to digital samplings, and defined in $\mathbb{R}^2$ or $\mathbb{R}^3$, it is connected with the spatial textures directly, and not via line sections. Its natural field of application is to textures having a limited number of phases, which can thus be reasonably modelled by sets. In this type of problem it has more flexibility than the classical Markov concept, since one phase only is required to be Markov.

G.4. Random germ models

We plan to end this section with a few words regarding point models considered by G. Matheron 1969, K. Krickeberg, 1974 and D. Ripley, 1976. In many applications, it is necessary to use random sets whose intersection with a bounded set contains a finite number of distinct points (nucleation points in crystal growth, for example). Call the set X a germ model if X is a countable set without accumulation points.

THEOREM XIII-10. *The proposition "X is a germ model" is an event with respect to the RACS X defined on $(\mathscr{F}, \sigma_f)$.*

In other words, the set of compact sets $K \in \mathscr{K}$ such that $X \cap K$ contains a finite number of distinct points, is measurable (belongs to the σ-algebra σ_f generated by $\mathscr{F}$). Now define a random germ set X by the triple $(\mathscr{F}, \sigma_f, P)$, with the following restriction: the event "X is a germ model" has probability 1. Thus Theorem 4 gives the basis for a counting measure applied to X, in the following manner:

THEOREM XIII-11: *If X is a random germ model, then the number $N(K)$ contained in every compact set is a random variable on $(\mathscr{F}, \sigma_f)$.*

This result guarantees, for example, the validity of the probabilities $P_n(B)$ calculated in Section F.

H. THE ROSE OF MODELS

To close this chapter, we propose a diagrammatic representation (Fig. XIII.25) of the relations between the main random sets.

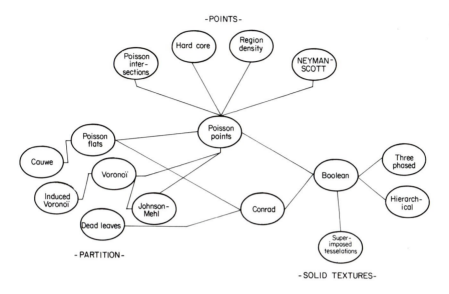

Figure XIII.25. The rose of models.

I. EXERCISES

I.1. Boolean model: a numerical application

The picture of Figure XIII.1a comes from choosing a germ density, $\theta = 300$ points/m², and disk diameter $= 2\,\mathrm{cm}$. To produce Figure XIII.1b, we turned the right half of the picture back onto the left half. Calculate the specific parameters A_A, U_A and N_A of both models.

I.2. Isolated primary grains (D. Stoyan, 1978)

Let X be a 2-D isotropic Boolean model with a random convex primary grain X' (area A', perimeter U', with a distribution function $F(dA', dU')$). Show that the proportion $\bar{\omega}$ of non-overlapping primary grains in X is given by:

$$\bar{\omega} = \int \exp\left\{ -\theta\left[A' + E(A') + \frac{U'E(U')}{2\pi} \right] \right\} \cdot F(dA', dU') \qquad (1)$$

Particularize this relation to the case where X' is a disk of a given radius r. Calculate the two proportions $\bar{\omega}$ (in number) and $\bar{\omega}'$ (in measure), of X occupied by the isolated disks. Compute it in the two cases of Exercise 1.

[Given a realization of a primary grain X' (A', U') centred at the point x, the probability that it misses the rest of X is

$$\exp\left\{-\theta\left[A + E(A') + U'\frac{E(U')}{2\pi}\right]\right\}$$

Randomizing X', we get (1). When X' is the disk of radius r, relation (1) becomes $\bar{\omega} = \exp(-\theta 4\pi r^2)$. The proportion $\bar{\omega}' =$ (specific area of the isolated grain)/(specific area of X); i.e.:

$$\bar{\omega}' = \frac{\theta\pi r^2 \exp(-4\theta\pi r^2)}{1 - \exp(-\theta\pi r^2)}$$

For Figure XIII.1a we find $\bar{\omega} = 0.69$ and $\bar{\omega}' = 0.72$, and for Figure XIII.1b, $\bar{\omega} = 0.47$ and $\bar{\omega}' = 0.51$ ($\bar{\omega}$ is immediately verifiable from the figure).]

I.3. Boolean model: a discrete approach (J. Serra, 1974)

(a) One knows from the general theorems of Chapter VI, that the probabilities $Q(B)$ of the Boolean model are digitalizable. However, these results are the direct consequences of the following argument. Start from a regular lattice of N points x_i, surrounding the origin O in $\mathbb{R}^2$. Each point serves as the centre of a primary grain with a very small probability $p = \theta/n$. The primary grains are deterministic compact sets X'^1 and B is a structuring element centred at the origin. Putting:

$$k_i(B) = 1 \quad \text{if } B \Uparrow X'^1_{x_i} \quad \text{and 0 if not,}$$

prove that the probability $Q^1(B)$, that B misses all the primary grains, satisfies:

$$\log Q^1(B) \simeq \frac{\theta}{n}\, \text{Mes}\,(X'^1 \oplus \check{B}) \tag{1}$$

$$\left[\log Q^1 = \sum_{i=1}^{N} \log\left[1 - \frac{\theta}{n}k_i(B)\right] \to \frac{\theta}{n}\,\text{Mes}\,(X'^1 \oplus \check{B}),\, \text{as } \frac{\theta}{n} \to 0\right]$$

(b) We now superimpose n similarly constructed random sets $(j = 1 \ldots n)$, the jth having deterministic primary grains X'^j. Putting $E[\text{Mes}\,(X'\oplus\check{B})] = (1/n)\sum_j \text{Mes}\,(X'^j \oplus \check{B})$, prove that:

$$Q(B) \to \exp\{-\theta E[\text{Mes}\,(X'\oplus B)]\}, \quad \text{as } n \to \infty, \theta \text{ fixed} \tag{2}$$

[Apply relation (1) to $\log Q = \sum_j \log Q^j$.]

Why does this type of proof not hold for connectivity numbers?

I.4. Convex Boolean model: induced grains (G. Matheron, 1967)

(a) Give a direct proof of relation XIII-14.

[Take a point $x_3 \in \mathbb{R}^3$, let ρ be its distance to the plane Π_ω, and x_2 its projection on Π_ω. Suppose that a grain placed at x_3 induces another one in Π_ω with a probability $\Pr(\rho)$. Summing up such events along a small cylinder C, normal to $\overline{\Pi_\omega}$ and with a base dx_2, we get:

$$\theta_2\, dx_2 = \theta_3 \int_C \Pr(\rho)\, dx_3 = \theta_3\, dx_2 \int_C \Pr(\rho)\, d\rho$$

(dx_2 and dx_3 are elementary volumes in Π_ω and $\mathbb{R}^3$ respectively). Note that $\int_C \Pr(\rho)\, d\rho$ is just the mathematical expectation $E[L(X'|\Delta_\omega)]$ of the projection of X' onto a normal to Π_ω. The proof for the passage from $\theta_3 \to \theta_1$ is similar.]

(b) Transpose the relationships VIII-37 to VIII-40 in terms of the Boolean model. Note that the covariance $C(h)$ is the same, measured in space or in section (restrict the question to isotropic X').

[The quantity C_0/q^2 of relation (XIII-9) is the same in $\mathbb{R}^3$, $\mathbb{R}^3 \cap \Pi$ or $\mathbb{R}^3 \cap \Delta$. Thus

$$\theta_3 K_{X'}(h) = \theta_2 K_{X' \cap \Pi}(h) = \theta_1 K_{X' \cap \Delta}(h).$$

Taking into account the result of part (a), we get for $h = 0$:

$$4E[V(X')] = E[L(X' \cap \Delta)] \cdot E[S(X')] \qquad (\mathbb{R}^3 \to \mathbb{R}^2)$$

$$2\pi E[V(X')] = E[A(X' \cap \Pi)] \cdot E[M(X')] \qquad (\mathbb{R}^3 \to \mathbb{R}^1)$$

$$\pi E[A(X')] = E[A(X' \cap \Delta)] \cdot E[U(X')] \qquad (\mathbb{R}^2 \to \mathbb{R}^1)$$

If we now differentiate the covariances K before putting $h = 0$, we find:

$$2\pi^2 E[S(X')] = E[M(X')] \cdot E[U(X' \cap \Pi)]\, (\mathbb{R}^3 \to R^2)]$$

Why are these results also valid for the individual analysis of Chapter VIII, Section E.4?

I.5. The Boolean model in one dimension (G. Matheron, 1969)

In $\mathbb{R}^1$, the convex primary grains are segments; let F denote their size distribution in number. After a Booleanization with a germ density θ, the size distribution of the grains and of the pores are F_1 and F_0 respectively.

(a) Show that $1 - F(h) = -K'(h)$, and that $K'(0) = -1$.

(b) Put $\Gamma(h) = \int_0^h [1 - F(x)]\, dx = K(0) - K(h)$. Show by a direct proof that $\Pr\{x + h \in X^C | x \in X^C\} = \exp[-\theta\Gamma(h)]$.

[Given there are n left ends of primary grains on $[0, h]$ {Prob. $[(\theta h)^n/n!]$ {exp $(-\theta h)$} }, then their distribution on $[0, h]$ is uniform. The probability that a given one of these primary grains misses the point h is, $(1/h) \int_0^h F(y) dy$. Hence:

$$\Pr(x + h \in X^C | x \in X^C) = \sum_n \frac{(\theta h)^n}{n!} \exp(-\theta h)$$

$$\times \left[\frac{1}{h} \int_0^h F(y) dy \right]^n = \exp\{(-\theta \Gamma(h)\}$$

(c) Prove that $\Pr(h \in X | 0 \in X^C) = 1 - \exp[-\theta \Gamma(h)]$, satisfies:

$$1 - \exp[-\theta \Gamma(h)] = \int_0^h \theta \exp[-\theta \Gamma(h - x)][1 - F(x)] dx$$

(d) Let Φ_1 and R be the Laplace transforms of f_1 and $\exp[-\theta \Gamma(h)]$ respectively. Show that if $m = \Gamma(\infty)$ is the mean length of the primary grain, the mean length m_1 of the Boolean grain is $m_1 = (1/\theta)[\exp(\theta m) - 1]$.

$$\left[R = \frac{1}{\lambda + \theta(1 - \Phi_1)}; \quad \text{as } \lambda \to 0, \text{ the } \frac{1}{\lambda} \text{ term of } R(\lambda) \text{ is } \frac{\exp(-\theta m)}{\lambda}. \right]$$

(e) Show that the range a is equal to $\theta(q^2/p)(m_1^2 + \sigma_1^2)$

$$\left[\text{From } m_0 = \frac{1}{\theta} \text{ and } m_1 = \frac{\exp(\theta m) - 1}{\theta}, \right.$$

$$\text{we have } q = \exp(-\theta m) = \frac{1}{1 + \theta m_1}$$

Hence

$$a = \frac{2}{pq} \int_0^\infty [C_{00}(h) - q^2] dh = \frac{2}{p} \int_0^\infty \{\exp[-\theta \Gamma(h)] - \exp(-\theta m)\} dh;$$

thus:

$$a = \frac{2}{p} \lim_{\lambda \to 0} \left[R(\lambda) - \frac{\exp(-\theta m)}{\lambda} \right] = \theta \frac{q^2}{p}(m_1^2 + \sigma_1^2) \right]$$

(f) Find a primary size distribution F, so that the Boolean set is a two state Markov process. (Note the paradox: we wish to recreate the lack of memory property of the Markov process by Booleanizing a primary random grain which does not have exponential length and is thus not Markov.)
[We need $\Phi_1 = a/(a + \lambda)$. Then the results of part e imply

$R = (a + \lambda)/\lambda \cdot 1/(a + \theta + \lambda)$; hence

$$\exp[-\theta \Gamma(h)] = \frac{a}{a+\theta} + \frac{\theta}{a+\theta} \exp[-(a+\theta)h].$$

Taking the logarithms and differentiating, we obtain:

$$1 - F(h) = \frac{(a+\theta)\exp[-(a+\theta)h]}{a + \theta \exp[-(a+\theta)h]}$$

A solution is indeed possible, giving the well known "logistic" law.]

I.6. Boolean model in $\mathbb{R}^3$: connectivity number (J. Serra, 1973)

Calculate the specific connectivity number N'_V of a 3-D Boolean model (convex primary grains without preferential orientation) by using the algorithms associated with the cubic grid of spacing (a); i.e.:

$$N'_V = \lim_{a \to 0} \frac{-Q\begin{bmatrix} \cdot \end{bmatrix} + 3Q\begin{bmatrix} 1 & 1 \\ 1 & 1 \end{bmatrix} - 3Q\begin{bmatrix} 1 & 1 \end{bmatrix} + q}{a^3}$$

where $Q[\cdot]$ is the probability that the cube (respectively the square, the segment) of size a is contained in the pores X^C of the model X. Compare the result with formula (XIII-28) and interpret.

[The cubic grid algorithm gives:

$$N'_V = \theta \exp[-E(V)]\left\{1 - \frac{3}{8\pi}\theta E(S) \cdot E(M) + \frac{3}{32}\theta^2 [E(S)]^3\right\}$$

The coefficients $1/4\pi$ and $\pi/96$ of relation XIII-28 have become $3/8\pi$ and $3/32$ respectively. Digitalization in 3-D has not relieved the problems encountered in two dimensions.]

I.7. Dead leaves tesselation (G. Matheron, 1969)

7.1. Intercepts distribution
Consider the intersection of the Dead leaves tesselation with straight line. We assume that we can distinguish between the disk-type: $\overline{}\!-\!\overline{}$, the scale-type: $\overline{}\!-\!\_$, and the relief-type: $\_\!-\!\_$, intercepts. Denote their respective specific number by $N_{L,d}$, $N_{L,s}$ and $N_{L,r}$; and their probability densities (in number) by f_d, f_s and f_r. Also denote by p_1 the probability that a given point x belongs to an intercept of type $i (i = d, s$ or $r)$ (i.e. the probability in measure), and by $\bar{\omega}_i = N_{L,i}/N_L$ the probability that a given intercept is of type i (probability in number).

(a) Show that by putting $K(O) = m$ and $K'(O) = -1$ in formulae (XIII-39 to XIII-41), we can work directly in $\mathbb{R}^1$. (Let $f(h)$ denote the probability density of the primary grain, with mean m).

(b) Prove that:

$$N_{L,r} \cdot f_r(h) = \frac{f(h)}{h+m} = \frac{K''(h)}{h+m}. \quad \text{Note the bias} \qquad (1)$$

Derive the relation $p_r + 2\bar{\omega}_r = 1$ (the intercepts of type r occupy a large proportion of the straight line, but they are less numerous than the intercepts of the other two types).

[A relief intercept with an origin in (O, dh_1) and an end in $(h, h+dh_2)$ occurs if: this configuration was already realized at time $-\delta t$, and has not been removed during the interval $-\delta t$ to O; or if: a primary grain of length h arrived at the origin between $-\delta t$ and O. Hence:

$$N_{L,r} f_r(h) = N_{L,r} f_r(h)[1 - \theta \delta t(h+m)] + \theta \delta t f(h)$$

By integration, we derive relation (1), and both

$$N_{L,r} = \int_0^\infty \frac{f(h)dh}{h+m}, \quad \text{and} \quad N_{L,r}(m_r + m) = 1 = p_r + 2\bar{\omega}r.]$$

(c) Prove that the disk size distribution satisfies:

$$N_{L,d} f_d(h) = 2\frac{K(h)}{(h+m)^3} \qquad (2)$$

[Consider the event: at time $-t$ a primary grain covered the segment of length h (Prob. $\theta dt K(h)$), then the segment has been missed by all later grains {Prob. $\exp[-\theta t(h+m)]$}, and, between $-t$ and O, a right extremity occurred in $(-dh_1, 0)$ and a left one in $(h, h+dh_2)$ (Prob. $\theta^2 t^2 dh_1 dh_2$). Hence:

$$N_{L,d} f_d(h) = \theta^3 K(h) \int_0^\infty t^2 \exp[-\theta t(h+m)]\, dt = \frac{2K(h)}{(h+m)^3}]$$

(d) Show that the number of disk and relief intercepts are the same.
[Integrate relation (1) and (2) by parts, to give:

$$N_{L,d} = N_{L,r} = \frac{1}{2m_0} + \int_0^\infty \frac{K'(h)}{(h+m)^2}\, dh.]$$

(e) Prove that the scale intercept density function satisfies $N_{L,s} f_s = -2K'(h)/(h+m)^2$.
[The proof is either by direct means, or by noticing that $\sum N_{L,i} f_i = N_L f_0$ (sum over $i = d, r, s$) and that $N_L f_0$ is the second derivative of $P_0(h)$, given by (XIII-39).]

7.2. *The connectivity number*

In this exercise, we calculate the connectivity number N_A resulting from a rectangular digitalization of the tesselation (the estimate is biased). From Exercise VI-4, we know that:

$$N_A = \lim_{\substack{a \to 0 \\ b \to 0}} \frac{Q\left(a\,\boxed{}^{b}\right) - Q\left(\,\boxed{}_a\,\right) - Q\left(\bullet\!\!-\!\!\overset{b}{-}\!\!-\!\!\bullet\right) + 1}{ab} \tag{1}$$

(a) Suppose that the primary grain is convex and without angular points. Prove that the probability $Q(a, b) = Q\left(a\,\boxed{}^{b}\right)$, that a small rectangle (a, b) parallel to the x and y axes is contained in just one class of the tesselation, is:

$$Q(a, b) = \frac{E[A(X')] - aK'_x(O) - bK'_y(O) + ab}{E[A(X')] + aK'_x(O) + bK'_y(O) + ab} \tag{2}$$

[Use Steiner's formula for the erosions and dilations (Ch. V), and apply relation (XIII-38).]

(b) Derive from (1) and (2) the expression for N_A. What happens in the isotropic case?

[(1) and (2) imply $N_A = [4K'_x(O)\cdot K'_y(O)]/[K(O)]^2$; in particular, when $K'_x(O) = K'_y(O) = (1/\pi)E[U(X')]$ we get $N_A = 4/m^2 = 1/m^2$. In words, the specific connectivity number is the inverse of the square of the mean intercept of the tesselation.]

I.8. A relief dead leaves model (G. Matheron, 1969)

Consider a relief dead leaves process (Sect. C.2b), where the primary grain is the random disk whose radius has a distribution function $\Phi(\cdot)$. The corresponding moments are $\mu_n = \int_0^\infty r^n \Phi(dr)$. Let $F(\cdot)$ denote the distribution function of the radius of the relief grain (with moments m_n).

(a) Starting from relation XIII-45, show that

$$\mu_n = \pi N_A (\mu_2 m_n + 2\mu_1 m_{n+1} + m_{n+2}) \tag{1}$$

(b) Derive a method for determining the original law Φ from N_A and $F(dr)$ (which are quantities actually measured).

[Using $m_1, \ldots, m_4$, relation (1) with $n = 1$ and $n = 2$ allow μ_1 and μ_2 to be estimated; and with $n = 3$, this estimation can be checked (i.e. tested to see whether the model correctly fits the phenomenon under study). Then (XIII-45) implies:

$$\Phi(dr) = \pi N_A [\mu_2 + 2\mu_1 r + r^2] F(dr).] \tag{2}$$

(c) Denoting the covariogram of the unit disk by $g(h)$, and by $P(h)$ the area proportion of the set after a linear erosion by a segment of length h, prove that:

$$P(h) = N_A \int_0^\infty r^2 g\left(\frac{h}{r}\right) \Phi(dr) \tag{3}$$

[Apply XIII-43. Note that this relation (3) can be used to test the model, since $P(h)$ is directly estimable from the image and $\Phi(dr)$ calculable from relation (2). Note also that these relief intercept distributions do not coincide with those of Exercise XIII-7.]

I.9. Multi-dead leaves model (D. Jeulin, 1979)

This model is constructed in a sequential way, like the dead leaves model, but with three differences:

(α) There are n different species of leaves, labelled $i\ (1 \ldots i \ldots n)$.
(β) At time $t = 0$, the space is empty and the process starts; we study it not only for $t = \infty$, but also after a finite deposit time t.
(γ) The leaves falling between t and $t + dt$ satisfy elementary Boolean models $X_i(t, t + dt)$ with densities $\theta_i(t)dt$ and primary grains $X_i'(t)$. They hide the former layers; define phase $Y_i(t)\ (i = 1 \ldots n)$ to be the *union* of all the leaves of type i at time t (and no longer the boundaries as in the dead leaves tessellation (C.2a)); let the background be denoted $Y_{n+1}(t)$ (i.e. $[\bigcup_i Y_i(t)]^c$). Note that the model turns out to be a Markov process, where the variable at time t is a stationary random set (as in Ex. 10).

(a) $K_i(h, t)$ denotes the geometric covariogram of $X_i'(t)$ and

$$f_i(h, t) = \theta_i(t)[2K_i(o, t) - K_i(h, t)]; \qquad f(h, t) = \sum_i f_i(h, t)$$

Prove that the proportion $p_i(t)$ of phase i is given by:

$$p_i(t) = \left[\exp - \int_0^t f(o, u)du \right] \times \int_0^t \frac{f_i(o, u)du}{\exp - \int_0^u f(o, v)dv} \tag{1}$$

[At time $u\ (0 \le u \le t)$ the origin O has been covered by i leaves, and by nothing between u and t.]
(b) When i varies from 1 to n and j varies from 1 to $n + 1$, prove that the covariances $C_{ij}(h, t)$ are given by the expressions:

$$C_{ij}(h, t) = \left[\exp - \int_0^t f(h, u)du \right] \int_0^t \left(\exp \int_0^u f(h, v)dv \right) A_{ij}(u)du \tag{2}$$

where $A_{ii}(u) = 2p_i(u)[K_i(o, u) - K_i(h, u)] + \theta_i(u)K_i(h, u)$.

$$(j \neq i, j \neq n+1); \; A_{i,j}(u) = p_i(u)[K_j(o, u) - K_j(h, u)]$$
$$+ p_j(u)[K_i(o, u) - K_i(h, u)]$$
$$A_{i,n+1}(u) = p_{n+1}(u)[K_i(o, u) - K_i(h, u)]$$

[for $C_{ii}(h, t)$: either both points O and h are covered by i leaves at time u, and by nothing afterwards, or one point only was already covered at time u, and only the second point is i-covered; no more leaves cover them between u and t. The other covariances are processed in the same way].

(c) What about the background $X_{n+1}(t)$?

[It is the background of a Boolean model with a covariance $\exp - \int_0^t f(h, u)du$]

(d) In what way are covariances (2) modified if the space is not empty at time $t = 0$, but is a $n+1$-phase stationary partition with covariances $C_{ij}(h, 0)$?
[A term $C_{ij}(h, 0)\exp - \int_0^{-t} f(h, u)du$ must be added; the two points O and h were in the proper configuration at $t = 0$, and have not been changed later.]

(e) Show that the model defined by relation $Y_i(t+dt) = Y_i(t) \cup [X_i(t, t+dt) \cap X_{n+1}(t)]$ is identical to the preceding [it is the same when seen from below!].

(f) Calculate the $C_{ij}(h)$ when the $\theta_i(t)$ and $K_i(h, t)$ are constant functions of the time.
[For $j \neq i \neq n+1$, we have

$$C_{ij}(h, t) = \alpha_{ij}\left[\frac{[1-\exp(-f(h)t)]f(o) - [1-\exp(-f(o)t)]f(h)}{f(o)f(h)[f(o)-f(h)]}\right]$$

with
$$\alpha_{ij} = f_i(o)f_j(h) + f_j(o)\cdot f_i(h) - 2f_i(o)\cdot f_j(o)$$

The ratios of covariances C_{ij} are *independent* of t. When $t = \infty$, $C_{ij} = \alpha_{ij}/f(o)f(h)$. If all the primary grains have the same distribution, the final structure $(t = \infty)$ is made up of n phases statistically identical. For example if $n = 2$, the closures of the blacks and the whites represent the *same* random set. The associated covariance is then:

$$C_{12}(h, \infty) = 1 - \frac{K(o)}{2K(o) - K(h)} \quad \text{(independent of } \theta\text{).]}$$

I.10. Hierarchical models (J. Serra, 1974)

By hierarchical models, we mean the following. During the time interval $(0, dt)$, we generate an initial Boolean model with density $\theta\,dt$ and primary grain X'_0; at the instant dt we generate a second elementary process that will interact with the first, etc. . . . . The resulting set at time t is denoted by X_t. We shall study two particular cases:

(i) Any primary grain X'_{t+dt} appearing during the interval $[t, t+dt]$ is eliminated if it touches X_t; this will generate a random set with separate grains (Fig. XIII. 26a).

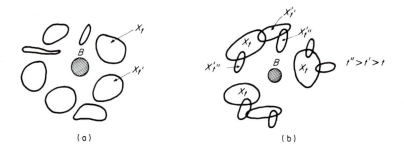

Figure XIII.26. Two types of hierarchical models.

(ii) Each primary grain X'_{t+dt} that *does not* touch X_t is suppressed, if not we take its union with X_t; this will generate a random set with clusters of interpenetrating crystals (as in a gypsum crystal for example).

(a) We first assume that X'_t is deterministic. Prove that the probability $Q_t(B)$ satisfies the relationship:

$$Q'_t(B) = -\theta \operatorname{Mes}(B \oplus \check{X}'_t) \cdot Q_t(X'_t) \qquad \text{(separate grains)} \qquad (1)$$

$$Q'_t(B) = -\theta \operatorname{Mes}(B \oplus \check{X}'_t) \cdot [1 - Q_t(X'_t)] \quad \text{(gypsum crystal model)} \qquad (2)$$

$(Q'_t = \text{derivative with respect to } t)$

[Separate grains: B is contained in the pores at time $t+dt$ (Prob. $Q_{t+dt}(B)$) if it is contained in the pores at time t (Prob. $Q_t(B)$) and if, in every $dx \in B \oplus \check{X}'_t$, no grain appears (Prob. $1 - \theta \, dx \, dt$) or one grain hits X_t (Prob. $\theta \, dt \, dx \, [1 - Q_t(X'_t)]$). In other words:

$$Q_{t+dt}(B) = Q_t(B) \exp(-\theta \, dt) \int_{\check{X}' \oplus B} Q_t(X') \, dx$$

which is just relation (1). The proof is the same for the gypsum crystal model.]

(b) Assume now that X'_t is random, and that its distribution in size and shape is governed by the law $F_t(d\lambda_1, \ldots d\lambda_n)$ where $\lambda_1 \ldots \lambda_n$ are n random parameters associated with X'_t (size, orientation, elongation . . .). Prove that:

$$Q'_t(B) = -\theta \int_{\lambda_1} \cdots \int_{\lambda_n} Q_t[X'_t(\lambda_1, \ldots, \lambda_n)]$$

$$\times F_t(d\lambda_1, \ldots, d\lambda_n) E[\text{Mes}(\check{X}'_t \oplus B)] \text{ (separate grain)} \qquad (3)$$

with a similar equation for the gypsum crystal model, obtained by replacing $Q_t(X')$ by $1 - Q_t(X')$, in the above integral.

[Divide the elementary Boolean model at time t into sub-classes, each corresponding to a set of given values for the λ_i's. For each sub-class, relation (1) can be applied. The number of sub-classes may be arbitrarily enlarged (infinite divisibility). Finally, superimpose the resulting models, by weighting with the probability $F_t(d\lambda_1 \ldots d\lambda_n)$ of the λ_i's relation (3) results.]

(c) Apply relation (3) to the case where X'_t is a random disk of radius r, with density $f_t(r)$, and take B to be the disk of radius R.

$$\left[Q'_t(R) = -\pi\theta \int_0^\infty Q_t(r) f_t(r)(r + R)^2 \, dr \right]$$

(d) Now assume that there are only two steps of the hierarchy. Starting from a first Boolean model (θ_1, X'_1), we generate a second one (θ_2, X'_2), and allow them to interact in one of the two ways described above. Calculate the probability $Q(B)$ of the resulting random set.

[The same proof as in (a) shows that for the separate grains model,

$$Q(B) = Q_1(B)\exp\{-\theta_2 Q_1(X'_2)E[\text{Mes}(B \oplus X'_2)]\}$$
$$= \exp\{-\theta_1 E[\text{Mes}(\check{X}'_1 \oplus B)] - \theta_2 E[\text{Mes}(\check{X}'_2 \oplus B)]$$
$$\times \exp[-\theta_1 E(\text{Mes}(X'_1 \oplus X'_2))]\}$$

(for the gypsum crystal model, replace $\exp[\ ]$ by $1 - \exp[\ ]$). Here the "separate grain" model is in fact a three phased model $(X_1, X_2$ and the background), where X_1 and X_2 are *disjoint* from each other.]

I.11. Superimposed tesselations (P. Delfiner, 1970)

The result proved in this exercise exhibits the existence of a domain of attraction for the Boolean model.

(a) Consider an arbitrary stationary random tesselation X of the space $\mathbb{R}^2$ or $\mathbb{R}^3$ where each class X' is almost surely bounded. If $N(B_x)$ denotes the random number of classes X' which hit B_x, prove that:

$$E[\text{Mes}(X' \oplus \check{B})] = E[N(B_x)] \cdot E[\text{Mes } X'] \qquad (1)$$

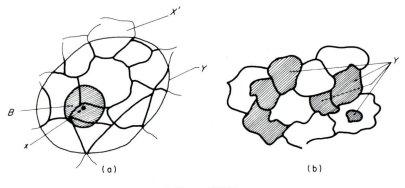

Figure XIII.27.

[Consider the intersection of a realization of X with a very large set Z. Ergodicity implies that:

$$E[N(B_x)] = \lim_{\text{Mes } Z \to \infty} \left[\frac{1}{\text{Mes } Z} \int_Z N(B_x)\,dx \right]$$

Label all the classes X' of the realization by an index $i \in I$. Then:

$$N(B_x) = \sum_{i \in I} v_i \quad \text{where} \quad \begin{cases} v_i = 1, & \text{if } B_x \Uparrow X_i' \\ v_i = 0, & \text{otherwise} \end{cases}$$

Or, equivalently, $N(B_x) = \sum_{i \in I} k_{X_i' \oplus \check{B}}(x)$ ($k_{X_i' \oplus \check{B}}$ is the indicator function of $\dot{X}_i' \oplus \check{B}$).
Thus

$$\int_Z N(B_x)\,dx = \sum_{i \in I} \int_Z k(x)_{X \oplus B}\,dx = \sum_{i \in I} \text{Mes}\left[(X_i' \oplus \check{B}) \cap Z \right]$$

Thus, as $\text{Mes } Z \to \infty$,

$$\sum_{i \in I} \text{Mes}\left[(X_i' \oplus \check{B}) \cap Z \right] \to E[N(Z)] \cdot E[\text{Mes}(X' \oplus \check{B})]$$

$$\to \frac{\text{Mes } Z}{E[\text{Mes } X']} \cdot E[\text{Mes}(X' \oplus \check{B})]$$

Relation (1) results.]

(b) Starting from a tesselation X, construct a two phased structure (i.e. a set) by assigning each class $X' \subset X$ to the phase "grains" with probability p, independently from one class to another. The topological closure of the classes assigned to the grains is a closed random set Y (Fig. XIII. 27b). Now $Q(B)$ denotes the probability that the compact set B is contained in the pores Y^C, in particular $Q\{O\} = 1 - p = q$.

Now superimpose n realizations of Y, and let Y_n denote the union of the grains of the n realizations. Y_n is in turn a stationary random closed set, with functional probabilities $Q_n(B)$ and with a porosity $Q_n\{O\} = q_n$.

(i) Prove that $Q_n(B) = [Q(B)]^n$, and that $Q \to 0$ as $n \to \infty$, except if $n \to \infty$, and $p \to 0$ in such a way that the product $np \to \theta (0 < \theta < \infty)$.

(ii) Assuming that $np \to \theta$ as $n \to \infty$, calculate the moments $Q_\infty(B)$ of the limit random set Y_∞. Prove that:

$$\lim_{n \to \infty} Q_n(B) = \exp\{-\theta E[N(B)]\} \tag{2}$$

where $N(B)$ is the number of classes of $X \cap B$.

[Let $\omega_v (v = 1, 2 \ldots)$ be the probability that $N(B) = v$, and let $G(s) = \sum_1^\infty \omega_v - s^v$ be the corresponding generating function. Given $N(B)$, the probability that $B \subset Y$ is $q^{N(B)}$, since the $N(B)$ classes hit by B must be pores. Unconditioning over v we find:

$$Q(B) = \sum_{N=1}^\infty \varpi_N q^N = G(q)$$

As $q \to 1 (p \to 0)$, $G(q) = 1 - pG'(1) + o(p)$, and noting that $G'(1) = E[N(B)]$, we obtain $G(q) = 1 - pE[N(B)] + o(p)]$. Now for $Q_n(B)$, we have:

$$\log[Q_n(B)] = n \log[Q(B)] = n \log[1 - pE(N(B)) + o(p)]$$

$$\sim -npE(N_B).$$

Taking the limit as $n \to \infty$, we find relation (2).]

(c) Prove that the limit random set Y_∞ is a Boolean set with a primary grain X' and a density $\theta' = \theta[\text{Mes } X']^{-1}$.

[According to Choquet's Theorem XIII, two random sets having the same functional probabilities, can be identified as the same. It only remains then to replace $E[N(B)]$ by its value given by relation (1). In particular, when X' is a Poisson tesselation, Y_∞ is the Poisson Boolean model.]

I.12. The Boole–Poisson Model (F. Conrad, 1971)

(a) Although F. Conrad constructed his model to describe aerial photographs of geological cracks, by using Poisson polygons and Boolean fissures, his "mixed" random set is extremely general. For simplicity, we will present it as it was originally. Start from an isotropic Poisson lines process in $\mathbb{R}^2$, with density λ, and denote it as X_1. Intersect each polygon Π of the tesselation with a stationary random set X_2, in the following special way. For each polygon of the given realization of X_1, use a *different* realization

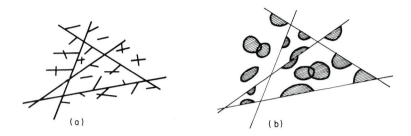

Figure XIII.28. Conrad's model.

X_2^* of X_2. The union of all these portions of X_2, together with X_1, make up the set we shall call Y. Take X_2 to be a Boolean model of segments (Fig. XIII.28a) with primary grain X_2' and density θ. If $Q_1(B)$, $Q_2(B)$ and $Q(B)$ denote the functional probabilities associated with X_1, X_2 and Y respectively, show that:

$$Q(B) = Q_1(B) \cdot Q_2(B) = \exp\{-\lambda U(B) - \theta E[A(X_2' \oplus \check{B})]\} \qquad (1)$$

where B is a connected compact set.

[When we know that B misses X_1 (i.e. is contained in a *single* polygon Π), it is equivalent to knowing that B misses $\Pi \cap X_2(\Pi)$, or to saying that B misses X_2:

$$Q(B) = \Pr\{B \subset Y^c\} = \Pr\{B \subset X_1^c, B \subset X_2^c\}$$
$$= \Pr\{B \subset X_1^c\} \cdot \Pr\{B \subset X_2^c \mid B \subset X_1^c\}$$
$$= Q_1(B) \cdot \Pr\{B \subset X_2^{*c} \mid B \subset X_1^c\}$$

Since X_1 and X_2^* are independent, formula (1) results.

Note that the first equality of relation (1) does not depend on the choice of the tesselation X_1 nor of the random set X_2. Another example is given in Chapter IX, Figure IX.12b.].

(b) Prove that formula (1) remains valid if we simply take the union of a realization of X_1 with a realization of X_2. The result seems to be in contradiction with Choquet's Theorem (XIII-5); the two random sets, namely Y as defined in a, and $X_1 \cup X_2$, exhibit the same functional probabilities $Q(B)$, but are clearly different. What is the explanation?

[In both models, relation (1) is satisfied only for B, a *connected* compact set. Thus, there is no contradiction with Choquet's theorem.]

I.13. Poisson lines in the plane
(R. E. Miles, 1962–G. Matheron, 1969)

We recall the definition of Poisson lines in $\mathbb{R}^2$. A moving axis Oz turns around the fixed origin O. When it sweeps the directions $(\alpha, \alpha + d\alpha)$, we realize on it an elementary Poisson point process, with a density $\lambda\,d\alpha$, where λ is a constant, and draw the straight lines normal to Oz going through the Poisson points. We then repeat the same operation, with independent Poisson point realizations, for all the directions α, obtaining a random tesselation X of the plane into convex polygons. Since λ does not depend on α, the tesselation is isotropic.

(a) Show that the point process induced by X on an arbitrary line Δ is Poisson with density 2λ.

   [The straight lines of X having a direction $\in (\alpha, \alpha + d\alpha)$ induce an elementary Poisson point process with density $\lambda \sin \alpha\,d\alpha$. After integration, we get

$$\int_0^\pi \lambda \sin \alpha\,d\alpha = 2\lambda]$$

(b) Consider the axis Oz turning around O, and for each orientation $\alpha + \pi/2$ of Oz, look at its intersection with the Poisson lines of direction $\in (\alpha, \alpha + d\alpha)$. When α varies we then generate a point process. Is it Poisson?

   [No! The point density expressed in polar coordinates (ρ, α), is λ/ρ. Hence the point process is not stationary and has an accumulation point at the origin.]

(c) Can we define Poisson lines by starting from Poisson points in $\mathbb{R}^2$, and through each point put a random line (where the directions of the lines are independently chosen from the uniform distribution)?

   [Definitely not! Calculate the number of lines hitting a given segment: it is almost surely infinite. This kind of approach, based on the Boolean model, necessarily requires the primary grains to be *bounded*.]

(d) Conditional upon the x-axis hitting a line of X between x and dx, what is the probability that this line has a direction $\in (\alpha, \alpha + d\alpha)$?

   [Divide the joint probability $\lambda \sin \alpha\,d\alpha\,dx$ by the probability $2\lambda\,dx$, that a line of X hits the segment $(x, x + dx)$. The conditional density $(\sin \alpha/2)$ $(O \le \alpha \le \pi)$ results. Note that the directions are not uniformly distributed on the interval $[O, \pi]$.]

(e) The average number N_A of polygons per unit area.

   Since the polygons are convex, the specific convexity number and the specific connectivity number coincide. Prove that this number is $N_A = \pi\lambda^2$.

   [Take a disk of radius R; assume that the disk hits two lines of X (a priori prob. $\frac{1}{2}(2\pi R\lambda)^2 \exp(-2\pi R\lambda)$. Show that the probability of the intersection of these two lines lying inside the disk, is 0.5: if the first intercept has length L, then the probability of the second line hitting it is $2L/2\pi R$ (Ex. X-11); also the conditional expectation of L is $\pi R/2$; we finally obtain 0.5. As $R \to 0$, the probability that the disk contains a vertex $\sim N_A \cdot \pi R^2$ by one method and also $\sim \frac{1}{2}(2\pi\lambda R)^2/2 = \pi^2\lambda^2 R^2$. Thus $N_A = \pi\lambda^2$.]

(f) Determine the number-weighted and the area-weighted means of the polygons areas ($E(A)$ and $\mathscr{M}(A)$ respectively).

$$[E(A) = (N_A)^{-1} = (\pi\lambda^2)^{-1}.$$

Let Y_0 be the polygon containing the origin. The probability that it also contains the infinitesimal zone $r\,dr\,d\alpha$, is $\exp(-2\lambda r)$. Thus

$$\mathscr{M}(A) = \int_0^{2\pi}\int_0^\infty \exp(-2\lambda r)r\,dr\,d\alpha = \pi/2\lambda^2$$

Notice that

$$\frac{\mathscr{M}(A)}{E(A)} - 1 = \frac{\operatorname{var} A}{[E(A)]^2} = \frac{\pi^2}{2} - 1 \simeq 4$$

which is the coefficient of variation (in number) of the polygon areas; it is quite high.]

(g) *The conditional distribution.*

Let U denote the perimeter of a Poisson polygon, and N its number of sides (or vertices). Conditional upon $N = n$, the densities of U are $f(u)$ (number-weighted polygons) and $g(u)$ (area-weighted polygons). Prove that:

$$g(u) = \lambda^n \frac{u^{n-1}}{(n-1)!}\exp(-\lambda u);$$

$$f(u) = \lambda^{n-2}\frac{u^{n-3}}{(n-3)!}\exp(-\lambda u) \qquad (n \geq 3) \tag{1}$$

[Consider the following event $\mathscr{E}$: "the polygon Y_0 of the Poisson tesselation with density $\lambda + \mu$, has n sides which all belong to the sub-tesselation with density λ." The *a priori* probability P_n that Y_0 has $N = n$ sides, does not depend on the density λ (self similarity of Poisson lines). Thus the probability of $\mathscr{E}$ is $[\lambda/(\lambda+\mu)]^n P_n$. Alternatively, this probability can be calculated by first conditioning on Y_0 having n sides, and then considering the probability that *this* polygon is not hit by any line of the random set with density μ; i.e.,

$$\Pr\{\mathscr{E}\} = P_n \int_0^\infty \exp(-\mu u)g(u)\,du,$$

so that $\int_0^\infty [\exp(-\mu u)]g(u)du = [\lambda/(\lambda+\mu)]^n$. Therefore, $g(u)$ is the function whose Laplace transform is $[\lambda/(\lambda+\mu)]^n$, namely the gamma density of relation (1). To find $f(u)$, we can either use the same approach by putting the origin at a vertex of the tesselation, or use the general formula (X, 27) which links laws in number with laws in measure.]

(h) Calculate the size distribution of the polygon, with respect to openings by disks.

[This size distribution is the probability that the origin belongs to $(Y_0)_{B(r)}$, or equivalently, the specific area of the opening of X^c by $B(r)$. We have:

$$\Pr\{O \in (Y_0)_{B(r)}\} = A_A[X^c_{B(r)}] = (1 + 2\pi\lambda r + \pi^2\lambda^2 r^2)\exp(-2\pi\lambda r)]$$

I.14 Poisson planes in $\mathbb{R}^3$
(R. E. Miles, 1967—G. Matheron, 1969)

As in $\mathbb{R}^2$, we construct Poisson planes X in R^3 by using a turning axis of direction ω. The planes of X with normal $\in(\omega, \omega + d\omega)$ (ω solid angle) induce on an axis of direction ω, a Poisson point process of density $\lambda_3\, d\omega$ (λ_3 constant) each direction being independent of the others. To avoid counting each plane twice, ω only ranges over the half unit sphere.

(a) Show that the number of planes hitting a compact convex set B, has a Poisson distribution with parameter $\lambda_3 M(B)$, ($M(B) = $ norm of B).

[The number of planes of direction ω hitting B is Poisson with parameter $\lambda_3 D^{(3)}(B, \omega)$. Where $D^{(3)}(B, \omega)$ is the width of B in direction ω. Integrate over all the directions and apply relation (V-14).]

(b) Show that on an arbitrary plane Π the planes of X induce a set of Poisson lines with density $\lambda_2 = (\pi/2)\lambda_3$.

[Let B be a planar compact convex set embedded in $\mathbb{R}^3$; according to relation V-15, its norm (in $\mathbb{R}^3$)is $\pi/2$ times its perimeter (in $\mathbb{R}^2$). The probability that a plane of the tesselation hits B is also the probability that the line intersected by this plane on Π, hits B.]

(c) Given that a plane of X hits the x-axis at x, show that the probability that this plane has an angle ω wih the x-axis, is $\sin 2\omega\, d\omega$ ($0 \le \omega \le \pi/2$).

(d) Let Y' and Y be two compact convex sets such that $Y' \subset Y$ and with norms M' and M respectively. Show that the probability that a plane of X hits Y', given that it hits Y, is M'/M.

(e) Prove that the density of polyhedra per unit volume is $N_V = (\pi/6)\lambda^3$.

[Given that the (random) planes X_1, X_2 and X_3 hit a ball $B(R)$, the probability that their common point lies inside the ball is $\pi^2/48$. The intersection $X_1 \cap B(R)$ is a disk C_1 of random radius $r = \sqrt{R^2 - x^2}$, which is uniformly distributed on $[0, R]$; thus $E(r^2) = \frac{2}{3}R^2$. Then impose the extra restrictions that X_2 and X_3 must hit C_1.]

(f) Calculate the mean v-weighted volume of the polyhedra $\mathcal{M}(V)$. Show that:

$$\frac{\mathcal{M}(V)}{F(V)} = \mathcal{M}(V) \cdot N_V = \frac{4\pi^2}{3} \simeq 13 \text{ and that } \frac{\text{var}(V)}{[E(V)]^2} = \frac{4\pi^2}{3} - 1 \simeq 12.$$

[Start from the probability $P(l)$ and integrate it over the space.]

(g) Calculate the size distribution of the polyhedra with respect to openings by $B(r)$.

$$\left[V[X^c_{B(r)}] = \exp(-4\pi\lambda_3 r)\left(1 - 4\pi\lambda_3 r + \frac{\pi^4}{6}\lambda_3^2 r^2 + \frac{2}{9}\pi^5\lambda_3^3 r^3 \right) \right]$$

(h) What happens when the density $\lambda_3(\omega)$ depends on ω; i.e. when the process is not isotropic?

[In (a) the parameter $\lambda_3 M(B)$ must be replaced by $\frac{1}{2}\int_\Omega \lambda_3(d\omega)D^{(3)}(\omega)$, etc.]

I.15. Poisson lines in $\mathbb{R}^3$ (id.)

Poisson lines in $\mathbb{R}^3$ are as follows: the lines having a direction $\in(\omega, \omega + d\omega)$ induce on a plane normal to ω, a Poisson point process with density λ_3 (independent of ω). Each direction is independent of the others. Prove that the number of lines hitting a compact convex set B is a Poisson variable with parameter $(\pi/2)\lambda_3 S(V)$. Is there a property similar to conditional invariance?

[Apply Cauchy's formula. Conditional invariance has no equivalent here.]

I.16. Random measures for Poisson lines (G. Matheron, 1970)

Consider in $\mathbb{R}^2$ the isotropic Poisson lines process X, with density λ. At each point x, we define the stationary random function $\mu(B_x) =$ "total length of the segments of $X \cap B_x$" where B is a given compact set with indicator function $k_B(x)$ and with geometric covariogram $K(h) = \int_{\mathbb{R}^2} k_B(x) \cdot k_B(x+h)dx$. We wish to calculate the mean, the variance and the covariance of the function $\mu(B_x)$. In order to solve the problem, define the measure μ associated with X as follows: μ is obtained by uniformly applying a unit mass per unit length to each line of the set X. Then $\mu(B_x)$ is just the regularized version of μ after taking a moving average with respect to $B(\mu(B) = \mu * k_B)$.

(a) Work in polar coordinates (ρ, α). Show that the Laplace transform associated with the infinitesimal zone $(\rho, \rho + d\rho; \alpha, \alpha + d\alpha)$, is $\exp\{-\lambda d\rho\, d\alpha[1 - e^{-\eta\Phi(\rho, \alpha)}]\}$, where η is the dummy variable, and $\Phi(\rho, \alpha) = \int_{D(\rho, \alpha)} k_B(r)dr$, is the integral of k_B along the line perpendicular to α, at the point (ρ, α). Then integrate over ρ and α.

[The Prob. that a Poisson line $D(\rho, \alpha)$ is in $(\rho, \rho + d\rho; \alpha, \alpha + d\alpha)$, is $\lambda d\rho\, d\alpha$; the associated (infinitesimal) Laplace transform is:

$$1 - \lambda d\rho\, d\alpha + \lambda d\rho\, d\alpha \exp\left[-\eta\Phi(\rho, \alpha)\right] \sim \exp\left\{-\lambda d\rho\, d\alpha[1 - e^{-\eta\Phi(\rho, \alpha)}]\right\}$$

To obtain the Laplace transforms $d\Gamma_\alpha(\eta)$ associated with all the lines perpendicular to α, multiply the (infinitesimal) transforms (Poisson independence):

$$d\Gamma_\alpha(\eta) = \exp\left\{-\lambda d\alpha \int_{-\infty}^{+\infty} \{1 - \exp[-\eta\Phi(\rho, \alpha)]\}d\rho\right\}$$

Similarly, we obtain the Laplace transform $\Gamma(\eta)$ of X, by multiplying the $d\Gamma_\alpha(\eta)$ associated with each direction, finally giving:

$$\Gamma(\eta) = \exp\left\{-\lambda \int_0^\pi d\alpha \int_{-\infty}^{+\infty} \{1 - \exp[-\eta\Phi(\rho, \alpha)]\}d\rho\right\}$$

(b) $\Gamma(\eta)$ is the Laplace transform of the random variable $\mu(B)$ (B fixed). Derive the mean and the variance of $\mu(B)$ via:

$$E[\mu(B)] = -[\log \Gamma(\eta)]^{(i)}_{\eta=0} \quad \mathrm{Var}[\mu(B)] = -[\log \Gamma(\eta)]^{(ii)}_{\eta=0}$$

Interpret the results in terms of random measures.
[The mean $E[\mu(B)]$ is:

$$E[\mu(B)] = \lambda \int_0^\pi d\alpha \int_{-\infty}^{+\infty} \Phi(\rho, \alpha) d\rho$$

$$= \lambda \int_0^\pi d\alpha \int_{-\infty}^{+\infty} \left[\int_{D(\alpha, \rho)} k_B(r) dr\right] d\rho = \lambda\pi \int_{\mathbf{R}^2} k_B(x) dx.$$

$E[\mu(B)] = \lambda\pi A(B)$ and the mean-measure is $m(dx) = \lambda\eta\, dx$.

For the variance, we have $\mathrm{Var}[\mu(B)] = \lambda \int_0^\pi d\alpha \int_{-\infty}^{+\infty} \Phi^2(\rho, \alpha)d\rho$; but

$$\int_{-\infty}^{+\infty} \Phi^2(\rho) d\rho = \int_{-\infty}^{+\infty} d\rho \int_{D(\rho, \alpha)} k_B(x_1)dx_1 \int_{D(\rho, \alpha)} k_B(x_2)dx_2$$

$$= \int_{D(\rho, \alpha)} dx_2 \int_{-\infty}^{+\infty} d\rho \int_{D(\rho, \alpha)} k_B(x_1)\cdot k_B(x_1 + x_2)dx_1$$

$$= \int_{D(O, \alpha)} K(x_2) dx_2 \quad (D(O, \alpha): \text{ line normal to } \alpha \text{ going}$$

through the origin).

Putting $x_2 = r$ and $\theta = \alpha + \pi/2$, and using the polar co-ordinates r, θ, we find:

$$\mathrm{var}[\mu(B)] = \lambda \int_0^\pi d\theta \int_{-\infty}^{+\infty} \frac{K(r)}{r} r\, dr = \lambda \int_{\mathbf{R}^2} \frac{K(x)dx}{r}$$

The covariance measure is $(\lambda/r) dr\, d\theta$. Notice that the function $1/r$ which could not be a covariance for a random function is perfectly acceptable for a random measure.]

I.17. Poisson slices in $\mathbb{R}^3$ (R. Miles, 1967, J. Serra, 1975)

Calculate the functional probabilities $Q(B)$ of the random set X obtained by dilating Poisson planes (density λ) with a ball of radius r.

(a) *Covariance*

Notice that the set X induces a Boolean model Y on the straight line going through points x and $x+h$; the primary grain is a segment. Calculate its geometric covariogram $K(h)$, and derive the covariance $C(h)$ of the pores of X.

[Let ω $(0 < \omega < \pi/2)$ be the direction of the Poisson plane whose intersection with the straight line produces a 1-D primary grain Y'. Its length $L(Y)'$, equals $2r/|\cos \omega|$, and ω has density $= \sin 2\omega$ on $(0, \pi/2)$ (Ex. XIII-14c). Thus:

$$K(h) = E[L(Y' \cap Y'_h)] = \int_0^{\pi/2} \left(\frac{2r}{\cos \omega} - |h| \right) \sin (2\omega)\, d\omega$$

$$= 4r - |h|; \quad \frac{2r}{|h|} \le 1$$

$$K(h) = E[L(Y' \cap Y'_h)] = \int_{\cos^{-1}(2r/h)}^{\pi/2} \left(\frac{2r}{\cos \omega} - |h| \right) \sin (2\omega)\, d\omega$$

$$= \frac{4r^2}{|h|}; \quad \frac{2r}{|h|} \le 1$$

The germ density on the straight line is $\pi\lambda$ (rel. XIII-49), so that the resulting pore covariance is finally:

$$C(h) = \exp\{ -\pi\lambda[2K(0) - K(h)] \}$$

$$= \begin{cases} \exp\{ -\pi\lambda[4r + |h|] \}; & 0 \le |h| \le 2r \\ \exp\{ -\pi\lambda[8r - (4r^2/|h|)] \}; & |h| \ge 2r \end{cases}$$

Note that:

(i) $(C(h)/C(0)) = \exp(-\pi\lambda|h|)$ is independent of r, for $|h| \le 2r$, and the slope $-\pi\lambda$ of the tangent at the origin allows estimation of λ.
(ii) $C(h)$ does not possess a finite range.
(iii) $C(0) = q = \exp(-4\pi\lambda r)$, allows estimation of r when λ is known.]

(b) *Linear probability $Q(l)$*

Using the conditional invariance, calculate the probability $Q(l)$, that a segment of length l is contained in the pores.

[Consider (in the initial tesselation) the polyhedron Y_0 containing the origin. The probability that Y_0 is not removed by spherical dilation is

$\exp(-4\pi\lambda r)$. Conditionally upon $Y_0 \ominus B(r) \neq \varnothing$, the probability that $Y_0 \ominus B(r)$ contains the segment ρ, is $\exp(-\lambda\pi)$. Therefore, $Q(l) = \exp\{-\lambda\pi[4r+l]\}$, which is exactly the covariance function for small $|h|$. This paradoxical result simply means that, if a grain appears between x and $x+h$, its length is necessarily $\geq 2r$. Notice that in order to judge the minimal thickness of the grains we use a *pore* erosion, and compare it to the *pore* covariance: the resulting criterion (i.e. $Q(l) \equiv C(l)$ for $l \leq l_0$) is more general and may be used for testing thickness independently of the above model.]

(c) *Erosion by a disk*

Prove that the probability that a disk of radius R is contained in the pores is $Q(R) = \exp\{-\lambda\pi[4r + \pi R]\}$.

[Same proof as in (b). It is also easy to obtain results for the pores *opened* by a disk.]

I.18. Random functions with a given variogram (J. Serra, 1969)

In order to simulate random structures, one often needs to start with realizations of random functions that possess an exponential or a linear variogram. The purpose of this exercise is to construct such functions from the Poisson lines process in $\mathbb{R}^2$.

(a) Let X be a random tesselation of $\mathbb{R}^2$ of Poisson polygons (density λ). Each polygon is independently given the state e_1 (Prob. p) or e_0 (Prob. $q = 1 - p$). Show that the random set induced on any straight line of the plane is a two state Markov process (see Ex. IX-15). Prove that the convexity number of the 2-D random set is $q^2 p\pi\lambda^2$.

(b) More generally, suppose that the number of states $e(t)$ is infinite, and that they are generated by a distribution with variance σ^2. The same procedure as in a) results in a random function $f(g)$. Show that its intersection with any straight line, generates a Markov process on the line, with (semi) variogram:

$$\gamma(h) = \tfrac{1}{2}E[f(y+h)-f(y)]^2 = \sigma^2[1 - \exp(-\lambda h)]$$

[The segment $[y, y+h]$ hits at least one boundary of the state with prob. $1 - \exp(-\lambda h)$. If this event occurs, then the two values $f(y+h)$ and $f(y)$ are independent, and their mean square difference is equal to $2\sigma^2$. Similarly, the covariance of the function is $\sigma^2 \exp(-\lambda h)$.]

(c) Return to the original realization X. Each line (labelled with an index i) divides the plane into two half planes which are given the two values $+z_i$ and $-z_i$ respectively; z_i's are independent and identically distributed as z is a random variable with mean m. The resulting value $f(y)$, at the point y is the

$\sum_i z_i$, summed over all lines of the process. Does the resulting function possess a covariance? Prove that its variogram is linear.

[Obviously $f(y) = \sum_i z_i$ is infinite. However $f(y+h) - f(y)$ is a Poisson random variable with parameter $m\lambda h$. The corresponding variogram is thus $\gamma(h) = \frac{1}{2}m\lambda h$.]

I.19. Triple edged tesselations

Prove that for any stationary tesselation in $\mathbb{R}^3$, where all the edges are triple (any edge is a common boundary of three classes), the specific number of grains f induced on a random section Π, is equal to the specific edge length L_V in $\mathbb{R}^3$, divided by four.

[The tesselation induces on Π a planar graph with f faces, e edges and v vertices per unit area. From Euler's relation $f + v = e$. Moreover, every edge is adjacent to two vertices and every vertex to three edges, thus $3v = 2e$, so that $2f = v$. According to the stereological formula (IV-36), v is precisely $L_V/2$.]

I.20 Random points on the unit sphere (J. Serra, 1974)

In metallography and geology, one often deals with orientations of crystals. Typically, one uses the technique of *stereographic projection*, which maps the northern hemisphere of the unit ball Ω (each point of Ω corresponds to a direction) onto the equatorial plane: if z_0 is the south pole and x a point in the northern hemisphere, then x is transformed into y, the intersection of the segment $[x, z_0]$ with the equatorial plane (Fig. XIII.29b). Poisson points on Ω are defined as in Euclidean space: the infinitesimal zone dx has one point with probability λdx, no points with probability $1 - \lambda dx$, and two disjoint zones are independent of each other. Assuming that the points are Poisson on the sphere, what can be said about their stereographic projection?

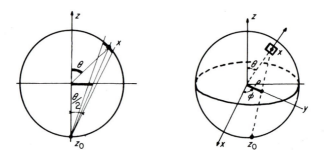

Figure XIII.29. (a) and (b) stereographic projection.

[First, suppose λ is a constant. The points falling on the surface element dx, around x, follow the Poisson distribution with parameter

$$d\lambda = \lambda\,dx = \lambda\,d\theta\,\sin\theta\,d\varphi$$

(θ, φ spherical co-ordinates of x (see Fig. XIII-29)). The induced process is obviously Poisson, but with a non constant density $\mu(\rho)$, satisfying the relation

$$d\lambda = \mu(\rho)\cdot\rho\,d\rho\,d\varphi \tag{1}$$

(ρ, φ = polar co-ordinates of the transform y of x). Noting that $\rho = \tan\theta/2$, we obtain:

$$\mu(\rho) = \frac{\lambda\,d\theta\,\sin\theta}{\rho\,d\rho} = \frac{4\lambda}{(1+\rho^2)^2} \tag{2}$$

Now consider the random number of points falling between the parallels of latitude, θ_1 and θ_2 ($\theta_2 > \theta_1$) and the meridians of longitude, φ_1 and φ_2 ($\varphi_2 > \varphi_1$). It is Poisson with parameter $\lambda(\varphi_2 - \varphi_1)(\cos\theta_1 - \cos\theta_2)$; the parameter M of the induced process of y's is therefore:

$$M(\varphi_1, \varphi_2, \theta_1, \theta_2) = \lambda(\varphi_2 - \varphi_1)\left[\frac{1-\rho_1^2}{1+\rho_1^2} - \frac{1-\rho_2^2}{1+\rho_2^2}\right]$$

More generally, if λ is a function $\lambda(\theta, \varphi)$ of the point x, equation (1) remains unchanged, and (2) becomes

$$\mu(\rho, \varphi) = \frac{4\lambda(2\arctan\rho, \varphi)}{(1+\rho^2)^2}$$

and

$$M(\varphi_1, \varphi_2, \rho_1, \rho_2) = 4\int_{\varphi_1}^{\varphi_2} dy \int_{\rho_1}^{\rho_2} \frac{\lambda(2\arctan\rho, \varphi)}{(1+\rho^2)^2}\,d\rho]$$

XIV. Modelling

The dialectic between criteria and models is far too difficult for us to resolve. In this concluding chapter, we would like to show how this discussion governs the questions of modelling. Although the subject itself could fill up a whole book, we will be satisfied with three glimpses. Firstly consider random set models: in our opinion, their major operational interest is to suggest new approaches to problems, to give ideas for criteria, etc. Such a *heuristic raison d'être* is contrasted with the more classical task of *prediction* using models (see Sect. A). Section B, very briefly, extends the question of modelling into the domain of mathematical models in morphology. We do not intend to give operational tools but simply to recall several models already studied in the book, in order to demonstrate the mathematical role played. We end with several modelling problems in pattern recognition (Sect. C). Here the notion of a model covers a broad spectrum and not all are equally accessible: the limits of mathematical morphology are reached!

A. JUDICIOUS USE OF PROBABILISTIC MODELS

A.1. Models and predicton

To construct a specific random set model for each new morphological problem is impossible (mathematical difficulties, incomplete knowledge of the actual physical process, etc. . . .). There is a danger here, since the physicist tends to use models which are already well worked, even when they are not really applicable to his problem. He should take care not to fall into this trap; Buffon's quotation (1752) is particularly apt here: "Lorsqu'on veut appliquer la géométrie et le calcul à des sujets de la physique trop compliqués, à des objets dont nous ne connaissons pas assez les propriétés pour pouvoir les mesurer, on est obligé dans tous ces cas de faire des suppositions toujours contraires à la Nature, de dépouiller le sujet de la plupart de ses qualités, d'en faire un être abstrait qui ne ressemble plus à l'être réel, et lorsqu'on a beaucoup raisonné et calculé sur les rapports et les propriétés de cet être abstrait, et qu'on est arrivé à une conclusion tout aussi abstraite, on croit avoir trouvé quelquechose de réel, et on transporte ce résultat idéal dans le sujet réel, ce qui produit une infinité de fausses conséquences et d'erreurs" (Buffon, *Hist. Nat., Discours d'intr.*, 1752).

In other words, prediction by using a random set model has a real meaning only if we do not try to deduce consequences too sophisticated for the hypotheses; as well, it must be possible to test it by sufficiently diverse experiments. Here we do not speak of "tests" in the statistical sense of the word (statistical inference for random sets is still in its infancy), but rather we construct falsifying experiments; that is, the emphasis is on proving false each hypothesis of the model in turn (Popper (1959)).

(a) Prediction aspects in Cauwe's model

Cauwe's model for the milling of rocks (analysed in Section E.1 of the last chapter) is a good example to take for discussion. Firstly, it demonstrates the idea of testing; i.e. of proving false. Cauwe based his study on the two criteria of linear and circular erosion. Measured on each specimen, they provided two experimental curves that he could match to their theoretical expressions under the hypotheses of Poisson planes or twin planes. The falsification of the first hypothesis, and the acceptance of the second one, resulted. Secondly, one is faced with the estimation of the parameters λ_2 and d_3 of the model. Treatments of both experimental curves provide two independent estimates of λ_2 and d_3; their respective ratios should be close to 1. This development is standard; what is not standard, is the decision to *ignore* the rock texture before milling:

(b) Implications of Cauwe's Law

It might seem rash to a petrologist that we have reached conclusions about relations between mineral texture and ease of milling, without even taking a look at the rocks. But should we? The milling circuit turns out to be an extraordinarily good ally in that it selects the "good" fissures (i.e. those that become active) from among the millions of microscopic cracks whose potential action is uncertain. From an operational point of view, how many times a rock is struck, and how many potential fissures it contains is not important; the only physical reality is whether a rock is broken (or not). The remarkable result is that the border of these breaks always generate the same structure, which is precisely the content of the law. We see many similarities with this and for example Ohm's Law: the milling conditions correspond to the potential difference (V); the ore corresponds to the resistance (R), the pair (λ, d) corresponds to the intensity (I), and most importantly the relation $V = IR$

corresponds to the Poisson doublet structure remaining unchanged over varying experimental conditions. Both laws concisely summarize the myriads of events that occur on a microscopic scale. The next question is one of delineating the domain of validity of the law; obviously, if the miller is filled with clay, Cauwe's law no longer holds. Little is known about where this distinction must be drawn.

(c) *Prediction and its limits*

An answer to Buffon's criticism, and an idea as to what is checkable in the model hypotheses, is contained in the above milling example. There, we were able to reply to Buffon's objections by submitting the model to judgement by the criteria. Was it really necessary to take a *stationary, ergodic, isotropic* process of twin planes, *independently* implanted, in order to generate fragments having the milled products' size distribution? Physically speaking, surely not, since there is absolutely no reason why the process of destruction of a rock will coincide, step by step, with the mathematical method for generating the random set (in reality the mill is bounded, the initial broken particles disturb the work of the balls of the mill, the elasticities of the small and large grains differ from each other . . .). The good fit between practical and theoretical data simply means that within the framework of the experimental limits, one factor emerges above all others: the particles are cracked into three parts and not into two. The role of the probabilistic theory was to "dress up" this idea to a level where size criteria can be used and the outcomes predicted. It is only relative to these criteria that Cauwe's model can be considered as an acceptable representation of pysical reality. If we now want to test the angularity of the milled particles for example, we would have to use different criteria, which probably would lead to the rejection of the doublet model.

Finally, the stationarity, ergodicity, etc . . . hypotheses which allow a theoretical formulation, look more like the catalysts of a chemical reaction than the elements of the reaction itself. They are difficult to verify from the limited number of bounded samples at the disposal of the experimenter. In essence they allow us the right to extrapolate the partial result we obtain to the whole object. Hence they are connected with chance occurences in the operating mode just as much as with the genetic process. Although we can sometimes proves the hypotheses to be false (see in Chapter VIII an example of a population of intercepts where the "mean chord" does not exist!), their level of generality results in us considering them more as methodologic choices than as checkable hypotheses. In an instructive book on probabilistic hypotheses, G. Matheron (1979) classifies them according to their degree of "vulnerability". Their importance from a theoretical point of view is to know whether acceptance or rejection in a given problem, introduces a substantial difference to the mathematical conclusions. Experimentally speaking, we usually cannot test them (in the statistical sense of the word), nor can we verify them on a single case study; we may only indirectly be able to get some idea of their "advisability", by looking at a large number of solved problems. We prefer therefore to say that the ergodic hypothesis (equally, the assumption that sets are closed, see Ch. III) is *useful* or *not useful, appropriate* or *not appropriate,* etc. . . ., than to say that it is *right* or *wrong.*

(d) *The key questions*
The lessons that can be learnt from the analysis of Cauwe's model are general to most probabilistic models. They can be summarized by the following questions, presented here in the order that they appeared in the discussion:

— Which criteria can be used to prove false each of the assumptions of the model?
— How can estimators for prediction purposes, be derived from the actual measurements?
— What part of the problem has been left untreated?
— What are the hypotheses that are not tested in the current study (and therefore are considered as methodological choices)?

We suggest that the reader asks and answers these four questions each time he wants to use, or to criticize, the application of a probabilistic model. Strictly speaking, the answers will not improve the quality of a given estimate. However they do provide a check-list of the possible weaknesses of the procedure, and can be equally applied when fitting a random set. As an exercise, we suggest that the reader apply this check list to the five global and local situations presented in Chapter VIII.

A.2. Models and heuristics

(a) *Models as descriptors*
The random set models of the previous chapter can be exploited in exactly the same way for example that an economist uses the function e^{at} when he claims that the electricity production of industrial countries follows an exponential law. If this were actually true, then in a few decades' time the world would collapse under the weight of the resulting electric power stations. However within a given time interval (whose upper limit is uncertain) the idea turns out to be operational; what he has captured is the *genesis* of the process (increment is proportional to the current value).

This way of thinking also appears in morphology. When the Boolean model is used for studying the firing of coke (Ch. XIII, Sect. B.8), or the spatial distribution of clay in soils (Fies, 1976), two steps occur: firstly a minimal experimental verification takes place (i.e. tested by linear or circular erosions); and secondly further investigation occurs based only on the primary grains and their density. In sum total then, the model appears here as an instrument for schematizing the structure, although we have less interest here in whether the proposed genesis is real, than we have in prediction. We are happy that the model is believable, the essential thing being that a very parsimonious description results, and that we have a way of choosing a few criteria from the many available.

(b) *Models as providers of criteria*

From a theoretical point of view, a random set model is a particularly well defined class of sets, which *all* satisfy a collection of formal properties. Indeed the spirit of Choquet's theorem (Ch. XIII, G.2) is exactly this, namely to know all the probabilities $\Pr \{ K \subset X^c \}$ for K compact is to give an exhaustive description of the random set. Unfortunately, this description adds a large number of unwanted properties to the ones that are really relevant to the problem under study. For example in Chapter X, Section E, we used the spherical model to provide an algorithm that relates size distributions of cross sections to the 3-D size distribution. In some cases this algorithm was equally acceptable for powders which were not at all spherical. Obviously all the properties of the spherical model (radii of curvature, sphericity) were not satisfied by the dust particles analysed in Chapter X, E.2. In general then, we start with a random set model, which defines a class of sets $\mathscr{E}$ satisfying a rather comprehensive set of properties $\mathscr{P}$ but we are in fact only interested in a subpart $\mathscr{P}'$ of $\mathscr{P}$ (see Fig. XIV.1). Then rather than try to verify the model $\mathscr{E}$, we profit more by looking for the domain of attraction of (more general) set models that satisfy $\mathscr{P}'$. As a general rule, the limits of this domain cannot be formally delineated (if they could be, then the domain would have been one starting point): simulations offer a practical way of gaining insight. This is illustrated by the following example.

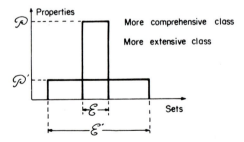

Figure XIV.1. The more a class of models is extensive, the less it is comprehensive.

(c) *Counts of overlapping cells*

Consider a histological thin section and suppose we want to count the 2-D number N_A of cells per unit area. Each cell is represented as a random set X' with an area A' (with average value $\overline{A'}$). Unfortunately, the cells overlap to a greater or lesser extent; how can we estimate N_A?

We will assume that the average $\overline{A'}$ is known, either by external biological information, or by measurements on the isolated cells of the specimen. Relationships between $\overline{A'}$ and N_A are easy in the following two simple cases:

(i) No overlapping. Then,

$$N_{A,1} = A_A/\overline{A'} \qquad \text{(XIV-1)}$$

(ii) The centres of the cells are Poisson. Then $X = \bigcup X'$ is a realization of a Boolean model, and relation (XIII-6) implies:

$$N_{A,2} = -\log\{(1 - A_A)/\overline{A'}\} \qquad \text{(XIV-2)}$$

(A_A = area proportion of X). Note that if there are few particles (N_A small) or if the cells are small ($\overline{A'}$ small), then there is a low probability of overlapping and the second expression tends towards the first. Also observe that these two estimates involve only area measurements (namely A_A) which are less sensitive to noise than the connectivity number algorithms. Question: are the two algorithms (XIV-1) and (XIV-2) very bad estimators when the cells *overlap* and have *non-Poisson* centres?

In cooperation with H. Digabel (1977), we tried to answer this question by simulating spatial distributions of cells on the texture analyser. The centres are realizations of Neyman–Scott models of the type (XIII, F.3), from which we remove those points with a neighbour at a distance less than d (hard core model). By exploiting the degrees of freedom of the simulations, we can generate either clusters, or spatial distributions more regular than the Poisson structure. Two types of size distribution are chosen; either constant size in $\mathbb{R}^2$ or in $\mathbb{R}^3$ (which induces variable sizes on sections). For each choice of simulation parameters, the number N of cells which fall in the measuring mask of the device is determined, as well as the areas A' of the N cells. For $\overline{A'}$ we take the mathematical expectation of the corresponding size distribution. The area proportion A_A is directly measured on the texture analyser; then algorithms (XIV-1) and (XIV-2) are applied, providing two estimates of N_A. Multiplying these two numbers by the area of the measuring mask we deduce two estimates N_1 and N_2 of the number of cells which can be compared with N.

Figure XIV.2 shows two of the simulations. The point process of centres is very similar, exhibiting small clusters. The size distribution is 2-D constant in (a) and 3-D constant in (b). Five different groups of parameters have been

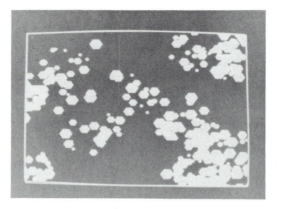

Figure XIV.2. Simulations of spatial distribution of cells. (The simulations have been performed by using hexagons for simplicity. Basically, this does not affect the validity of the results.)

chosen for each of them; ten independent realizations are built up in order to average N_1 and N_2, and reduce their statistical fluctuations. The mean results are shown in the table in Fig. XIV.3.

	Point Process						$\dfrac{N-N_1}{N \times 100}$	$\dfrac{N-N_2}{N \times 100}$
	Clusters	Hard core						
		distance d	Cell diameters	N	N_1	N_2		
	weak	8	2-D constant = 10	365	318	370	6·9	− 1·3
	weak	15	2-D constant = 9	179	166	176	7·3	0·2
Simulations	strong·	15	2-D constant = 9	156	168	155	5·1	0·6
	weak	15	3-D constant = 9	186	177	186	4·8	0
	strong	15	3-D constant = 9	191	173	182	9·4	4·7

Figure XIV.3. Table of simulation results.

In every case, the estimate N_2 derived from the Boolean model (XIV-2) is better than N_1, and leads to relative errors of only a few percent (we believe the choices of simulations were sufficiently realistic: subsequently, algorithm N_2 was applied several times on real cell populations and gave satisfactory results). Clearly, the domain of attraction of relationship (XIV-2) is much vaster than the Boolean model. The above results extend it to convex particles with a narrow relative variance of sizes, whose centres cluster much as the cells in situ. Probably, the algorithm remains valid for other classes of objects (non convex, other size distributions . . .). If need be, its suitability for such classes could be tested by a simulation approach similar to the one just presented.

(d) *Other examples of heuristics in models*
Several examples have already appeared earlier in the book, but are collected here for completeness. A model can be used for testing an algorithmic approximation, as we will now see. The algorithm (X-32) which gave the variance of the star is complicated to apply on digital images, and it was replaced by the simplified algorithm (X-33). However in the case of Poisson polygons (the model) they can both be calculated, and lead to very similar results. But the large variety of shapes present in Poisson polygons could in effect be used as a simulation of general convex particles, thus justifying the appearence of the approximation in these more general cases.

Models are also sometimes used for constructing counter examples that are often useful in delineating domains of attraction. In Chapter IX, Section D.1 we showed that a pseudo-periodic covariance did not correspond to a periodic structure, while in Section D.2, we showed that a knowledge of the probabilities of order two may still result in a completely unreliable characterization of structures. Another of the many examples is that of the conditional simulations of random functions (A. Journel, 1978); i.e. simulations with a certain covariance that takes fixed values at given points.

B. MATHEMATICAL MODELS IN MORPHOLOGY

Buffon's criticism should be directed not only at stochastic models, but also at mathematical formalization in general. The primary purpose of mathematical models is to provide relationships between possible measurements and properties for certain classes of sets. An example is offered by the perimeter, which in the convex ring can be considered as a function of the number of intercepts. But there is a second purpose (which answers Buffon's criticism), namely to investigate the domains of attraction of the "good" relationships established within the framework of the given model. The approach here will be different from that of the previous section.

Sometimes, the method can be developed in such a way that the experimental sets can be exactly fitted to those of the model. For example, no experiment is able to distinguish between an object and its topological closure (the notion of a topological closure is a property of a set, i.e. a model, and not of an experimental figure). Therefore, the intersection topology is constructed so that it is "blind" to the differences between two sets having the same closure.

However, more often we are not able to fit the exact degree of generality of the model with the range of textures under study. Then one approach is to build two different models, each representing extreme situations, and verify that they lead to equivalent key relationships. It is this approach that we have systematically exploited in the book. In Chapter V, the regular model and the convex ring lead to virtually the same parameters; in Chapter VI, the four theorems on digitalization, identify the continuous and discrete approaches etc. We are well aware that such a technique can be lengthy, since it doubles the number of proofs. However we have often found it to be the only one capable of judging the domain of validity of algorithms and criteria.

C. PATTERN RECOGNITION MODELLING

C.1. Structural Description and Pattern Recognition

Any quantitative image analysis lies somewhere on an axis, the two opposing poles of which are pattern recognition and structural description (Fig. XIV.4). In the first of these two, we know *a priori* a collection of archetypes which we use to match the shape of the object under study (chromosomes, letters of alphabet); here the problem is theoretically well defined. The other extreme case corresponds to a need for description independent of archetypes, for example when one wants to correlate microscopic structures and physical properties (see the study on Caucasian dolerites in Ch. X, Sect. C, and on wood anatomy in Ch. IX, Sect. E). Often one is somewhere in between, where one wants to use the structural description for classifying according to patterns which are not yet defined. For example, the type of cancerous cells seen on a cervical smear is not completely specified by the physiology (Ch. XI, Sect.?), nor are the number of neurone layers in the brain cortex (Ch. X, Sect. C). In this case, the structural analysis is used *both* for defining the archetypes

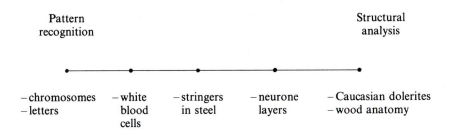

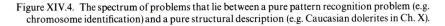

Pattern recognition				Structural analysis
– chromosomes – letters	– white blood cells	– stringers in steel	– neurone layers	– Caucasian dolerites – wood anatomy

Figure XIV.4. The spectrum of problems that lie between a pure pattern recognition problem (e.g. chromosome identification) and a pure structural description (e.g. Caucasian dolerites in Ch. X).

and subsequently for classifying additional samples. Sometimes we iterate the procedure by returning to the defining stage and repeating. Symbolically then, these auto-learning procedures correspond to oscillations on the axis of Figure XIV.4, converging towards a limit point somewhere between the two poles. (The relaxation labelling methods of O. Faugeras, A. Rosenfeld, Zucker (1977) belong to this category.)

C.2. Can a computer have feelings?

For problems which only involve classification of objects based solely on their morphology, pattern recognition generally gives fairly good results. But what a difference there is between this limited success, and the global claims of those schools of pattern recognition that want to substitute the computer for human perception. This "scientific" preconception comes from the decision that all perception can be reduced to observable geometric parameters, and that the spatial information (in the sense of Wiener's information theory) induces the meaning of the object. It is as if the concept of "A, typed on a machine", that of "house" and that of "friendly behaviour" could be recognized from images, by the *same* kind of geometrical methods. Alas in psychology this is not the situation at all. The mind gives a meaning to all images coming to the consciousness; i.e. adds something to the external information, not in order to transmit the image to other destinations, but to take it for itself. The consciousness is able to build up its own automatic circuits (recognition of letters, etc) so as to unload that which is routine, but that does not mean we can reduce all its activity to the low-level parts that are easily translated into external algorithms. In particular the consciousness only ever discharges its activity in a provisional manner, and always reserves the right to make a deeper interpretation as soon as its automatic circuits meet an ambiguity that purely geometric interpretations cannot resolve. Consider the following example. The text that you, the reader, are now scanning could just as easily be "read" by a computer. (In fact, does the computer read, or does it just decant the letters elsewhere, as one decants whisky from a barrel to a bottle? Does the bottle taste the whisky?) But imagine now that you are reading a handwritten version of this book, and that you come across the following:

Figure XIV.5.

After a frown and a slight hesitation, you will restart your reading and concentrate on the subsequent reading. You resort for an instant to the consciousness in order to call on your prior knowledge, then the mind goes back onto automatic pilot. What computer could perform such feats?

Suppose now we overdo our present position, and present you with the following sentence: "Pattern recognition is a big joke; it will surely lead to promising results".

In this case, the totality of the context helps in its understanding, so that between "surely" and "lead" you will immediately add the negation "not". By so doing you will have generated information from *nothing* strictly observable, and will have completely inverted the textual meaning of a portion of the sentence. Computers which can wittingly take such initiatives are way off in the future. The semantic world is not constructed in a piecemeal fashion, but each situation needs to be integrally surveyed in order to understand every portion of it (R. Ruyer, 1966).

C.3. "Pravda" or "Istina" again

According to the Chinese proverb, a poet either looks at, or is looked at, by the moon. So the criteria, as we have already said in Chapter I, are a panoply of well finished tools, and nothing more; with them one "grasps" the world "The poet looks at the moon". An already existing tool is sharpened by the acquired experience of auto-learning methods, but these can never produce a *new* tool from the object.

To interpret what has been grasped by the criteria, the morphologist then constructs a mathematical-type semantic world which is essentially distinct from human perception, but works in a somewhat similar way. "The poet is looked at by the moon". A group of points suggests a random points model, exactly as a smile suggests a friendly attitude. Neither the idea of a friendly attitude nor that of random points can be stated in *purely* geometric terms; they are transpatial and linked to the physical world by (different) inter-mediaries. When the brain interprets a smile it transcodes geometric inform-ation, just as the statistician infers the hypothesis of random implantation from measurements on the distance between points.

We can see a demonstration of the separation between the semantics in the quantitative (mathematical-type) world and in human perception, in their different performances. If the computer that will thrill to the softness of a smile is not yet ready, then equally the human eye is totally incapable of recognizing whether points are Poisson (just as it is unable to comprehend superposition of scales, estimation of specific perimeters, averaging one hundred images, etc . . .). It may happen that an image receives the same interpretation from the two semantic worlds; for example a given photograph can lead to the concept of "aerial photograph of a village" in both systems. Unfortunately this happy parallelism does not occur in a systematic way. The semantics in the quantitative world are developed in such a way that one can access them via *logical* paths, which in practice are necessarily *finite* (even if one can potentially construct as many random functions and models as one likes). These two characteristics markedly contrast this world to that of human perception.

As an illustration of the above discussions, we will consider the well-known psychological test of Rorschach. Ten plates are taken, each showing large ink spots. Presented to a subject, they provoke a series of interpretations which help in a diagnosis of his psychology. To an outside observer, the subject speaks about himself via the ink spots, as if the spots were looking at him. Now imagine the same test given to a large computer which contains in its memories all the pattern recognition models ever written. It will find several hundred possible interpretations (a butterfly, chords having an exponential distribution, the profile of a witch on her broomstick, symmetries of order two, etc . . .), none of which carry any information at all, since determining psychoses belongs strictly to the human perception world and not to the other.

Quantitative methods of image analysis will never exactly copy the work of the brain. It is not worth rejoicing or mourning over such an apparent failure. The inventor (genius though he was) who wanted to reproduce the exact movement of a bird's wings in order to fly, fell down badly. Much later (using other principles) aeroplanes reached performances in flight that were previously unimaginable.

Quantitative image analysis will never recognize that a flower is beautiful; it will do completely different things, and in some senses much more.

Appendix

THE DUALITY WITH RESPECT TO THE COMPLEMENTATION

One of the principal themes of this book was the duality w.r. to the complementation, which appears naturally when we deal with sets. In the present appendix, we briefly review the major situations where this notion is invoked. The operation Ψ which transforms a set X into $\Psi(X)$ also works on the background, since X^c becomes $[\Psi(X)]^c$. This latter transformation is called dual Ψ^* of the former one Ψ, and we have by definition:

$$\Psi^*(X) = [\Psi(X^c)]^c \tag{1}$$

Ψ may depend on a set or a scalar parameter, and X may represent several sets. For example if Ψ denotes the set union, relation (1) yields the famous Morgan's theorem:

$$\left[\bigcup_{i \in I} X_i\right]^c = \bigcap_{i \in I} (X_i)^c \quad I: \text{finite or not integer}$$

The notion of duality is also valid for a logical relation $\mathcal{R}$. Then $X \mathcal{R}^* Y = X^c \mathcal{\bar{R}} Y$ where $\mathcal{\bar{R}}$ is the negation of the relation. For example the dual form of $B \subset X$ is $B \Uparrow X$. Finally the duality applies in the description of the remarkable features in an object, and for example make the connected components correspond with the holes of a set. By duality, an extensive set transformation becomes anti-extensive, but the properties of being increasing, homotopic and idempotent remain unchanged (see table below).

Initial notions	*Dual notions*
Set X, its caps, isthmuses and islands	Set X^c, its gulfs, channels and lakes
union $X \cup Y \begin{cases} X, Y \text{ arguments} \\ Y \text{ parameter} \end{cases}$	intersection $X \cap Y$ set difference X/Y
erosion $X \ominus B$, opening $(X \ominus \check{B}) \oplus B$ (B par.)	dilation $X \oplus B$, closing $(X \oplus \check{B}) \ominus B$
Hit or Miss $X \circledast T (T = (T_1, T_2)$ par.)	Hit or Miss $X \circledast T^*$ with $T^* = (T_2, T_1)$
Thinning $X \bigcirc T$ (T par.)	thickening $X \odot T^*$
Cond. thickening (resp. dilation) $X \odot T; Y$ (X, Y arguments, $T = (T_1, T_2)$ parameter)	Cond. thinning (resp. erosion) $X \bigcirc T^*; Y$
closed set; space $\mathscr{F}$ of all closed sets ($\mathbb{R}^n$) $X_t \to X$; $\overline{\mathrm{Lim}} (X_i)$; $X_i \uparrow X$	open set; space $\mathscr{G}$ of all open sets ($\mathbb{R}^n$) $X_i \to X$; $\underline{\mathrm{Lim}} (X_i)$; $X_i \downarrow X$
upper semi-continuity $\overline{\mathrm{Lim}} \, \Psi(X_i) \subset \Psi(X)$	lower semi-continuity $\underline{\mathrm{Lim}} \, \Psi(X_i) \supset \Psi(X)$
homotopic sets (or functions)	homotopic sets (or functions)
Specific Euclidean $\begin{cases} \textit{Area } A_A(X); \text{ perimeter } U_A(X) \\ \text{connectivity number } N_A(x) \\ \text{convexity number } N_A^+(X) \end{cases}$	equal $\begin{cases} 1 - A_A(X^c); \, U_A(X^c) \\ -N_A(X^c) \\ N_A^-(X^c) \end{cases}$ respectively
function $f(x)$; picture $p(x)(0 \le p \le m)$	funtion $-f(x)$; picture $m - p(x)$
Section X_t	Section X_{-t}^- (function); X_{m-t}^- (pictures)
Sup (f, g) (f, g functions or pictures)	inf (f, g)
$-(f \oplus g)$ (functions)	equals $-f \oplus g$
$m - (f \ominus g)$ (pictures)	equals $(m - f) \oplus g$
For functions and pictures $\begin{cases} \text{maximum, summits} \\ \text{divides, watersheds} \\ \text{ridges, saddles} \end{cases}$	become $\begin{cases} \text{minimum, sinks} \\ \text{channels, hills} \\ \text{ruts, saddles} \end{cases}$ respectively
Variograms: $\gamma_X(h); \gamma_f(h); \gamma_p(h)$	equal $\gamma_{X^c}(h); \gamma_{-f}(h); \gamma_{m-p}(h)$ resp.

Bibliography

The literature presented below does not pretend to cover the whole field of Image Analysis. It only refers to documents associated with the new morphological approach. The essential ideas appear in the papers of Fontainebleau School, plus a small representative set of books, including among others:

— H. HADWIGER (1957), E. F. HARDING and D. G. KENDALL (1974), L. A. SANTALO (1976) for integral geometry and geometrical probability.
— G. MATHERON (1975) for the mathematical bases of Morphology.
— E. E. UNDERWOOD (1970) and E. R. WEIBEL (1980) for stereology applied to material sciences and to biology respectively.
— R. O. DUDA and P. E. HART (1972), A. ROSENFELD and A. C. KAK (1976), W. K. PRATT (1978), for the problems involved in picture processing.
— JAMES and JAMES (1976) for a general dictionary of Mathematics.

In this text, several references dating from the eighteenth and nineteenth centuries are given. They can all be found in R. E. MILES and J. SERRA (1978), where digests of the main studies, from Euler through to Minkowski inclusive, are proposed in their original typographies.

As often happens with new theories, some concepts have been "borrowed" from Mathematical Morphology and renamed by other workers (the notion of an opening probably holds the record for this). It would not be worthwhile mentioning these pilferings except that they can lead to a distorted presentation of the method. For example, the notion of an opening loses all meaning when it is separated from that of size distribution and of convexity. To avoid problems of this sort we recommend that readers whose first contact with Mathematical Morphology has been with these plagiarized versions, go back to the original sources as indicated in the text.

Adler, R. J. (1977). Some erratic patterns generated by the planar Wiener process. *Adv. in Appl. Prob.* (March 1978) **10**, 22–27.
Adler, J. R. (1979). "The Geometry of Random Fields" University of New South Wales.
Ahuja, N. (1979). Unpublished Ph.D. thesis. (Univ. Maryland.)
Alberny, R., Serra, J. and Turpin, M. (1969). Use of covariograms for Dendrite arm-spacing measurements. *Trans. AIME* **245**, 55–59.
Andrews, H. C. (1978). Tutorial and Selected Papers in Digital Image Processing. *IEEE Transactions*, New York.
Avtandilov, G. G. (1980). Vedenie v kolitchestvennuiju patologitcheskuiju morphologiyu, Moscow, Meditsina, 214 pp.
Avtandilov, G. G., Iavlutchanski, N. I. and Gubenko, V. G. (1981). Systemanaïa Stereometria v izutcheni pathologitcheskovo processa, Moscow, Meditsina, 189 pp.

Bachacou, J. and Decourt, N. (1976). Etude de la compétition dans des plantations régulières à l'aide de variogrammes. *Annales des sciences forestières* **33**, 177–198.

Baddeley, A. (1979). Absolute curvatures in integral geometry. *Math. Proc. Camb. Phil. Soc.* (1980) **88**, 45–58.

Baddeley, A. (1980). A limit theorem for statistics of spacial data. *Adv. Appl. Prob.* **12**, 447–461.

Barbery, G., Huyet, G., Bloise, R and Laurenceau, H. (1977). Etude théorique et simulation de la reconstruction de la composition volumétrique de population de particules composites à partir de mesures réalisées sur des sections. *In* J. L. Chermant (ed.), 203–212.

Bartlett, M. S. (1963). The spectral analysis of point processes. *J. Roy. Stat. Soc. B*, **25**, 264–296.

Bartlett, M. S. (1976). "The Statistical Analysis of Spatial Pattern." Chapman and Hall, London.

Baudin, M. (1979). Microearthquakes and point process. C. M. M., int. rep. 50 pp.

Benzecri, J. P. (1974). "L'analyse des données." Dunod, Paris.

Bernroider, G. and Adam, H. (1979). Synthesized point information sampling. Fifth Int. Cong. ICCS. (To be published in *Mikroscopie*.)

Bertram, J. F. and Rogers, A. W. (1979). A morphometric study of human bronchial epithelium. Light Microscopic Measurements. Int. report. The Medical School, Flinders University of South Australia.

Beucher, S. (1977). "Random Processes Simulations on the Texture Analyser." Lecture Notes in Biomathematics, No. 23. Springer-Verlag.

Beucher, S. and Lantuéjoul, Ch. (1979). On the change of space in image analysis. 5th Congress of the ICCS, (To be published in *Mikroscopie*.)

Beucher, S. and Lantuéjoul, Ch. (1979). Use of watersheds in contour detection. Int. Workshop on Image Processing, CCETT. Rennes, France.

Binet, J. L. (1977). "La cytologie quantitative." PUF, Paris.

Binet, J. L., Meyer, F., Merle-Beral, H., Bodega, E., Dighiero, G. and Lesty, C. (1979). Fine chromatin analysis by contrast algorithms in normal and CLL lymphocytes. Fifth Int. Congress of Stereol. (To be published in *Mikroscopie*.)

Bisconte, J. C. and Margules, S. (1979). Continuous quantitative analysis of cultured living cells by image analysis. Fifth Int. Cong. of ICCS. (To be published in *Mikroscopie*.)

Blasckhe, W. V. (1936). "Vorlesungen über integral geometrie ." Teubner, Leipzig.

Blanc, M. (1979). Contribution de la morphometrie à l'étude de matériaux polycristallins. Ph.D. thesis. Paris School of Mines, 180 pp.

Blum, H. (1962). An Associative machine for dealing with the visual fields and some of its biological implications. *Biol. Prot. and Synth. Syst.* **1**, 244–260.

Blum, H. (1973). *Biological Shape and Bol.* **38**, 205–287.

Bodziony, J. (1981). On the relationship between basic stereological characteristics. *Stereol. Iugosl.* **3**, suppl. 1, pp. 97–102.

Bolender, R. P. (1979). An analysis of stereological reference systems used to interpret drug induced changes in biological membranes. *Mikroscopie* **37**, suppl. 165–172.

Bookstein, F. L. (1978). (Ed.). Lecture Notes in Biomathematics, No. 24. "The Measurement of Biological Shape and Shape Change." Springer-Verlag, Berlin, Heidelberg, New York.

Bouchon, J. (1979). Structure des peuplements forestiers. *Ann. Sci. Forest.* **36** (3), 175–209.

Bowie, J. E. and Young, I. T. (1977). An analysis technique for biological shape. *II–III, Acta Cytol.* **21**.

Bridgmann, W. (1927). "The Logic of Modern Physics." New York.

Brugal, G., Garbay, C., Giroud, F. and Adelh, A. (1979). A double scanning microphotometer for image analysis: hardware, software and biomedical applications. J. Histochem. Cytochem. **27**, 114–152

Cahn, J. W. (1972). The generation and characterization of shape. Special supplement to *Advances in Applied Probability*.

Calabi, L. (1965). A study of the skeleton of plane figures. *Parke Math. Lab. Inc.*

Calabi, L. and Hartnett, W. E. (1966). Skeletons of gray-scale pictures. Parke Math. Lab. inc. 6 pp.

Castan, S. and Leloup, J. (1977). Extraction d'une forme visuelle d'un bruit aléatoire. *GRETSI.* Nice 26–30, Avril 1977.

Castan, S. and Nabonne, A. (1975). Un algorithme de squelettisation opérant par tests locaux sur des suites de voisinages successifs. Colloque National sur le traitement du signal et ses applications. Nice 16–21, Juin 1975.

Cauchy, A. (1841). Note sur divers théorèmes relatifs à la rectification des courbes et à la quadrature des surfaces. *C. R. Acad. Sci. Paris,* **13**, 1060–1065.

Cauwe, Ph. (1973). Morphologie mathématique et fragmentation. Ph.D. Thesis. Univ. of Louvain, Belgium. Vol. I, 177 pp.; Vol. II, 103 pp.

Celeski, J. C., Jeulin, D. and Schneider, M. (1977). Etude quantitative du réseau poreux de trois cokes moulés, *IRSID R. E.,* 433.

Chermant, J. L. and Coster, M. (1977). Quantitative analysis of particle growth in TiC–Ni alloys. *J. of Micr.* **111**, part. 3, 269–282.

Chermant, J. L. (1978). (ed.) Quantitative analysis of microstructures in material sciences, biology and medicine, Special issue of *Practical Metallography*, Riederer–Verlag, Stuttgart.

Chinkarenko, V. C. and Tchernouk, A. M. (1979). Application of the texture analyser for investigations on microcirculation. (In Russian.) *Bull. Exp. Biol. and Medicine.* **5**, 497–500.

Choquet, G. (1953). Theory of capacities. *Ann. Inst. Fourier V.* 131–295.

Choquet, G. (1966). "Topology." Academic Press.

Cochran, W. G. (1977). "Sampling Techniques." Wiley, New York.

Coleman, R. (1979). The distribution of the sizes of spheres from observations through a thin slice. Proceedings of the Fifth Int. Congress of Stereology. (To be published in *Mikroscopie*.)

Coleman, R. (1979). An introduction to mathematical stereology. Memoirs n° 3, Department of theoretical statistics, University of Aarhus.

Conrad, F. and Jacquin, C. (1972). Représentation d'un réseau bi-dimensionnel de failles par un modèle probabiliste. Int. Report Centre de Morphologie Mathématique Fontainebleau, France.

Coster, M. and Deschanvres, A. (1978). Fracture, objet fractal et morphologie mathématique. *Spec. issue. Pract. Metal.* **8**, 61–73.

Cowan, R. (1978). The use of the ergodic theorems in random geometry. *Supp. Adv. Appl. Prob.* **10**, 47–57.

Cox, D. C. (1962). "Renewal Processes." Methuen, London.

Cressie, N. (1978). A strong limit theorem for random sets. *Suppl. Adv. Appl. Prob.* **10**, 36–46.

Crofton, M. W. (1868). On the theory of local probability, applied to straight lines drawn at random in a plane, the method used being also extended to the proof of certain new theorems in the integral calculus. *Philos. Trans. Roy. Soc., London,* **158**, 181–189.

Cruz-Orive, L. M. (1976). Correction of stereological parameters from biased samples on nucleated particle phases. I: Nuclear volume fraction; II: Specific surface area. *J. of Micr.* **106**, part 1, 1–32.

Cruz-Orive, L. M. (1978). Particle size-shape distribution: the general spheroïd problem II. Stochastic model and practical guide. *J. of Micr.* **112**, part 2, 153–168.

Davies, E. R. and Plummer, A. P. N. (1980). Thinning Algorithms and their role in image processing. BPRA Conference on Pattern Recognition, Oxford, 40 pp.

Davy, P. J. (1977). Aspects of random set theory. *Adv. appl. probability,* **10**, 28–35.

Davy, P. J. and Miles, R. E. (1977). Sampling theory for opaque spatial specimens. *J. R. Statist. soc.* **B.39**, 56–65.

Delfiner, P. (1970). Le schéma booléen-Poissonien. Int. Report, Centre de Morphologie Mathématique Fontainebleau, France.

Delfiner, P. (1971). Etude morphologique des milieux poreux et automatisation des mesures en plaques minces. Ph.D. thesis, University of Nancy.

Deltheil, R. (1926). "Probabilités géométriques". Gauthier-Villars, Paris.

Digabel, H., Mariana, J. C., Mauleon, P. and Serra, J. (1973). Etude de la distribution spatiale des ovogonies dans l'ovaire d'embryon de ratte à l'aide de l'analyseur de textures. *Annales de Biologie Animale.* **13**, No. hors série, 115–125.

Digabel, H., and Lantuéjoul, Ch. (1978). "Iterative Algorithms." In J. L. Chermant (ed.).

Van Driel-Kulker, A. M. J., Meyer, F. and Ploem, J. S. (1980). Automated analysis of cervical specimens using the T. A. S. *Microscopica acta*, suppl. **4**, 73–81.

Dubreil, P. and Dubreil-Jacotin, M. L. (1964). "Leçons d'algèbre moderne." Dunod, Paris.

Duda, R. O. and Hart, P. E. (1972). "Pattern Classification and Scene Analysis." Wiley Interscience publication.

Duff, M. J. B. and Watson, D. M. (1974). The cellular logic array image processor. Dept of Physics and Astronomy, University College London.

Duff, M. J. B., Watson, D. M., Fountain, T. J. and Shaw, G. K. (1973). A cellular logic array for image processing. *Patt. Recog.* **5**, 229–247.

Elias, H. and Weibel, E. R. (1967). "Quantitative methods in Morphology." Springer-Verlag, Berlin.

Exner, H. E. (1979). An introduction to the development and effects of voïds in sintered materials. *J. of Micr.* **116**, part 1, 25–38.

Faugeras, O. and Pratt, W. K. (1978). Visual discrimination of stochastic textures. *IEEE Trans. on Systems, Man and Cybernetics.*

Federer, H. (1969). "Geometric Measure Theory." Springer-Verlag, Berlin.

Fejes Toth, L. (1953). "Lagerungen in Der Ebene auf der Kugel und im Raum." Springer-Verlag.

Feller, W. (1960). "An Introduction to Probability Theory." Wiley. (Vol. I, 509 pp.; Vol. II, 625 pp.)

Fies, J. C. (1980). Etude des écoulements en milieux saturés et non saturés, en relation avec l'espace poral. (Contrat D. G. R. S. T. No. 79-4-0074, technical report.)

Fischmeister, H. F. (1972). Applications of quantitative microscopy in material engineering *J. of Micro.* **95**, part 1, 119–143.

Flint, E. (1957). "Principles of Metallography." (In Russian.) (Ed.) Mir, Moscow. 237 pp.

Flook, A. G. (1978). The use of dilation logic on the quantimet to achieve fractal dimension characterisation of textured and structured profiles. *Powder Technology* **21**, 295–298.

Frossard, E. (1978). Caracterisation pétrographique et propriétés mécaniques des sables. Ph.D. thesis, Paris School of Mines.

Fu, K. S. (1976). "Digital Pattern Recognition." Springer Verlag, 234 pp.

Galadkowskaja, G. A. (1968). O vlianii tektonicheskikh processov na formirovanie ingernerno geologicheskikh svoïstu gornikh parod. Voproçi ingenernoï geologi i gruntovedenia. Iz. vo. M. G. OU.

Gamow, G. (1950). "One, Two, Three . . . Infinity". Viking Press, N. Y.

Giger, H. and Hadwiger, H. (1968). Uber Treffzahlwahrscheinlichkeiten in Eiköperfeld. Z. Wahrscheinlichkeitstheorie verw. Geb. 10, 329–334.

Gilbert, E. N. (1961). Random subdivisions of space into crystals. Ann. Math. Statist. 33, 958 (1962).

Goetcharian, V. (1980). From binary to grey level tone image processing by using fuzzy logic concepts. Patt. Recog. 12, 7–15.

Goetcherian, V. (1980). Parallel image processes and real-time texture analysis. Thesis Doctor of Philosophy, University College, London.

Golay, M. J. E. (1969). Hexagonal parallel pattern transformation. IEEE Trans. Comput. C-18, 733–740.

Granlund, G. H. (1978). In search of a general picture processing operator. Computer Graphic and Image Processing 8, 155–173.

Grant, G., Reid, A. F. and Zuiderwyk, M. (1979). Simplified size and shape description of ore particles as measured by automated sem. International Powder Bulk Solids Conf., Philadelphia, 1979. Processings, 168–175.

Grant, G. and Reid, A. F. (1979). A fast and precise boundary tracing algorithms for image analysis of objects of any shape. Fifth Int. congress of Stereology. (To be published in Mikroscopie.)

Gray, S. B. (1971). Local properties of binary images in two dimensions. IEEE Transactions on Computer, C-20, No. 5.

Greco, A., Jeulin, D. and Serra, J. (1979). The use of the texture analyser to study sinter structure. J. of Micr. 116, part 2.

Grenander, U. (1973). statistical geometry: A tool for pattern analysis. Bull. of the Amer. Math. Soc. 79, No. 5.

Grolier, J. and Hucher, M. (1973). Construction de mosaïques pour l'analyse des textures subpolyèdriques. Bull. Soc. fr. Mineral. Cristallogr. 93, 105–122.

Gundersen, H. J. (1977). Notes on the estimation of the numerical density of arbitrary profiles: the edge effect. J. of Micro. 111, part 2, 219–223.

Haas, A., Matheron, G. and Serra, J. (1967). Morphologie mathématique et granulométries en place. Ann. Mines. XI, 736–753 and XII, 767–782.

Haas, A. (1968). Contribution à l'étude de la structure des grains d'un métal. rapport interne IRSID.

Hadwiger, H. (1957). "Vorslesungen über Inhalt, Oberfläche und Isoperimetrie." Springer, Berlin.

Hadwiger, H. (1959). Normale Körper im euklidischen Raum und ibre topologischen und metrischen Eigenschaften. Math. Zeitschr. 71, 124–140.

Hammer, J. (1977). Unsolved Problems Concerning lattice points (Pitman, London Melbourne.)

Haralick, R. M. (1978) Scene matching methods. (In issues in Digital Image Processing (ed. by R. M. Haralick and J. C. Simon), series E: Applied Science, No. 34.)

Hasofer, A. M. (1977). Upercrossings of random fields. Adv. in Appl. prob. March 1978, No. 10, 14–21.

Hausdorff, F. (1914). Grundzüge der Mengenlehre. Leipzig: Viet and Co. (1914) 3 (Gekûrzte) Aufl. Berlin, Leipzig: Gôschens Lehrbücherei (1935).

Hersant, T., Jeulin, D. and Parniere, P. (1976–1977). Basic notions of mathematical morphology used in quantitative metallography. (*Newsletter in Stereology*, Aug. 76, Dec. 77, edited by Gesellschaft für Kernforschung M. B. H. Karlsruhe.)

De Hoff, R. T. and Rhines, F. N. (1968). "Quantitative Microscopy." McGraw-Hill, New York.

Horalek, V. (1980). On evaluating exposures containing both extracted and sectioned particles. *Mikroskopic* 37, suppl., 90–93.

Hunzider, O., Schultz, U., Serra, J. and Walliser, Ch. (1976). Morphometric analysis of neurons in different depths of the cat's brain cortex after hypoxia. *N. B. S. Spec. publ.* No. 431, 203–206.

Jacod, J. and Joathon, P. (1971). Use of random genetic models in the study of sedimentary processes. *J. of Assoc. for Math. Geol.* 3, No. 3, 265–279.

James and James (1976). "Mathematics Dictionary." Van Nostrand, New York.

Jeulin, D. (1979). Multicomponent random models for the description of complex microstructures. Fifth Congress of ICCS, (To be published in *Mikroscopie*.)

Jeulin, P. and D. (1981). Synthesis of rough surfaces by random morphological models. *Stereol. Iugosl.* 3, Suppl. 1, 239–246.

Johnson, W. A. and Mehl, R. F. (1939). Reaction kinetics in processes of nucleation and growth. *Trans. A.I.M.M.E.* 135, 416.

Jeulin, D. (1976). Automatic recognition of non-metallic stringers in steel. *N.B.S.* Special pub. 431, 265–268.

Kalisnik, M. (1979). A stereological access to the quantitative morphological analysis of surfaces. Fifth Int. Congress of Stereology. (To be published in *Mikroscopie*.)

Kandinsky (1926). Punkt und linie zu fläche. Beitrag zur Analyse der Malerischen Elemente, München, A. Langen, 180 pp.

Kaufmann, A. (1975). "Theory of the Fuzzy Subsets." Academic Press.

Kendall, M. G. and Moran, P. A. P. (1963). "Geometrical Probability." Griffin, London.

Kendall, D. G. (1974). Foundations of a theory of random sets. *In* "Stochastic Geometry." (E. F. Harding and D. G. Kendall., eds.) Wiley.

Kesarev, V. C. (1978). Quantization of the architecture of the human brain. (*In* Russian.) *Acc. med. SSSR* 12, 29–36.

Kingman, J. (1972). Regenerative phenomena.

Klein, J. C. and Serra, J. (1971). The texture analyser. *J. of Micr.* 95, part 2, April 1973, 349–356.

Klein, J. C. (1976). Conception et réalisation d'une unité logique pour l'analyse quantitative d'images. Thesis, University of Nancy, France, 124 p.

König, D. and Stoyan, D. (1979). Stereological formulas through random marked point processes. Fifth Int. Congress for Stereol. Salzburg.

Kramer, H. P. and Bruckner, J. B. (1975). Iterations of non-linear transformations for enhancement on digital images. *Patt. Recog.* 7, 53–58.

Krekule, I., Hladis, K., Pilny, J., Blazek, V., Muzik, K. and Opatrna, J. (1979). Comparison of some optical and electronic methods in image analysis of biological structure. Fifth Int. Congress of Stereol. (To be published in *Mikroscopie*.)

Krickberg, K. (1977). Statistical problems on point processes. Conferences au Centre Banach Varsovie, September 76.

Kruse, B. (1973). A parallel picture processing machine. *IEEE Trans. Comput.* C-22, p. 1075.

Kulpa, Z. (1977). Area and perimeter measurement of blobs in discrete binary pictures. *Comp. Graph. and Im. Proc.* 6, 434–451.

Lantuéjoul, Ch. (1977). Sur le modèle de Johnson-Mehl généralisé. *Rapport Interne Centre de Morph. Math.* Fontainebleau, France.

Lantuéjoul, Ch. (1978). Détection automatique des lignes de défauts dans les systèmes eutectiques lamellaires. *Rapport interne.* Centre de Morphologie Mathématique Fontainebleau, France.

Lantuéjoul, Ch. (1978). La squelettisation et son application aux mesures topologiques des mosaïques polycristallines. Ph.D. thesis, Paris School of Mines, 83 pp.

Lantuéjoul, Ch. (1978). Grain dependence test in a polycristalline ceramic, *In* J. L. Chermant, ed.), 40–50.

Lantuéjoul, Ch. (1980). Skeletonization in quantitative metallography. *In* Issues of Digital Image Proc. (R. M. Haralick and J. C. Simon, eds.) Sijthoff and Noordhoff.

Levialdi, S. (1971). Parallel pattern processing. *IEEE Trans. Syst. Man. Cybern. SMC-1,* 292–296.

Levy, P. (1965). Processus stochastiques et mouvement brownien. Gauthier Villard, Paris, 438 pp.

Lewis, P. A. W. (Ed.) (1972). "Stochastic Point Processes." Wiley, New York.

Little, D. V. (1974). A third note on recent research in geometrical probability. *Adv. Appl. Prob.* **6**, 103–130.

Mandelbrot, B. B. (1977). "Fractals: Form, Chance, Dimension." W. H. Freeman and Co., San Francisco and London.

Marbeau, J. P. (1973). Etude structurale d'un peuplement en régénération. Int. report Centre Morph. Math. Fbleau (France).

Mase, S. (1980). Asymptotic properties of stereological estimators of volume fraction for stationary randoms sets. Technical Report, Hiroshima University, Japan.

Matern, B. (1960). Spatial Variation. Meddelanden från Statens. skogsforskningsinstitut Band 49. NR 5.

Matheron, G. (1965). Les variables régionalisées et leur estimation. (Doctorate thesis.) *Appl. Sci.*, Masson, Paris.

Matheron, G. (1967). Eléments pour une théorie des milieux poreux. Masson, Paris.

Matheron, G. (1969). Théorie des ensembles aléatoires. Ecole des Mines de Paris.

Matheron, G. (1969). Cours de Processus stochastiques, Ecole d'Eté, C. M. M., 78 pp.

Matheron, G. (1971). Les polyèdres poissoniens isotropes. 3° Symposium Européen sur la Fragmentation, Cannes 5–8 October 1971.

Matheron, G. (1971). Random sets theory and its applications to stereology. *J. of Micr.* **95**, part 1, Feb. 1972, 15–23.

Matheron, G. (1972). Quelques aspects de la montée. Note interne. Centre de Morphologie Mathématique Fontainebleau, France.

Matheron, G. (1972). Ensembles fermés aléatoires, ensembles semi-markoviens et polyèdres poissoniens. *Adv. in Appl. Prob.* **4**, No. 3, 306–377.

Matheron, G. (1975). "Random Sets and Integral Geometry." Wiley, New York.

Matheron, G. (1976). La formule de Crofton pour les sections épaisses. *J. Appl. Prob.* **13**, 707–713.

Matheron, G. (1978). "Estimer et Choisir. Essai sur la pratique des Probabilités." (To be published.)

Matheron, G. (1978). Quelques propriétés topologiques du squelette. Rapport Interne Centre de Morphologie Mathématique Fontainebleau, France.

Matheron, G. (1979). L'émergence de la loi de Darcy. Int. report, C.M.M., 87 pp.

Meijering, J. L. (1953). Interface area, edge length, and number of vertices in crystal aggregates with random nucleation. Philips Research Reports 8, 270–290.

Mendelsonn, M. L. and Mayall, B. H. (1974). Chromosome identification by image analysis and quantitative cytochemistry. "Human Chromosomes Methodologie." Academic Press.

Meyer, F. (1977). "Contrast Feature Extraction." (*In* J. L. Chermant, ed.).

Meyer, F. (1978). Iterative image transformation for an automatic screening of cervical smears. *J. of Histochemistry and Cytochemistry* **27**, No. 1, 128–135.

Meyer, F. (1979). Cytologie Quantitative et Morphologie Mathématique. Thèse docteur ingénieur, Ecole des Mines de Paris, 4 Mai 1979.

Meyer, F. (1980). Quantitative analysis of the chromatin of lymphocytes an essai on comparative structuralism. *Blood Cells* **6**, 159–172.

Meyer, F. and Van Driel-Kulker, A. (1979). Automatic screening of papanicolaou stained cervical smears with the T. A. S. Leitz Symposium on Quantitative Morphology and Image Analysis. (To be published in "Acta Microscopica".)

Mignot, J. (1978). Etude morphologique des fibres conjonctives sur biopsies ou pièces opératoires. Lecture Notes in Biomath. Springer-Verlag, Berlin, No. 23, 77–84.

Miles, R. E. (1964). Random Polytopes. Ph.D. thesis, Cambridge, England.

Miles, R. E. (1969). Poisson Flats in Euclidean Spaces. Part I. *Adv. Appl. Prob.* **1**, 211–237.

Miles, R. E. (1972). The random division of space. Special supplement to *Adv. in Appl. Prob.* 265 pp.

Miles, R. E. (1974). On the elimination of edge effects in planar sampling. *In* "Stochastic Geometry." Wiley and Sons.

Miles, R. E. and Davy, P. (1976). Precise and general conditions for the validity of a comprehensive set of stereological fundamental formulae. *J. of Micr.*, **107**, part 3, 211–226.

Miles, R. E. and Davy, P. (1977). On the choice of quadrats in stereology. *J. of Micr.* **110**, part 1, 27–44.

Miles, R. E. and Serra, J. (1978). (ed.) Geometrical probabilities and biological structures. Lecture Notes in Biomathèmatics, No. 23. Springer-Verlag.

Millier, C., Poissonnet, M. and Serra, J. (1970). Morphologie Mathématique et Sylviculture. 3 rs Conference, Advisory Group of Forest Statisticians, Jouy-en-Josas, INRA Publications 73–2, 287–307.

Minkowski, H. (1903). Volumen and Oberfläche. *Math. Ann. Vol.* **57**, 447–495.

Minsky, M. L. and Papert, S. (1970). "Perceptions, Introduction to Computational Geometry." M.I.T. Press, Cambridge, Mass.

Modestino, J. W., Fries, R. W. and Vickers, A. L. (1981). Stochastic Image Models Generated by Random Tesselations of the Plane in Image Modelling, pp. 301–325. Academic Press, London and New York.

Montanari, G. U. (1970). On limit properties in digitalization schemes. *J. of the Assoc. for Comp. Mach.* **17**, No. 2, 348–360.

Moran, P. A. P. (1966). Measuring the length of a curve. *Biometrika* **53**, 3 and 4, 359–364.

Moran, P. A. P. (1972). The probabilistic basis of stereology. *Suppl. Adv. Appl. Prob.* 69–91

Motzkin, Th. (1935). Sur quelques propriétés caractéristiques des ensembles bornés non convexes. *Atti Acad. Naz. Lincei* **21**, 773–779.

Nawrath, R. and Serra, J. (1979). Quantitative image analysis. 1. Theory and instrumentation; 2. Applications using sequential transformations. *Microscopa Acta* **82**, No 2, 101–111 and 113–128.

Neumann, J. V. (1951). The general logical theory of automata. In "Cerebral

Mechanisms in Behavior." (The Hixon Symposium, L. A. Jeffress, ed.) Wiley, New York.

Neveu, J. (1964). "Calcul des probabilités." Masson, Paris.

Northrop, E. P. (1944). Riddles in mathematics. D. Van Nostrand, N.Y. 225 pp.

Ondracek, G. and Spieler, K. (1976). Gasätzen—ein neues Kontrastierungsverfahren in der Gefügepräparation. Gesellschaft für Kernforschung M. B. H. Karlsruhe.

Pavel, M. (1970). Fondements mathématiques de la reconnaissance des structures. Hermann, Paris, 256 pp.

Pavlidis, T. (1977). "Structural Pattern Recognition." Springer, Berlin, Heidelberg, New York.

Pettijohn, F. J. (1970). "Sedimentary Rocks." Harper and Row, New York.

Piaget, J. (1967). La Psychologie de l'Intelligence. Armand Collin.

Piaget, J. and Ulmo, J. (1967). "Epistémologie de la Physique" (dans Logique et Connaissance Scientifique, Collection de la Pléiade).

Piefke, F. (1978). Masze von Halbräumen und hyperebenen im euklid'schen raum. Braunschweig, Sept. 1978, Thesis University of Braunschweig.

Ploem, J. S. and Van Driel-Kulker, A. (1980). "Automated Analysis of Papanicolaou stained Cervical Cells using a Television based Analysis Systems." Leytas. (To be published.)

Pont, J. C. (1975). Topologie algébrique des origines à Poincaré, PUF.

Popper, K. P. (1959). "The Logic of Scientific Discovery." Hutchinson Co, London.

Pratt, W. K. (1978). "Digital Image Processing." J. Wiley, New York.

Preston, K. (1961). The cellscan system. A leukocyte pattern analyzer. In "Proc. WJCC".

Preston, K. (1970). "Feature Extraction by Golay Hexagonal Pattern Transforms." IEEE Symposium on Feature Extraction and Selection in Pattern Recognition, Argonne, III.

Prewitt, J. M. S. (1978). On some application of pattern recognition and image processing to cytology. "Cytogenetics and Histology." Uppsala University.

Rasson, J. P. (1979). Estimation de formes convexes du plan. *Statistiques et Analyse des Données* 31–46.

Riley, J. A. (1965). Plane graphs and their skeletons, TM-1-60429. Parke Math. Lab. Inc.

Rink, M. and Schopper, J. R. (1978). On the application of image analysis to formation evaluation. Presented at the Fifth European SPWLA Logging Symposium, Paris, pp. 20–21.

Ripley, B. D. (1977). Modelling spatial patterns. *J. R. Statist. Soc.* **B.39**.

Roach, S. A. (1968). The Theory of Random Clumping. Methuenen, London, 100 pp.

Roberts, F. D. K. (1967). A Monte-Carlo solution of a two-dimensional unstructured cluster problem. *Biometrika.* **54**, 3 and 4, p. 625.

Rodenacker, K., Gais, P. and Jütting, U. (1981). Segmentation and measurement of the texture in digitized images. *Stereol. Iugosl.* **3**, suppl., 165–174.

Rosenfeld, A. and Pfaltz, J. L. (1966). Sequential operations in digital picture processing. *J. of the Assoc. for Comp. Mach.* 13–14, 471.

Rosenfeld, A. (1970). Connectivity in digital pictures. *J. of the Assoc. for Comp. Mach.* **17**, No. 1, 146–160.

Rosenfeld, A. (1973). Arcs and curves in digital pictures. *J. of the Assoc. for Comp. Mach.*, **20**, No. 1, 81–87.

Rosenfeld, A. and Kak, A. C. (1976). "Digital Picture Processing." Academic Press, New York, London.

Rosenfeld, A. (1978). Iterative methods in image analysis. *Patt. Recog.*, **10**, 181–187.
Rosenfeld, A. and Davis, L. S. (1978). Iterative Histogram Modification. *IEEE Transactions on Systems, Man, and Cybernetics*, **SMC 8**, No. 4.
Rosenfeld, A. (ed.) (1981). "Image Modeling" Academic Press, London and New York, 445 pp.
Rutovitz, D. (1969) Centremere finding: some shape descriptors for small chromosome outlines. *Mach. Intelligence* **5**, 435–462.
Ruyer, R. (1966) "La Cybernétique et l'origine de l'information." Flammarion, Paris.
Sáltykov, S. A. (1970). "Stereometric Metallography." Metallurgizdat, Moscow.
Sandler, S. S. (1975). "Picture Processing and Reconstitution". Lexington Books, D. C. Heatland C°, lex–Mass. 158 pp.
Santalo, L. A. (1953). "Introduction to Integral Geometry." Hermann, Paris.
Santalo, L. A. (1976). "Integral Geometry and Geometric Probability." Addison Wesley, Reading, Mass.
Saxl, I. (1981). Overview of contemporary theoretical approaches in stereology. *Stereol. Iugosl.* **3**, suppl. 1, 19–41.
Schachter, B. J., Rosenfeld, A. and Davis, L. S. (1978). Random mosaïc models for textures. *IEEE Transactions on Systems, Man, and Cybernetics*, Vol. SMC-8, n° 9.
Schwartz, L. (1967). "Mathematics for the Physical Sciences." Hermann, Paris.
Sergueev, E. M. (1971). Grountovédénié. Université de Moscou.
Serra, J. (1965). L'analyse des textures par la géométrie aléatoire. *Compte-rendu du Comité Scientifique de l'IRSID*.
Serra, J. "Patents of the Texture Analyser." First applications in France, No. 1-449-052, July 2, 1965; No. 70-21-322, June 10, 1970; No. 7521925, June 23, 1975. (Patented in Austria, Belgium, Canada, England, Fed. Germany, Japan, Sweden, U.S.A.)
Serra, J. (1967). Buts et réalisation de l'analyseur de textures. *Revue de l'Industrie Minérale*, **49**.
Serra, J. (1968). Morphologie mathématique et genèse des concrétions carbonatées des minerais de fer de Lorraine. *Sedimentology*, 183–208.
Serra, J. (1968). Les structures gigognes: morphologie mathématique et interprétation Métallogénique. *Mineral Deposita (Berl.)* **3**, 135–154.
Serra, J. (1969). Introduction à la Morphologie Mathématique. (Cahiers du Centre de Morphologie Mathématique, Booklet No. 3, 160 pp. E.N.S.M.P.)
Serra, J. (1972). Stereology and structuring elements. *J. of Micr*, part 1, 93–103. (Third Intern. Congress for Stereol., Berne, 26–31 Aug. 1971.)
Serra, J. and Muller, W. (1973). Das Problem der Auflösung und der durch Messlogik bedingten Fehler in Rahm der quantitative bild analyse. (*Leitz Mitteilungen*, suppl. 1–4, 125–136.
Serra, J. (1974). Aggregats de points et Morphologie. Rapport interne Centre de Morphologie Mathématique, Fontainebleau, France.
Serra, J., Mariaux, A. and Peray, O. (1976). The texture analyser measurements and Physical anisotropy of tropical woods. *N.B.S. Spec. pub.* No. 431, 257–260.
Serra, J. (1975). Morphologie pour les fonctions "à peu près en tout ou rien". Techn. report. Centre de Morphologie Mathématique Fontainebleau, France.
Serra, J. (1976). Lectures on Image Analysis by Mathematical Morphology. Sponsored by Tokyo Noko University, September 1976.
Serra, J. and Kolomenski, E. N. (1976). La Quantification en Pétrologie. *Bulletin de l'Association Intern. de Géologie de l'Ingénieur* No. 13, parts I and II, 83–88; 89–97. Krefeld.

Serra, J. (1978). En matière d'introduction. *In* "Buffon Symposium." Lectures Notes in Biomathematics. (Co-author: R. Miles.) Springer-Verlag.

Serra, J. (1980). The Boolean model and random sets. *Comp. Graph. and Im. Proc.* **12**, 99–126.

Simon, J. C. and Quinqueton, J. (1978). On the use of a peano scanning in image processing. *In* Issues of "Digital Image Processing." (R. M. Haralick and J. C. Simon, eds). Series E: *Applied Science*, No. 34.

Sklansky, J. (1972). Measuring concavity on a regular mosaïc. *IEEE Transactions on Computers.* **C-21**.

Slansky, J. (1980). Biomedical image analysis. *In* "Issues in Digital Image Processing." Sijhoff and Noordhoff, Series E, n° 34, pp. 291–312.

Smith, C. S. (1948). Grains, phases and interfaces: an interpretation of microstructure. *Trans. Am. Inst. Min. Metall. Engnrs* **175**, 15.

Snyder, D. L. (1975). "Random Point Processes." Wiley, London, 485 pp.

Solomon, H. R. (1953). Distribution of the measure of a random two dimensional set. *Ann. Math. Stat.* **24**, 650–656.

Sommerville, D. M. (1958). An Introduction to the Geometry in *n* Dimensions." Dover Publ., New York.

Steiner, J. (1840). Uber parallele flächen. Montsber, Akd. Wiss. Berlin, 2, pp. 114–118.

Sternberg, S. R. (1979). "Parallel Architecture for Image Processing." Proceedings of Third International IEEE Compsac, Chicago.

Sternberg, S. R. (1980). "Cellular computer and biomedical image processing." To be published in lecture notes in bio-Mathematics, Springer-Verlag.

Stienen, H. (1979). The sectioning of randomely dispersed particles. A (computer) simulation. Fifth Int. Congress of Stereol. (To be published in *Mikroscopie*.)

Stoyan, D. (1978). On some qualitative properties of the boolean model of stochastic geometry. *Bergakademie Freibert Sektion Mathematik* DDR, 92, Freiberg.

Stoyan, D. (1979). Proof of some fundamental formulas of stereology without Poisson-type assumptions. *Bergakademie Freibert sektion Mathematik DDR*, 92, Freiberg.

Streit, F. (1976). On methods and problems of geometrical stochastics. *Bull. of the ISI* **46(2)**, 600–605.

Stroeven, P. (1977). The analysis of fibre distributions in fibre reinforced materials. *J. of Micr.* **111**, part 3, 283–296.

Struik, D. J. (1961). "Lectures on Classical Differential Geometry." Addison Wesley, Cambridge, Mass.

Switzer, P. (1967). Mapping a geographically correlated environment. "Statistical Ecology." Vol. 1, Univ. Pemm. Press.

Takahashi, T. and Suwa, N. (1978). Stereological and Topological Analysis of Corrhotic Livers Lectures Notes in Biomathematics, Springer-Verlag, No. 23.

Tcherniavski, K. S. (1977). Stereologia v metallovedenii, Moscow, Metallurgia, 279 pp.

Timcak, G. M. (1973). A probabilist definition of the one and two dimensional "size". Proc. Pore Symp. RILEM/IUPAC, Prague, Vol. 3, pp. A33–A45, Academia.

Ullmann, J. R. (1973). "Pattern Recognition Techniques." Butterworths, London, 412 pp.

Underwood, E. E. (1970). "Quantitative Stereology." Addison Wesley, Reading, Mass.

Underwood, E. E. (1972). The stereology of projected images. *J. Microsc.* **95**, 25.

Vollath, D. (1981). PACOS. A software controlled image analysis system. *Stereol. Iugosl.* **3**, suppl. 1, 211–217.

Voronin, Yu. A. (ed.) (1977). Primenenie matematitcheskikh metodov u E.V.M. pri poïskakh i razvedke polesnikh skonaemikh. A. N. S. S. S. R., Sibirskoe otdele, 156 pp.

Voss, K., Simon, H. and Wenzelides, K. (1981). Hardware and software of a system for automatic microscopical picture analysis and its application in medecine. *Stereol. Iugosl.* **3**, suppl. 1, 139–143.

Vurpillot, E. (1963). "L'organisation perceptive, son rôle dans l'évolution des illusions optico-géométriques" Vrin, Paris, 186 pp.

Wadell, H. (1932). Volume shape and roundness of rock particles. *J. Geol.* **40**, 443–451.

Wadell, H. (1935). Volume shape and roundness of quartz particles. *J. Geol.* **43**, 250–280.

Watson, G. (1973). "Mathematical Morphology." Tech. Report No. 21, Dept of Stat., Princeton Univ., New Jersey.

Weibel, E. R. (1980). "Practical Methods for Biological Morphometry. Vol. I: Stereological Methods." Academic Press.

Weszka, J. S. (1978). A survey of threshold selection techniques. *Computer Graphics and Image Processsing* **7**, 259–265.

Wicksell, S. D. (1925). The corpuscle problem: case of spherical particles. *Biometrika* **17**, 84.

Zucker, S. W., Hummel, R. A. and Rosenfeld, A. (1977). An application of relaxation labeling to line and curve enhancement, *IEEE Trans. on Comp.* **C-26**, 394–403.

Subject Index

A

Accumulation point, 68
Adherent point, 71
Algorithm, 20–24
Alignments of points, 541–543
Alternating capacity, 548
Angular measures, 119
Anamorphoses, 435
Angularity, 140, 144, 385
  and size distribution, 330, 362–364
Anisotropy, 271, 283–286, 301–304
  fast characterization, 226
Antiextensivity, *see also* Extensivity, 254, 270, 319
Area
  digital, 35
  digitalization, 220
  Euclidean, 104
  weighted representation, 210
Asymptote
  for the covariance, 282
  for fractal sets, 150
Automorphism, 167
Average, 254–262
  infinite, 263

B

Balls, 262
  covariance of, 310
  Boolean, 499
  size distribution, 350–352
  volume of, 124

Basis
  digital, 168, 175
  topological, 67
Bezout's identity, 168
Bias, 245, 248
Borel sets, 114
Boolean disks, 482, 553
  pore size distributions, 361
Boolean functions, 469
Boolean sets, 485–502, 581
  in one dimension, 555
Boole-Poisson model, 565–566
Boundary, 13, 69, 142, 151
  digital, 183, 198, 200, 392
  genus of, 135, 199
Bounded set, 219, 236, 244
Brownian motion trajectories, 543–544

C

C-additivity, 109, 548, 550
  and sup-additivity, 465
Cauwe's model, 28, 520–524, 577–579
Cauchy formula, 104, 137
Cells with nuclei, estimation, 255
Channel, 447, 456
Change of scale (principle)
  compatibility under, 9, 151
  digital, 202
  and convexity, 96–98
  and size distributions, 334, 357–358
Choquet's theorem, 242, 548, 565
Chord, *see* Linear size distribution
Circularity factor, 336–339